VOLUME ONE HUNDRED AND THIRTY TWO

CURRENT TOPICS IN DEVELOPMENTAL BIOLOGY

Organ Development

CURRENT TOPICS IN DEVELOPMENTAL BIOLOGY

"A meeting-ground for critical review and discussion of developmental processes"
A.A. Moscona and Alberto Monroy (Volume 1, 1966)

VOLUME ONE HUNDRED AND THIRTY TWO

Current Topics in DEVELOPMENTAL BIOLOGY

Organ Development

Edited by

DENEEN M. WELLIK
Department of Cell & Regenerative Biology,
University of Wisconsin,
Madison, WI, United States

ELSEVIER
ACADEMIC PRESS
An imprint of Elsevier

Academic Press is an imprint of Elsevier
50 Hampshire Street, 5th Floor, Cambridge, MA 02139, United States
525 B Street, Suite 1650, San Diego, CA 92101, United States
The Boulevard, Langford Lane, Kidlington, Oxford OX5 1GB, United Kingdom
125 London Wall, London, EC2Y 5AS, United Kingdom

First edition 2019

ISBN: 978-0-12-810489-7
ISSN: 0070-2153

Publisher: Zoe Kruze
Acquisition Editor: Fiona Pattison
Editorial Project Manager: Shellie Bryant
Production Project Manager: Denny Mansingh
Cover Designer: Miles Hitchen

Typeset by SPi Global, India

Contents

Contributors

Haley Abbott
Department of Cell and Developmental Biology, University of Michigan Medical School, Ann Arbor, MI, United States

Madeline G. Andrews
Department of Neurobiology; Neuroscience Graduate Program, University of California, Los Angeles, CA, United States

Nadean L. Brown
Department of Cell Biology & Human Anatomy, University of California Davis School of Medicine, Davis, CA, United States

Samantha J. Butler
Department of Neurobiology; Eli and Edythe Broad Center of Regenerative Medicine and Stem Cell Research, University of California, Los Angeles, CA, United States

Terence D. Capellini
Department of Human Evolutionary Biology, Harvard University; Broad Institute of Harvard and MIT, Cambridge, MA, United States

Ondine Cleaver
Department of Molecular Biology and Center for Regenerative Science and Medicine, University of Texas Southwestern Medical Center, Dallas, TX, United States

Taylor Cool
Institute for the Biology of Stem Cells, Program in Molecular, Cell, and Developmental Biology, University of California, Santa Cruz, Santa Cruz, CA, United States

Corinna Cozzitorto
Max Delbrück Center for Molecular Medicine (MDC), Berlin, Germany

Edward Daniel
Department of Molecular Biology and Center for Regenerative Science and Medicine, University of Texas Southwestern Medical Center, Dallas, TX, United States

E. Camilla Forsberg
Institute for the Biology of Stem Cells, Department of Biomolecular Engineering, University of California, Santa Cruz, Santa Cruz, CA, United States

Ankur Garg
Department of Pediatrics, University of California, San Diego, La Jolla, CA, United States

Deborah L. Gumucio
University of Michigan, Cell and Developmental Biology Department, Ann Arbor, MI, United States

Saher Sue Hammoud
Department of Human Genetics; Department of Obstetrics and Gynecology; Department of Urology, University of Michigan Medical School, Ann Arbor, MI, United States

Stacey S. Huppert
Division of Gastroenterology, Hepatology & Nutrition, Cincinnati Children's Hospital Medical Center; Department of Pediatrics, University of Cincinnati College of Medicine, Cincinnati, OH, United States

Makiko Iwafuchi-Doi
Department of Pediatrics, University of Cincinnati College of Medicine; Division of Developmental Biology, Cincinnati Children's Hospital Medical Center, Cincinnati, OH, United States

Gabrielle Kardon
Department of Human Genetics, University of Utah, Salt Lake City, UT, United States

Jennifer Kong
Department of Neurobiology; Neuroscience Graduate Program, University of California, Los Angeles, CA, United States

Hailey Larose
Department of Human Genetics, University of Michigan Medical School, Ann Arbor, MI, United States

Lei Lei
Department of Cell and Developmental Biology; Department of Biomedical Engineering, University of Michigan Medical School, Ann Arbor, MI, United States

Gabriel Manske
Department of Human Genetics, University of Michigan Medical School, Ann Arbor, MI, United States

Joel B. Miesfeld
Department of Cell Biology & Human Anatomy, University of California Davis School of Medicine, Davis, CA, United States

Bennett G. Novitch
Department of Neurobiology; Eli and Edythe Broad Center of Regenerative Medicine and Stem Cell Research, University of California, Los Angeles, CA, United States

Kyriel M. Pineault
Department of Cell & Regenerative Biology, University of Wisconsin, Madison, WI, United States

Elizabeth M. Sefton
Department of Human Genetics, University of Utah, Salt Lake City, UT, United States

Licia Selleri
Program in Craniofacial Biology, Department of Orofacial Sciences; Program in Craniofacial Biology, Department of Anatomy, Eli and Edythe Broad Center of Regeneration Medicine & Stem Cell Research, University of California, Institute of Human Genetics, San Francisco, CA, United States

Adrienne Niederriter Shami
Department of Human Genetics, University of Michigan Medical School, Ann Arbor, MI, United States

Pragya Sidhwani
Division of Biological Sciences, University of California, San Diego, CA, United States

Jane Y. Song
Program in Cell and Molecular Biology Program, University of Michigan, Ann Arbor, MI, United States

Francesca M. Spagnoli
Max Delbrück Center for Molecular Medicine (MDC), Berlin, Germany; Centre for Stem Cell and Regenerative Medicine, King's College London, London, United Kingdom

Pengfei Sui
Department of Pediatrics, University of California, San Diego, La Jolla, CA, United States

Xin Sun
Department of Pediatrics; Department of Biological Sciences, University of California, San Diego, La Jolla, CA, United States

Jamie M. Verheyden
Department of Pediatrics, University of California, San Diego, La Jolla, CA, United States

Katherine D. Walton
University of Michigan, Cell and Developmental Biology Department, Ann Arbor, MI, United States

Sha Wang
University of Michigan, Cell and Developmental Biology Department, Ann Arbor, MI, United States

Deneen M. Wellik
Department of Cell & Regenerative Biology, University of Wisconsin, Madison, WI, United States

Deborah Yelon
Division of Biological Sciences, University of California, San Diego, CA, United States

Lisa R. Young
Division of Pulmonary Medicine, Center for Childhood Lung Research, Department of Pediatrics, Vanderbilt University Medical Center, Nashville, TN, United States

Mariel Young
Department of Human Evolutionary Biology, Harvard University, Cambridge, MA, United States

Preface

Organogenesis is one of the most fascinating aspects of embryonic development. It is the generation of these structures that initiate after the embryo has created germ layers and begun segmenting into spatially distinct regions that permit an organism to think, see, breathe, eat, move, and reproduce. A review how organs initiate, grow, and maintain and repair themselves form the basis of this volume. While it is not possible to be comprehensive, this volume includes a review of the formation of endodermal organs, such as the lung (Dr. Xin Sun, PhD and colleagues), gut (Katherine Walton, PhD), pancreas (Francesca Spagnoli, MD, PhD) and liver (Stacey Huppert, PhD). Mesodermal organs reviewed include the hematopoietic system (Camilla Forsberg, PhD), muscle and muscle connective tissue (Gabrielle Kardon, PhD), the vascular system (Ondine Cleaver, PhD), the pelvis and scapula (Licia Selleri, MD, PhD and Terrance Capellini, PhD), the heart (Deborah Yelon, PhD), and the limb musculoskeletal system (Deneen Wellik, PhD). The ectodermal organs are represented by reviews of the spinal cord (Samantha Butler, PhD and Bennett Novitch, PhD), the eye (Nadean Brown, PhD), and germ cells (Sue Hammoud, PhD and Lei, PhD).

It is of note that all of the reviews do not focus solely on the organogenesis processes associated with embryonic development, nor do they focus solely on embryology and genetics. While a vast amount of our current knowledge continues to be impacted by our ability to manipulate both the genome and the embryo itself, the field has expanded to include genomics, systems biology, and biophysical, structural and biomechanical studies into the existing body of knowledge. Further, continued work in the field has begun to reveal that organogenesis, and the study of this basic knowledge is highly relevant to postnatal and adult organ biology and is critical for understanding organ maintenance, organ response to injury and disease, and the organ's ability (or inability) to heal and regenerate itself after insults. Many of the same factors and pathways that are critical for the initial development and early growth of our organs continue to play fundamental roles throughout life. Further, these original pathways and processes are additionally influenced by new cell types (i.e., the microbiome), new molecular factors, and the immune system. Current work in the field is using a combination of basic knowledge from many fields to begin to dissect the more complex interplay that impact adult organ systems in health and disease.

A Note About the Authors

The thirteen chapters included in this volume are contributed by leaders in their respective fields, scientists that promote biomedical science and education through their own outstanding research programs, peer-review of grants and manuscripts, administrative service, teaching, and outreach at their own institutions and around the world. Notably, at least one of the corresponding authors for each chapter is female, highlighting the broad and essential contributions of female scientists to the fields of organogenesis, and serving as a reminder of the gender-neutral power of genetics to reveal fundamental scientific truths.

DENEEN M. WELLIK, PHD
Professor and Chair
Department of Cell & Regenerative Biology
University of Wisconsin-Madison
Madison, WI, United States

Stacey S. Huppert, PhD., Associate Professor, Dept. of Pediatrics, Cinncinnati Children's

Xin Sun, PhD, Professor, Dept. of Pediatrics, UCSD

Katherine Walton, PhD., Assistant Professor, Dept. Cell & Devel. Biol., Michigan Medicine

Nadean Brown, PhD, Professor, Cell Biol. & Human Anat., UC Davis

E. Camilla Forsberg, PhD, Professor, Biomolecular Engineering, UC Santa Cruz

S. Sue Hammoud, PhD, Assistant Professor, Dept. Human Genetics, Michigan Medicine

Lei, PhD, Assistant Professor, Dept. Cell & Devel. Biol., Michigan Medicine

Ondine Cleaver, PhD, Professor, Dept. Mol. Biology, UT Southwestern

Terrance D. Capellini, PhD, Associate Professor, Human Evolutionary Biol., Harvard University

Licia Selleri, MD/PhD, Professor, Dept. Oralfacial Sciences, UCSF

Samantha J. Butler, PhD, Associate Professor, Dept. Neurobiology, UCLA

Deb Yelon, Professor and Vice Chair, Sect Cell & Devel. Biol., Sect Cell & Devel. Biol.

Francesca Spagnoli, MD/PhD, Professor, Max Delbruck Center, Max Delbruck Center

Max Delbruck Center, Professor, Dept. Human Genetics, University of Utah

CHAPTER ONE

Chasing Mavericks: The quest for defining developmental waves of hematopoiesis

Taylor Cool[a], E. Camilla Forsberg[b,*]
[a]Institute for the Biology of Stem Cells, Program in Molecular, Cell, and Developmental Biology, University of California, Santa Cruz, Santa Cruz, CA, United States
[b]Institute for the Biology of Stem Cells, Department of Biomolecular Engineering, University of California, Santa Cruz, Santa Cruz, CA, United States
*Corresponding author: e-mail address: cforsber@ucsc.edu

Contents

Abstract

Hematopoiesis is the process by which mature blood and immune cells are produced from hematopoietic stem and progenitor cells (HSCs and HSPCs). The last several decades of research have shed light on the origin of HSCs, as well as the heterogeneous pools of fetal progenitors that contribute to lifelong hematopoiesis. The overarching concept that hematopoiesis occurs in dynamic, overlapping waves throughout development, with each wave contributing to both continuous and developmentally limited cell types, has been solidified over the years. However, recent advances in

Current Topics in Developmental Biology, Volume 132
ISSN 0070-2153
https://doi.org/10.1016/bs.ctdb.2019.01.001

our ability to track the production of hematopoietic cells in vivo have challenged several long-held dogmas on the origin and persistence of distinct hematopoietic cell types. In this review, we highlight emerging concepts in hematopoietic development and identify unanswered questions.

1. Introduction

Hematopoietic stem cells (HSCs) are blood stem cells capable of self-renewal and differentiation into lineage-restricted progenitor cells that ultimately make mature cell types with highly specialized functions. HSCs were first described in the 1960s as cells capable of giving rise to spleen colonies containing clonally multipotent cells (Becker, McCulloch, & Till, 1963; Wu, Till, Siminovitch, & McCulloch, 1968). These studies laid the groundwork for our initial understanding of blood stem cells and defined them as cells with combined self-renewal and multilineage hematopoietic differentiation capability upon transplantation into a suitable host. Subsequently, Spangrude et al. demonstrated that 30 HSCs were sufficient to rescue half a population of lethally irradiated mice, indicating the potent reconstitution capacity of these cells (Spangrude, Heimfeld, & Weissman, 1988). Since then, the field of molecular and cellular biology has advanced rapidly, allowing for extremely sensitive detection of donor-derived cells to assess the clonal capacity of distinct subsets of hematopoietic stem and progenitor cells (HSPCs).

While transplantation of HSPCs into conditioned hosts has long been the "gold standard" in the field, in situ fate mapping, barcoding, and RNA sequencing (RNA-seq) have become increasingly common to interrogate HSC identity and function. There are pros and cons to each of these techniques. Transplantation assays test the inherent potential of stem and progenitor cells to produce different cell types. However, the conditioning regimens typically used to ensure robust reconstitution may affect the lineage capacity and/or preference of the incoming cells as well as the clearance of resident cell types, thus altering the perceived potential of the transplanted cell. Although reconstitution assays remain an essential tool to assess HSPC in vivo lineage potential and the therapeutic potential of cell subpopulations for regenerative medicine, HSPC transplantation might not reflect the cell ontogeny of normal hematopoiesis. Fate mapping using permanent genetic marking allows for tracking of HSPC differentiation in situ without disrupting normal hematopoiesis. Yet, it is difficult to reliably pinpoint the initiation of labeling to only HSCs, since different progenitors may transiently express the gene responsible for the labeling. In situ barcoding assays allow us to determine contributions of single clones to the hematopoietic system (Busch et al., 2015; Pei et al., 2017; Sun et al., 2014) (reviewed in Dharampuriya et al., 2017). However, interpretation of the outcomes is complex and relies heavily on mathematical modeling, and the conclusions have been challenged (Sawai et al., 2016). In addition to these functional assays, RNA-seq of bulk populations or single cells is heavily utilized. This is an extremely useful, yet relatively simple, tool to identify differences between cells. Although the data collected from RNA-seq are descriptive and not without caveats (Arzalluz-Luque, Devailly, Mantsoki, & Joshi, 2017; Griffiths, Scialdone, & Marioni, 2018; Herring, Chen, McKinley, & Lau, 2018), they provide a valuable database of genes to further investigate. Regardless of caveats, these assays have collectively proven extremely beneficial in furthering our understanding of hematopoiesis.

Increasingly sophisticated versions of these techniques have led to reconsideration of some long-held dogmas of hematopoiesis. Vast heterogeneity in stem and progenitor populations throughout development and in adulthood has been revealed, and the idea that the adult HSC pool is derived from one small pool of cells, as previously believed, has been challenged (Ganuza, Hall, et al., 2017; Pei et al., 2017; Yu et al., 2016). Accumulating evidence suggests that embryonic progenitors that precede lifelong, engraftable HSCs contribute to distinct cell types that persist with little replenishment from the adult HSC pool (Gomez Perdiguero et al., 2015; Hoeffel & Ginhoux, 2015). The extent to which different HSCs, and their fetal precursors, contribute to hematopoiesis during development and over the lifespan is under debate (Sawai et al., 2016). Even the concept that definitive, repopulation-capable HSCs persist for life has been challenged (Beaudin et al., 2016). In order to solidify emerging views and continue to gain deeper insights into the biology of HSCs, many labs focus their attention to the developmental origin of HSCs and the lifelong contribution of discrete waves of hematopoiesis. Important goals include gaining an actionable understanding of the biology of hematopoietic disease, and how to generate and culture HSCs in vitro for therapeutic use. In this review, with emphasis on lessons learned from the mouse, we aim to summarize the current perspective on the contributions of developmental waves of HSPCs to the specification and maintenance of lifelong hematopoiesis, and to pinpoint questions that remain unanswered.

2. HSC emergence: Where, when, and how?

2.1 Cellular origins of HSCs: Hemangioblast versus hemogenic endothelium

The mechanisms of HSC emergence have been under debate for decades. One hypothesis posited that hematopoietic and endothelial lineages arise through a specialized bi-potent progenitor known as the hemangioblast (Fig. 1). This term was initially used to describe a group of mesodermal cells from the primitive streak that contained endothelial and hematopoietic cells (Murray, 1932). One study demonstrated that a subset of embryonic stem cell (ESC)-derived progenitors were capable of producing both endothelial and hematopoietic lineages (Lacaud, Robertson, Palis, Kennedy, & Keller, 2001). The authors concluded that this progenitor population was an in vitro equivalent of the hemangioblast. Using a brachyury-GFP

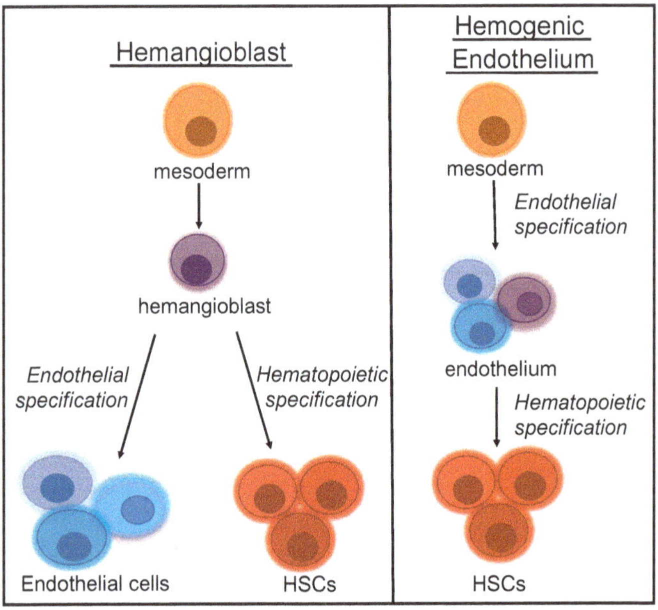

Fig. 1 Hemangioblast versus hemogenic endothelium schematic. Two proposed models of HSC emergence in the embryo. One model proposes that a specialized bi-potent precursor named the hemangioblast gives rise to both endothelial cells and HSCs. An alternative model proposes that mesoderm gives rise first to endothelium, a small proportion of which can give rise to HSCs. In this schematic, the purple cell represents the subset of hemogenic endothelial cells. The different shades represent heterogeneity within both lineages.

transgene, Huber and colleagues identified a subset of Flk1+brachyury+ cells that modeled the mesodermal-to-hemangioblast transition (Huber, Kouskoff, Fehling, Palis, & Keller, 2004) providing further ex vivo support for the existence of a hemangioblast.

Ueno and colleagues tested the hemangioblast concept in vivo by making tetrachimeric mouse embryos that clonally expressed different fluorescent proteins (Ueno & Weissman, 2006). The dissimilar proportions of colored endothelial and hematopoietic cells within individual blood islands argued against a shared bi-potent progenitor that contributes equally to both lineages. Another group used a temporally-restricted fate mapping model, VE-cadherin-CreERT2, to selectively label endothelial cells and thereby distinguish between HSC emergence from the endothelium or underlying mesenchyme (Zovein et al., 2008). By injecting pregnant mothers at E9.5 with tamoxifen, they aimed to label aorta-gonad-mesonephros (AGM)-derived HSCs, while evading the earlier VE-cadherin expression of the yolk sac (YS). By surveying additional models (SM22α-Cre and Myocardin-Cre) that labeled mesenchymal cells and their progeny, they ruled out a direct mesenchyme-to-HSC transition and concluded that only the endothelial lineage was capable of generating HSCs. Similarly supporting an endothelial ancestry, Oberlin et al. demonstrated that the majority of adult BM HSCs are derived from a VE-cadherin positive precursor (Oberlin, El Hafny, Petit-Cocault, & Souyri, 2010). However, VE-cadherin may not be entirely restricted to the endothelium. One study demonstrated that fetal HSCs express VE-cadherin at E13.5, but lose expression by E16.5 (Kim, Yilmaz, & Morrison, 2005); thus, the observed labeling of hematopoietic cells in this model may not signify an endothelial origin.

An alternative hypothesis to the bi-potent hemangioblast proposed that mesoderm first gives rise to endothelial cells, some of which possess the capacity to generate HSCs (Fig. 1). In support of HSC emergence after endothelial specification, Lancrin and colleagues identified a hemogenic endothelial precursor intermediate that was capable of giving rise to the first hematopoietic cells (Lancrin et al., 2009). Another group, using in vitro time-lapse imaging, concluded that both ESC-derived mesodermal cells and mesodermal cells isolated from E7.5 embryos were able to give rise to hematopoietic cells through an endothelial intermediate, supporting the endothelial origin of HSCs (Eilken, Nishikawa, & Schroeder, 2009). Boisset and colleagues used in vivo time-lapse imaging to visualize HSC emergence from endothelium in the AGM region (Boisset et al., 2010). Similar findings were made simultaneously in zebrafish: two groups concluded that HSCs emerge from ventral aortic hemogenic endothelial cells prior to the onset of blood circulation (Bertrand et al., 2010; Kissa & Herbomel, 2010). Thus, HSC emergence appears similar in zebrafish and mouse, although de Brujin et al. concluded that HSC activity was limited to a few cells within the endothelial layer of the dorsal aorta in mouse, as opposed to the ventral aorta in fish (de Bruijn et al., 2002).

More recently, a group used single cell isolation and RNA-seq to identify molecular signatures of pre-HSCs (Zhou, Li, et al., 2016). They concluded that Rictor, a component of the mTORC pathway, is required for the emergence of HSCs from endothelial cells, a finding that supports the notion that HSCs arise from endothelial cells (Figs. 1 and 2). Ganuza and colleagues performed ex vivo cultures of murine cells from the midgestation aorta, vitelline artery, and umbilical artery and found that only a very small percentage of the cells had in vitro functional properties of hemogenic endothelium (Ganuza, Hadland, et al., 2017). Further, they suggested that heterogeneity within the hemogenic endothelial population might contribute to distinct populations with differential capacities to specify HSCs (Fig. 2). This finding suggests that several types of precursors contribute to HSPCs. This idea is novel and striking given that some labs have believed for decades that one specific type of precursor contributes

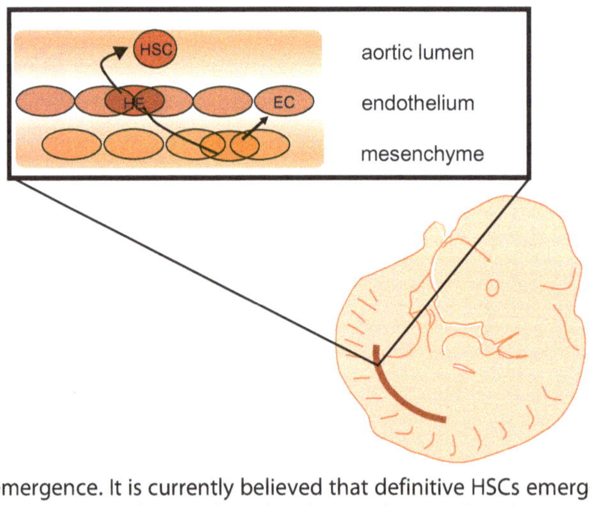

Fig. 2 HSC emergence. It is currently believed that definitive HSCs emerge in the AGM region of E10.5 murine embryos. There has been substantial evidence supporting the notion that a subset of mesenchyme-derived endothelial cells (ECs), called hemogenic endothelium (HE), has the potential to give rise to HSCs. Such *de novo* HSC specification in the AGM may contribute to lifelong hematopoiesis in parallel with YS-derived hematopoietic cells.

to the first HSCs. While no definitive conclusion has been made regarding the exact properties of the in vivo precursors to HSCs, it is clear that endothelial and hematopoietic lineages are closely linked (Figs. 1 and 2). The concepts of the hemangioblast versus the endothelial-to-hematopoietic transition are quite similar and may not be mutually exclusive: in both cases, mesoderm cells give rise to endothelial and hematopoietic lineages (Figs. 1 and 2). Several reviews discuss the hemangioblast versus hemogenic endothelial origin of HSCs in greater detail (Boisset & Robin, 2012; de Bruijn, 2014; Lacaud & Kouskoff, 2017; Nishikawa, 2012).

2.2 Are HSCs specified multiple times in distinct locations?

Tracking the ontogeny of HSCs has proven challenging due to the different temporal and spatial emergence of blood cells. It has been proposed that HSCs emerge de novo in either the YS, placenta, and/or AGM, and eventually seed the fetal liver (FL) and adult bone marrow (BM). Early in vitro and in vivo colony forming assays (Huang & Auerbach, 1993; Liu & Auerbach, 1991; Moore & Metcalf, 1970; Perah & Feldman, 1977; Toles, Chui, Belbeck, Starr, & Barker, 1989) suggested that HSCs emerge in the YS, migrate to the FL, and finally to the BM, where they reside in adulthood and contribute to lifelong hematopoiesis. In support of this model, Samokhvalov et al. demonstrated that Runx1 + cells, labeled with a single tamoxifen injection at E7.5 in a MER-cre-MER model, were able to develop into fetal myeloid and lymphoid progenitors and contribute to adult hematopoiesis (Samokhvalov, Samokhvalova, & Nishikawa, 2007). They concluded that these cells were able to colonize the umbilical cord, AGM, and ultimately the FL, suggesting that at least some definitive HSCs might be descendants of hematopoietic cells first specified in the YS. More recent data from their group substantiate these findings by providing evidence that a subset of extraembryonic Runx1+ Gata1− cells in the E7.5 embryo migrate to intraembryonic sites prior to the onset of circulation (initiation of a heartbeat) (Tanaka et al., 2014). Similar findings were made using a Lyve1-cre mouse model (Lee et al., 2016). These results are quite striking as they suggest that systemic circulation is not required for HSC migration through the embryo.

However, it has also been suggested that HSCs are generated de novo in the AGM, challenging an exclusive YS origin of definitive HSCs (Medvinsky & Dzierzak, 1996). Using an in vitro organ culture system to tease apart the hematopoietic potential of cells residing in the YS and AGM region, the authors concluded that the AGM is the source of definitive

HSCs that ultimately seed the adult BM. Taoudi and colleagues reported that VE-cadherin + CD45 + pre-definitive HSCs in the AGM region give rise to the first definitive HSC (Taoudi et al., 2008). This finding contradicted the YS origin of HSCs and suggested that HSCs are generated de novo in the AGM rather than migrating from the YS. Several other studies have also reported de novo generation of HSCs in the AGM region (de Bruijn, Speck, Peeters, & Dzierzak, 2000; Durand & Dzierzak, 2005; Peeters et al., 2009) or in the placenta, independent of circulation (Gekas et al., 2010; Rhodes et al., 2008).

Many labs have utilized Cre lineage tracing to directly assess the emergence of HSCs. Using an inducible Stem Cell Leukemia (SCL)-specific Cre line, one group addressed whether HSCs are generated de novo at each site during fetal development (Gothert et al., 2005). They injected tamoxifen into pregnant mothers at E10.5 in order to label AGM HSCs and analyzed the BM of adult progeny for labeled HSCs. As ~10% of the adult BM HSCs were labeled, at least some of the fetal HSCs appear to seed the adult BM. However, this result also revealed either inefficient HSC labeling (at this time point) or that additional HSCs may be specified after this time point. To address the possibility of de novo generation of HSCs at later time points, they compared the percent EYFP + cells in the adult BM of mice dosed at E10.5 and E11.5 with the percent EYFP + cells in the BM of mice transplanted with previously marked E14.5 FL cells. The authors reasoned that de novo generation of HSCs would have led to a decrease in EYFP + cells in the BM of the transplanted mice. They found that mice transplanted with E14.5 FL HSCs had a similar percentage of labeled HSCs in their adult BM compared to the embryos treated with tamoxifen at E10.5, thus arguing against de novo generation beyond E10.5. However, it is important to note that the labeling efficiency of CreER models during embryonic development is frequently incomplete, and that the origins of the remaining 90% of unlabeled cells in this model remain unknown.

The lack of consensus on the anatomical location(s) of the emergence of the first HSC(s) is likely, in part, due to models and methods being used. One issue is the length of time tamoxifen persists in the system. For example, Nishikawa's group suggested that YS cells migrate to colonize the embryo; however, labeling of intraembryonic cells could potentially be explained by tamoxifen remaining in the system and labeling cells within the embryo that express Runx1 at a later time point. In addition, most groups use different Cre models, independently, that undoubtedly label divergent cell types. A thorough side-by-side comparison of the Cre lines that are specific to early

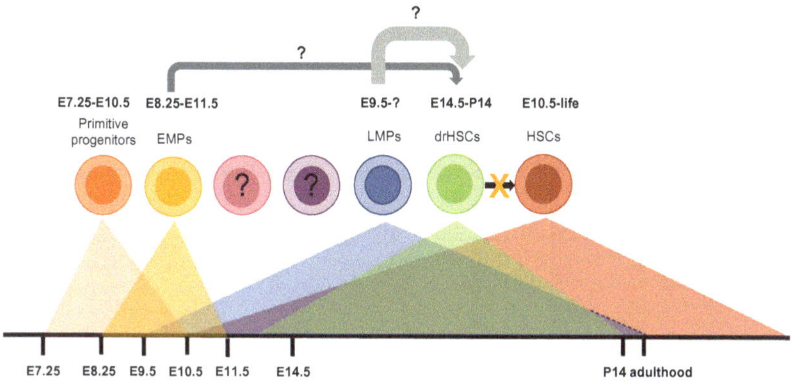

Fig. 3 Fetal hematopoiesis occurs in waves. Several distinct progenitors exist at different developmental stages *in situ*. Primitive erythroid precursors can be isolated from E7.25 until E10.5. Erythro-myeloid progenitors (EMPs) exist from E8.25 until E11.5. Lympho-myeloid progenitors (LMPs) appear around E9.5. Engraftable, developmentally restricted (dr)HSCs can be detected from E12.5 until postnatal day 14 (P14). Definitive HSCs that contribute to lifelong hematopoiesis are identifiable from E10.5 onward. It remains to be determined whether the drHSC is derived from EMPs, LMPs, or an independent precursor. It is clear that drHSCs do not give rise to definitive HSCs. Whether other previously unidentified HSPCs exist during this timeframe remains to be determined.

hematopoietic development may help in our data analysis and interpretations, and ultimately better define hematopoietic development (Abram, Roberge, Hu, & Lowell, 2014). Additionally, transplantation assays may not be sufficient to define which "HSCs" persist in situ (Fig. 3). It has been demonstrated that embryonic HSCs engraft better in neonates compared to adults (Arora et al., 2014; Yoder, Hiatt, & Mukherjee, 1997). Additionally, evidence is accumulating that cells that do not engraft long term upon transplantation can persist for long periods of time in situ (Busch et al., 2015; Schoedel et al., 2016; Sun et al., 2014). Models that do not rely on niche perturbations or culturing of cells may provide us with better insights to the development of the blood system in situ (Ganuza, Hall, et al., 2017).

3. Molecular regulation of HSC specification

3.1 What are the molecular drivers of HSC specification and maintenance?

The molecular drivers of hematopoiesis change several times from a developing embryo to a fully formed vertebrate (Table 1). Efforts focused on understanding the different genetic programs that regulate HSC specification,

Table 1 Molecular regulators of HSC specification.

Stage	Molecular regulator	Deficit when deleted	Required for primitive hematopoiesis?	Required for definitive hematopoiesis?	References
Primitive hematopoiesis	Scl	No YS hematopoiesis. Disrupted capillary formation	Yes	No	Visvader, Fujiwara, and Orkin (1998), Robb et al. (1995), Shivdasani, Mayer, and Orkin (1995), Tanaka et al. (2000), Mikkola et al. (2003), Schlaeger, Mikkola, Gekas, Helgadottir, and Orkin (2005), and Org et al. (2015)
Primitive hematopoiesis	Cpb	Defective blood vessel formation	Yes	?	Tanaka et al. (2000)
Primitive hematopoiesis	Runx1	Absence of fetal liver hematopoiesis	Yes	No	Okuda, van Deursen, Hiebert, Grosveld, and Downing (1996), Chen, Yokomizo, Zeigler, Dzierzak, and Speck (2009), and Tober, Yzaguirre, Piwarzyk, and Speck (2013)
Primitive	Sox17	Reduced hemogenic endothelium and HSCs in AGM region	Yes	No	Clarke et al. (2013) and Kim, Saunders, and Morrison (2007)
Primitive to definitive transition	ATF4	FL hematopoiesis impaired. HSC self-renewal impaired	Yes	Yes?	Masuoka and Townes (2002) and Zhao et al. (2015)
Primitive to definitive transition	Cbfb	Nascent HSCs stuck in AGM (cannot enter circulation)	No	Yes	Bresciani et al. (2014) and Sasaki et al. (1996)

Primitive to definitive transition	Gata-2	Blocked HSC formation in AGM. Adult HSCs have impaired proliferation	No	Yes	Ling et al. (2004), Mehta et al. (2017), Rodrigues et al. (2005), and Tsai et al. (1994)
Primitive to definitive transition	C-myb	Blocked migration from AGM. Conditional deletion in adults leads to impaired self-renewal	Yes	Yes	Emambokus et al. (2003), Lieu and Reddy (2009), Mucenski et al. (1991), and Zhang, Jin, Li, Qin, and Wen (2011)
Primitive to definitive transition	C/EBPa	Impaired granulopoiesis. Enhanced repopulation activity of FL HSCs. Conditional deletion in adults leads to increased proliferation of HSCs	Yes	Yes	Ye et al. (2013)
Definitive hematopoiesis	Tet/Etv6	Impaired megakaryopoiesis	No	Yes	Hock et al. (2004)
Definitive hematopoiesis	E2A	Blocked B cell development leading to postnatal death. HSC pool maintenance perturbed. Impaired myelolymphoid/myeloerythroid progenitor maturation	No	Yes	Semerad, Mercer, Inlay, Weissman, and Murre (2009) and Zhuang, Soriano, and Weintraub (1994)
Definitive hematopoiesis	Bmi-1	Impaired hematopoiesis postnatally. Reduced hematopoietic numbers. Increased adipocytes in bone marrow	No	Yes	van der Lugt et al. (1994) and Park et al. (2003)

A summary of the different genes involved in HSC specification and hematopoietic maintenance throughout development and the lifespan.

expansion, and maintenance have identified distinct transcription factors involved in primitive erythropoiesis. Several studies have demonstrated that SCL/TAL1 is required for primitive hematopoiesis and specification of HSCs (Robb et al., 1995; Shivdasani et al., 1995; Visvader et al., 1998) and divergence of cardiac and hematopoietic lineages during development (Org et al., 2015). SCL is required for the initial specification of the hematopoietic system, but not for long-term HSC colonization in the FL (Schlaeger et al., 2005), and conditional deletion of SCL in adult mice showed that SCL is dispensable for adult HSC maintenance (Mikkola et al., 2003). Similarly, the transcriptional regulators Rbtn2 (LMO2) (Warren et al., 1994) and CBP (Tanaka et al., 2000) appear to be required for primitive hematopoiesis and specification of HSCs, but not for HSC maintenance.

Like SCL, Runx1 is a transcription factor and an essential regulator of hematopoietic specification. Okuda et al. showed that Runx1 is required for FL hematopoiesis (Okuda et al., 1996). Mice with a homozygous mutation of Runx1 exhibited normal YS erythropoiesis, but eventually died around E12.5 due to failed FL hematopoiesis. Importantly, endothelial-specific loss of Runx1 precluded HSC formation (Chen et al., 2009), whereas hematopoietic deletion of Runx1 allowed normal EMP and HSC formation (Ichikawa et al., 2004). This supports HSC derivation from endothelial cells in a Runx1-dependent process. To further define the window of Runx1 dependency, Tober et al. performed timed deletions of Runx1 from E7.5 until E13.5 (Tober et al., 2013). When Runx1 was deleted between E9.5 and E11.5, recipients had no engraftment, while deletion after E11.5 led to high levels of engraftment. The authors concluded that Runx1 is required for EMP and HSC formation up until E10.5 and E11.5, respectively. In addition to Runx1, ATF4 (Masuoka & Townes, 2002; Zhao et al., 2015), Cbfb (Bresciani et al., 2014; Sasaki et al., 1996), Gata2 (Ling et al., 2004; Mehta et al., 2017; Rodrigues et al., 2005; Tsai et al., 1994), Sox17 (Clarke et al., 2013; Kim et al., 2007), and C-Myb (Emambokus et al., 2003; Lieu & Reddy, 2009; Mucenski et al., 1991; Zhang et al., 2011) have also been shown to be indispensable for normal mouse FL hematopoiesis (Table 1). Some of these factors also regulate adult HSC maintenance and function, while some are dispensable for adult hematopoiesis (Table 1). Likewise, a unique set of genes are required for the maintenance of adult HSCs (Hock et al., 2004; Semerad et al., 2009; van der Lugt et al., 1994; Ye et al., 2013; Zhuang et al., 1994). Interestingly, Sox17 and Lin28 are highly expressed in FL, but not adult, HSCs, and can be used to reprogram adult HSCs into cells that have regained at least some of the properties of fetal HSCs (He, Kim, Lim, & Morrison, 2011; Lang, Shi, & Chin, 2013; Yuan, Nguyen, Liu, Kanellopoulou, & Muljo, 2012). These studies have identified several key regulators of primitive and definitive hematopoiesis and have given us insight into the development and maintenance of distinct stages of HSCs (Table 1). However, the intricate network of genes being regulated by these factors remains under investigation, with the hopes of understanding the coordinated actions of key hematopoietic pathways that contribute to lifelong hematopoiesis.

4. Phenotypic and functional heterogeneity within the HSPC pool

4.1 Extensive transcriptome heterogeneity of cell populations

For several decades, the field has believed that one small pool of HSCs gives rise to a relatively homogeneous HSC pool and ultimately the blood system (Kumaravelu et al., 2002; Medvinsky & Dzierzak, 1996; Medvinsky, Rybtsov, & Taoudi, 2011; Morrison, Hemmati, Wandycz, & Weissman, 1995; Spangrude et al., 1988). Numerous studies have identified subsets of hematopoietic precursors and HSPCs using single-cell (sc)RNA-seq (Guo et al., 2013; Kowalczyk et al., 2015; Macaulay et al., 2016; Moignard, Woodhouse, Fisher, & Göttgens, 2013; Wilson et al., 2015). In an impressive analyses of early mesoderm specification, one group assayed over 3900 cells from developing embryos spanning E7.0 and E8.5 (Moignard et al., 2015). This revealed three cell clusters within the embryo: primitive streak/neural plate associated cells, a mixed group containing primitive streak/neural plate and hemogenic endothelial associated cells, and a final group consisting of hematopoietic associated cells. Computational analysis identified Sox7 as a potential regulator of early blood divergence. An inducible Sox7 overexpression model validated this prediction. Embryos overexpressing Sox7 displayed a clear reduction in primitive erythroid cells, indicating that downregulation of Sox7 is required for early blood cell commitment. Similarly, Scialdone et al. used bulk and scRNA-seq to interrogate the endothelial and hematopoietic specification from E6.5-E8.5 embryos (Scialdone et al., 2016). Gene expression data from wild-type embryos and SCL mutants indicated that SCL regulates several genes associated with early blood specification, including Sox7.

In parallel to these embryonic studies, several groups have addressed heterogeneity within the adult HSPC pool using RNA-seq (Kowalczyk et al., 2015; Wilson et al., 2015). One group performed scRNA-seq of adult HSCs and their direct descendants, MPP1 cells, and identified subsets of quiescent and "active" HSCs, based on differential expression of cell cycle genes, DNA replication genes, and myeloid associated genes (Yang et al., 2016). The vast heterogeneity identified by these studies has prompted investigation by our lab and others as to whether other distinct stem and progenitor cells exist throughout the lifespan. Analogous scRNA-seq analysis of fetal stem and progenitor populations may shed light on previously unidentified populations in the developing embryo.

4.2 Functional heterogeneity of hematopoietic progenitor populations

To address the functional consequences of the revealed heterogeneity, many labs have utilized transplantation, barcoding, and lineage tracing with complex reporter systems. Interestingly, experiments that investigated the capacity of HSCs isolated from different developmental time points and tissues revealed extensive heterogeneity that changes throughout development and the lifespan (Benz et al., 2012). This revelation suggested the existence of several distinct clones of HSCs and that pools are clonally selected based on supporting niches and the needs of the developing mouse. It also prompted the creation of more complex systems to address the clonal capacity of individual HSCs. Ganuza et al. utilized the Confetti system with several Cre lines to conclude, with the help of mathematical modeling, that several hundred developmental precursors contribute to long-term hematopoiesis (Ganuza, Hall, et al., 2017). Unlike previous estimations of HSC numbers in the developing embryo, this model did not rely on ex vivo culture or transplantation of embryonic cells. Also using a multicolor reporter strategy, Yu et al. reported that the murine hematopoietic system is composed of persistent and non-persistent clones, with a small population of "dominant" clones that persist and give rise to most of the blood system while others completely disappear (Yu et al., 2016). Such a non-persistent, developmentally restricted (dr)HSC that retain serial, multilineage reconstitution capability was recently identified in FL (Beaudin et al., 2016). Pei et al., using an artificial DNA recombination locus, "polylox," were able to tag thousands of single progenitor cells in the embryo and attribute the cell production capacity of individual clones (Pei et al., 2017).

Like Ganuza and Yu, they concluded that the embryo is made up of many clones of HSCs and that some clones are more productive than others. Similar results have been made in the zebrafish (Henninger et al., 2017). While the extent of heterogeneity within the embryonic and adult HSC pools is still unclear, there is mounting consensus that many more clones than previously believed contribute to hematopoiesis during development and throughout the lifespan.

4.3 Do all HSCs persist?

There is conflicting evidence regarding the persistence of HSCs. One hypothesis is that all HSCs, endowed with self-renewal capability by definition, persist throughout the lifespan. As discussed above, the Nishikawa and Inlay/Mikkola groups have provided evidence that hematopoietic cells specified in the YS can persist for life (Lee et al., 2016; Samokhvalov et al., 2007). However, there is also accumulating evidence for the converse model, with some "HSC" populations extinguishing over time (Yu et al., 2016). Recently, our lab identified a developmentally-restricted HSC (drHSC) that does not persist into adulthood in unperturbed mice, supporting the notion that not all pools of HSCs endure for life (Beaudin et al., 2016). We identified this drHSC via use of the $Flk2^{Cre}/Rosa^{mTmG}$, also known as "FlkSwitch," mice (Boyer, Beaudin, & Forsberg, 2012; Boyer, Schroeder, Smith-Berdan, & Forsberg, 2011). Although functional HSCs by serial transplantation assays, the drHSCs only exist perinatally (Figs. 3 and 4). The irreversible, Cre-mediated genetic deletion that marks the drHSCs (but not adult HSCs) means that drHSCs cannot be precursors of adult HSCs, thereby questioning the idea that hematopoiesis is one linear progression through development. The temporal restriction of drHSCs to a perinatal window also defies the principle that engraftable HSCs persist and repopulate the entire blood system throughout the lifespan. The drHSCs efficiently gave rise to unique subsets of cells, including innate-like B and T lymphocytes, supporting the concept of developmental waves that comprise both unique lineage capacities and varying degrees of overlap in lineage potential (Fig. 3). The ability to prospectively isolate drHSCs will enable new insights to the mechanisms regulating HSC persistence and lineage potential and extend recent work in the field that has highlighted the vast heterogeneity of hematopoietic progenitors (Table 2) (Eaves, 2015).

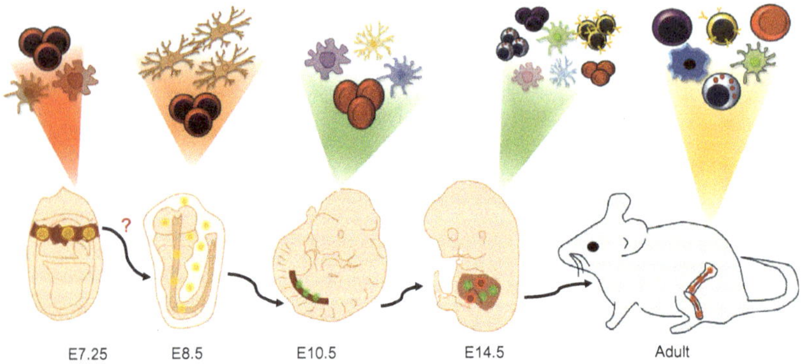

Fig. 4 Progenitors at different time points give rise to both overlapping and distinct subsets of cells. A simplified model depicting key features of developmental hematopoiesis. Around E7, cells of the yolk sac blood islands give rise to primitive macrophages and to primitive erythrocytes that are nucleated and express fetal globins. At E8.5, EMPs contribute to the first wave of tissue resident macrophages, including microglia in the developing brain. By E10.5, the first engraftable HSCs emerge in the AGM region of the embryo proper and initiate production of traditional mature blood cells (erythrocytes, platelets, granulocytes, monocytes/macrophages, B cells, T cells, and NK cells. At E14.5, developmentally restricted (dr)HSCs and persisting HSCs coexist in the fetal liver; drHSCs efficiently produce innate-like and traditional lymphoid cells. After birth, definitive HSCs primarily reside in the bone marrow and contribute to "traditional" mature blood cells for the lifespan of the animal.

Table 2 Summary of different hematopoietic stem and progenitor cells (HSPCs) that contribute to development of the blood system.

Progenitor	Stage	Persists throughout life?	Lineage bias	References
Primitive progenitors	E7.25-E10.5	No	Myeloid (primitive red blood cells)	Haar and Ackerman (1971)
Erythromyeloid progenitors (EMPs)	E8.25-E11.5	No	Myeloid (erythrocytes, megakaryocytes, macrophages)	Palis, Robertson, Kennedy, Wall, and Keller (1999)
Lymphomyeloid progenitors (LMPs)	E9.5s life	ND	Lymphoid, GM	Boiers et al. (2013)
Developmentally-restricted HSCs (drHSCs)	E12.5-P14	No	Lymphoid (innate-like lymphocytes)	Beaudin et al. (2016)
Definitive HSCs	E10.5-life	Yes	Clone dependent?	Medvinsky et al. (2011) and Ottersbach and Dzierzak (2005)

5. Purpose of developmental waves

5.1 Immune system layering

The very first hematopoietic cells produced during embryogenesis, prior to the emergence of reconstituting HSCs, are nucleated erythrocytes and macrophages (Table 2, Figs. 3 and 4). Erythrocytes are responsible for carrying oxygen throughout the developing embryo, and macrophages play important roles in tissue development and homeostasis (Blander, 2017; Gomez Perdiguero & Geissmann, 2013; Kierdorf, Prinz, Geissmann, & Gomez Perdiguero, 2015; Nicolás-Ávila, Hidalgo, & Ballesteros, 2018). The specification of these cells is intriguing since they do not appear to derive from the classical hierarchy that defines later stages of hematopoiesis. Similarly, tissue resident macrophages (TrMacs) have been the topic of substantial interest in recent years due to their reported pre-HSC origin (Lichanska & Hume, 2000; Mizoguchi, Takahashi, Takeya, Naito, & Morioka, 1992; Sorokin, Hoyt, Blunt, & McNelly, 1992). TrMacs have been widely implicated in disease, from neurodegeneration and autoimmune disorders to cancer (Honold & Nahrendorf, 2018; Nicolás-Ávila et al., 2018; Sevenich, 2018; Yin, Valin, Dixon, & Leavenworth, 2017; Zhou, Ding, et al., 2016). Understanding how TrMacs develop and how they are maintained long-term will provide a better understanding of several diseases and may uncover potential therapeutic targets.

In the last decade, several groups have utilized fate mapping to identify distinct subsets of TrMacs that arise from YS progenitors and "self-renew" in their respective tissues (Epelman et al., 2014; Ghigo et al., 2013; Ginhoux et al., 2010; Hashimoto et al., 2013; Hoeffel et al., 2015, 2012; Kierdorf et al., 2013; Schulz et al., 2012; Yona et al., 2013) (reviewed in Hoeffel & Ginhoux, 2018; Lavin, Mortha, Rahman, & Merad, 2015). Neither resident brain macrophages (microglia) nor skin macrophages (Langerhans cells) appear to be maintained by BM-derived progenitors (Ajami, Bennett, Krieger, Tetzlaff, & Rossi, 2007; Merad et al., 2002). Instead, microglia and Langerhans cells are likely of fetal origin, without significant contribution from adult HSCs at steady state (Fig. 4). Several groups have since reported BM-independent maintenance of additional TrMacs, including Kupffer cells of the liver, alveolar macrophages of the lung, cardiac macrophages of the heart, and red pulp macrophages of the spleen (Fig. 4) (Epelman et al., 2014; Hashimoto et al., 2013; Yona et al., 2013). While most groups agree that many TrMacs have a fetal origin, the mechanisms that regulate seeding of these cells in distinct developmental waves and tissues require further investigation. Surprisingly, we have found that tissue monocytes express IL7rα and contribute to TrMacs in several tissues (Beaudin, McCann, Leung, Worthington, & Forsberg, unpublished data). This finding was unexpected, as IL7rα is restricted to the lymphoid lineage in canonical hematopoiesis (Schlenner et al., 2010). Additional unanticipated regulators of tissue resident immune cells will likely be uncovered in the future.

Mast cells have also been suggested to have fetal origins. In 1983, one group injected YS- or FL-derived mast precursor cells into adult mice that lack mast cells to determine the site of precursor activity (Hayashi, Sonoda, & Kitamura, 1983). They demonstrated that the precursor activity was limited to the YS, arguing that these likely do not originate from definitive HSCs. Very recently, Gentek et al. used a VE-cadherin fate mapping model to assess the ontogeny of mast cells (Gentek et al., 2018). They concluded that mast cells largely originate from YS progenitors, but that adult definitive HSCs also contribute. They also demonstrated that YS-derived mast cells are transcriptionally distinct from their adult-derived counterparts. This finding is supportive of multiple waves of HSPCs contributing to tissue resident myeloid cells.

In addition to TrMacs and mast cells, several types of innate-like lymphocytes have been shown to have primarily fetal origins. One such cell type, B1a cells that predominantly exist in serous cavities, has become highly scrutinized (Beaudin & Forsberg, 2016). The Herzenberg group showed that fetal progenitor cells have a unique capability of generating Ly-1+ B cells (B1 cells) (Hayakawa, Hardy, Herzenberg, & Herzenberg, 1985). They concluded that only neonatal, not adult, cells had the capacity to generate B1a cells upon transplantation into irradiated recipients. This finding was further corroborated by additional studies (Barber, Montecino-Rodriguez, & Dorshkind, 2011; Ghosn et al., 2012). However, the conclusion that FL HSCs are incapable of contributing to B1a cells in an irradiated adult host (Ghosn et al., 2016) has been convincingly contested (Beaudin et al., 2016). The source of the divergent conclusions likely stem from strict phenotypic (Ghosn) versus functional (Beaudin and Kristensen) definitions of fetal HSCs. Several recent reviews provide more in depth discussion (Beaudin & Forsberg, 2016; Hadland & Yoshimoto, 2018; Hardy & Hayakawa, 2015; Herzenberg, 2015; Yoshimoto, 2015). While the cellular precursors of B1a cells remain under debate, there is general consensus on a primarily fetal origin of the B1a lineage.

Although the exact contribution of each wave to the developing blood system remains under investigation, there is clearly substantial overlap in hematopoietic cell production between waves. It appears that the earlier waves contribute to innate immune cells that provide basic immunity and aid in tissue formation and organization, while later perinatal waves contribute to innate and adaptive immune cells that prepare the fetus for exposure to environmental pathogens after birth. How early life exposures affect the emergence and expansion of different immune cells, and how that influence immune cell function in later life, is an important avenue of investigation.

6. Conclusion and future directions

As highlighted in this review, hematopoiesis undergoes continual molecular and functional changes throughout development and the lifespan. It seems likely that several clones of HSCs arise independently, in temporally and spatially regulated waves (Fig. 4 and Table 2). Exactly how many HSCs emerge and persist in the embryo, and their relative lifelong contribution to the blood system remains under debate. Cre-lox, barcoding, and transplantation tracking systems have, together with RNA-seq, improved our understanding of blood development, yet some questions remain difficult to address. It is unclear which waves are essential and how much functional overlap exists between them. Rigorous testing would require specifically deleting one lineage without perturbing the others, but that will depend on development of more precise and efficient tools. The novel genetic regulators identified by RNA-seq, in conjunction with reconstitution assays, fate mapping models, and innovative microscopy techniques, promise to provide an increasingly comprehensive understanding of blood system development. Quantitative assessment of individual HSC output, in situ and upon transplantation, is also needed to provide deeper insight into the lifespan and differentiation capacity of HSCs. Additionally, understanding the amazing epigenetic memory of cells, such as the drHSCs that maintain lineage biased output upon serial transplantation, may allow us to better utilize, and potentially engineer, HSCs for therapeutic use, and will likely also shed light on the drivers of blood and immune disease. While our understanding of hematopoiesis has expanded immensely, the spatial and temporal regulation of hematopoietic development warrant continued investigation.

Acknowledgments

We thank Dr. Smrithi Rajendiran, Atesh Worthington, Eric Martin, and Stephanie Smith-Berdan for insightful comments on the manuscript. E.C.F. is supported by NIH/NIDDK (R01DK100917), NIH/NHLBI (R01HL115158), by American Cancer Society Research Scholar Award (RSG-13-193-01-DDC), and an American Asthma Foundation Research Scholar award.

Author contributions

TM and ECF co-wrote the manuscript.

Competing interests

There are no conflicts of interest.

References

Abram, C. L., Roberge, G. L., Hu, Y., & Lowell, C. A. (2014). Comparative analysis of the efficiency and specificity of myeloid-Cre deleting strains using ROSA-EYFP reporter mice. *Journal of Immunological Methods*, *408*, 89–100. https://doi.org/10.1016/j.jim.2014.05.009.

Ajami, B., Bennett, J. L., Krieger, C., Tetzlaff, W., & Rossi, F. M. V. (2007). Local self-renewal can sustain CNS microglia maintenance and function throughout adult life. *Nature Neuroscience*, *10*, 1538–1543. https://doi.org/10.1038/nn2014.

Arora, N., Wenzel, P. L., McKinney-Freeman, S. L., Ross, S. J., Kim, P. G., Chou, S. S., et al. (2014). Effect of developmental stage of HSC and recipient on transplant outcomes. *Developmental Cell*, *29*, 621–628. https://doi.org/10.1016/j.devcel.2014.04.013.

Arzalluz-Luque, Á., Devailly, G., Mantsoki, A., & Joshi, A. (2017). Delineating biological and technical variance in single cell expression data. *The International Journal of Biochemistry & Cell Biology*, *90*, 161–166. https://doi.org/10.1016/j.biocel.2017.07.006.

Barber, C. L., Montecino-Rodriguez, E., & Dorshkind, K. (2011). Reduced production of B-1-specified common lymphoid progenitors results in diminished potential of adult marrow to generate B-1 cells. *Proceedings of the National Academy of Sciences of the United States of America*, *108*, 13700–13704. https://doi.org/10.1073/pnas.1107172108.

Beaudin, A. E., Boyer, S. W., Perez-Cunningham, J., Hernandez, G. E., Derderian, S. C., Jujjavarapu, C., et al. (2016). A transient developmental hematopoietic stem cell gives rise to innate-like B and T cells. *Cell Stem Cell*, *19*, 768–783. https://doi.org/10.1016/j.stem.2016.08.013.

Beaudin, A. E., & Forsberg, E. C. (2016). To B1a or not to B1a: Do hematopoietic stem cells contribute to tissue-resident immune cells? *Blood*, *128*, 2765–2769. https://doi.org/10.1182/blood-2016-10-697813.

Becker, A. J., McCulloch, E. A., & Till, J. E. (1963). Cytological demonstration of the clonal nature of spleen colonies derived from transplanted mouse marrow cells. *Nature*, *197*, 452–454.

Benz, C., Copley, M. R., Kent, D. G., Wohrer, S., Cortes, A., Aghaeepour, N., et al. (2012). Hematopoietic stem cell subtypes expand differentially during development and display distinct lymphopoietic programs. *Cell Stem Cell*, *10*, 273–283.

Bertrand, J. Y., Chi, N. C., Santoso, B., Teng, S., Stainier, D. Y. R., & Traver, D. (2010). Haematopoietic stem cells derive directly from aortic endothelium during development. *Nature*, *464*, 108–111. https://doi.org/10.1038/nature08738.

Blander, J. M. (2017). The many ways tissue phagocytes respond to dying cells. *Immunological Reviews*, *277*, 158–173. https://doi.org/10.1111/imr.12537.

Boiers, C., Carrelha, J., Lutteropp, M., Luc, S., Green, J. C., Azzoni, E., et al. (2013). Lymphomyeloid contribution of an immune-restricted progenitor emerging prior to definitive hematopoietic stem cells. *Cell Stem Cell*, *13*(5), 535548.

Boisset, J.-C., & Robin, C. (2012). On the origin of hematopoietic stem cells: Progress and controversy. *Stem Cell Research*, *8*, 1–13. https://doi.org/10.1016/j.scr.2011.07.002.

Boisset, J.-C., van Cappellen, W., Andrieu-Soler, C., Galjart, N., Dzierzak, E., & Robin, C. (2010). In vivo imaging of haematopoietic cells emerging from the mouse aortic endothelium. *Nature*, *464*, 116–120. https://doi.org/10.1038/nature08764.

Boyer, S. W., Beaudin, A. E., & Forsberg, E. C. (2012). Mapping differentiation pathways from hematopoietic stem cells using Flk2/Flt3 lineage tracing. *Cell Cycle*, *11*, 3180–3188. https://doi.org/10.4161/cc.21279.

Boyer, S. W., Schroeder, A. V., Smith-Berdan, S., & Forsberg, E. C. (2011). All hematopoietic cells develop from hematopoietic stem cells through Flk2/Flt3-positive progenitor cells. *Cell Stem Cell*, *9*, 64–73. https://doi.org/10.1016/j.stem.2011.04.021.

Bresciani, E., Carrington, B., Wincovitch, S., Jones, M., Gore, A. V., Weinstein, B. M., et al. (2014). CBF and RUNX1 are required at 2 different steps during the development of hematopoietic stem cells in zebrafish. *Blood*, *124*, 70–78. https://doi.org/10.1182/blood-2013-10-531988.

Busch, K., Klapproth, K., Barile, M., Flossdorf, M., Holland-Letz, T., Schlenner, S. M., et al. (2015). Fundamental properties of unperturbed haematopoiesis from stem cells in vivo. *Nature*, *518*, 542–546. https://doi.org/10.1038/nature14242.

Chen, M. J., Yokomizo, T., Zeigler, B. M., Dzierzak, E., & Speck, N. A. (2009). Runx1 is required for the endothelial to haematopoietic cell transition but not thereafter. *Nature*, *457*, 887–891. https://doi.org/10.1038/nature07619.

Clarke, R. L., Yzaguirre, A. D., Yashiro-Ohtani, Y., Bondue, A., Blanpain, C., Pear, W. S., et al. (2013). The expression of Sox17 identifies and regulates haemogenic endothelium. *Nature Cell Biology*, *15*, 502–510. https://doi.org/10.1038/ncb2724.

de Bruijn, M. (2014). The hemangioblast revisited. *Blood*, *124*, 2472–2473. https://doi.org/10.1182/blood-2014-09-597674.

de Bruijn, M. F., Ma, X., Robin, C., Ottersbach, K., Sanchez, M.-J., & Dzierzak, E. (2002). Hematopoietic stem cells localize to the endothelial cell layer in the midgestation mouse aorta. *Immunity*, *16*, 673–683.

de Bruijn, M. F., Speck, N. A., Peeters, M. C., & Dzierzak, E. (2000). Definitive hematopoietic stem cells first develop within the major arterial regions of the mouse embryo. *The EMBO Journal*, *19*, 2465–2474. https://doi.org/10.1093/emboj/19.11.2465.

Dharampuriya, P. R., Scapin, G., Wong, C., John Wagner, K., Cillis, J. L., & Shah, D. I. (2017). Tracking the origin, development, and differentiation of hematopoietic stem cells. *Current Opinion in Cell Biology*, *49*, 108–115. https://doi.org/10.1016/j.ceb.2018.01.002.

Durand, C., & Dzierzak, E. (2005). Embryonic beginnings of adult hematopoietic stem cells. *Haematologica*, *90*, 100–108.

Eaves, C. J. (2015). Hematopoietic stem cells: Concepts, definitions, and the new reality. *Blood*, *125*, 2605–2613. https://doi.org/10.1182/blood-2014-12-570200.

Eilken, H. M., Nishikawa, S.-I., & Schroeder, T. (2009). Continuous single-cell imaging of blood generation from haemogenic endothelium. *Nature*, *457*, 896–900. https://doi.org/10.1038/nature07760.

Emambokus, N., Vegiopoulos, A., Harman, B., Jenkinson, E., Anderson, G., & Frampton, J. (2003). Progression through key stages of haemopoiesis is dependent on distinct threshold levels of c-Myb. *The EMBO Journal*, *22*, 4478–4488. https://doi.org/10.1093/emboj/cdg434.

Epelman, S., Lavine, K. J., Beaudin, A. E., Sojka, D. K., Carrero, J. A., Calderon, B., et al. (2014). Embryonic and adult-derived resident cardiac macrophages are maintained through distinct mechanisms at steady state and during inflammation. *Immunity*, *40*, 91–104. https://doi.org/10.1016/j.immuni.2013.11.019.

Ganuza, M., Hadland, B., Chabot, A., Li, C., Kang, G., Bernstein, I., et al. (2017). Murine hemogenic endothelial precursors display heterogeneous hematopoietic potential ex vivo. *Experimental Hematology*, *51*, 25–35.e6. https://doi.org/10.1016/j.exphem.2017.04.006.

Ganuza, M., Hall, T., Finkelstein, D., Chabot, A., Kang, G., & McKinney-Freeman, S. (2017). Lifelong haematopoiesis is established by hundreds of precursors throughout mammalian ontogeny. *Nature Cell Biology*, *19*, 1153–1163. https://doi.org/10.1038/ncb3607.

Gekas, C., Rhodes, K. E., Vanhandel, B., Chhabra, A., Ueno, M., & Mikkola, H. K. A. (2010). Hematopoietic stem cell development in the placenta. *The International Journal of Developmental Biology*, *54*, 1089–1098. https://doi.org/10.1387/ijdb.103070cg.

Gentek, R., Ghigo, C., Hoeffel, G., Bulle, M. J., Msallam, R., Gautier, G., et al. (2018). Hemogenic endothelial fate mapping reveals dual developmental origin of mast cells. *Immunity*, *48*, 1160–1171.e5. https://doi.org/10.1016/j.immuni.2018.04.025.

Ghigo, C., Mondor, I., Jorquera, A., Nowak, J., Wienert, S., Zahner, S. P., et al. (2013). Multicolor fate mapping of Langerhans cell homeostasis. *The Journal of Experimental Medicine*, *210*, 1657–1664. https://doi.org/10.1084/jem.20130403.

Ghosn, E. E. B., Waters, J., Phillips, M., Yamamoto, R., Long, B. R., Yang, Y., et al. (2016). Fetal hematopoietic stem cell transplantation fails to fully regenerate the B-lymphocyte compartment. *Stem Cell Reports*, *6*, 137–149. https://doi.org/10.1016/j.stemcr.2015.11.011.

Ghosn, E. E. B., Yamamoto, R., Hamanaka, S., Yang, Y., Herzenberg, L. A., Nakauchi, H., et al. (2012). Distinct B-cell lineage commitment distinguishes adult bone marrow hematopoietic stem cells. *Proceedings of the National Academy of Sciences of the United States of America*, *109*, 5394–5398. https://doi.org/10.1073/pnas.1121632109.

Ginhoux, F., Greter, M., Leboeuf, M., Nandi, S., See, P., Gokhan, S., et al. (2010). Fate mapping analysis reveals that adult microglia derive from primitive macrophages. *Science*, *330*, 841–845. https://doi.org/10.1126/science.1194637.

Gomez Perdiguero, E., & Geissmann, F. (2013). Myb-independent macrophages: A family of cells that develops with their tissue of residence and is involved in its homeostasis. *Cold Spring Harbor Symposia on Quantitative Biology*, *78*, 91–100. https://doi.org/10.1101/sqb.2013.78.020032.

Gomez Perdiguero, E., Klapproth, K., Schulz, C., Busch, K., de Bruijn, M., Rodewald, H.-R., et al. (2015). The origin of tissue-resident macrophages: When an erythro-myeloid progenitor Is an erythro-myeloid progenitor. *Immunity*, *43*, 1023–1024. https://doi.org/10.1016/j.immuni.2015.11.022.

Gothert, J. R., Gustin, S. E., Hall, M. A., Green, A. R., Göttgens, B., Izon, D. J., et al. (2005). In vivo fate-tracing studies using the Scl stem cell enhancer: Embryonic hematopoietic stem cells significantly contribute to adult hematopoiesis. *Blood*, *105*, 2724–2732. https://doi.org/10.1182/blood-2004-08-3037.

Griffiths, J. A., Scialdone, A., & Marioni, J. C. (2018). Using single−cell genomics to understand developmental processes and cell fate decisions. *Molecular Systems Biology*, *14*, e8046. https://doi.org/10.15252/msb.20178046.

Guo, G., Luc, S., Marco, E., Lin, T.-W., Peng, C., Kerenyi, M. A., et al. (2013). Mapping cellular hierarchy by single-cell analysis of the cell surface repertoire. *Cell Stem Cell*, *13*, 492–505. https://doi.org/10.1016/j.stem.2013.07.017.

Haar, J. L., & Ackerman, G. A. (1971). A phase and electron microscopic study of vasculogenesis and erythropoiesis in the yolk sac of the mouse. *The Anatomical Record*, *170*, 199–223. https://doi.org/10.1002/ar.1091700206.

Hadland, B., & Yoshimoto, M. (2018). Many layers of embryonic hematopoiesis: New insights into B-cell ontogeny and the origin of hematopoietic stem cells. *Experimental Hematology*, *60*, 1–9. https://doi.org/10.1016/j.exphem.2017.12.008.

Hardy, R. R., & Hayakawa, K. (2015). Perspectives on fetal derived CD5$^+$ B1 B cells. *European Journal of Immunology*, *45*, 2978–2984. https://doi.org/10.1002/eji.201445146.

Hashimoto, D., Chow, A., Noizat, C., Teo, P., Beasley, M. B., Leboeuf, M., et al. (2013). Tissue-resident macrophages self-maintain locally throughout adult life with minimal contribution from circulating monocytes. *Immunity*, *38*, 792–804. https://doi.org/10.1016/j.immuni.2013.04.004.

Hayakawa, K., Hardy, R. R., Herzenberg, L. A., & Herzenberg, L. A. (1985). Progenitors for Ly-1 B cells are distinct from progenitors for other B cells. *The Journal of Experimental Medicine*, *161*, 1554–1568.

Hayashi, C., Sonoda, T., & Kitamura, Y. (1983). Bone marrow origin of mast cell precursors in mesenteric lymph nodes of mice. *Experimental Hematology*, *11*, 772–778.

He, S., Kim, I., Lim, M. S., & Morrison, S. J. (2011). Sox17 expression confers self-renewal potential and fetal stem cell characteristics upon adult hematopoietic progenitors. *Genes & Development*, *25*, 1613–1627. https://doi.org/10.1101/gad.2052911.

Henninger, J., Santoso, B., Hans, S., Durand, E., Moore, J., Mosimann, C., et al. (2017). Clonal fate mapping quantifies the number of haematopoietic stem cells that arise during development. *Nature Cell Biology*, *19*, 17–27. https://doi.org/10.1038/ncb3444.

Herring, C. A., Chen, B., McKinley, E. T., & Lau, K. S. (2018). Single-cell computational strategies for lineage reconstruction in tissue systems. *Cellular and Molecular Gastroenterology and Hepatology*, *5*, 539–548. https://doi.org/10.1016/j.jcmgh.2018.01.023.

Herzenberg, L. A. (2015). Layered evolution in the immune system: A view from history. *Annals of the New York Academy of Sciences*, *1362*, 1–5. https://doi.org/10.1111/nyas.12795.

Hock, H., Meade, E., Medeiros, S., Schindler, J. W., Valk, P. J. M., Fujiwara, Y., et al. (2004). Tel/Etv6 is an essential and selective regulator of adult hematopoietic stem cell survival. *Genes & Development*, *18*, 2336–2341. https://doi.org/10.1101/gad.1239604.

Hoeffel, G., Chen, J., Lavin, Y., Low, D., Almeida, F. F., See, P., et al. (2015). C-Myb(+) erythro-myeloid progenitor-derived fetal monocytes give rise to adult tissue-resident macrophages. *Immunity*, *42*, 665–678. https://doi.org/10.1016/j.immuni.2015.03.011.

Hoeffel, G., & Ginhoux, F. (2015). Ontogeny of tissue-resident macrophages. *Frontiers in Immunology*, *6*, 486. https://doi.org/10.3389/fimmu.2015.00486.

Hoeffel, G., & Ginhoux, F. (2018). Fetal monocytes and the origins of tissue-resident macrophages. *Cellular Immunology*, *330*, 5–15. https://doi.org/10.1016/j.cellimm.2018.01.001.

Hoeffel, G., Wang, Y., Greter, M., See, P., Teo, P., Malleret, B., et al. (2012). Adult Langerhans cells derive predominantly from embryonic fetal liver monocytes with a minor contribution of yolk sac-derived macrophages. *The Journal of Experimental Medicine*, *209*, 1167–1181. https://doi.org/10.1084/jem.20120340.

Honold, L., & Nahrendorf, M. (2018). Resident and monocyte-derived macrophages in cardiovascular disease. *Circulation Research*, *122*, 113–127. https://doi.org/10.1161/CIRCRESAHA.117.311071.

Huang, H., & Auerbach, R. (1993). Identification and characterization of hematopoietic stem cells from the yolk sac of the early mouse embryo. *Proceedings of the National Academy of Sciences of the United States of America*, *90*, 10110–10114.

Huber, T. L., Kouskoff, V., Fehling, H. J., Palis, J., & Keller, G. (2004). Haemangioblast commitment is initiated in the primitive streak of the mouse embryo. *Nature*, *432*, 625–630. https://doi.org/10.1038/nature03122.

Ichikawa, M., Asai, T., Saito, T., Yamamoto, G., Seo, S., Yamazaki, I., et al. (2004). AML-1 is required for megakaryocytic maturation and lymphocytic differentiation, but not for maintenance of hematopoietic stem cells in adult hematopoiesis. *Nature Medicine*, *10*, 299–304. https://doi.org/10.1038/nm997.

Kierdorf, K., Erny, D., Goldmann, T., Sander, V., Schulz, C., Perdiguero, E. G., et al. (2013). Microglia emerge from erythromyeloid precursors via Pu.1- and Irf8-dependent pathways. *Nature Neuroscience*, *16*, 273–280. https://doi.org/10.1038/nn.3318.

Kierdorf, K., Prinz, M., Geissmann, F., & Gomez Perdiguero, E. (2015). Development and function of tissue resident macrophages in mice. *Seminars in Immunology*, *27*, 369–378. https://doi.org/10.1016/j.smim.2016.03.017.

Kim, I., Saunders, T. L., & Morrison, S. J. (2007). Sox17 dependence distinguishes the transcriptional regulation of fetal from adult hematopoietic stem cells. *Cell*, *130*(3), 470–483. https://doi.org/10.1016/j.cell.2007.06.011.

Kim, I., Yilmaz, O. H., & Morrison, S. J. (2005). CD144 (VE-cadherin) is transiently expressed by fetal liver hematopoietic stem cells. *Blood*, *106*, 903–905. https://doi.org/10.1182/blood-2004-12-4960.

Kissa, K., & Herbomel, P. (2010). Blood stem cells emerge from aortic endothelium by a novel type of cell transition. *Nature*, *464*, 112–115. https://doi.org/10.1038/nature08761.

Kowalczyk, M. S., Tirosh, I., Heckl, D., Rao, T. N., Dixit, A., Haas, B. J., et al. (2015). Single-cell RNA-seq reveals changes in cell cycle and differentiation programs upon aging of hematopoietic stem cells. *Genome Research*, *25*, 1860–1872. https://doi.org/10.1101/gr.192237.115.

Kumaravelu, P., Hook, L., Morrison, A. M., Ure, J., Zhao, S., Zuyev, S., et al. (2002). Quantitative developmental anatomy of definitive haematopoietic stem cells/long-term repopulating units (HSC/RUs): Role of the aorta-gonad-mesonephros (AGM) region and the yolk sac in colonisation of the mouse embryonic liver. *Development*, *129*, 4891–4899.

Lacaud, G., & Kouskoff, V. (2017). Hemangioblast, hemogenic endothelium, and primitive versus definitive hematopoiesis. *Experimental Hematology*, *49*, 19–24. https://doi.org/10.1016/J.EXPHEM.2016.12.009.

Lacaud, G., Robertson, S., Palis, J., Kennedy, M., & Keller, G. (2001). Regulation of hemangioblast development. *Annals of the New York Academy of Sciences*, *938*, 96–107. discussion 108.

Lancrin, C., Sroczynska, P., Stephenson, C., Allen, T., Kouskoff, V., & Lacaud, G. (2009). The haemangioblast generates haematopoietic cells through a haemogenic endothelium stage. *Nature*, *457*, 892–895. https://doi.org/10.1038/nature07679.

Lang, J.-Y., Shi, Y., & Chin, Y. E. (2013). Reprogramming cancer cells: Back to the future. *Oncogene*, *32*, 2247–2248. https://doi.org/10.1038/onc.2012.349.

Lavin, Y., Mortha, A., Rahman, A., & Merad, M. (2015). Regulation of macrophage development and function in peripheral tissues. *Nature Reviews. Immunology*, *15*, 731–744. https://doi.org/10.1038/nri3920.

Lee, L. K., Ghorbanian, Y., Wang, W., Wang, Y., Kim, Y. J., Weissman, I. L., et al. (2016). LYVE1 Marks the divergence of yolk sac definitive hemogenic endothelium from the primitive erythroid lineage. *Cell Reports*, *17*, 2286–2298. https://doi.org/10.1016/j.celrep.2016.10.080.

Lichanska, A. M., & Hume, D. A. (2000). Origins and functions of phagocytes in the embryo. *Experimental Hematology*, *28*, 601–611.

Lieu, Y. K., & Reddy, E. P. (2009). Conditional c-myb knockout in adult hematopoietic stem cells leads to loss of self-renewal due to impaired proliferation and accelerated differentiation. *Proceedings of the National Academy of Sciences of the United States of America*, *106*, 21689–21694. https://doi.org/10.1073/pnas.0907623106.

Ling, K.-W., Ottersbach, K., van Hamburg, J. P., Oziemlak, A., Tsai, F.-Y., Orkin, S. H., et al. (2004). GATA-2 plays two functionally distinct roles during the ontogeny of hematopoietic stem cells. *The Journal of Experimental Medicine*, *200*, 871–882. https://doi.org/10.1084/jem.20031556.

Liu, C. P., & Auerbach, R. (1991). In vitro development of murine T cells from prethymic and preliver embryonic yolk sac hematopoietic stem cells. *Development, 113*, 1315–1323.
Macaulay, I. C., Svensson, V., Labalette, C., Ferreira, L., Hamey, F., Voet, T., et al. (2016). Single-cell RNA-sequencing reveals a continuous spectrum of differentiation in hematopoietic cells. *Cell Reports, 14*, 966–977. https://doi.org/10.1016/j.celrep.2015.12.082.
Masuoka, H. C., & Townes, T. M. (2002). Targeted disruption of the activating transcription factor 4 gene results in severe fetal anemia in mice. *Blood, 99*, 736–745.
Medvinsky, A., & Dzierzak, E. (1996). Definitive hematopoiesis is autonomously initiated by the AGM region. *Cell, 86*, 897–906.
Medvinsky, A., Rybtsov, S., & Taoudi, S. (2011). Embryonic origin of the adult hematopoietic system: Advances and questions. *Development, 138*, 1017–1031. https://doi.org/10.1242/dev.040998.
Mehta, C., Johnson, K. D., Gao, X., Ong, I. M., Katsumura, K. R., McIver, S. C., et al. (2017). Integrating enhancer mechanisms to establish a hierarchical blood development program. *Cell Reports, 20*, 2966–2979. https://doi.org/10.1016/j.celrep.2017.08.090.
Merad, M., Manz, M. G., Karsunky, H., Wagers, A., Peters, W., Charo, I., et al. (2002). Langerhans cells renew in the skin throughout life under steady-state conditions. *Nature Immunology, 3*, 1135–1141. https://doi.org/10.1038/ni852.
Mikkola, H. K. A., Klintman, J., Yang, H., Hock, H., Schlaeger, T. M., Fujiwara, Y., et al. (2003). Haematopoietic stem cells retain long-term repopulating activity and multipotency in the absence of stem-cell leukaemia SCL/tal-1 gene. *Nature, 421*, 547–551. https://doi.org/10.1038/nature01345.
Mizoguchi, S., Takahashi, K., Takeya, M., Naito, M., & Morioka, T. (1992). Development, differentiation, and proliferation of epidermal Langerhans cells in rat ontogeny studied by a novel monoclonal antibody against epidermal Langerhans cells, RED-1. *Journal of Leukocyte Biology, 52*, 52–61.
Moignard, V., Woodhouse, S., Fisher, J., & Göttgens, B. (2013). Transcriptional hierarchies regulating early blood cell development. *Blood Cells, Molecules & Diseases, 51*, 239–247. https://doi.org/10.1016/j.bcmd.2013.07.007.
Moignard, V., Woodhouse, S., Haghverdi, L., Lilly, A. J., Tanaka, Y., Wilkinson, A. C., et al. (2015). Decoding the regulatory network of early blood development from single-cell gene expression measurements. *Nature Biotechnology, 33*, 269–276. https://doi.org/10.1038/nbt.3154.
Moore, M. A., & Metcalf, D. (1970). Ontogeny of the haemopoietic system: Yolk sac origin of in vivo and in vitro colony forming cells in the developing mouse embryo. *British Journal of Haematology, 18*, 279–296.
Morrison, S. J., Hemmati, H. D., Wandycz, A. M., & Weissman, I. L. (1995). The purification and characterization of fetal liver hematopoietic stem cells. *Proceedings of the National Academy of Sciences of the United States of America, 92*, 10302–10306.
Mucenski, M. L., McLain, K., Kier, A. B., Swerdlow, S. H., Schreiner, C. M., Miller, T. A., et al. (1991). A functional c-myb gene is required for normal murine fetal hepatic hematopoiesis. *Cell, 65*, 677–689.
Murray, P. D. F. (1932). The development in vitro of the blood of the early chick embryo. *Proceedings of the Royal Society B: Biological Sciences, 111*, 497–521. https://doi.org/10.1098/rspb.1932.0070.
Nicolás-Ávila, J. A., Hidalgo, A., & Ballesteros, I. (2018). Specialized functions of resident macrophages in brain and heart. *Journal of Leukocyte Biology, 104*, 743–756. https://doi.org/10.1002/JLB.6MR0118-041R.
Nishikawa, S. (2012). Hemangioblast: An in vitro phantom. *Wiley Interdisciplinary Reviews: Developmental Biology, 1*, 603–608. https://doi.org/10.1002/wdev.38.

Oberlin, E., El Hafny, B., Petit-Cocault, L., & Souyri, M. (2010). Definitive human and mouse hematopoiesis originates from the embryonic endothelium: A new class of HSCs based on VE-cadherin expression. *The International Journal of Developmental Biology, 54,* 1165–1173. https://doi.org/10.1387/ijdb.103121eo.

Okuda, T., van Deursen, J., Hiebert, S. W., Grosveld, G., & Downing, J. R. (1996). AML1, the target of multiple chromosomal translocations in human leukemia, is essential for normal fetal liver hematopoiesis. *Cell, 84,* 321–330.

Org, T., Duan, D., Ferrari, R., Montel-Hagen, A., Van Handel, B., Kerenyi, M. A., et al. (2015). Scl binds to primed enhancers in mesoderm to regulate hematopoietic and cardiac fate divergence. *The EMBO Journal, 34,* 759–777. https://doi.org/10.15252/embj.201490542.

Ottersbach, K., & Dzierzak, E. (2005). The murine placenta contains hematopoietic stem cells within the vascular labyrinth region. *Developmental Cell, 8,* 377–387.

Palis, J., Robertson, S., Kennedy, M., Wall, C., & Keller, G. (1999). Development of erythroid and myeloid progenitors in the yolk sac and embryo proper of the mouse. *Development, 126,* 5073–5084.

Park, I. K., Qian, D., Kiel, M., Becker, M. W., Pihalja, M., Weissman, I. L., et al. (2003). Bmi-1 is required for maintenance of adult self-renewing haematopoietic stem cells. *Nature, 423,* 302–305.

Peeters, M., Ottersbach, K., Bollerot, K., Orelio, C., de Bruijn, M., Wijgerde, M., et al. (2009). Ventral embryonic tissues and Hedgehog proteins induce early AGM hematopoietic stem cell development. *Development, 136,* 2613–2621. https://doi.org/10.1242/dev.034728.

Pei, W., Feyerabend, T. B., Rössler, J., Wang, X., Postrach, D., Busch, K., et al. (2017). Polylox barcoding reveals haematopoietic stem cell fates realized in vivo. *Nature, 548,* 456–460. https://doi.org/10.1038/nature23653.

Perah, G., & Feldman, M. (1977). In vitro activation of the in vivo colony-forming units of the mouse yolk sac. *Journal of Cellular Physiology, 91,* 193–199. https://doi.org/10.1002/jcp.1040910205.

Rhodes, K. E., Gekas, C., Wang, Y., Lux, C. T., Francis, C. S., Chan, D. N., et al. (2008). The emergence of hematopoietic stem cells is initiated in the placental vasculature in the absence of circulation. *Cell Stem Cell, 2,* 252–263. https://doi.org/10.1016/j.stem.2008.01.001.

Robb, L., Lyons, I., Li, R., Hartley, L., Köntgen, F., Harvey, R. P., et al. (1995). Absence of yolk sac hematopoiesis from mice with a targeted disruption of the scl gene. *Proceedings of the National Academy of Sciences of the United States of America, 92,* 7075–7079.

Rodrigues, N. P., Janzen, V., Forkert, R., Dombkowski, D. M., Boyd, A. S., Orkin, S. H., et al. (2005). Haploinsufficiency of GATA-2 perturbs adult hematopoietic stem-cell homeostasis. *Blood, 106,* 477–484. https://doi.org/10.1182/blood-2004-08-2989.

Samokhvalov, I. M., Samokhvalova, N. I., & Nishikawa, S. (2007). Cell tracing shows the contribution of the yolk sac to adult haematopoiesis. *Nature, 446,* 1056–1061. https://doi.org/10.1038/nature05725.

Sasaki, K., Yagi, H., Bronson, R. T., Tominaga, K., Matsunashi, T., Deguchi, K., et al. (1996). Absence of fetal liver hematopoiesis in mice deficient in transcriptional coactivator core binding factor beta. *Proceedings of the National Academy of Sciences of the United States of America, 93,* 12359–12363.

Sawai, C. M., Babovic, S., Upadhaya, S., Knapp, D. J. H. F., Lavin, Y., Lau, C. M., et al. (2016). Hematopoietic stem cells are the major source of multilineage hematopoiesis in adult animals. *Immunity, 45,* 597–609. https://doi.org/10.1016/j.immuni.2016.08.007.

Schlaeger, T. M., Mikkola, H. K. A., Gekas, C., Helgadottir, H. B., & Orkin, S. H. (2005). Tie2Cre-mediated gene ablation defines the stem-cell leukemia gene (SCL/tal1)-dependent window during hematopoietic stem-cell development. *Blood, 105,* 3871–3874. https://doi.org/10.1182/blood-2004-11-4467.

Schlenner, S. M., Madan, V., Busch, K., Tietz, A., Läufle, C., Costa, C., et al. (2010). Fate mapping reveals separate origins of T cells and myeloid lineages in the thymus. *Immunity*, *32*, 426–436. https://doi.org/10.1016/j.immuni.2010.03.005.

Schoedel, K. B., Morcos, M. N. F., Zerjatke, T., Roeder, I., Grinenko, T., Voehringer, D., et al. (2016). The bulk of the hematopoietic stem cell population is dispensable for murine steady-state and stress hematopoiesis. *Blood*, *128*, 2285–2296. https://doi.org/10.1182/blood-2016-03-706010.

Schulz, C., Perdiguero, E. G., Chorro, L., Szabo-Rogers, H., Cagnard, N., Kierdorf, K., et al. (2012). A lineage of myeloid cells independent of Myb and hematopoietic stem cells. *Science*, *336*, 86–90. https://doi.org/10.1126/science.1219179.

Scialdone, A., Tanaka, Y., Jawaid, W., Moignard, V., Wilson, N. K., Macaulay, I. C., et al. (2016). Resolving early mesoderm diversification through single-cell expression profiling. *Nature*, *535*, 289–293. https://doi.org/10.1038/nature18633.

Semerad, C. L., Mercer, E. M., Inlay, M. A., Weissman, I. L., & Murre, C. (2009). E2A proteins maintain the hematopoietic stem cell pool and promote the maturation of myelolymphoid and myeloerythroid progenitors. *Proceedings of the National Academy of Sciences of the United States of America*, *106*, 1930–1935. https://doi.org/10.1073/pnas.0808866106.

Sevenich, L. (2018). Brain-resident microglia and blood-borne macrophages orchestrate central nervous system inflammation in neurodegenerative disorders and brain cancer. *Frontiers in Immunology*, *9*, 697. https://doi.org/10.3389/fimmu.2018.00697.

Shivdasani, R. A., Mayer, E. L., & Orkin, S. H. (1995). Absence of blood formation in mice lacking the T-cell leukaemia oncoprotein tal-1/SCL. *Nature*, *373*, 432–434. https://doi.org/10.1038/373432a0.

Sorokin, S. P., Hoyt, R. F., Blunt, D. G., & McNelly, N. A. (1992). Macrophage development: II. Early ontogeny of macrophage populations in brain, liver, and lungs of rat embryos as revealed by a lectin marker. *The Anatomical Record*, *232*, 527–550. https://doi.org/10.1002/ar.1092320410.

Spangrude, G. J., Heimfeld, S., & Weissman, I. L. (1988). Purification and characterization of mouse hematopoietic stem cells. *Science*, *241*, 58–62.

Sun, J., Ramos, A., Chapman, B., Johnnidis, J. B., Le, L., Ho, Y.-J., et al. (2014). Clonal dynamics of native haematopoiesis. *Nature*, *514*, 322–327. https://doi.org/10.1038/nature13824.

Tanaka, Y., Naruse, I., Hongo, T., Xu, M., Nakahata, T., Maekawa, T., et al. (2000). Extensive brain hemorrhage and embryonic lethality in a mouse null mutant of CREB-binding protein. *Mechanisms of Development*, *95*, 133–145.

Tanaka, Y., Sanchez, V., Takata, N., Yokomizo, T., Yamanaka, Y., Kataoka, H., et al. (2014). Circulation-independent differentiation pathway from extraembryonic mesoderm toward hematopoietic stem cells via hemogenic angioblasts. *Cell Reports*, *8*, 31–39. https://doi.org/10.1016/j.celrep.2014.05.055.

Taoudi, S., Gonneau, C., Moore, K., Sheridan, J. M., Blackburn, C. C., Taylor, E., et al. (2008). Extensive hematopoietic stem cell generation in the AGM region via maturation of VE-cadherin + CD45 + pre-definitive HSCs. *Cell Stem Cell*, *3*, 99–108. https://doi.org/10.1016/j.stem.2008.06.004.

Tober, J., Yzaguirre, A. D., Piwarzyk, E., & Speck, N. A. (2013). Distinct temporal requirements for Runx1 in hematopoietic progenitors and stem cells. *Development*, *140*, 3765–3776. https://doi.org/10.1242/dev.094961.

Toles, J. F., Chui, D. H., Belbeck, L. W., Starr, E., & Barker, J. E. (1989). Hemopoietic stem cells in murine embryonic yolk sac and peripheral blood. *Proceedings of the National Academy of Sciences of the United States of America*, *86*, 7456–7459.

Tsai, F. Y., Keller, G., Kuo, F. C., Weiss, M., Chen, J., Rosenblatt, M., et al. (1994). An early haematopoietic defect in mice lacking the transcription factor GATA-2. *Nature*, *371*, 221–226. https://doi.org/10.1038/371221a0.

Ueno, H., & Weissman, I. L. (2006). Clonal analysis of mouse development reveals a polyclonal origin for yolk sac blood islands. *Developmental Cell*, *11*, 519–533. https://doi.org/10.1016/j.devcel.2006.08.001.

van der Lugt, N. M., Domen, J., Linders, K., van Roon, M., Robanus-Maandag, E., te Riele, H., et al. (1994). Posterior transformation, neurological abnormalities, and severe hematopoietic defects in mice with a targeted deletion of the bmi-1 proto-oncogene. *Genes & Development*, *8*, 757–769.

Visvader, J. E., Fujiwara, Y., & Orkin, S. H. (1998). Unsuspected role for the T-cell leukemia protein SCL/tal-1 in vascular development. *Genes & Development*, *12*, 473–479.

Warren, A. J., Colledge, W. H., Carlton, M. B., Evans, M. J., Smith, A. J., & Rabbitts, T. H. (1994). The oncogenic cysteine-rich LIM domain protein rbtn2 is essential for erythroid development. *Cell*, *78*, 45–57.

Wilson, N. K., Kent, D. G., Buettner, F., Shehata, M., Macaulay, I. C., Calero-Nieto, F. J., et al. (2015). Combined single-cell functional and gene expression analysis resolves heterogeneity within stem cell populations. *Cell Stem Cell*, *16*, 712–724. https://doi.org/10.1016/j.stem.2015.04.004.

Wu, A. M., Till, J. E., Siminovitch, L., & McCulloch, E. A. (1968). Cytological evidence for a relationship between normal hemotopoietic colony-forming cells and cells of the lymphoid system. *The Journal of Experimental Medicine*, *127*, 455–464.

Yang, J., Tanaka, Y., Seay, M., Li, Z., Jin, J., Garmire, L. X., et al. (2016). Single cell transcriptomics reveals unanticipated features of early hematopoietic precursors. *Nucleic Acids Research*, *45*, 1281–1296. https://doi.org/10.1093/nar/gkw1214.

Ye, M., Zhang, H., Amabile, G., Yang, H., Staber, P. B., Zhang, P., et al. (2013). C/EBPa controls acquisition and maintenance of adult haematopoietic stem cell quiescence. *Nature Cell Biology*, *15*, 385–394. https://doi.org/10.1038/ncb2698.

Yin, J., Valin, K. L., Dixon, M. L., & Leavenworth, J. W. (2017). the role of microglia and macrophages in CNS homeostasis, autoimmunity, and cancer. *Journal of Immunology Research*, *2017*, 1–12. https://doi.org/10.1155/2017/5150678.

Yoder, M. C., Hiatt, K., & Mukherjee, P. (1997). In vivo repopulating hematopoietic stem cells are present in the murine yolk sac at day 9.0 postcoitus. *Proceedings of the National Academy of Sciences of the United States of America*, *94*, 6776–6780.

Yona, S., Kim, K.-W., Wolf, Y., Mildner, A., Varol, D., Breker, M., et al. (2013). Fate mapping reveals origins and dynamics of monocytes and tissue macrophages under homeostasis. *Immunity*, *38*, 79–91. https://doi.org/10.1016/j.immuni.2012.12.001.

Yoshimoto, M. (2015). The first wave of B lymphopoiesis develops independently of stem cells in the murine embryo. *Annals of the New York Academy of Sciences*, *1362*, 16–22. https://doi.org/10.1111/nyas.12612.

Yu, V. W. C., Yusuf, R. Z., Oki, T., Wu, J., Saez, B., Wang, X., et al. (2016). Epigenetic memory underlies cell-autonomous heterogeneous behavior of hematopoietic stem cells. *Cell*, *167*, 1310–1322.e17. https://doi.org/10.1016/j.cell.2016.10.045.

Yuan, J., Nguyen, C. K., Liu, X., Kanellopoulou, C., & Muljo, S. A. (2012). Lin28b reprograms adult bone marrow hematopoietic progenitors to mediate fetal-like lymphopoiesis. *Science*, *335*, 1195–1200. https://doi.org/10.1126/science.1216557.

Zhang, Y., Jin, H., Li, L., Qin, F. X.-F., & Wen, Z. (2011). cMyb regulates hematopoietic stem/progenitor cell mobilization during zebrafish hematopoiesis. *Blood*, *118*, 4093–4101. https://doi.org/10.1182/blood-2011-03-342501.

Zhao, Y., Zhou, J., Liu, D., Dong, F., Cheng, H., Wang, W., et al. (2015). ATF4 plays a pivotal role in the development of functional hematopoietic stem cells in mouse fetal liver. *Blood*, *126*, 2383–2391. https://doi.org/10.1182/blood-2015-03-633354.

Zhou, Z., Ding, M., Huang, L., Gilkeson, G., Lang, R., & Jiang, W. (2016). Toll-like receptor-mediated immune responses in intestinal macrophages; implications for mucosal immunity and autoimmune diseases. *Clinical Immunology*, *173*, 81–86. https://doi.org/10.1016/j.clim.2016.09.005.

Zhou, F., Li, X., Wang, W., Zhu, P., Zhou, J., He, W., et al. (2016). Tracing haematopoietic stem cell formation at single-cell resolution. *Nature*, *533*, 487–492. https://doi.org/10.1038/nature17997.

Zhuang, Y., Soriano, P., & Weintraub, H. (1994). The helix-loop-helix gene E2A is required for B cell formation. *Cell*, *79*, 875–884.

Zovein, A. C., Hofmann, J. J., Lynch, M., French, W. J., Turlo, K. A., Yang, Y., et al. (2008). Fate tracing reveals the endothelial origin of hematopoietic stem cells. *Cell Stem Cell*, *3*, 625–636. https://doi.org/10.1016/j.stem.2008.09.018.

Further reading

Davies, L. C., Jenkins, S. J., Allen, J. E., & Taylor, P. R. (2013). Tissue-resident macrophages. *Nature Immunology*, *14*, 986–995. https://doi.org/10.1038/ni.2705.

Lavin, Y., & Merad, M. (2013). Macrophages: Gatekeepers of tissue integrity. *Cancer Immunology Research*, *1*, 201–209. https://doi.org/10.1158/2326-6066.CIR-13-0117.

CHAPTER TWO

Signals and forces shaping organogenesis of the small intestine

Sha Wang[†], Katherine D. Walton[*,†], Deborah L. Gumucio
University of Michigan, Cell and Developmental Biology Department, Ann Arbor, MI, United States
*Corresponding author: e-mail address: kdwalton@med.umich.edu

Contents

Abstract

The adult gastrointestinal tract (GI) is a series of connected organs (esophagus, stomach, small intestine, colon) that develop via progressive regional specification of a continuous tubular embryonic organ anlage. This chapter focuses on organogenesis of the small intestine. The intestine arises by folding of a flat sheet of endodermal cells into a tube of highly proliferative pseudostratified cells. Dramatic elongation of this tube is driven by rapid epithelial proliferation. Then, epithelial-mesenchymal crosstalk and

[†]equal contribution

Current Topics in Developmental Biology, Volume 132
ISSN 0070-2153
https://doi.org/10.1016/bs.ctdb.2018.12.001

physical forces drive a stepwise cascade that results in convolution of the tubular surface into finger-like projections called villi. Concomitant with villus formation, a sharp epithelial transcriptional boundary is defined between stomach and intestine. Finally, flask-like depressions called crypts are established to house the intestinal stem cells needed throughout life for epithelial renewal. New insights into these events are being provided by *in vitro* organoid systems, which hold promise for future regenerative engineering of the small intestine.

1. Generation and growth of the intestinal tube

1.1 Embryonic origins of the murine small intestine

We focus this chapter primarily on development of the mouse small intestine, since the majority of work has been done in that system. Around embryonic day 6 (E6), the mouse embryo is cup-shaped and ready for gastrulation. The cup is composed of two cellular layers: (1) epiblast cells, which will give rise to the embryo proper, line the inner surface and are surrounded by (2) a layer of squamous cells known as the embryonic visceral endoderm (emVE). As gastrulation initiates, presumptive definitive endoderm (DE) cells travel through the primitive streak and intercalate into the emVE layer. Early lineage tracing using injected dyes or markers and histological analyses suggested that DE cells insert contiguously, displacing the emVE cells laterally, so that these extraembryonic cells are eventually completely eliminated from the portion of the developing endodermal sheet that will give rise to the embryonic gastrointestinal tract (reviewed in Viotti, Foley, & Hadjantonakis, 2014). In those studies, cell movements were inferred from analysis of multiple samples harvested at different times after labeling. However, more recent work, using broadly expressed lineage tracing tools, combined with live imaging of developing embryos, has revealed the actual dynamics of cellular movements and showed that the entering DE cells disperse rather than displace the emVE cells (Kwon, Viotti, & Hadjantonakis, 2008). Indeed, scattered emVE cells can be found in the completely formed gut tube as late as 15 somites (E8.75), and perhaps even later. However, it is not known whether these emVE-derived cells or their progeny have characteristics or roles that are distinct from DE cells, or how long they persist in the growing animal.

1.2 Shaping the intestinal tube

As the definitive endodermal sheet is being established (E7.0–8.0), the embryo is elongating along the anterior/posterior axis, head and cardiac regions are being defined and left-right asymmetry is initiating. Interestingly, the developing endodermal layer influences all of these processes (Martinez Barbera et al., 2000; Nascone & Mercola, 1995; Saund et al., 2012; Schultheiss, Xydas, & Lassar, 1995; Viotti, Niu, Shi, & Hadjantonakis, 2012; Withington, Beddington, & Cooke, 2001). In the case of left-right patterning, two groups showed that loss of the transcription factor *Sox17* in the endoderm perturbs both gut development and the acquisition of left-right asymmetry (Saund et al., 2012; Viotti et al., 2012). In the endoderm of *Sox17* mutant mice, Connexin 43 (CX43), a gap junctional protein, is missing (Viotti et al., 2012), resulting in defective intercellular transport between endoderm cells. Further investigations led both groups to conclude that the planar transmission of signals across gap junctional connections within the endoderm acts as a conduit to transfer a left-right asymmetry signal, generated in the node, to the lateral plate mesoderm.

Around E8.0, the flat endodermal sheet becomes indented at its anterior end and, shortly thereafter, its posterior end, forming the anterior and caudal intestinal portals (Fig. 1). Careful cell tracing studies have revealed the intricate morphogenic movements of endodermal cells during the formation and extension of these portals as the endodermal sheet is wrapped into a tube between E8.0 and E9.0 (Franklin et al., 2008; Tremblay & Zaret, 2005). Even before this complex cellular choreography establishes the gut tube proper,

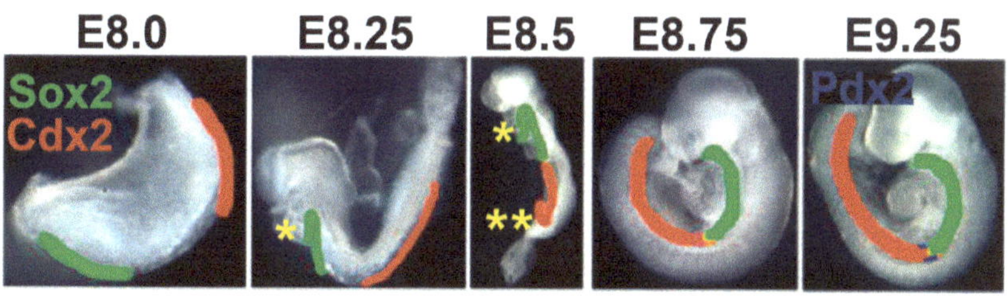

Fig. 1 Expression domains of CDX2, SOX2, and PDX1 during gut tube closure and formation of the anterior and caudal intestinal portals. At E8.0, SOX2 (green) and CDX2 (red) are expressed in the anterior- and posterior-most endoderm, respectively. At E8.25, the anterior intestinal portal is observed in the SOX2 domain (*) and at E8.5, the caudal intestinal portal (**) begins to form within the CDX2 domain. Between E8 and E8.75, the two domains extend toward each other as the endoderm wraps into a tube. The two domains meet at E8.5 and cells at the boundary co-express SOX2 and CDX2 (yellow). By E9.25, the boundary appears sharper in whole mount tissue (but see text) and PDX1 expression is seen in cells in the future pyloric region, extending from both sides of this boundary (blue). *Images were modified with permission from Sherwood, R. I., Chen, T. Y., & Melton, D. A. (2009). Transcriptional dynamics of endodermal organ formation.* Developmental Dynamics, 238, *29–42.*

anterior-posterior (A-P) patterning is initiating, regionalizing the gut into foregut, midgut and hindgut territories (Zorn & Wells, 2009). Soluble signals from surrounding mesodermal tissues are responsible, at least in part, for A-P patterning; an important role for FGF and WNT signals in posterior specification has been documented (Dessimoz, Opoka, Kordich, Grapin-Botton, & Wells, 2006; Sherwood, Maehr, Mazzoni, & Melton, 2011). By E7.75, the transcription factors *Sox2* and *Cdx2* are activated in the most anterior and most posterior endoderm, respectively (Fig. 1). As the portals move toward the middle of the embryo, the expression domains of these factors also move toward one another. At E9.5, after turning of the embryo completes gut tube formation, a distinct *Sox2-Cdx2* border is visible in whole mount tissue, in the region of the future pylorus (Sherwood, Chen, & Melton, 2009). Transcriptomic studies have revealed a number of additional transcription factors and cell surface proteins that show dynamic and region-specific expression patterns during this critical morphogenic process, and likely participate in establishment of the regional pattern (Sherwood et al., 2009).

1.3 Formation of the convoluted intestinal tract

The midgut first forms a hairpin loop at E10.5 (Fig. 2A, D, and J), which extends toward the ventral side of the embryo (Onouchi, Ichii, Otsuka, Hashimoto, & Kon, 2013) (Fig. 2B, E, E′, and K). As the gut tube actively elongates, it herniates into the umbilicus (Fig. 2B, E, and E′). At the same time, it also rotates 90° counterclockwise, around the axis of the superior mesenteric artery, so that its proximal end is to the right and distal end is to the left (Fig. 2B) (Kaufman, 1992; Kluth, Jaeschke-Melli, & Fiegel, 2003). A cecum bulge also forms more posteriorly, which separates the small and large intestine (Fig. 2K). The midgut continues to elongate rapidly while coiling into loops over the next 5 days outside of the body (Fig. 2E, E′, F, F′, G, G′, and L). Retraction of the gut back into the body cavity begins proximally, possibly driven by the muscular contraction of the duodenojejunal region (Cervantes, Yamaguchi, & Hebrok, 2009) (Fig. 2C and H). During this retraction, the gut undergoes a further counterclockwise rotation (Fig. 2C), which adjusts the position of the future small intestine with respect to the future large intestine (Davis et al., 2008). By E16.5, the entire gut has returned to the abdominal cavity with a properly convoluted pattern (Fig. 2I, I′, and M).

The complex process of herniation, looping of the growing intestine outside of the body and ordered return of the midgut to the body cavity

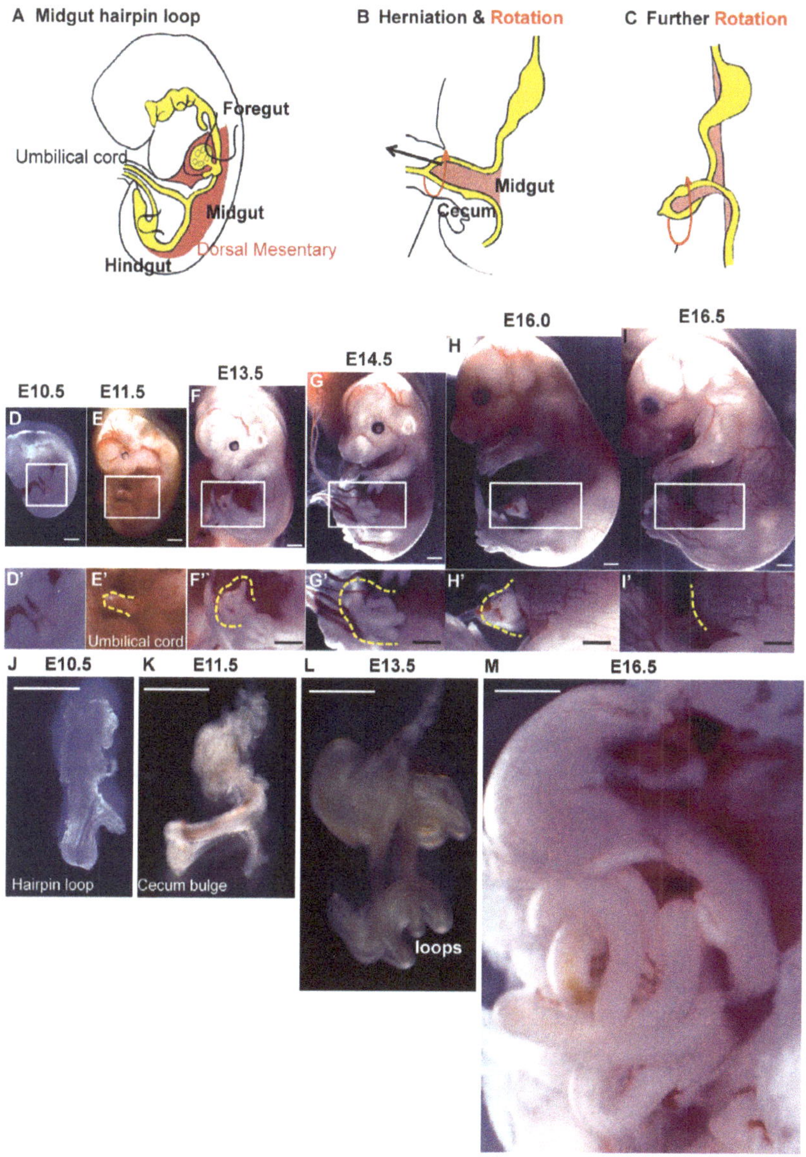

Fig. 2 Intestinal hernia, rotation, looping, and retraction during development. (A-C) Schematic illustration of intestinal hairpin loop formation (A), herniation (B) and rotation (B, C) in amniotes. As the midgut elongates, it protrudes into the umbilical cord and forms an intestinal hernia. As it protrudes, the midgut loop makes a counterclockwise rotation. Outside of the body, the gut elongates dramatically; looping/coiling occurs as the result of differential growth of the intestine and the attached mesentery. As the gut returns to the body cavity, it makes a further rotation. (D-I) Mouse embryos at E10.5–16.5. (D′-I′) Higher magnifications of the white boxed regions in (D-I), respectively (herniation and retraction). Yellow dashed line outlines the intestine located outside of the body cavity. (J-L) Dissected mouse GI tracts at E10.5 (hairpin loop), E11.5 (cecum bulge formation), E13.5 (looping). (M) Convoluted GI tract in an opened abdominal cavity of an E16.5 mouse embryo, showing intestinal loops. Scalebars are 1 mm.

are critical for proper intestinal function after birth. Intestinal malrotation, a congenital mis-positioning of the intestine within the peritoneal cavity (Applegate, Anderson, & Klatte, 2006), is a relatively common human birth defect (1 in 500 births) (Torres & Ziegler, 1993). Since the intestine is attached along its length to the dorsal mesentery, which provides the blood supply, abnormal rotation can cause ischemia of bowel segments, bowel necrosis or volvulus (Applegate et al., 2006).

The characteristic intestinal loops that form while the intestine is herniated allow intestinal length to extend well beyond the linear length of the embryo, while establishing an ordered configuration in the body cavity (Nerurkar, Mahadevan, & Tabin, 2017). Interestingly, looping is driven by mechanical buckling, due to elongation of the intestine against the constraint of the attached dorsal mesentery (Arnaout & Stainier, 2011; Savin et al., 2011). These two tissues have different growth rates, with the intestine outpacing the mesentery. As a result, the elongating intestine stretches the mesentery, which pulls against the intestine, compacting it into regular loops. The intestinal looping pattern (number and size of loops) is also species-specific and this observation led Nerurkar et al. to discover that BMP2, expressed by the dorsal mesentery, controls the rate of growth of this tissue (Nerurkar et al., 2017). Ectopic expression of BMP2 in the chick mesentery does not alter intestinal growth but further slows the growth of the mesentery, which enhances the growth rate gap between the two tissues and causes more tension on the mesentery, resulting in more and tighter loops (Nerurkar et al., 2017). Thus, through its differential effect on growth of two attached tissues, BMP signals tune the tensile forces that control looping patterns.

1.4 Intestinal elongation

One of the most amazing aspects of the small intestine is its length, a feature that contributes to the extensive absorptive surface area necessary for its function. The human intestine attains 40% of its adult length before birth (Weaver, Austin, & Cole, 1991). Failure of proper elongation of fetal intestine (congenital short bowel syndrome) is often fatal (Hasosah, Lemberg, Skarsgard, & Schreiber, 2008; van der Werf, Halim, Verheij, Alves, & Hofstra, 2015).

The intestinal tube begins to actively elongate immediately after its formation. Elongation can be divided into two distinct phases, based on the absence (Phase I) or presence (Phase II) of villi (Fig. 3). During Phase I (E10.5–14.5),

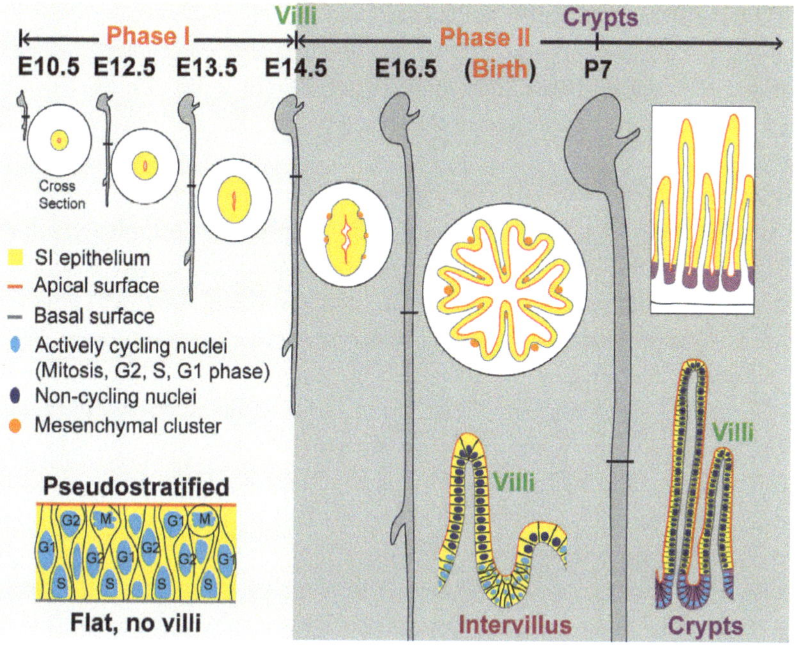

Fig. 3 Small intestinal elongation and its epithelial configuration in Phase I and Phase II. Schematics of intestinal lengthening (drawn to scale) are shown at top. Accompanying circular diagrams represent intestinal cross sections at the point of the horizontal line in top figures. Bottom diagrams show detail of cell shape in the epithelium. In Phase I, the epithelium is pseudostratified with a small flat apical surface (red line) and nearly all cells are actively cycling (blue nuclei, cell cycle phases are noted). In Phase II, beginning at E14.5, mesenchymal clusters form beneath the epithelium (orange) and villi begin to emerge. Cells on villi stop cycling (black nuclei) and change shape, becoming columnar. Cells in the intervillus regions remain proliferative (blue nuclei). Around P7, crypts (purple) form and house proliferating cells.

the epithelial tube is pseudostratified (Grosse et al., 2011) with a flat lumenal surface (Kohlnhofer, Thompson, Walker, & Battle, 2016; Walton, Freddo, Wang, & Gumucio, 2016). Phase II begins at E14.5. During this phase, finger-like villi emerge at the luminal surface, convoluting and expanding the apical surface area (Freddo et al., 2016; Walton, Freddo, et al., 2016; Walton et al., 2012). In addition to differences in epithelial structure, epithelial proliferation patterns differ in Phase I and II. Proliferating cells are distributed throughout the Phase I epithelium, but once villi emerge in Phase II, epithelial cells on top of the villi withdraw from the cell cycle and proliferation becomes confined to intervillus regions (Noah, Donahue, & Shroyer, 2011; Walton, Mishkind, Riddle, Tabin, & Gumucio, 2018).

During Phase I, the mouse small intestinal length, measured from pylorus (the opening from the stomach to the small intestine) to cecum (which marks the transition from small intestine to large intestine), increases more than 11-fold. Early studies led to the proposal that this rapid elongation was driven by convergent-extension movements, as multiple cell layers of a stratified epithelium converge into one layer, generating length by sacrificing girth (Cervantes et al., 2009; Matsumoto, Hashimoto, Yoshioka, & Otani, 2002; Reed et al., 2009). However, later work established that Phase I intestinal epithelium is composed of a single pseudostratified cell layer (Fig. 3) (Grosse et al., 2011; Wang, Cebrian, Schnell, & Gumucio, 2018). Additionally, the girth of the epithelial tube increases as it elongates as well, further ruling out convergent-extension as a mechanism of lengthening (Wang et al., 2018). Wang et al. found that all or nearly all epithelial cells are actively cycling, with cell cycle times of approximately 16 h. This uniformly rapid proliferation therefore appears to be the predominant driver for Phase I midgut growth.

As with other pseudostratified epithelia, the Phase I intestinal epithelium undergoes interkinetic nuclear migration (IKNM), in which nuclei synthesize DNA near the basal surface, migrate apically for mitosis and then return to basal to repeat the cycle (Fig. 4) (Grosse et al., 2011; Wang et al., 2018). Because of this constant nuclear motion, nuclei are staggered along the apical-basal axis. By combining high-resolution 3D confocal imaging and 2D live imaging, Wang et al. found that, as the nucleus moves apically to divide, the cell remains connected to the basal surface via a basal process. During mitosis, the basal process splits in a basal-to-apical direction. Subsequently, one of the two basal processes is often lost or retracted and the other is inherited by one of the two daughter cells. This means that one nascent daughter is born without a basal connection, though it does retain an apical link to its sister. During G1, the nuclei of both daughter cells return to the basal surface, but they do so in two distinct modes. In Mode I, one nucleus returns faster than the other. The daughter that inherits the basal process uses this basally connected process as a "conduit" to quickly return its nucleus to the basal side. In contrast, its sister actively projects a new process to establish a basal pathway and uses that path to return its nucleus; nuclear return by "pathfinding" takes twice as long as the "conduit" strategy. In Mode II, despite the fact that one daughter possesses a basal process, it does not use it to return its nucleus. Instead, both daughters utilize the "pathfinding" strategy, and in this case, both return their nuclei at a similar (slow) pace (Wang et al., 2018) (Fig. 4). Whether the mode of nuclear return or the

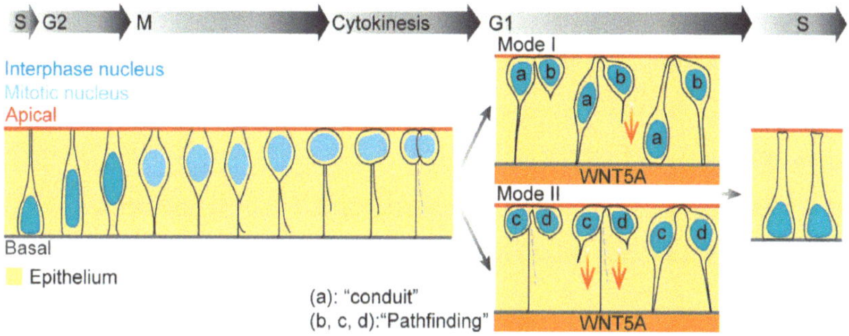

Fig. 4 The choreography of IKNM-associated cell division in Phase I small intestinal epithelium. As the nucleus migrates apically for mitosis, the cell maintains a basal connection *via* a thin filament, termed the basal process. This process splits into two, but only one of the two remains intact and is inherited by one daughter cell, leaving the other daughter disconnected basally. During G1, daughter pairs return their nuclei to the basal side in two distinct modes. In Mode I, the daughter with the basal process uses it as a "conduit" for a quick nuclear return; return of the other daughter's nucleus is slower as this cell must generate a filopodium to build a new pathway to return its nucleus—the "pathfinding" strategy. In Mode II, both daughters utilize "pathfinding" and return their nuclei basally at a similar pace. Pathfinding sometimes fails in the absence of the mesenchymal cue, WNT5A. The two daughters remain apically connected throughout mitosis. Finally, two daughters separate from each other (they are not side-by-side in the epithelium) and prepare for the next round of division. *Adapted from Fig. 7L from Wang, S., Cebrian, C., Schnell, S., & Gumucio, D. L. (2018). Radial WNT5A-guided post-mitotic filopodial pathfinding is critical for midgut tube elongation.* Developmental Cell, 46, *173–188.e3.*

asymmetric inheritance of the basal process has any functional consequence, with respect to stem cell potential, is currently unknown.

The critical importance of the "pathfinding" strategy for rapid Phase I elongation is revealed by analysis of cell behavior in *Wnt5a* knockout mice (Wang et al., 2018) which have severely shortened intestines (Cervantes et al., 2009). During Phase I growth, WNT5A is expressed by the mesenchymal cells that underlie and surround the intestinal epithelium. In its absence, filopodial pathfinding is impaired. Though daughter cells that retain a basal process can still return their nuclei by the conduit strategy, "pathfinding" daughter cells are impaired in their ability to build a new filopodial path. Their nuclei often remain at the apical side for prolonged periods, and eventually, some of these cells undergo apoptosis. The continual loss of approximately 10% of epithelial cells during this phase is sufficient to explain the intestinal length deficit seen in these mice (Wang et al., 2018). Of note, without WNT5A, some "pathfinding" daughters do manage to return basally, indicating the likely presence of other guidance cues, which could also play critical roles for Phase I midgut elongation.

ROR2 is a known WNT5A receptor and is expressed in both epithelial and mesenchymal cells in the developing intestine. Loss of *Ror2* also results in a shortened intestine, but to a lesser extent (by E13.5, *Wnt5a*$^{-/-}$ small intestines are only 28% of the wild-type length; *Ror2*$^{-/-}$ small intestines are about 63% of wild-type length) (Cervantes et al., 2009; Yamada et al., 2010). It is possible that ROR2 acts as a WNT5A sensor for "pathfinding" daughter cells to sense the gradient of WNT5A and extend filopodia toward to the basal surface. Also of note, mutations in FLNA (Filamin A), an actin-binding protein that acts downstream of WNT5A/ROR2 in filopodia formation (Nishita et al., 2006), have been identified in patients with congenital short bowel syndrome (van der Werf et al., 2013). A next important goal is to further probe the potential involvement of a WNT5A-ROR2-FLNA pathway in intestinal elongation, which might shed a light on the pathogenesis of congenital short bowel syndrome.

Several other genes and signaling pathways have also been reported to affect intestinal lengthening during Phase I, including GATA4 (Kohlnhofer et al., 2016), Hedgehog (Hh) (Mao, Kim, Rajurkar, Shivdasani, & Mcmahon, 2010), and FGF9 (Geske, Zhang, Patel, Ornitz, & Stappenbeck, 2008). GATA4 is an epithelially-expressed transcription factor required for epithelial cell proliferation, acting in part by regulating transcription of cell cycle mediators. Eliminating GATA4 from the midgut endoderm reduces intestinal length to 84% of control length by E14.5 (Kohlnhofer et al., 2016). In addition, perturbing the growth of the underlying mesenchyme can also hinder intestinal elongation, though in this case, it is hard to know whether simply reducing the amount of mesenchyme, or the perturbed signaling that necessarily results, is the cause (or both). Elimination of Shh and Ihh from the early intestinal endoderm (factors that signal from the epithelium to the underlying mesenchyme, in part to induce its proliferation) reduces intestinal length to approximately 10% of controls by E12.5 (Kim, Kim, Mao, Rowan, & Shivdasani, 2011; Mao et al., 2010; Ramalho-Santos, Melton, & McMahon, 2000). Similarly, loss of *Fgf9* limits mesenchymal fibroblast proliferation and reduces small intestinal length to 87% of its normal length by E14.5 (Geske et al., 2008).

Midgut elongation continues after villus initiation, but the underlying mechanisms are less well understood. Dissection of Phase II elongation is more complicated because the epithelium is no longer a set of relatively homogeneous proliferating pseudostratified cells. Cells at the tips of emerging villi become shorter and wider and cease cycling; thus, proliferative cells are confined to intervillus regions. It is not clear how these changes in cell shape and proliferative patterns affect lengthening. The programs that regulate proliferation also differ in the two phases. Canonical Wnt signals are required for epithelial proliferation in Phase II, but not in Phase I, when canonical Wnt activity is suppressed by ID2 expression (Chin et al., 2016; Korinek et al., 1998; Nigmatullina et al., 2017). Indeed, loss of epithelial β-catenin does not affect lengthening during Phase I but reduces the intestinal length to about 83% of control length between E14.5 and E15.5, the beginning of Phase II (Chin et al., 2016). Further investigation is needed to dissect what additional factors may influence Phase II elongation.

2. Patterning and regionalization

2.1 The pylorus: The anterior intestinal boundary

Embryonically, the pyloric region is an amazing hub of organogenesis, since epithelial and mesenchymal cells of this region give rise to stomach, intestine, spleen, pancreas, liver and gall bladder; complex signaling crosstalk among all of these organs is critical for development of this region (reviewed in Udager, Prakash, & Gumucio, 2010). In the adult epithelium, the pyloric boundary between stomach and intestine is particularly sharp; stomach cells sit directly adjacent to intestinal cells. Establishment of this dramatic tissue-specific border occurs progressively during development.

As discussed above, by E9.5, a distinct SOX2/CDX2 boundary can be seen at the pyloric border by immunofluorescence, marking the boundary between the future gastric and intestinal domains (Sherwood et al., 2009). However, while this boundary appears sharp when whole mount tissue is examined, stained transverse sections of the pyloric border region, examined at higher resolution, reveal a somewhat diffuse boundary at the cellular level (Li et al., 2009). At E14.5, SOX2 expression is high in stomach and a gradient of expression fades off into intestinal territory; similarly, CDX2 is expressed robustly in the intestine and its expression declines a short distance into the stomach. Expression of the intestine-specific structural protein, VILLIN, shows a similar "fuzzy" border at E14.5, but at E16.5, the intestine-stomach boundary is precisely demarcated for both VILLIN and CDX2 (SOX2 expression recedes to the forestomach) (Li et al., 2009). Several additional intestine-specific transcription factors (*Hnf4γ*, *Creb3l3* and *Tcfec*) also exhibit sharp anterior expression borders in the E16.5 pyloric epithelium (Li et al., 2009). Thus, it appears that the formation of a very precise anterior intestinal boundary occurs rather late in development, as the intestine begins to elaborate its characteristic finger-like villi.

Interestingly, at E14.5, transcriptomic analysis of developing gastric and intestinal epithelium and mesenchyme reveals only a handful of differences in gene expression (Li et al., 2009). These primarily represent clear regional patterning of *Sox2* and *Cdx2*, as well as several Hox and Parahox genes that are known to be regionally expressed. However, at E16.5, hundreds of intestinal genes are coordinately upregulated in the intestine and the vast majority of these are expressed in the epithelial, not the mesenchymal compartment. Gene ontology analysis indicates that the protein products of these genes have roles in metabolism and absorption, critical intestinal functions. Therefore, this transcriptional burst, which occurs concomitantly with villus development, appears to represent the acquisition of intestinal character or "intestinalization" (Li et al., 2009).

2.2 Signaling crosstalk: Endoderm/mesoderm interactions in gut tube patterning

Elegant tissue recombination studies performed over several decades by Dr. Nicole LeDouarin and colleagues produced solid evidence for the critical role of signaling crosstalk between the endoderm and its underlying mesoderm in patterning the intestinal domain. In practice, these experiments involve enzymatic separation of epithelium (prior to villus emergence) from mesenchyme and recombination with heterologous tissues from different regions of the GI tract or even with tissues from heterologous species (e.g., chick-quail). Grafts were derived from mouse, rat or chick intestine and were incubated in a vascular-rich environment such as under the skin of immunodeficient mice, under the rat kidney capsule, on the chick chorioallantoic membrane, or *ex vivo* in an appropriate culture system. The literature derived from these studies is extensive and thought-provoking. An excellent review interweaves the early literature with some of the more recent data (Grapin-Botton, 2005). Here, a few specific findings from these grafting experiments are highlighted.

First, pre-villus chick or rat endoderm cultured *in vitro* without underlying mesoderm will develop into a columnar epithelium with a well-polarized intestine-like apical surface that is studded with microvilli and express some intestine-specific enzyme markers (reviewed in Mizuno & Yasugi, 1990). While this suggests the presence of at least some "intestinal identity" of the endoderm, villi do not form in these cultures. In contrast, when endodermal pieces are recombined with isolated mesenchyme, villi are observed

(Duluc, Freund, Leberquier, & Kedinger, 1994). In fact, mesenchyme of many varieties will do. Even grafts composed of E14 rat intestinal endoderm and skin fibroblasts will form villi and express intestinal differentiation markers (Kedinger et al., 1998). However, the regional source of the endoderm exerts some dominance in the outcome. For example, when E14 rat endoderm from the proximal jejunum is combined with rat mesenchyme isolated from the distal ileum (or *vice versa*), the recombinants give rise to villus structures that express enzymes appropriate to the endodermal rather than the mesodermal portion of the graft (Duluc et al., 1994). Thus, while mesenchymal signals are critical for villus morphogenesis, the intestinal endoderm responds molecularly to mesenchymal signals by obeying a region-specific pre-pattern that is at least partially set before villi arise.

Second, small intestinal mesoderm has a powerful ability to instruct endoderm derived from other embryonic organ domains to become intestinal. When mesenchyme from pre-villus chick small intestine is combined with 5-day endoderm from esophagus, proventriculus, gizzard or colon, the grafts always take on an intestinal morphology (Duluc et al., 1994; Gumpel-Pinot, Yasugi, & Mizuno, 1978). In rodents, rat intestinal mesenchyme can instruct rat colon endoderm (but not lung or stomach endoderm) to form villi (Duluc et al., 1994). However, potentially due in part to its pre-patterned state, small intestinal endoderm tends to be rather resistant to heterologous instructions provided by mesoderm from other regions of the GI tract (Gumpel-Pinot et al., 1978).

Third, much of the mesenchymal instructional information appears to lie in a sub-epithelial fibroblast population. When 5.5-day chick gizzard endoderm (or E14 rat intestinal endoderm) is wrapped by a confluent sheet of pericryptal fibroblasts isolated from the intestine of postnatal rats and grafted into the chick coelomic cavity, the endoderm develops villi and expresses intestinal enzymes (Fritsch, Simon-Assmann, Kedinger, & Evans, 1997; Haffen, Lacroix, Kedinger, & Simon-Assmann, 1983). Some grafts also exhibited surrounding smooth muscle layers and express biochemical markers of intestine (Fritsch et al., 1997). Strikingly, a rat intestinal cell line, IEC-17, can be substituted for the endoderm; when wrapped with E14 rat intestinal mesenchyme and grafted under the rat kidney capsule, 10% of the grafts formed villi and exhibited an epithelium with typical intestinal cell types and enzymatic signatures of the intestine, with a well-developed surrounding muscular layer (Kedinger et al., 1986).

Finally, distinct molecular determinants seem to underlie structural morphogenesis and cytodifferentiation. For example, in grafts of E14 mouse lung endoderm with mouse intestinal mesenchyme, villi form successfully, but the epithelial component expresses lung-specific surfactants instead of intestinal enzymes (Duluc et al., 1994). Nevertheless, the degree of morphogenic re-organization of both endoderm (villus formation) and mesoderm (intestinal muscle formation) in these recombination studies provides substantial evidence for bi-directional crosstalk between the two tissue layers. Indeed, the very fact that intestinal mesenchyme can instruct lung endoderm to generate villi (Duluc et al., 1994) supports a back and forth molecular cascade that is self-organizing at the local (villus) level. Newer molecular data that support this hypothesis are discussed below.

2.3 Regionalization within the small intestine: Duodenum, jejunum, and ileum

While a major job of the small intestine is absorption and metabolism, different parts of the intestine are specialized for distinct purposes. The duodenum secretes a variety of enzymes which, in association with pancreatic and bile secretions, serve to break down food so that nutrients can be absorbed by the jejunum, while the ileum specializes in vitamin B_{12} and bile acid absorption. The complete transcriptional network that controls these differences in gene expression has not been fully elucidated, but some important findings have emerged.

First, both the chick and the mouse have a well-organized Hox code throughout the gut, with overlapping domains from anterior to posterior (Sakiyama, Yokouchi, & Kuroiwa, 2001; Sekimoto et al., 1998). Most of these Hox genes are mesenchymally expressed in the intestinal domain, though a handful (e.g., *Hoxd13*) are epithelial. In general, while loss of some of these regionally-restricted Hox genes affects the development of the domain in which that gene is expressed, homeotic transformations of regional fates are not seen (Beck, Tata, & Chawengsaksophak, 2000; Boulet & Capecchi, 1996; Zacchetti, Duboule, & Zakany, 2007). Interestingly, the positions of the anterior borders of expression of several Hox genes seem to correspond to anatomical sphincters or distinctive regions in the gut and there is functional evidence that Hox genes play a role in the morphogenesis of the pyloric, ileocecal and anal sphincters (Kondo, Dolle, Zakany, & Duboule, 1996; Zakany & Duboule, 1999).

The dynamic expression pattern of the caudal transcription factor, CDX2, during establishment of the gut anlage is discussed above. Once the gut tube is completed, CDX2 marks the stomach-intestine border, but its expression in the intestine is graded, with highest levels in the posterior ileum and colon (Silberg, Swain, Suh, & Traber, 2000). Indeed, loss of CDX2 early in development results in emergence of a squamous esophageal-like epithelium in the ileum and colon (Gao, White, & Kaestner, 2009). In *Cdx2* mutant tissue, a large number of intestinal genes are downregulated, while genes normally expressed in esophagus are upregulated, indicative of an anterior transformation. Interestingly, only subtle transient changes in the intestinal Hox code are noted in *Cdx2* mutants, reinforcing the idea that the Hox code may not be a primary determinant of regional patterning in the gut (Gao et al., 2009). Additionally, only minor morphological and transcriptional changes are seen in the duodenum of *Cdx2* null mice (Gao et al., 2009); thus, *Cdx2* function is not required to maintain intestinal character in the proximal intestine.

One of the most interesting regionalization phenotypes in Cdx2 mutants arises in the proximal colon of adult mice that are heterozygous for this gene. These animals develop polyps in tissue where the untargeted allele of *Cdx2* is ectopically silenced (Beck, Chawengsaksophak, Waring, Playford, & Furness, 1999; Chawengsaksophak, James, Hammond, Kontgen, & Beck, 1997). In the center of the polyp, the colonic epithelium takes on squamous, esophageal character. Amazingly, directly surrounding these lesions, a regular progression of regional gut identities can be seen, with concentric rings of tissue exhibiting gastric, followed by intestinal morphologies and gene expression patterns. Since analysis of *Cdx2* mutant ileum revealed dramatic changes in both Wnt and Hedgehog signaling (Gao et al., 2009), it will be interesting to determine whether such changes contribute to these interesting patterns.

Regionalization of the proximal duodenum is highly dependent upon proper expression of the Parahox factor, PDX1. *Pdx1* expression extends from the antral stomach to the mid-duodenum and includes the region of the developing pancreas and bile duct (Offield et al., 1996). In mutants lacking *Pdx1*, pancreatic buds form, but fail to grow out, and islets and acini are not established (Offield et al., 1996). In the proximal duodenum, villi do not form; instead, the surface epithelium is composed of cuboidal cells that resemble (morphologically and molecularly) bile duct epithelium. Finally, Brunner's glands, the submucosal glands characteristic of the duodenum, are reduced and structurally abnormal.

Regional patterning of the jejunum and ileum requires activity of GATA family transcription factors. GATA4 is expressed in the duodenum and jejunum, but absent from the ileum, while GATA6 is present throughout the intestine (Battle et al., 2008; Beuling et al., 2012; Bosse et al., 2006;

Fang, Olds, & Sibley, 2006). GATA4 is both necessary and sufficient to determine jejunal fate, since its loss results in expansion of ileal genes into the jejunal territory (Kohlnhofer et al., 2016; Walker, Thompson, Kohlnhofer, Faber, & Battle, 2014), whereas ectopic expression of GATA4 in the ileum (where it is normally absent) leads to activation of jejunal genes in the ileal domain (Thompson et al., 2017). On the other hand, *Gata6* loss has no effect on jejunal fate, but when this factor is deleted from the ileum, colonic markers are expressed in this domain (Beuling et al., 2011; Walker et al., 2014), consistent with an important role for GATA6 in suppressing colonic fate in the ileum.

3. Villus development

3.1 Signals and physical forces driving formation of the villi

Villus morphogenesis requires synchronized development of the epithelium and mesenchyme. This coordination is achieved by secreted signaling molecules as well as physical forces generated by tissue expansion and cell shape changes. At the top of the signaling hierarchy is the Hh signaling pathway which plays critical roles in multiple aspects of intestinal development in sea urchins, cnidarians, *Drosophila*, and vertebrates (Hoch & Pankratz, 1996; Matus, Magie, Pang, Martindale, & Thomsen, 2008; Sukegawa et al., 2000; Walton, Warner, Hertzler, & Mcclay, 2009). Loss of Hh signaling results in a plethora of intestinal malformations including duodenal stenosis, abnormal innervation, loss of smooth muscle, loss of villi, malrotation, reduced length, altered radial patterning and imperforate anus (Madison et al., 2005; Mo et al., 2001; Ramalho-Santos et al., 2000; Sukegawa et al., 2000).

In the small intestine, Hh signaling is paracrine. Shh and Ihh are produced and secreted from the nascent intestinal epithelium, even before gut tube closure (Bitgood & McMahon, 1995; Echelard et al., 1993). Hh signals are received by a variety of mesenchymal cell types including serosa, smooth muscle, nerves, endothelial cells, sub-epithelial mesenchymal cells, fibroblasts, and myofibroblasts (Kolterud et al., 2009). Constitutive activation of Hh signaling results in a massive expansion of the mesenchyme (Mao et al., 2010) and a large increase in smooth muscle differentiation (Zacharias et al., 2011), while conditional loss of embryonic Hh signaling, beginning at E10.5, results in reduced mesenchymal proliferation (Mao et al., 2010), especially of myofibroblasts and smooth muscle (Kosinski et al., 2010;

Madison et al., 2005; van Dop et al., 2009; Zacharias et al., 2011). Though the epithelium does not receive Hh signals, reduced Hh signaling causes spatial disruption of epithelial proliferative zones (normally restricted to intervillus regions during fetal stages). This indirect effect appears to be due to alterations in the signaling network downstream of Hh, including reduced Bmp signaling and increased Wnt signaling (Kosinski et al., 2010; Madison et al., 2005). Interestingly, a network of Hh responsive, FoxL1-positive subepithelial telocytes have recently been shown to secrete Wnt ligands and define the epithelial stem cell niche in adults (Aoki et al., 2016; Degirmenci, Valenta, Dimitrieva, Hausmann, & Basler, 2018; Shoshkes-Carmel et al., 2018). These cells may be, at least in part, the mesenchymal cells responsible for these epithelial alterations. It will be important to determine when these telocytes arise during development.

In mammals, villi emerge as domes from a flat epithelial tube and grow directly into finger-like projections (Fig. 5) (Walton et al., 2018). Villus emergence begins around E14.5 in mice when tight aggregations of Hh responsive mesenchymal cells aggregate into "clusters" (Karlsson, Lindahl, Heath, & Betsholtz, 2000; Walton et al., 2012). Such clusters were first noted in early electron microscopic studies of villus development (Mathan, Moxey, & Trier, 1976) and later shown to be hubs of PDGF signaling (Karlsson et al., 2000). Cluster cells express PDGFRα, while the epithelial cells above them express PDGFA ligand. Loss of either the receptor or the ligand does not affect the formation of the initial clusters but does suppress the proliferative activity of PDGFRα-positive mesenchymal cells so that formation of subsequent clusters is inhibited, leading to sparsely spaced and aberrantly shaped villi (Karlsson et al., 2000).

Since PDGF signaling does not affect cluster initiation, other signals are clearly necessary. Analysis of Hh signal transduction prior to villus emergence revealed that scattered sub-epithelial mesenchymal cells express Hh target genes (Walton et al., 2012). At E14.5, in the most proximal intestine, these mesenchymal cells begin to aggregate to form clusters next to the basement membrane of the epithelium. Hh signaling is critical for cluster formation since inhibition of this pathway abolishes cluster formation, while Hh pathway augmentation increases cluster size (Madison et al., 2005; Walton et al., 2012). Furthermore, cluster formation is required for villus emergence (Walton, Whidden, et al., 2016). Downstream from Hh, Bmp signals sent by and received by cells within the clusters themselves regulate cluster size and spacing through a self-organizing Turing field mechanism. Progressively reducing the level of Bmp signaling changes

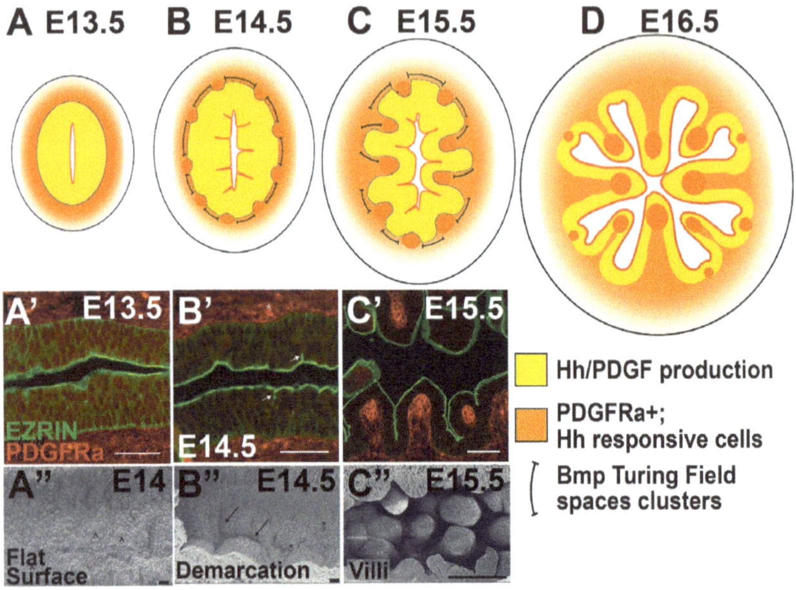

Fig. 5 Villus formation. (A) At E13.5, the intestinal epithelium (yellow) is a tube with a flat lumenal surface that uniformly secretes Hh and PDGF signals (yellow) to subepithelial mesenchymal cells (orange). (A′) The flat apical surface is highlighted by EZRIN staining (green) and the even distribution of PDGF responsive cells is noted by PDGFRa staining (red). (A″) SEM image showing the flat lumenal surface at E13.5. (B) By E14.5 in the anterior duodenum, Hh/PDGF responsive mesenchymal cells begin to aggregate to form clusters (orange dots) in a regularly spaced field; spacing is determined by a Bmp-dependent Turing field, in which formation of one cluster prevents the initiation of another cluster in the immediate area. (B′) Mesenchymal clusters result in uneven compression in the epithelium, leading to membrane invaginations that demarcate villi, here shown as T-shaped extensions of the apical surface (arrows). Apical demarcations (arrows) form between mesenchymal clusters. (B″) SEM reveals demarcations of initial villi at the apical surface (arrows). (C and C′) Villi emerge at the site of each cluster; clusters remain associated with tips of emerging villi. (C″) Villus domes are obvious in SEM. (D) By E16.5, villi are longer and additional rounds of cluster formation occur in intervillus regions; villi are present throughout the small intestine. *Panels (A′-C″): Reproduced from Freddo, A. M., Shoffner, S. K., Shao, Y., Taniguchi, K., Grosse, A. S., Guysinger, M. N., et al. (2016). Coordination of signaling and tissue mechanics during morphogenesis of murine intestinal villi: A role for mitotic cell rounding.* Integrative Biology: Quantitative Biosciences from Nano to Macro, 8, *918–928 with permission from The Royal Society of Chemistry.*

the pattern of discrete clusters (spots) into a continuous line of connected clusters (stripes) (Walton, Whidden, et al., 2016).

Hh-responsive mesenchymal clusters are signaling centers that express a myriad of soluble factors, including BMP ligands and modifiers

(BMP2,4,5,7, BMP1, TSG1, Noggin, FLSTL1) (Walton, Whidden, et al., 2016) as well as WNT5A (Wang et al., 2018). In addition to serving as patterning signals for the clusters themselves, these molecules feed back to the overlying epithelium, causing cells directly above the clusters to change shape and withdraw from the cell cycle. At this time, it is not known which signal or combination of signals produces these effects. Nevertheless, epithelial cells above the clusters, initially tall and pseudostratified, shorten and widen (without changing volume), becoming more columnar in shape (Freddo et al., 2016; Walton, Whidden, et al., 2016). During initial epithelial cell shortening, the apical surface remains flat and the cluster, which is adhered to the basement membrane, is drawn up into a shallow alcove in the epithelium (Fig. 5C). The focal widening of epithelial cells over clusters that accompanies cell shortening appears to place pressure on surrounding epithelial cells located in regions between clusters (Freddo et al., 2016). Within these regions, specialized cell divisions cause formation of rapid, deep membrane invaginations (Fig. 5B and B′), effectively carving out the villus domains and allowing the first villus "bumps" to emerge (Fig. 5B″) (Freddo et al., 2016). A similar mechanism of rapid membrane invagination, triggered by dividing cells, is also seen during *Drosophila* tracheal invagination (Kondo & Hayashi, 2013). In both cases, intraepithelial physical forces act on a tall, pseudostratified epithelium to fold it quickly (in the scale of minutes) in a patterned manner.

As villi emerge, cell proliferation at the tips is suppressed, perhaps due to BMP signals from the underlying clusters, which remain closely associated with the tips of emerging villi. Thus, proliferation becomes restricted to the intervillus regions. Additional rounds of villus formation occur, with new clusters forming beneath the proliferative, expanding intervillus epithelium; in the mouse, four rounds of cluster-driven villus formation are detected between E14.5 and E18.5 (Walton et al., 2012).

Curiously, the mechanisms of villus formation are quite different in the chick, though similar signaling molecules are used. In the mouse, the signaling cascade downstream of Hh-driven cluster formation provides the patterning cue that sets the stage for villus demarcation (Freddo et al., 2016; Walton et al., 2012; Walton, Whidden, et al., 2016). In contrast, the chick uses physical forces for patterning, driven by the sequential formation of smooth muscle layers in the deep mesenchyme, which confine the growing epithelium (Shyer, Huycke, Lee, Mahadevan, & Tabin, 2015; Shyer et al., 2013). A step-wise progression of epithelial bending results, which starts with length-wise ridges caused by formation of the circular muscle and progresses to compression of the ridges into zig-zags as the longitudinal muscle forms. The deep zig-zag folds are proposed to "trap" epithelial Hh ligands, resulting in the formation of clusters at the most compressed points (Shyer et al., 2015, 2013). These clusters express BMP ligands which serve to suppress WNT signaling in the epithelium overlying the clusters, simultaneously restricting WNT signals to the intervillus regions.

This coordination of muscle layer development with epithelial bending observed in the developing chick intestine is not observed in mouse or humans (Walton, Whidden, et al., 2016), though physical confinement and coordinated contractions may be important for villus outgrowth. Also, in the mouse, the pseudostratified epithelium does not transduce WNT signals prior to cluster-induced villus emergence. Rather, epithelial WNT signaling is activated in the intervillus regions as villi emerge (Chin et al., 2016; Nigmatullina et al., 2017). The signaling program that drives cell proliferation in the early pseudostratified mouse epithelium is unknown.

3.2 Mesenchymal factors controlling villus development

Several mesenchymally-expressed transcription factors have been shown to play a role in villus development. Genetic deletion of the homeodomain transcription factor *Nkx2–3* (a Hh target) results in reduced gut mesenchyme, impaired intervillus localization of epithelial proliferation and delayed villus formation, specifically in the jejunum (Pabst, Zweigerdt, & Arnold, 1999). The few villi that do form in these mutants are short and misshapen. Interestingly, at adult stages, proliferation is increased and both villus and crypt morphology are greatly disturbed; villi are abnormally long and branched, with increased villus core vasculature, while crypts extend deeper and are disorganized with increased fissioning. Mice mutant for *FoxL1/Fkh-6*, another direct target of Hh signaling (Madison, McKenna, Dolson, Epstein, & Kaestner, 2009), display a phenotype very similar to that of *Nkx 2–3* mutants (Kaestner, Silberg, Traber, & Schütz, 1997; Katz et al., 2004). Yet, *Nkx 2–3* mutants have normal levels of *FoxL1* expression and *FoxL1* mutants have normal levels of *Nkx 2–3* (Pabst et al., 1999). In both cases, no changes in *Shh* or *Ihh* expression are detectable, but marked downregulation of *Bmp2* and *Bmp4* is seen in the jejunum where villus defects are noted. Loss of *FoxF* genes ($FoxF2^{-/-}$ or the double heterozygote, $FoxF1^{+/-}$; $FoxF2^{+/-}$) also exhibits reduced Bmp4 expression, perturbed patterning of proliferative regions and ectopic WNT signaling along the villi (Ormestad et al., 2006). These findings underline the critically important role of mesenchymal cells in the proper patterning and emergence of villi. Of note, all of these mesenchymal factors (*Nkx2–3, FoxF* and *FoxL*) are Hh target genes. Similar phenotypes are also observed in mice transgenically expressing low levels of a secreted form of Hhip, an inhibitor of Hh signaling (Madison et al., 2005). Thus, all of these transcription factors may function together downstream of epithelial Hh signals to control proper villus morphogenesis.

Epimorphin/Syntaxin2 is highly expressed in the sub-epithelial mesenchyme during villus morphogenesis. Epimorphin is a t-SNARE protein which is involved in the targeting and fusion of intracellular transport vesicles for exocytosis in neural and polarized epithelial cells. Thus, *Epimorphin* might play a role in regulating secretion of cluster morphogens (such as Bmps or Bmp antagonists). In support of this notion, increased expression of *Epimorphin* results in increased *Bmp4* expression (Fritsch et al., 2002) and genetic deletion of *Epimorphin* leads to decreased *Bmp4* expression (Wang et al., 2006). In addition, Bmps are known to inhibit Wnt signaling in intestinal crypts (He et al., 2004) and consistent with a loss of Bmp expression, expansion of epithelial Wnt signal transduction (increased nuclear β-catenin) is noted in *Epimorphin* null mice; villi are longer, crypts are deeper and increased crypt fissioning is observed (Wang et al., 2006). Co-culture of rat intestinal endoderm with myofibroblasts transfected with *Epimorphin* engrafted into the chick coelomic cavity results in ectopic formation of villi. This is Epimorphin dependent, since inhibition of Epimorphin expression in the myofibroblasts prevents villus morphogenesis. Finally, treatment of these co-cultures with the Bmp antagonist, Noggin, partially blocks the ability of Epimorphin to induce villus emergence, suggesting that a Bmp signaling feed-back loop may be at least partially responsible for the effects of Epimorphin on the epithelium (Fritsch et al., 2002).

3.3 Epithelial factors important in villus emergence

The initiation of villus emergence entails both morphological change and local cell differentiation. Morphologically, the epithelium is initially composed of tall, thin, uniformly proliferative pseudostratified cells. But as villi form, the epithelium directly above mesenchymal clusters transitions to a shorter, wider columnar shape. This localized shape change simultaneously decreases the stiffness of the epithelial layer and increases its area (the wider cells take up more space), resulting in a pattern of epithelial buckling and villus emergence over each cluster. Simultaneously, these same cells cease to divide and begin to differentiate, while epithelial cells located between the clusters remain pseudostratified and highly proliferative.

In addition to Shh and Ihh secreted by the epithelium, multiple epithelial transcription factors have been shown to contribute to these events of epithelial differentiation, villus morphogenesis and homeostasis (Chin, Hill, Aurora, & Spence, 2017). However, the role of the KLF5-FOXA1-ELF3 transcriptional networks is particularly noteworthy, since it appears to control the transition from pseudostratified epithelium to villus emergence. *Klf5* (Kruppel-like factor 5) is expressed in the embryonic endoderm that gives rise to the intestinal epithelium and expression continues during and after villi emergence (Dong et al., 2009; McConnell et al., 2011). *Foxa1* is active in the pseudostratified epithelium, but suppressed at E14.5 when villi form (Bell et al., 2013), while *Elf3* (ETS-related Transcription Factor 3) is activated at E14.5 (Ng et al., 2002). Deletion of epithelial *Klf5* expression prior to villus morphogenesis results in decreased intestinal length, poor epithelial differentiation and dramatically impaired villus morphogenesis (Bell et al., 2013). Loss of *Elf3* strongly mirrors this *Klf5* phenotype (Ng et al., 2002). Importantly, in Klf5 mutants, expression of *FoxA1* is expanded and expression of *Elf3* is suppressed. *FoxA1* is a repressor of epithelial differentiation, while *Elf3* is an important driver of differentiation. Bell et al. determined that KLF5 is required to suppress *FoxA1* activity at E14.5, as villus formation initiates. In its absence, *FoxA1*, a repressor of *Elf3*, is overexpressed and this impedes villus formation and epithelial differentiation (Bell et al., 2013). This leaves one important conundrum: since KLF5 is present before and after villus formation, while FOXA1 is only suppressed at the point of villus formation and not before, what temporal trigger limits the repressive effect of KLF5 on FOXA1 to E14.5 and after?

Finally, it is important to consider the role that the basement membrane likely plays in villus development. Basement membranes in tissues help maintain tissue integrity and compartmentalization as well as promote differentiation and provide a scaffold for cell migration. These roles may be conveyed through physical forces or through cell signaling. Indeed, as villi form, mesenchymal clusters interact closely with the overlying basement membrane (Walton et al., 2012). Expression of several laminin isoforms is tightly controlled and developmentally regulated (Lefebvre, Sorokin, Kedinger, & Simon-Assmann, 1999; Miner, 1998; Perreault, Vachon, & Beaulieu, 1995; Simon-Assmann et al., 1998). Loss of laminin a5 from the intestine results in an increased production of laminin a1 and a4, laminins typically expressed at high levels in the colon. This switch in regional laminins is accompanied by structural and fate alterations of the epithelium to resemble colon rather than small intestinal villi. Whether this effect is the result of structural differences, changes in cell signaling, or changes in localization of sub-epithelial mesenchymal cell types that affect positional information is unclear.

4. Making crypts

Crypts are flask-like indentations of the epithelium into the underlying mesenchyme which serve as critical architectural units for housing the stem cells that ensure the continuous renewal of epithelial cells. In both mouse and human, crypts form in the intervillus regions after villi have formed. Human crypts are established early in gestation (weeks 11–12) (Trier & Moxey, 1979), while in the mouse, crypts only begin to form in the first week after birth (Dehmer et al., 2011). After observing a relatively constant distance between the bottom of intervillus regions/crypts and the smooth muscle layer at E17, post-natal day 3 (P3) and P8, Calvert et al. proposed that crypt formation occurs by "zipping up" the villus base, from bottom to top, driven by mesenchymal cell movements. This hypothesis held until Sumigray et al. carefully characterized the formation of nascent crypt structures, using 3D confocal imaging coupled with transcriptomic analysis to reveal and describe a dynamic process of crypt invagination (Sumigray, Terwilliger, & Lechler, 2018).

Sumigray et al. showed that CD44v6 is robustly expressed in crypt progenitor units throughout the formation of crypts; thus, they used CD44v6 to trace the structure of developing crypts step by step (Fig. 6A-D). With this marker, they sorted developing crypt cells from villus cells between P0 and P10 and performed RNA-seq to identify key regulators of crypt morphogenesis. They observed that, at P0, the CD44v6 positive compartment is a flat sheet of intervillus cells. Interestingly, CD44v6 expression becomes restricted to a circular pattern within the intervillus area, even before crypt invagination begins, marking the nascent crypt region (Fig. 6A, inset). Invagination of this compartment generates a cup-like structure (Fig. 6B and F). By analysis of their transcriptomic datasets, the authors found that a large subset of myosin II-associated contractility genes are enriched during crypt initiation and confirmed that Myosin II-driven apical constriction initiates this invagination process. Additionally, the transcriptomic analysis revealed several novel crypt-enriched and villus-enriched markers that will be useful in future analyses.

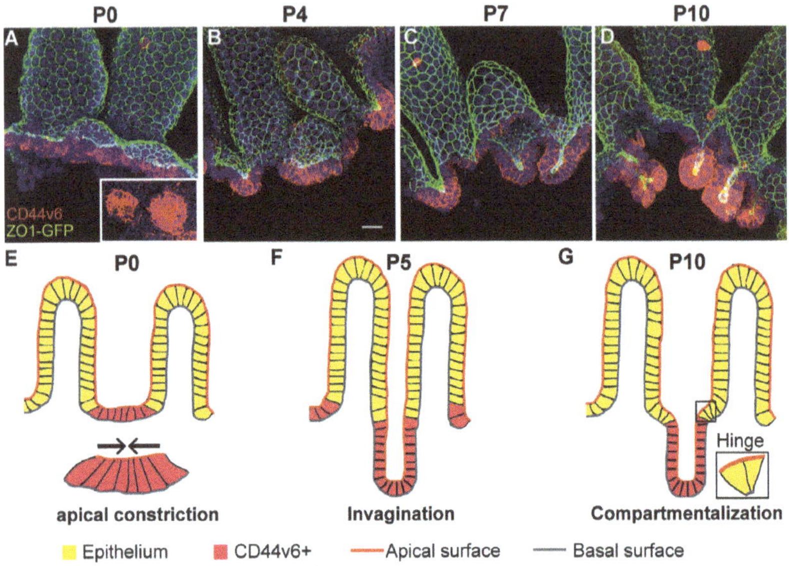

Fig. 6 Crypt development. (A-D) Whole mount staining of CD44v6 (magenta) in epithelium of ZO1-GFP labeled mouse intestines at P0, 4, 7 and 10. (E-G) Schematic summary of crypt morphogenesis. (E) The CD44v6+ compartment starts within the flat sheet of intervillus cells at P0. Myosin II-drives apical constriction of CD44v6 positive cells. (F) Forming crypts invaginate into the underlying mesenchyme. (G) Cells at the bend between the crypt and villus adopt a wedge-like shape—the "hinge" region—which morphologically separates crypts from villi. *Panels (A-D): Reproduced with permission from Sumigray, K. D., Terwilliger, M., & Lechler, T. (2018). Morphogenesis and compartmentalization of the intestinal crypt.* Developmental Cell, 45, *183–197.e185 (Fig. B). Panels (E-G): Adapted with permission from the graphical abstract in Sumigray, K. D., Terwilliger, M., & Lechler, T. (2018). Morphogenesis and compartmentalization of the intestinal crypt.* Developmental Cell, 45, *183–197.e185.*

By P3, the authors noted a clear bend between the emerging crypts and nearby villi (Fig. 6B and F). By P7, this defined "hinge" is formed from wedge-shaped cells that are expanded apically and constricted basally (Fig. 6C, D, and G). Genetic ablation of Rac1 impairs hinge formation and results in disorganized villus spacing. In part, Rac1 appears to function by downregulating α6/β4 integrins; these integrins make up hemidesmosomes, which attach these cells to the underlying basement membrane, perhaps so that the necessary basal constriction can be accomplished. In the future, it will be interesting to know how the CD44v6 population is first restricted to initiate crypt formation and what role the interactions between the epithelium and mesenchyme play in crypt formation, patterning, and maintenance.

5. Organoid systems to study intestinal development

The term "intestinal organoid" is a generic term for engineered intestinal tissues (epithelium only or epithelium with combinations of supporting mesenchyme) that are grown in a 3D matrix (Matrigel or various versions of hydrogels) and provided with essential factors to support intestinal stem cells and epithelial cell progeny. The first iterations of organoid culture were started from epithelial cells isolated from primary small intestinal tissue (Barker et al., 2010; Ootani et al., 2009; Sato et al., 2009); however, organoids can now be started from hESCs or iPSCs and driven toward an endodermal fate using specifically-timed treatments with combinations of growth factors (Spence et al., 2011). Many different forms/variations of engineered intestinal tissues using different starting material and different culture conditions have been developed. To further clarify these variations, a standard of nomenclature has been proposed by the National Institutes of Health Intestinal Stem Cell Consortium (Stelzner et al., 2012). "Enteroids" refer to multi-lobulated structures that develop from enterospheres, a spherical structure that forms from isolated epithelial cells that round off to form cysts, and contain only epithelium while organoids contain both epithelium and mesenchyme.

The ability to engineer 3D intestinal tissues in a dish holds great promise for future treatment of patients with shortened intestines or loss of absorptive surface area. However, currently, while organoids or enteroids readily generate crypt-like structures, finger-like villi only develop after transplantation into an animal host that provides additional growth factors, immune cells, vasculature, and perhaps nerve cells (Watson et al., 2014). Despite this deficit, organoids have been extremely useful for a variety of studies, including examining the role of specific genes in cell fate decisions and intestinal morphogenesis (Carulli et al., 2015; Chin et al., 2016), defining the role of the microbiome in maturation of the absorptive surface (Foulke-Abel et al., 2014; Hill et al., 2017), examining the role of specific cell types in intestinal morphogenesis (Schlieve et al., 2017; Workman et al., 2017), drug screening (Baumann, 2017; Liu et al., 2016), and studying patient specific diseases (VanDussen et al., 2015; Vlachogiannis et al., 2018). Two excellent recent reviews cover molecular and cellular development of intestinal organoids (Dedhia, Bertaux-Skeirik, Zavros, & Spence, 2016; Nakamura & Sato, 2018).

Interestingly, a 3D setting is not required to observe and investigate the organization of the crypt-villus axis. Two dimensional "enteroid monolayers," generated by plating the endodermal portion of isolated crypts on Matrigel-coated plates, develop dense foci of proliferative cells that express stem cell markers encircled by non-proliferating cells that express markers of differentiated enterocytes, goblet cells and enteroendocrine cells (Thorne et al., 2018). Paneth cells are also present and are found in the center of the dense foci. In such a monolayer setting, ectopic addition of WNT or BMP has the expected effect of increasing or decreasing (respectively) the zone of proliferation. Importantly, the focal organization of cells within the monolayer sets up its own feedback controls: the main source of WNT is the Paneth population at the center of the foci, while the main source of BMP is the surrounding differentiated cells. Indeed, treatment of such cultures by addition of WNT3a expands the proliferative compartment initially, but the subsequent increase in differentiated cells leads to increased BMP production that then feeds back to suppress proliferation. Such feedback can be blocked by LDN, an inhibitor of BMP receptors (Thorne et al., 2018). Since *in vivo* loss of Paneth cells does not perturb the crypt proliferation or maintenance of stem cells (Durand et al., 2012), it is clear that these interepithelial signaling loops, while powerful, are not sufficient for normal homeostasis. Nevertheless, these findings point to a remarkable self-regulating activity of the epithelial cells, which is likely critical in maintaining an organized "crypt-villus" axis.

6. Summary

In 300 BC, Aristotle wrote the following about embryonic development: "*It is possible, then, that A should move B and B should move C, that, in fact, the case should be the same as with the automatic machines shown as curiosities. For the parts of such machines while at rest have a sort of potentiality of movement in them, and when any external force puts the first into motion, immediately the next is moved in actuality… in these automatic machines, the external force moves the parts in a certain sense (not by touching any part at the moment but by having touched one previously).*"

Certainly, establishment and regional patterning of the endodermal gut tube, its dramatic elongation and looping, and decoration of its lumenal surface with villi and then crypts, follows a type of "automatic" sequence suggested by Aristotle, which is rife with cascading inductions. Modifications of the tube anlage occur in a precisely timed temporal sequence, at a distinct regional address. Though we still know little about the temporal triggers, we are learning that the "external force" includes not only physical forces *per se*, derived from tissue growth, constraint, or cell shape change, but soluble signals expressed by subsets of cells that act in an instructive manner on other cells. Together, these complex and precisely delivered stimuli result in a fully differentiated and functional intestinal tube. While development of the muscular features and enteric nervous system of the gut have not been discussed here in detail, recent studies provide more information about those aspects, which are also essential to adult intestinal function (Cotton et al., 2017; Gurdziel, Vogt, Walton, Schneider, & Gumucio, 2016; Huang et al., 2013; Kosinski et al., 2010; Mao et al., 2010; Nagy & Goldstein, 2017; Zacharias et al., 2011; reviewed in Heanue & Pachnis, 2007; Nagy & Goldstein, 2017; Obermayr, Hotta, Enomoto, & Young, 2013; Sasselli, Pachnis, & Burns, 2012; Uesaka, Young, Pachnis, & Enomoto, 2016). Importantly, secreted signals from endoderm are important for smooth muscle differentiation, while signals from the mesenchyme are believed to guide the influx of enteric nerves.

For the future, it remains clear that one of the most important clinical goals of research in this area is to learn how to produce or regenerate functional intestine that can successfully absorb nutrients in patients with intestinal failure or short bowel syndrome. Answers to several of the intriguing biological questions that are raised in this chapter will help move us closer to that goal by revealing key control factors. Further experimentation with organoid/enteroid systems holds promise for not only dissecting these critical signals but also for assembling functional villus/crypt units *in vitro*.

References

Aoki, R., Shoshkes-Carmel, M., Gao, N., Shin, S., May, C. L., Golson, M. L., et al. (2016). Foxl1-expressing mesenchymal cells constitute the intestinal stem cell niche. *Cellular and Molecular Gastroenterology and Hepatology*, *2*, 175–188.

Applegate, K. E., Anderson, J. M., & Klatte, E. C. (2006). Intestinal malrotation in children: A problem-solving approach to the upper gastrointestinal series. *Radiographics*, *26*, 1485–1500.

Arnaout, R., & Stainier, D. Y. R. (2011). Developmental biology: Physics adds a twist to gut looping. *Current Biology*, *21*, R854–R857.

Barker, N., Huch, M., Kujala, P., van de Wetering, M., Snippert, H. J., van Es, J. H., et al. (2010). Lgr5(+ve) stem cells drive self-renewal in the stomach and build long-lived gastric units in vitro. *Cell Stem Cell*, *6*, 25–36.

Battle, M. A., Bondow, B. J., Iverson, M. A., Adams, S. J., Jandacek, R. J., Tso, P., et al. (2008). GATA4 is essential for jejunal function in mice. *Gastroenterology*, *135*, 1676–1686.e1.

Baumann, K. (2017). Stem cells: Colonic organoids for drug testing and colorectal disease modelling. *Nature Reviews. Molecular Cell Biology*, *18*, 467.

Beck, F., Chawengsaksophak, K., Waring, P., Playford, R. J., & Furness, J. B. (1999). Reprogramming of intestinal differentiation and intercalary regeneration in Cdx2 mutant mice. *Proceedings of the National Academy of Sciences of the United States of America*, *96*, 7318–7323.

Beck, F., Tata, F., & Chawengsaksophak, K. (2000). Homeobox genes and gut development. *BioEssays*, *22*, 431–441.

Bell, S. M., Zhang, L., Xu, Y., Besnard, V., Wert, S. E., Shroyer, N., et al. (2013). Kruppel-like factor 5 controls villus formation and initiation of cytodifferentiation in the embryonic intestinal epithelium. *Developmental Biology*, *375*, 128–139.

Beuling, E., Aronson, B. E., Tran, L. M., Stapleton, K. A., ter Horst, E. N., Vissers, L. A., et al. (2012). GATA6 is required for proliferation, migration, secretory cell maturation, and gene expression in the mature mouse colon. *Molecular and Cellular Biology*, *32*, 3392–3402.

Beuling, E., Baffour-Awuah, N. Y., Stapleton, K. A., Aronson, B. E., Noah, T. K., Shroyer, N. F., et al. (2011). GATA factors regulate proliferation, differentiation, and gene expression in small intestine of mature mice. *Gastroenterology*, *140*, 1219–1229.e1–2.

Bitgood, M. J., & McMahon, A. P. (1995). Hedgehog and Bmp genes are coexpressed at many diverse sites of cell-cell interaction in the mouse embryo. *Developmental Biology*, *172*, 126–138.

Bosse, T., Piaseckyj, C. M., Burghard, E., Fialkovich, J. J., Rajagopal, S., Pu, W. T., et al. (2006). Gata4 is essential for the maintenance of jejunal-ileal identities in the adult mouse small intestine. *Molecular and Cellular Biology*, *26*, 9060–9070.

Boulet, A. M., & Capecchi, M. R. (1996). Targeted disruption of hoxc-4 causes esophageal defects and vertebral transformations. *Developmental Biology*, *177*, 232–249.

Carulli, A. J., Keeley, T. M., Demitrack, E. S., Chung, J., Maillard, I., & Samuelson, L. C. (2015). Notch receptor regulation of intestinal stem cell homeostasis and crypt regeneration. *Developmental Biology*, *402*, 98–108.

Cervantes, S., Yamaguchi, T. P., & Hebrok, M. (2009). Wnt5a is essential for intestinal elongation in mice. *Developmental Biology*, *326*, 285–294.

Chawengsaksophak, K., James, R., Hammond, V. E., Kontgen, F., & Beck, F. (1997). Homeosis and intestinal tumours in Cdx2 mutant mice. *Nature*, *386*, 84–87.

Chin, A. M., Hill, D. R., Aurora, M., & Spence, J. R. (2017). Morphogenesis and maturation of the embryonic and postnatal intestine. *Seminars in Cell & Developmental Biology*, *66*, 81–93.

Chin, A. M., Tsai, Y. H., Finkbeiner, S. R., Nagy, M. S., Walker, E. M., Ethen, N. J., et al. (2016). A dynamic WNT/beta-CATENIN signaling environment leads to WNT-independent and WNT-dependent proliferation of embryonic intestinal progenitor cells. *Stem Cell Reports*, *7*, 826–839.

Cotton, J. L., Li, Q., Ma, L., Park, J. S., Wang, J., Ou, J., et al. (2017). YAP/TAZ and hedgehog coordinate growth and patterning in gastrointestinal mesenchyme. *Developmental Cell*, *43*, 35–47.e34.

Davis, N. M., Kurpios, N. A., Sun, X., Gros, J., Martin, J. F., & Tabin, C. J. (2008). The chirality of gut rotation derives from left-right asymmetric changes in the architecture of the dorsal mesentery. *Developmental Cell*, *15*, 134–145.

Dedhia, P. H., Bertaux-Skeirik, N., Zavros, Y., & Spence, J. R. (2016). Organoid models of human gastrointestinal development and disease. *Gastroenterology*, *150*, 1098–1112.

Degirmenci, B., Valenta, T., Dimitrieva, S., Hausmann, G., & Basler, K. (2018). GLI1-expressing mesenchymal cells form the essential Wnt-secreting niche for colon stem cells. *Nature*, *558*, 449–453.

Dehmer, J. J., Garrison, A. P., Speck, K. E., Dekaney, C. M., Van Landeghem, L., Sun, X., et al. (2011). Expansion of intestinal epithelial stem cells during murine development. *PLoS One*, *6*, e27070.

Dessimoz, J., Opoka, R., Kordich, J. J., Grapin-Botton, A., & Wells, J. M. (2006). FGF signaling is necessary for establishing gut tube domains along the anterior-posterior axis in vivo. *Mechanisms of Development*, *123*, 42–55.

Dong, B., Horie, T., Denker, E., Kusakabe, T., Tsuda, M., Smith, W., et al. (2009). Tube formation by complex cellular processes in Ciona intestinalis notochord. *Developmental Biology, 330*, 237–249.

Duluc, I., Freund, J. N., Leberquier, C., & Kedinger, M. (1994). Fetal endoderm primarily holds the temporal and positional information required for mammalian intestinal development. *Journal of Cell Biology, 126*, 211–221.

Durand, A., Donahue, B., Peignon, G., Letourneur, F., Cagnard, N., Slomianny, C., et al. (2012). Functional intestinal stem cells after Paneth cell ablation induced by the loss of transcription factor Math1 (Atoh1). *Proceedings of the National Academy of Sciences of the United States of America, 109*, 8965–8970.

Echelard, Y., Epstein, D. J., St-Jacques, B., Shen, L., Mohler, J., McMahon, J. A., et al. (1993). Sonic hedgehog, a member of a family of putative signaling molecules, is implicated in the regulation of CNS polarity. *Cell, 75*, 1417–1430.

Fang, R., Olds, L. C., & Sibley, E. (2006). Spatio-temporal patterns of intestine-specific transcription factor expression during postnatal mouse gut development. *Gene Expression Patterns, 6*, 426–432.

Foulke-Abel, J., In, J., Kovbasnjuk, O., Zachos, N. C., Ettayebi, K., Blutt, S. E., et al. (2014). Human enteroids as an ex-vivo model of host-pathogen interactions in the gastrointestinal tract. *Experimental Biology and Medicine (Maywood, N.J.), 239*, 1124–1134.

Franklin, V., Khoo, P. L., Bildsoe, H., Wong, N., Lewis, S., & Tam, P. P. (2008). Regionalisation of the endoderm progenitors and morphogenesis of the gut portals of the mouse embryo. *Mechanisms of Development, 125*, 587–600.

Freddo, A. M., Shoffner, S. K., Shao, Y., Taniguchi, K., Grosse, A. S., Guysinger, M. N., et al. (2016). Coordination of signaling and tissue mechanics during morphogenesis of murine intestinal villi: A role for mitotic cell rounding. *Integrative Biology: Quantitative Biosciences From Nano to Macro, 8*, 918–928.

Fritsch, C., Simon-Assmann, P., Kedinger, M., & Evans, G. S. (1997). Cytokines modulate fibroblast phenotype and epithelial-stroma interactions in rat intestine. *Gastroenterology, 112*, 826–838.

Fritsch, C., Swietlicki, E. A., Lefebvre, O., Kedinger, M., Iordanov, H., Levin, M. S., et al. (2002). Epimorphin expression in intestinal myofibroblasts induces epithelial morphogenesis. *The Journal of Clinical Investigation, 110*, 1629–1641.

Gao, N., White, P., & Kaestner, K. (2009). Establishment of intestinal identity and epithelial-mesenchymal signaling by Cdx2. *Developmental Cell, 16*, 588–599.

Geske, M. J., Zhang, X. Q., Patel, K. K., Ornitz, D. M., & Stappenbeck, T. S. (2008). Fgf9 signaling regulates small intestinal elongation and mesenchymal development. *Development, 135*, 2959–2968.

Grapin-Botton, A. (2005). Antero-posterior patterning of the vertebrate digestive tract: 40 years after Nicole Le Douarin's PhD thesis. *International Journal of Developmental Biology, 49*, 335–347.

Grosse, A. S., Pressprich, M., Curley, L. B., Margolis, B., Hildebrand, J. D., & Gumucio, D. L. (2011). Cell dynamics in fetal intestinal epithelium: Implications for intestinal growth and morphogenesis. *Development (Cambridge, England), 138*, 4423–4432.

Gumpel-Pinot, M., Yasugi, S., & Mizuno, T. (1978). [Differentiation of the endodermal epithelium associated with the splanchnic mesoderm]. *Comptes rendus hebdomadaires des séances de l'Académie des sciences. Série D: Sciences naturelles, 286*, 117–120.

Gurdziel, K., Vogt, K. R., Walton, K. D., Schneider, G. K., & Gumucio, D. L. (2016). Transcriptome of the inner circular smooth muscle of the developing mouse intestine: Evidence for regulation of visceral smooth muscle genes by the hedgehog target gene, cJun. *Developmental Dynamics, 245*, 614–626.

Haffen, K., Lacroix, B., Kedinger, M., & Simon-Assmann, P. M. (1983). Inductive properties of fibroblastic cell cultures derived from rat intestinal mucosa on epithelial differentiation. *Differentiation*, *23*, 226–233.

Hasosah, M., Lemberg, D. A., Skarsgard, E., & Schreiber, R. (2008). Congenital short bowel syndrome: A case report and review of the literature. *Canadian Journal of Gastroenterology = Journal Canadien de Gastroenterologie*, *22*, 71–74.

He, X. C., Zhang, J., Tong, W. G., Tawfik, O., Ross, J., Scoville, D. H., et al. (2004). BMP signaling inhibits intestinal stem cell self-renewal through suppression of Wnt-beta-catenin signaling. *Nature Genetics*, *36*, 1117–1121.

Heanue, T. A., & Pachnis, V. (2007). Enteric nervous system development and Hirschsprung's disease: Advances in genetic and stem cell studies. *Nature Reviews. Neuroscience*, *8*, 466–479.

Hill, D. R., Huang, S., Nagy, M. S., Yadagiri, V. K., Fields, C., Mukherjee, D., et al. (2017). Bacterial colonization stimulates a complex physiological response in the immature human intestinal epithelium. *eLife*, *6*.

Hoch, M., & Pankratz, M. J. (1996). Control of gut development by fork head and cell signaling molecules in Drosophila. *Mechanisms of Development*, *58*, 3–14.

Huang, H., Cotton, J. L., Wang, Y., Rajurkar, M., Zhu, L. J., Lewis, B. C., et al. (2013). Specific requirement of Gli transcription factors in Hedgehog-mediated intestinal development. *The Journal of Biological Chemistry*, *288*, 17589–17596.

Kaestner, K. H., Silberg, D. G., Traber, P. G., & Schütz, G. (1997). The mesenchymal winged helix transcription factor Fkh6 is required for the control of gastrointestinal proliferation and differentiation. *Genes & Development*, *11*, 1583–1595.

Karlsson, L., Lindahl, P., Heath, J. K., & Betsholtz, C. (2000). Abnormal gastrointestinal development in PDGF-A and PDGFR-(alpha) deficient mice implicates a novel mesenchymal structure with putative instructive properties in villus morphogenesis. *Development*, *127*, 3457–3466.

Katz, J. P., Perreault, N., Goldstein, B. G., Chao, H. H., Ferraris, R. P., & Kaestner, K. H. (2004). Foxl1 null mice have abnormal intestinal epithelia, postnatal growth retardation, and defective intestinal glucose uptake. *American Journal of Physiology. Gastrointestinal and Liver Physiology*, *287*, G856–G864.

Kaufman, M. H. (1992). *The atlas of mouse development.* New York: Elsevier Academic Press.

Kedinger, M., Duluc, I., Fritsch, C., Lorentz, O., Plateroti, M., & Freund, J. N. (1998). Intestinal epithelial-mesenchymal cell interactions. *Annals of the New York Academy of Sciences*, *859*, 1–17.

Kedinger, M., Simon-Assmann, P. M., Lacroix, B., Marxer, A., Hauri, H. P., & Haffen, K. (1986). Fetal gut mesenchyme induces differentiation of cultured intestinal endodermal and crypt cells. *Developmental Biology*, *113*, 474–483.

Kim, T. H., Kim, B. M., Mao, J., Rowan, S., & Shivdasani, R. A. (2011). Endodermal Hedgehog signals modulate Notch pathway activity in the developing digestive tract mesenchyme. *Development*, *138*, 3225–3233.

Kluth, D., Jaeschke-Melli, S., & Fiegel, H. (2003). The embryology of gut rotation. *Seminars in Pediatric Surgery*, *12*, 275–279.

Kohlnhofer, B. M., Thompson, C. A., Walker, E. M., & Battle, M. A. (2016). GATA4 regulates epithelial cell proliferation to control intestinal growth and development in mice. *Cellular and Molecular Gastroenterology and Hepatology*, *2*, 189–209.

Kolterud, A., Grosse, A. S., Zacharias, W. J., Walton, K., Kretovich, K. E., Madison, B. B., et al. (2009). Paracrine Hedgehog signaling in stomach and intestine: New roles for hedgehog in gastrointestinal patterning. *Gastroenterology*, *137*, 618–628.

Kondo, T., Dolle, P., Zakany, J., & Duboule, D. (1996). Function of posterior HoxD genes in the morphogenesis of the anal sphincter. *Development*, *122*, 2651–2659.

Kondo, T., & Hayashi, S. (2013). Mitotic cell rounding accelerates epithelial invagination. *Nature*, *494*, 125–129.

Korinek, V., Barker, N., Moerer, P., van Donselaar, E., Huls, G., Peters, P. J., et al. (1998). Depletion of epithelial stem-cell compartments in the small intestine of mice lacking Tcf-4. *Nature Genetics*, *19*, 379–383.

Kosinski, C., Stange, D., Xu, C., Chan, A. S., Ho, C., Yuen, S. T., et al. (2010). Indian hedgehog regulates intestinal stem cell fate through epithelial-mesenchymal interactions during development. *Gastroenterology*, *139*, 893–903.

Kwon, G. S., Viotti, M., & Hadjantonakis, A. K. (2008). The endoderm of the mouse embryo arises by dynamic widespread intercalation of embryonic and extraembryonic lineages. *Developmental Cell*, *15*, 509–520.

Lefebvre, O., Sorokin, L., Kedinger, M., & Simon-Assmann, P. (1999). Developmental expression and cellular origin of the laminin alpha2, alpha4, and alpha5 chains in the intestine. *Developmental Biology*, *210*, 135–150.

Li, X., Udager, A., Hu, C., Qiao, X. T., Richards, N., & Gumucio, D. L. (2009). Dynamic patterning at the pylorus: Formation of an epithelial intestine-stomach boundary in late fetal life. *Developmental Dynamics*, *238*, 3205–3217.

Liu, F., Huang, J., Ning, B., Liu, Z., Chen, S., & Zhao, W. (2016). Drug discovery via human-derived stem cell organoids. *Frontiers in Pharmacology*, *7*, 334.

Madison, B. B., Braunstein, K., Kuizon, E., Portman, K., Qiao, X. T., & Gumucio, D. L. (2005). Epithelial hedgehog signals pattern the intestinal crypt-villus axis. *Development*, *132*, 279–289.

Madison, B. B., McKenna, L. B., Dolson, D., Epstein, D. J., & Kaestner, K. H. (2009). FoxF1 and FoxL1 link hedgehog signaling and the control of epithelial proliferation in the developing stomach and intestine. *The Journal of Biological Chemistry*, *284*, 5936–5944.

Mao, J., Kim, B., Rajurkar, M., Shivdasani, R., & Mcmahon, A. (2010). Hedgehog signaling controls mesenchymal growth in the developing mammalian digestive tract. *Development*, *137*, 1721–1729.

Martinez Barbera, J. P., Clements, M., Thomas, P., Rodriguez, T., Meloy, D., Kioussis, D., et al. (2000). The homeobox gene Hex is required in definitive endodermal tissues for normal forebrain, liver and thyroid formation. *Development*, *127*, 2433–2445.

Mathan, M., Moxey, P. C., & Trier, J. S. (1976). Morphogenesis of fetal rat duodenal villi. *The American Journal of Anatomy*, *146*, 73–92.

Matsumoto, A., Hashimoto, K., Yoshioka, T., & Otani, H. (2002). Occlusion and subsequent re-canalization in early duodenal development of human embryos: Integrated organogenesis and histogenesis through a possible epithelial-mesenchymal interaction. *Anatomy and Embryology*, *205*, 53–65.

Matus, D. Q., Magie, C. R., Pang, K., Martindale, M. Q., & Thomsen, G. H. (2008). The Hedgehog gene family of the cnidarian, Nematostella vectensis, and implications for understanding metazoan Hedgehog pathway evolution. *Developmental Biology*, *313*, 501–518.

McConnell, B. B., Kim, S. S., Yu, K., Ghaleb, A. M., Takeda, N., Manabe, I., et al. (2011). Kruppel-like factor 5 is important for maintenance of crypt architecture and barrier function in mouse intestine. *Gastroenterology*, *141*, 1302–1313, 1313.e1–6.

Miner, J. H. (1998). Developmental biology of glomerular basement membrane components. *Current Opinion in Nephrology and Hypertension*, *7*, 13–19.

Mizuno, T., & Yasugi, S. (1990). Susceptibility of epithelia to directive influences of mesenchymes during organogenesis: Uncoupling of morphogenesis and cytodifferentiation. *Cell Differentiation and Development*, *31*, 151–159.

Mo, R., Kim, J. H., Zhang, J., Chiang, C., Hui, C. C., & Kim, P. C. (2001). Anorectal malformations caused by defects in sonic hedgehog signaling. *The American Journal of Pathology*, *159*, 765–774.

Nagy, N., & Goldstein, A. M. (2017). Enteric nervous system development: A crest cell's journey from neural tube to colon. *Seminars in Cell & Developmental Biology*, *66*, 94–106.

Nakamura, T., & Sato, T. (2018). Advancing intestinal organoid technology toward regenerative medicine. *Cellular and Molecular Gastroenterology and Hepatology*, *5*, 51–60.

Nascone, N., & Mercola, M. (1995). An inductive role for the endoderm in Xenopus cardiogenesis. *Development*, *121*, 515–523.

Nerurkar, N. L., Mahadevan, L., & Tabin, C. J. (2017). BMP signaling controls buckling forces to modulate looping morphogenesis of the gut. *Proceedings of the National Academy of Sciences of the United States of America*, *114*, 2277–2282.

Ng, A. Y., Waring, P., Ristevski, S., Wang, C., Wilson, T., Pritchard, M., et al. (2002). Inactivation of the transcription factor Elf3 in mice results in dysmorphogenesis and altered differentiation of intestinal epithelium. *Gastroenterology*, *122*, 1455–1466.

Nigmatullina, L., Norkin, M., Dzama, M. M., Messner, B., Sayols, S., & Soshnikova, N. (2017). Id2 controls specification of Lgr5(+) intestinal stem cell progenitors during gut development. *The EMBO Journal*, *36*, 869–885.

Nishita, M., Yoo, S. K., Nomachi, A., Kani, S., Sougawa, N., Ohta, Y., et al. (2006). Filopodia formation mediated by receptor tyrosine kinase Ror2 is required for Wnt5a-induced cell migration. *The Journal of Cell Biology*, *175*, 555–562.

Noah, T. K., Donahue, B., & Shroyer, N. F. (2011). Intestinal development and differentiation. *Experimental Cell Research*, *317*, 2702–2710.

Obermayr, F., Hotta, R., Enomoto, H., & Young, H. M. (2013). Development and developmental disorders of the enteric nervous system. *Nature Reviews. Gastroenterology & Hepatology*, *10*, 43–57.

Offield, M. F., Jetton, T. L., Labosky, P. A., Ray, M., Stein, R. W., Magnuson, M. A., et al. (1996). PDX-1 is required for pancreatic outgrowth and differentiation of the rostral duodenum. *Development*, *122*, 983–995.

Onouchi, S., Ichii, O., Otsuka, S., Hashimoto, Y., & Kon, Y. (2013). Analysis of duodenojejunal flexure formation in mice: Implications for understanding the genetic basis for gastrointestinal morphology in mammals. *Journal of Anatomy*, *223*, 385–398.

Ootani, A., Li, X., Sangiorgi, E., Ho, Q., Ueno, H., Toda, S., et al. (2009). Sustained in vitro intestinal epithelial culture within a Wnt-dependent stem cell niche. *Nature Medicine*, *15*, 701–706.

Ormestad, M., Astorga, J., Landgren, H., Wang, T., Johansson, B. R., Miura, N., et al. (2006). Foxf1 and Foxf2 control murine gut development by limiting mesenchymal Wnt signaling and promoting extracellular matrix production. *Development*, *133*, 833–843.

Pabst, O., Zweigerdt, R., & Arnold, H. H. (1999). Targeted disruption of the homeobox transcription factor Nkx2-3 in mice results in postnatal lethality and abnormal development of small intestine and spleen. *Development*, *126*, 2215–2225.

Perreault, N., Vachon, P. H., & Beaulieu, J. F. (1995). Appearance and distribution of laminin A chain isoforms and integrin alpha 2, alpha 3, alpha 6, beta 1, and beta 4 subunits in the developing human small intestinal mucosa. *The Anatomical Record*, *242*, 242–250.

Ramalho-Santos, M., Melton, D. A., & McMahon, A. P. (2000). Hedgehog signals regulate multiple aspects of gastrointestinal development. *Development*, *127*, 2763–2772.

Reed, R. A., Womble, M. A., Dush, M. K., Tull, R. R., Bloom, S. K., Morckel, A. R., et al. (2009). Morphogenesis of the primitive gut tube is generated by Rho/ROCK/myosin II-mediated endoderm rearrangements. *Developmental Dynamics*, *238*, 3111–3125.

Sakiyama, J., Yokouchi, Y., & Kuroiwa, A. (2001). HoxA and HoxB cluster genes subdivide the digestive tract into morphological domains during chick development. *Mechanisms of Development*, *101*, 233–236.

Sasselli, V., Pachnis, V., & Burns, A. J. (2012). The enteric nervous system. *Developmental Biology*, *366*, 64–73.

Sato, T., Vries, R., Snippert, H., Van De Wetering, M., Barker, N., Stange, D., et al. (2009). Single Lgr5 stem cells build crypt–villus structures in vitro without a mesenchymal niche. *Nature, 459*, 262–265.

Saund, R. S., Kanai-Azuma, M., Kanai, Y., Kim, I., Lucero, M. T., & Saijoh, Y. (2012). Gut endoderm is involved in the transfer of left-right asymmetry from the node to the lateral plate mesoderm in the mouse embryo. *Development, 139*, 2426–2435.

Savin, T., Kurpios, N. A., Shyer, A. E., Florescu, P., Liang, H. Y., Mahadevan, L., et al. (2011). On the growth and form of the gut. *Nature, 476*, 57–62.

Schlieve, C. R., Fowler, K. L., Thornton, M., Huang, S., Hajjali, I., Hou, X., et al. (2017). Neural crest cell implantation restores enteric nervous system function and alters the gastrointestinal transcriptome in human tissue-engineered small intestine. *Stem Cell Reports, 9*, 883–896.

Schultheiss, T. M., Xydas, S., & Lassar, A. B. (1995). Induction of avian cardiac myogenesis by anterior endoderm. *Development, 121*, 4203–4214.

Sekimoto, T., Yoshinobu, K., Yoshida, M., Kuratani, S., Fujimoto, S., Araki, M., et al. (1998). Region-specific expression of murine Hox genes implies the Hox code-mediated patterning of the digestive tract. *Genes to Cells, 3*, 51–64.

Sherwood, R. I., Chen, T. Y., & Melton, D. A. (2009). Transcriptional dynamics of endodermal organ formation. *Developmental Dynamics, 238*, 29–42.

Sherwood, R. I., Maehr, R., Mazzoni, E. O., & Melton, D. A. (2011). Wnt signaling specifies and patterns intestinal endoderm. *Mechanisms of Development, 128*, 387–400.

Shoshkes-Carmel, M., Wang, Y. J., Wangensteen, K. J., Toth, B., Kondo, A., Massassa, E. E., et al. (2018). Subepithelial telocytes are an important source of Wnts that supports intestinal crypts. *Nature, 557*, 242–246.

Shyer, A. E., Huycke, T. R., Lee, C., Mahadevan, L., & Tabin, C. J. (2015). Bending gradients: How the intestinal stem cell gets its home. *Cell, 161*, 569–580.

Shyer, A. E., Tallinen, T., Nerurkar, N. L., Wei, Z., Gil, E. S., Kaplan, D. L., et al. (2013). Villification: How the gut gets its villi. *Science, 342*, 212–218.

Silberg, D. G., Swain, G. P., Suh, E. R., & Traber, P. G. (2000). Cdx1 and cdx2 expression during intestinal development. *Gastroenterology, 119*, 961–971.

Simon-Assmann, P., Lefebvre, O., Bellissent-Waydelich, A., Olsen, J., Orian-Rousseau, V., & De Arcangelis, A. (1998). The laminins: Role in intestinal morphogenesis and differentiation. *Annals of the New York Academy of Sciences, 859*, 46–64.

Spence, J. R., Mayhew, C. N., Rankin, S. A., Kuhar, M. F., Vallance, J. E., Tolle, K., et al. (2011). Directed differentiation of human pluripotent stem cells into intestinal tissue in vitro. *Nature, 470*, 105–109.

Stelzner, M., Helmrath, M., Dunn, J. C., Henning, S. J., Houchen, C. W., Kuo, C., et al. (2012). A nomenclature for intestinal in vitro cultures. *American Journal of Physiology. Gastrointestinal and Liver Physiology, 302*, G1359–G1363.

Sukegawa, A., Narita, T., Kameda, T., Saitoh, K., Nohno, T., Iba, H., et al. (2000). The concentric structure of the developing gut is regulated by Sonic hedgehog derived from endodermal epithelium. *Development, 127*, 1971–1980.

Sumigray, K. D., Terwilliger, M., & Lechler, T. (2018). Morphogenesis and compartmentalization of the intestinal crypt. *Developmental Cell, 45*, 183–197.e5.

Thompson, C. A., Wojta, K., Pulakanti, K., Rao, S., Dawson, P., & Battle, M. A. (2017). GATA4 is sufficient to establish jejunal versus ileal identity in the small intestine. *Cellular and Molecular Gastroenterology and Hepatology, 3*, 422–446.

Thorne, C. A., Chen, I. W., Sanman, L. E., Cobb, M. H., Wu, L. F., & Altschuler, S. J. (2018). Enteroid monolayers reveal an autonomous WNT and BMP circuit controlling intestinal epithelial growth and organization. *Developmental Cell, 44*, 624–633.e4.

Torres, A. M., & Ziegler, M. M. (1993). Malrotation of the intestine. *World Journal of Surgery, 17*, 326–331.

Tremblay, K. D., & Zaret, K. S. (2005). Distinct populations of endoderm cells converge to generate the embryonic liver bud and ventral foregut tissues. *Developmental Biology, 280*, 87–99.

Trier, J. S., & Moxey, P. C. (1979). Morphogenesis of the small intestine during fetal development. *Ciba Foundation Symposium*, 3–29.

Udager, A., Prakash, A., & Gumucio, D. L. (2010). Dividing the tubular gut: Generation of organ boundaries at the pylorus. *Progress in Molecular Biology and Translational Science, 96*, 35–62.

Uesaka, T., Young, H. M., Pachnis, V., & Enomoto, H. (2016). Development of the intrinsic and extrinsic innervation of the gut. *Developmental Biology, 417*, 158–167.

van der Werf, C. S., Halim, D., Verheij, J. B., Alves, M. M., & Hofstra, R. M. (2015). Congenital short bowel syndrome: From clinical and genetic diagnosis to the molecular mechanisms involved in intestinal elongation. *Biochimica et Biophysica Acta, 1852*, 2352–2361.

van der Werf, C. S., Sribudiani, Y., Verheij, J. B. G. M., Carroll, M., O'Loughlin, E., Chen, C. H., et al. (2013). Congenital short bowel syndrome as the presenting symptom in male patients with FLNA mutations. *Genetics in Medicine, 15*, 310–313.

van Dop, W. A., Uhmann, A., Wijgerde, M., Sleddens-Linkels, E., Heijmans, J., Offerhaus, G. J., et al. (2009). Depletion of the colonic epithelial precursor cell compartment upon conditional activation of the hedgehog pathway. *Gastroenterology, 136*, 2195–2203.e1–7.

VanDussen, K. L., Marinshaw, J. M., Shaikh, N., Miyoshi, H., Moon, C., Tarr, P. I., et al. (2015). Development of an enhanced human gastrointestinal epithelial culture system to facilitate patient-based assays. *Gut, 64*, 911–920.

Viotti, M., Foley, A. C., & Hadjantonakis, A. K. (2014). Gutsy moves in mice: Cellular and molecular dynamics of endoderm morphogenesis. *Philosophical Transactions of the Royal Society of London. Series B, Biological Sciences, 369*, 1–11.

Viotti, M., Niu, L., Shi, S. H., & Hadjantonakis, A. K. (2012). Role of the gut endoderm in relaying left-right patterning in mice. *PLoS Biology, 10*, e1001276.

Vlachogiannis, G., Hedayat, S., Vatsiou, A., Jamin, Y., Fernandez-Mateos, J., Khan, K., et al. (2018). Patient-derived organoids model treatment response of metastatic gastrointestinal cancers. *Science, 359*, 920–926.

Walker, E. M., Thompson, C. A., Kohlnhofer, B. M., Faber, M. L., & Battle, M. A. (2014). Characterization of the developing small intestine in the absence of either GATA4 or GATA6. *BMC Research Notes*, 7, 902.

Walton, K. D., Freddo, A. M., Wang, S., & Gumucio, D. L. (2016). Generation of intestinal surface: An absorbing tale. *Development, 143*, 2261–2272.

Walton, K. D., Kolterud, A., Czerwinski, M. J., Bell, M. J., Prakash, A., Kushwaha, J., et al. (2012). Hedgehog-responsive mesenchymal clusters direct patterning and emergence of intestinal villi. *Proceedings of the National Academy of Sciences of the United States of America, 109*, 15817–15822.

Walton, K. D., Mishkind, D., Riddle, M. R., Tabin, C. J., & Gumucio, D. L. (2018). Blueprint for an intestinal villus: Species-specific assembly required. *Wiley Interdisciplinary Reviews: Developmental Biology*, 7(4), e317.

Walton, K., Warner, J., Hertzler, P., & Mcclay, D. (2009). Hedgehog signaling patterns mesoderm in the sea urchin. *Developmental Biology, 331*, 26–37.

Walton, K. D., Whidden, M., Kolterud, A., Shoffner, S., Czerwinski, M. J., Kushwaha, J., et al. (2016). Villification in the mouse: Bmp signals control intestinal villus patterning. *Development, 143*, 427–436.

Wang, S., Cebrian, C., Schnell, S., & Gumucio, D. L. (2018). Radial WNT5A-guided post-mitotic filopodial pathfinding is critical for midgut tube elongation. *Developmental Cell, 46*, 173–188.

Wang, Y., Wang, L., Iordanov, H., Swietlicki, E. A., Zheng, Q., Jiang, S., et al. (2006). Epimorphin(-/-) mice have increased intestinal growth, decreased susceptibility to dextran sodium sulfate colitis, and impaired spermatogenesis. *The Journal of Clinical Investigation, 116*, 1535–1546.

Watson, C. L., Mahe, M. M., Munera, J., Howell, J. C., Sundaram, N., Poling, H. M., et al. (2014). An in vivo model of human small intestine using pluripotent stem cells. *Nature Medicine, 20*, 1310–1314.

Weaver, L. T., Austin, S., & Cole, T. J. (1991). Small intestinal length: A factor essential for gut adaptation. *Gut, 32*, 1321–1323.

Withington, S., Beddington, R., & Cooke, J. (2001). Foregut endoderm is required at head process stages for anteriormost neural patterning in chick. *Development, 128*, 309–320.

Workman, M. J., Mahe, M. M., Trisno, S., Poling, H. M., Watson, C. L., Sundaram, N., et al. (2017). Engineered human pluripotent-stem-cell-derived intestinal tissues with a functional enteric nervous system. *Nature Medicine, 23*, 49–59.

Yamada, M., Udagawa, J., Matsumoto, A., Hashimoto, R., Hatta, T., Nishita, M., et al. (2010). Ror2 is required for midgut elongation during mouse development. *Developmental Dynamics, 239*, 941–953.

Zacchetti, G., Duboule, D., & Zakany, J. (2007). Hox gene function in vertebrate gut morphogenesis: The case of the caecum. *Development, 134*, 3967–3973.

Zacharias, W. J., Madison, B. B., Kretovich, K. E., Walton, K. D., Richards, N., Udager, A. M., et al. (2011). Hedgehog signaling controls homeostasis of adult intestinal smooth muscle. *Developmental Biology, 355*, 152–162.

Zakany, J., & Duboule, D. (1999). Hox genes and the making of sphincters. *Nature, 401*, 761–762.

Zorn, A. M., & Wells, J. M. (2009). Vertebrate endoderm development and organ formation. *Annual Review of Cell and Developmental Biology, 25*, 221–251.

CHAPTER THREE

Consider the lung as a sensory organ: A tip from pulmonary neuroendocrine cells

Ankur Garg[a], Pengfei Sui[a], Jamie M. Verheyden[a], Lisa R. Young[b], Xin Sun[a,c,*]

[a]Department of Pediatrics, University of California, San Diego, La Jolla, CA, United States
[b]Division of Pulmonary Medicine, Center for Childhood Lung Research, Department of Pediatrics, Vanderbilt University Medical Center, Nashville, TN, United States
[c]Department of Biological Sciences, University of California, San Diego, La Jolla, CA, United States
*Corresponding author: e-mail address: xinsun@ucsd.edu

Contents

Abstract

While the lung is commonly known for its gas exchange function, it is exposed to signals in the inhaled air and responds to them by collaborating with other systems including immune cells and the neural circuit. This important aspect of lung physiology led us to consider the lung as a sensory organ. Among different cell types within the lung that mediate this role, several recent studies have renewed attention on pulmonary neuroendocrine cells (PNECs). PNECs are a rare, innervated airway epithelial cell type that accounts for $<1\%$ of the lung epithelium population. They are enriched at airway branch points. Classical *in vitro* studies have shown that PNECs can respond to an array of aerosol stimuli such as hypoxia, hypercapnia and nicotine. Recent *in vivo* evidence suggests an essential role of PNECs at neuroimmunomodulatory sites of action,

Current Topics in Developmental Biology, Volume 132
ISSN 0070-2153
https://doi.org/10.1016/bs.ctdb.2018.12.002

releasing neuropeptides, neurotransmitters and facilitating asthmatic responses to allergen. In addition, evidence supports that PNECs can function both as progenitor cells and progenitor niches following airway epithelial injury. Increases in PNECs have been documented in a large array of chronic lung diseases. They are also the cells-of-origin for small cell lung cancer. A better understanding of the specificity of their responses to distinct insults, their impact on normal lung function and their roles in the pathogenesis of pulmonary ailments will be the next challenge toward designing therapeutics targeting the neuroendocrine system in lung.

1. Overview: Lung as a sensory organ

The lung is essential for survival at first breath. Every minute in an adult human, 5–8 L of air flows in and out of the lung for oxygen intake and carbon dioxide output. Respiration is facilitated by two functional compartments in the lung—the conducting airways and gas-exchange alveoli. It is estimated that an adult lung on average is composed of 480 million alveoli, with a remarkable surface area of $80\,m^2$ (Ochs et al., 2004; Wiebe & Laursen, 1995).

Aside from its gas exchange function, an overshadowed role of the lung is its ability to sense different aerosol inputs and interpret them into distinct physiological responses. The intake air could carry allergens, pollutants or pathogens, and vary in moisture, pressure, or O_2/CO_2 levels. The lung responds to these challenges through coordinating with other systems including the immune and neuronal systems. Given the large surface area and constant exposure, the lung is a primary organ in tune with the aerosol environment. In this review, we will consider the lung as a sensory organ.

Pulmonary neuroendocrine cells (PNECs) (Fig. 1) were first identified in 1954 and are innervated airway epithelial cells that are enriched at branch point junctions (Feyrter, 1954; Kuo & Krasnow, 2015). PNECs have long been implicated as a sensory cell type in the lung. Classical *in vitro* studies have shown that the isolated primary PNECs or transformed PNEC cell lines can respond to O_2/CO_2/nicotine in culture (Cattaneo, Codignola, Vicentini, Clementi, & Sher, 1993; Lauweryns, Cokelaere, Deleersynder, & Liebens, 1977; Schuller, 1994; Schuller, Nylen, Park, & Becker, 1990). However, they have not attracted much attention partly because of their rarity: they account for ~0.5% of lung epithelial population in humans (Boers, den Brok, Koudstaal, Arends, & Thunnissen, 1996). Recent studies demonstrated *in vivo* roles of PNECs and led to renewed interest in these cells. It has been shown that the PNECs are key regulators of the immune

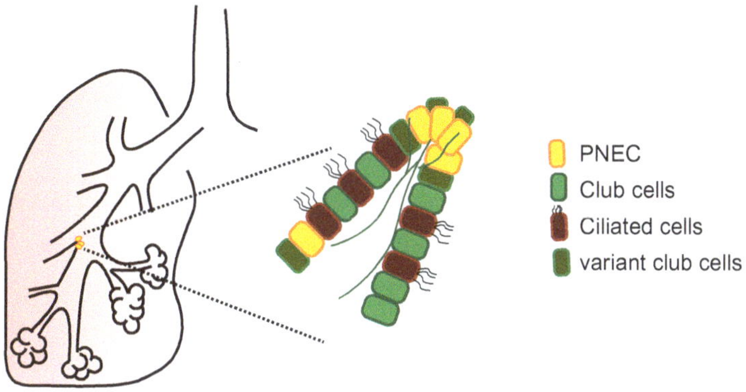

Fig. 1 Pulmonary neuroendocrine cells are rare, innervated epithelial cells that are enriched at airway branch points.

milieu and are essential for allergen-induced asthmatic responses (Branchfield et al., 2016; Sui et al., 2018). Following injury, PNECs can act as progenitors, expand and differentiate into a small number of lost epithelial cells in the airway (Reynolds, Giangreco, Power, & Stripp, 2000; Song et al., 2012). In addition, they can act as a niche for overlaying variant club cells in their role as progenitors (Guha et al., 2012; Reynolds et al., 2000). Experimental evidence suggests that PNECs are the tumor cell of origin for small cell lung cancer, the most aggressive type of lung cancer (Park et al., 2011; Song et al., 2012; Sutherland et al., 2011). In addition, PNEC pathology has been noted in a large array of lung-associated chronic diseases including COPD, fibrosis and pulmonary hypertension (Alshehri, Cutz, Banzhoff, & Canny, 1997; Heath et al., 1990; Johnson, Wobken, & Landrum, 1988). In this review, we will integrate classical and recent findings, and synthesize the central roles of PNECs as sentinels of the lung.

2. PNEC lineage origin and specification

The origin of PNECs can be traced back to the gills of fish (Hockman et al., 2017), demonstrating evolutionary conservation. Based on functional similarity to glomus cells in carotid bodies, PNECs were initially thought to be derived from neural crest cells. However, recent genetic lineage tracing data demonstrated that PNECs are derived from the endoderm (Hockman et al., 2017; Kuo & Krasnow, 2015; Song et al., 2012). Based on their expression of *Ascl1*, a PNEC-fate defining gene, PNECs are the first specified cell type to appear in the respiratory epithelium, detectable at embryonic day (E)12.5 in mice (Kuo & Krasnow, 2015). They initiate as solitary cells and cluster into neuroepithelial bodies (NEBs) by E15.5, with preferential enrichment near airway branch points (Kuo & Krasnow, 2015; Noguchi, Sumiyama, & Morimoto, 2015). In mice, expression of CGRP, a PNEC neuropeptide and differentiation marker, was clearly detected at E15.5 (Kuo & Krasnow, 2015). In humans, PNECs containing serotonin can be detected at 8 weeks of gestation (Van Lommel & Lauweryns, 1997). By 10 weeks, PNECs containing bombesin (also known as GRP) are also observed.

Live imaging of explanted developing mouse lung showed that solitary PNECs migrate toward branch point junctions to form NEBs (Kuo & Krasnow, 2015; Noguchi et al., 2015). One study showed that PNECs temporarily downregulate junctional proteins such as ZO-1 and E-Cadherin and upregulate mesenchymal transcription factor, Snail (Kuo & Krasnow, 2015). They lose attachment to basal lamina and crawl over and around neighboring epithelial cells in a behavior coined as "cell slithering." As they meet, these cells resume the expression of adhesion and polarity proteins, re-establish epithelial junctions to form NEBs, and start to differentiate shortly after. A parallel study, however, detected attachment to basement membrane of all PNECs as they undergo directed migration to cluster (Noguchi et al., 2015). Data from our laboratory indicate that SLIT-Roundabout (ROBO) signaling is essential for PNEC clustering into NEBs (Branchfield et al., 2016).

Recent lineage tracing using *Cgrp*^{creERT2} knockin mice, which is more sensitive at labeling PNECs than the use of anti-CGRP antibody, showed labeling in both PNECs and a small percentage of alveolar type 1 and type 2 (AT1 and AT2) cells (Song et al., 2012). Lineage labeled alveolar cells were only observed if Cre activation by tamoxifen was carried out before E15.5, while activation after E15.5 only labeled PNECs. This could be due to a low level of CGRP expression in a subset of the AT1/2 precursors. Alternatively, this could be due to an early and transient lineage sharing between PNEC and a small subset of alveolar epithelium (Song et al., 2012).

Ascl1 plays a critical role in PNEC specification. Inactivation of *Ascl1* leads to disruption of PNEC development without affecting other cell types in the lung epithelium (Ito et al., 2000; Sui et al., 2018). Conversely, ectopic expression of human *ASCL1* in the airway epithelium using *Scgb1a1* promoter in mice during development resulted in dysplasia of distal airway epithelium and progressive bronchiolarization of the alveoli with increased club-like and ciliated-like cells, but no additional PNECs (Linnoila et al., 2000).

It is possible that overexpression driven by *Scgb1a1* is too late for airway epithelial cells to transdifferentiate into PNECs. Alternatively, this could suggest that ASCL1 is not sufficient to specify PNECs without potential partners such as E12/E47 (Linnoila et al., 2000).

A plethora of evidence suggests that appropriate Notch activity in early airway development is critical for the specification and proper number of PNECs, as Notch signaling inhibits *Ascl1* expression (Chen et al., 1997; Sasai, Kageyama, Tagawa, Shigemoto, & Nakanishi, 1992). Inactivation of *Hes1* in the lung epithelium, a positive Notch effector, led to upregulation of *Ascl1* and increased PNECs (Ito et al., 2000). Consistent with this, lack of Notch signaling by deleting *Pofut1* or *Rbpjk* in the lung epithelium resulted in the expansion of PNECs and an absence of secretory cell fate (Tsao et al., 2009). A more significant increase of PNECs was found in *Notch1*; *Notch2*; *Notch3* triple mutants compared to *Rbpjk* mutant, suggesting that other downstream mediators may be involved in this process (Morimoto, Nishinakamura, Saga, & Kopan, 2012). Conversely, transgenic expression of human Notch intracellular domain (ICN1) driven by CGRP promoter led to a reduction in PNEC differentiation, marked by a decline in CGRP- and PGP9.5-positive NE cells (Shan, Aster, Sklar, & Sunday, 2007). Aside from PNECs, precise levels of Notch signaling are critical for the formation of other cell types in the lung epithelium (Morimoto et al., 2012). How this balance of Notch is achieved and how Notch integrates with other genetic control of PNEC formation remains poorly understood.

3. PNEC innervation

Innervation of PNECs varies across species in terms of timing, abundance, pattern and types of sensory afferent/efferent nerves. In developing rabbit lungs, a majority of PNECs/NEBs are innervated that connect *via* mucosal, submucosal and intercorpuscular neural connections (Pan, Yeger, & Cutz, 2004). However, in adult rabbits, only one-third of NEBs are innervated (Lauweryns, Van Lommel, & Dom, 1985). In adult rats, ~58% PNECs/NEBs are innervated (Larson, Schelegle, Hyde, & Plopper, 2003). Recent studies in mice have suggested that only NEBs but not solitary PNECs are innervated (Kuo & Krasnow, 2015). In humans, at least a subset of the solitary PNECs are innervated, in addition to NEBs (Gu et al., 2014).

Ultrastructural analysis in rabbits revealed that PNECs are innervated by two types of nerves: sensory afferent neurons with mitochondria-rich, synaptic vesicle-poor nerve endings, as well as efferent neurons with mitochondria-poor and small agranular synaptic vesicles-rich nerve endings (Lauweryns et al., 1985). Studies in mouse, rat and rabbit lungs have shown that PNECs are innervated by afferent neurons with cell bodies in the vagal ganglion as well as possibly in the dorsal root ganglia (Adriaensen et al., 1998; Lauweryns et al., 1985; Van Lommel, Lauweryns, & Berthoud, 1998). These nerve fibers express a number of markers including P2X3 receptors for ATP, CGRP, P2ry1 and Calbindin (Adriaensen et al., 1998; Brouns, Adriaensen, Burnstock, & Timmermans, 2000; Brouns et al., 2003; Brouns, Van Genechten, Scheuermann, Timmermans, & Adriaensen, 2002; Chang, Strochlic, Williams, Umans, & Liberles, 2015). However, further investigation is needed on how PNECs and nerves interact with each other to command their function.

4. PNECs in lung development

Since PNECs are the first cell type to be specified in the lung epithelium, it has been speculated that they may play a role in the lung developmental steps that follows. The role of PNEC function in lung growth was studied in *ex vivo* fetal and newborn lungs across mammalian species from mouse to human. Gastrin-releasing peptide (GRP), a mammalian homolog of amphibian bombesin, is exclusively expressed by PNECs, and its receptor was found to be expressed in both epithelium and mesenchymal components in the lung (Wang, Yeger, & Cutz, 1996). Bombesin treatment in lung organ culture from mouse, baboon and human, and *via* intraperitoneal injection *in utero* in mice resulted in the increase in either SPC/DNA/protein content of alveolar type 2 cells, DNA content of mesenchymal cells and airway branching (Emanuel et al., 1999; King, Torday, & Sunday, 1995; Sunday, Hua, Dai, Nusrat, & Torday, 1990). These findings raised the possibility that PNECs act through their products to promote lung growth.

This conclusion from gain-of-function experiments is not supported by *in vivo* loss-of-function data. As mentioned above, inactivation of *Ascl1* led to absence of PNECs, but these mutant lungs are normal in branching pattern and size as well as the differentiation of other major epithelial cells, including airway club and ciliated cells and alveolar AT1 and AT2 cells (Ito et al., 2000; Sui et al., 2018). These results indicate that PNECs are not required for primary aspects of lung development.

5. PNEC function as progenitors and progenitor niches

PNECs have been shown to function both as progenitor cells and progenitor cell niches. Pioneer studies using naphthalene, which ablates club cells, showed that variant (naphthalene-resistant) club cells reside apical to PNECs and both are EdU-positive. Moreover, CGRP- and SCGB1A1-double positive cells are observed around PNEC clusters, suggesting that PNECs are able to differentiate into club cells to function as stem cells (Reynolds et al., 2000). This is confirmed using CGRP-creERT2 line to lineage label PNECs, which showed that in the naphthalene-induced injury model, PNECs can differentiate into club and ciliated cells (Song et al., 2012). However, genetic ablation of PNECs by diphtheria toxin expression under the control of CGRP promoter did not affect club cell regeneration following naphthalene. More detailed study revealed that activation of Notch signaling, but not Hippo/Hh signaling, drives transdifferentiation of PNECs, with interplay of Polycomb repressive complex 2 and IL-6/Stat3 pathways (Yao et al., 2018).

PNECs also serve as niche for progenitor cells. PNECs preferentially localized to the areas of label retaining cells (LRCs) (Engelhardt, 2001). The development of variant club cells, as mentioned above, is highly dependent on PNECs. In *Ascl1* mutant mice which lack PNECs, the variant club cells are also ablated (Guha et al., 2012). Further studies showed that variant club cell development is guided by Notch signaling. In RBP-J mutant mice, although PNEC number is increased, there is a lack of variant club cells. No variant club cells are detected in *Notch1*; *Notch2*; *Notch3* triple mutant mice which have increased number of PNECs (Guha et al., 2012; Morimoto et al., 2012). PNECs also express Dll1 and Dll3, suggesting its potential to act through these ligands to activate Notch signaling in variant club cells and promote their fate (Post, Ternet, & Hogan, 2000; Verckist et al., 2017).

6. PNEC function in response to airway inputs

Current evidence suggests that PNECs function primarily through their secreted products. Upon activation, PNECs release small neuropeptides (CGRP, bombesin) and neurotransmitters (serotonin, GABA, ATP) (Brouns et al., 2000; Cho, Chan, & Cutz, 1989; Cutz et al., 1993; Emanuel et al., 1999; Lauweryns, Cokelaere, & Theunynck, 1973). These molecules may either act locally in autocrine and paracrine fashion or signal through neurons that innervate them (Cattaneo et al., 1993; Cutz, Pan, Yeger, Domnik, & Fisher, 2013; Schuller, 1994). PNECs have been implicated in a number of physiological processes primarily by *in vitro* data, while *in vivo* evidence is emerging.

6.1 As an immune modulator

PNEC hyperplasia and increased peptide content have been documented in several inflammatory pediatric and adult lung disorders. The associations led to speculation that PNECs may act as an immune modulator. Whether PNEC dysregulation acts as a cause or consequence of the symptoms remains debated.

Chronic inflammation appears in premature infants where truncation of alveolar development, exposure to hyperoxia and mechanical ventilation culminate into bronchopulmonary dysplasia (BPD) (Davidson & Berkelhamer, 2017; Hwang & Rehan, 2018). Increased levels of bombesin-like peptide have been shown to be associated with a 10-fold increase in the incidence of BPD in premature infants (Cullen et al., 2002). In mice, intratracheal instillation of bombesin led to mast cell recruitment, suggesting a link to inflammation (Subramaniam et al., 2003).

Using mouse genetics, our results demonstrated an immune regulatory role of PNECs *in vivo*. In the loss-of-function *Robo* mutant where PNECs failed to cluster, there is increased production of neuropeptides. This led to increased immune cell infiltration in the neonatal lung, causing matrix remodeling and alveolar simplification (Branchfield et al., 2016).

In a separate study, our recent findings demonstrated that PNECs play a significant role in amplifying allergen-induced asthmatic responses (Sui et al., 2018). In *Shhcre;Ascl1* mutants which lack PNECs, the response to allergen is significantly blunted. Further data indicated that upon allergen challenge, PNECs release CGRP to activate ILC2s, which in turn elicited downstream type 2 immune responses. PNECs also release neurotransmitter GABA to induce goblet-cell metaplasia. These data demonstrated that *in vivo*, PNECs and ILC2s, both enriched at airway branch points, form a neuroimmune module. Together, they function to amplify the asthmatic response.

6.2 Activation by hypoxia, carbon dioxide and acid

Function of PNECs as intrapulmonary chemoreceptors, similar to central and peripheral chemoreceptors (e.g., carotid bodies), was described nearly five decades ago (Lauweryns & Cokelaere, 1973; Lauweryns et al., 1977; Lauweryns, Cokelaere, Lerut, & Theunynck, 1978). The ultrastructural analysis of rabbit PNECs revealed basally localized dense core vesicles containing serotonin, and the release of serotonin can be triggered by hypoxic/hypercapnic challenge (Lauweryns & Cokelaere, 1973; Lauweryns et al., 1977). The same group also performed cross-circulation studies in animals which showed that hypoxia but not hypoxemia, caused the increase in the release of secretory vesicles (Lauweryns et al., 1978). Complementary studies demonstrated that hypoxia leads to the release of serotonin from isolated PNECs and PNECs in cultured lung slices (Cutz, Fu, & Yeger, 2004; Cutz et al., 1993).

Various *in vitro* biochemical studies using human and animal-derived PNECs, fresh lung slices and SCLC tumor cell lines have shed light on the mechanism of how PNECs can sense the intraluminal levels of oxygen/carbon dioxide/pH (acidosis) (Burleson, Mercer, & Wilk-Blaszczak, 2006; Cutz et al., 1993; Fu, Nurse, Wong, & Cutz, 2002; Jonz, Fearon, & Nurse, 2004; Lauweryns et al., 1977; Youngson, Nurse, Yeger, & Cutz, 1993). It was proposed that during low oxygen conditions, reduced H_2O_2 formation by NADPH oxidase keeps oxygen-sensitive K^+ channels in closed state. As a result, voltage-gated Ca^{2+} channels open, followed by the influx of extracellular Ca^{2+}, triggering the release of neurotransmitter. This chemotransmission is also further modulated by the level of extracellular ATP (Fu, Nurse, & Cutz, 2004).

The idea that PNECs play a role in hypoxia sensing is also supported by findings from gp91^{phox} (Cybb or Nox2) loss-of-function mice. These oxidase-deficient mice exhibited abnormal breathing patterns: reduced tidal volume, increased breathing frequency under both normoxia and hypoxia, and reduced peak minute ventilation under hypoxia (Kazemian, Stephenson, Yeger, & Cutz, 2001). These oxidase-deficient mice showed normal hypoxia sensing control in cell types such as carotid body glomus cells, adrenomedullary chromaffin cells and pulmonary artery smooth muscle cells (Archer et al., 1999; He et al., 2002; Thompson, Farragher, Cutz, & Nurse, 2002). However, it was shown that the PNECs from these mutants are insensitive to hypoxia as measured by *in vitro* patch clamping (Fu, Wang, Nurse, Dinauer, & Cutz, 2000; Kazemian et al., 2001). Despite these lines of evidence, the *in vivo* physiological impact of PNECs during hypoxia requires further study.

Several studies have shown that changes in CO_2/H^+ levels can activate the release of serotonin/CGRP using either primary cells or *ex vivo* organ culture (Ebina, Hoyt, McNelly, Sorokin, & Linnoila, 1997; Lauweryns et al., 1977; Schuller, 1994). This response is dependent on the increase in the intracellular Ca^{2+} levels and was significantly reduced in the presence of inhibitors for carbonic anhydrase as well as voltage-gated calcium channels (Livermore et al., 2015). Similar observations were made in neuroendocrine cells derived from gills of zebrafish, suggesting functional conservation during evolution (Qin, Lewis, & Perry, 2010).

6.3 Activation by nicotine

PNECs express nicotinic acetylcholine receptors. In cultured SCLCs or fetal hamster PNECs, nicotine stimulated the release of serotonin and this release was inhibited by specific ganglionic nicotinic antagonists as well as voltage-gated calcium channel blockers (Cattaneo et al., 1993; Codignola et al., 1994; Plummer, Sheppard, & Schuller, 2000; Sheppard, Williams, Plummer, & Schuller, 2000). In a neonatal rabbit model, ultrastructural analysis of NEBs showed that after intratracheal nicotine challenge, there was an increase in the exocytosis of dense core vesicles (Lauweryns et al., 1977). This nicotine-induced serotonin release was suggested to be downstream of activated MAPK pathway and phosphorylation of c-Myc (Jull, Plummer, & Schuller, 2001).

It has been shown that nicotine has mitogenic effects on PNECs. This effect was enhanced by increased levels of CO_2 and was blocked by the inhibitors of protein kinase C and receptors of 5-HT/bombesin. This led to the speculation that the mitogenic effect underlies the frequent association of SCLC to smoking (Plummer et al., 2000; Schuller, 1994). However, a causal relationship has not been demonstrated.

6.4 Activation by stretch

Freshly isolated rabbit PNECs can release cytoplasmic serotonin in response to cyclic sinusoidal stretch, mimicking the intrauterine environment. This effect is blocked by inhibitors of mechano-sensitive channels (Pan, Copland, Post, Yeger, & Cutz, 2006). Similarly, in a mouse lung slice model, mechanical stimulation through application of hypo-osmotic solution resulted in increased Ca^{2+} within NEBs (Lembrechts et al., 2012). Interestingly, Piezo2, the mechano-transduction channel for proprioception, is expressed in both of the vagal sensory nerves and PNECs in the lung, indicating that PNECs are capable of sensing mechanical stretch. However, inactivating *Piezo2* in vagal sensory nerves, but not in PNECs, is essential for respiratory transition at birth (Nonomura et al., 2017). Mouse mutants with no PNECs survive birth and live into adulthood (Sui et al., 2018). Thus, the *in vivo* importance of stretch sensation in PNECs remains unclear.

7. PNECs in chronic lung diseases

PNECs have been implicated in a wide array of lung diseases (Table 1). The pathologies are often observed as an increase in PNEC number and/or products (Linnoila, 2006). It has not been established whether the PNEC pathologies are a consequence of the diseases or are causative to symptoms. Given that PNECs primarily function through their secreted products, and secreted factors are more readily targeted using pharmacological agents, there is a strong incentive to understand PNEC involvement in disease.

PNEC-associated pediatric lung-diseases include neuroendocrine cell hyperplasia of infancy (NEHI) and congenital diaphragmatic hernia (CDH). NEHI is a rare disease that was so named based on the increased PNECs in lung biopsies in the absence of other prominent histologic findings (Deterding et al., 2005). In NEHI lungs, there is a prominence of solitary bombesin-positive PNECs in the distal airways, as well as increase in the numbers and size of NEBs (Deterding et al., 2005; Young et al., 2011). Alveolar development and surfactant protein expression are normal, airways are structurally patent, and inflammation is relatively lacking in both lung sections and bronchoalveolar lavage (Deterding et al., 2005; Deutsch et al., 2007; Young et al., 2011). NEHI typically presents in otherwise healthy infants who were asymptomatic at birth but exhibit gradual onset of chronic tachypnea, hypoxemia, and crackles on chest auscultation in the first year of life. Physiologic abnormalities can be measured by pulmonary function tests (PFTs), demonstrating reduced compliance, marked air-trapping, but normal FEV1/FVC ratio (or FEV0.5/FVC ratio in infants) with small airway flow obstruction without bronchodilator response (Kerby et al., 2013; Young et al., 2011). While pulmonary symptoms and hypoxemia gradually improve over time, long-term oxygen use is required, and abnormal lung function persists into adolescence and adulthood in at least some cases (Deterding et al., 2005; Nevel, Garnett, Schaudies, & Young, 2018; Nevel et al., 2016). There is no treatment for NEHI beyond supportive care.

Congenital diaphragmatic hernia (CDH) is a birth defect with a high mortality rate due to pulmonary hypoplasia, pulmonary hypertension and heart failure (Kardon et al., 2017). Studies have shown an increased number of bombesin-positive PNECs in CDH patient lung biopsies (Asabe et al., 1999; Ijsselstijn et al., 1997). A similar increase of PNECs has been found in a rat model of chemical nitrofen-induced CDH and this increase was correlated with the dysregulation of Notch signaling

Table 1 Pediatric and adult disorders associated with abnormalities with pulmonary neuroendocrine system.

	Symptoms	References
Rare pediatric disorders		
Congenital diaphragmatic hernia (CDH)	PNEC hyperplasia, Increased bombesin-like peptide expression	Gaxiola, Varon, and Valladolid (2009), Asabe, Tsuji, Handa, Kajiwara, and Suita (1999), Ijsselstijn, Gaillard, de Jongste, Tibboel, and Cutz (1997)
Congenital central hypoventilation syndrome	Atrophy of carotid bodies and compensatory increase in PNECs	Cutz, Ma, Perrin, Moore, and Becker (1997)
Wilson-Mikity syndrome	Increase in the number of PNECs	Hoepker et al. (2008), Gillan and Cutz (1993)
Neuroendocrine cell hyperplasia of infancy (NEHI)	Increase in the number and size of PNECs	Deterding, Pye, Fan, and Langston (2005), Young et al. (2011)
Sudden infant death syndrome (SIDS)	PNEC hyperplasia/ hypertrophy	Gillan, Curran, O'Reilly, Cahalane, and Unwin (1989), Perrin, McDonald, and Cutz (1991), Aita et al., 2000, Cutz, Perrin, Pan, Haas, and Krous (2007), Cutz, Yeger, & Pan, 2007)
Congenital pneumonia	PNEC hyperplasia	Saad, Heffelfinger, and Stanek (2003)
Interstitial pneumonia	PNEC hyperplasia	Jiramethee, Erasmus, Nogee, and Khoor (2017), Chatterjee, Kamimoto, Dunn, Mittadodla, and Joshi (2016), Ito et al. (2002), Reyes, Majo, Perich, and Morell (2007)
Common pediatric disorders		
Bronchopulmonary dysplasia	Increased immunoreactivity of neurotransmitters of PNECs	Johnson, Lock, Elde, and Thompson (1982), Johnson, Kulik, Lock, Elde, and Thompson (1985), Johnson and Wobken (1987), Johnson, Anderson, and Burke (1993)

Table 1 Pediatric and adult disorders associated with abnormalities with pulmonary neuroendocrine system.—cont'd

	Symptoms	**References**
Cystic fibrosis	Increase in the number of serotonin immunoreactivity in PNECs	Johnson et al. (1988)
Rare adult disorders		
Diffuse idiopathic pulmonary neuroendocrine cell hyperplasia (DIPNECH)	PNEC hyperplasia with no other pre-existing lung disease	Aguayo et al. (1992), Koliakos, Thomopoulos, Abbassi, Duc, and Christodoulou (2017)
Common adult disorders		
Asthma	Increase in bombesin and CGRP-positive PNECs	Carmichael and Zacher (2005), Sui et al. (2018)
Chronic obstructive pulmonary disorder (COPD)	Idiopathic diffuse hyperplasia of bombesin-producing PNECs	Alshehri et al. (1997)

(IJsselstijn, Perrin, de Jongste, Cutz, & Tibboel, 1995; Santos et al., 2007). In a genetic mouse model of CDH where the *Roundabout* receptor genes were inactivated, we found increase in PNEC products and increased baseline immune cells in the lung, leading to remodeling of matrix and simplification of alveoli, which resemble the symptoms in CDH patients (Branchfield et al., 2016; Kantarci & Donahoe, 2007).

There are several additional rare pediatric disorders with PNEC pathology. Congenital central hypoventilation syndrome (CCHS) patients present with failed autonomic nervous control of respiration. In CCHS patients, there is an atrophy of the carotid bodies and increased PNECs (Cutz et al., 1997). Wilson-Mikity syndrome, prevalent in low-weight newborn babies, is associated with cystic interstitial emphysema causing chronic obstructive lung disease. This disease has been linked to increased PNECs in the airways (Gillan & Cutz, 1993). Sudden infant death syndrome (SIDS) is also associated with expansion of PNEC/NEB population in numbers as well as size. This has been postulated to arise in response to chronic hypoxia, however, this is difficult to test (Aita et al., 2000;

Cutz, Perrin, et al., 2007; Cutz, Yeger, & Pan, 2007; Gillan et al., 1989; Perrin et al., 1991). In addition, increase of PNECs has been reported in rare pneumonia such as congenital pneumonia and interstitial pneumonia (Chatterjee et al., 2016; Ito et al., 2002; Jiramethee et al., 2017; Reyes et al., 2007; Saad et al., 2003)

A more prevalent pediatric condition that is linked to PNECs is BPD. BPD rate is inversely proportional to gestation age at birth, and plagues ~40% of infants born prior to or at 28 weeks of gestation. BPD lungs show respiratory deficiencies which often persist into adulthood (Johnson et al., 1993). BPD patients display an increase of PNEC secretory products including serotonin, CGRP and bombesin (Johnson et al., 1993, 1985, 1982; Johnson & Wobken, 1987). Many BPD patients develop pulmonary hypertension, but it remains to be tested whether the increase in PNEC products may contribute to hypertension. Likewise, several studies have implicated PNECs in pediatric asthma. Increased PNECs and PNEC secretions are found in pediatric asthma patient lungs (Cutz, Yeger, & Pan, 2007; Sui et al., 2018). In older patients with cystic fibrosis, there is an increase in serotonin-positive bronchioles, suggesting a possible role for PNECs in the progression of this disease (Johnson et al., 1988).

In the adult, both rare and common pathologies have also been linked to PNECs (Table 1). Rare diseases include diffuse idiopathic pulmonary neuroendocrine cell hyperplasia (DIPNECH) (Aguayo et al., 1992; Alshehri et al., 1997; Cutz, Perrin, et al., 2007; Cutz, Yeger, & Pan, 2007; Johnson et al., 1988; Sui et al., 2018). DIPNECH is a rare condition defined by the pathological findings of increased PNECs. While the pathology is similar to that of NEHI, there is no connection between NEHI and DIPNECH clinically. DIPNECH patients have adult-onset, irreversible airflow obstruction and airway fibrosis, typically with no prior history of smoking (Aguayo et al., 1992). It is unclear how the PNEC increase occurs and if it is a cause for the symptoms.

PNEC pathology has also been documented in a majority of prevalent adult lung diseases, including chronic obstructive pulmonary disease (COPD), bronchitis and pulmonary hypertension (Carmichael & Zacher, 2005; Johnson et al., 1988; Sui et al., 2018). In study of smokers, alteration of PNECs precedes the onset of chronic lung disease, suggestive of a causative relationship (Aguayo et al., 1989). Based on the vasomodulatory properties of PNEC products such as 5-HT/CGRP, PNECs may play a role in the pathogenesis of pulmonary hypertension (Cutz, Yeger, & Pan, 2007). It is intriguing that PNEC abnormalities are found in diseases as diverse as asthma and pulmonary hypertension. It is possible that PNECs may only contribute to symptoms in a subset of these diseases. Alternatively, heterogeneity within PNECs may underlie differential responses. Further effort investigating PNECs contribution to these diseases is urgently needed.

8. Pulmonary neuroendocrine cells in cancer

Other than developmental and adult chronic disorders, PNECs are also the cell of origin for lung carcinomas including carcinoids, small cell lung cancer (SCLC) and large cell neuroendocrine carcinoma (LCNC) (Fisseler-Eckhoff & Demes, 2012; Gazdar et al., 2015; Gorshtein et al., 2012; Park et al., 2011). Carcinoids are low/intermediate grade tumors that are well-differentiated PNECs, whereas small cell and large cell lung carcinomas present with poorly differentiated PNECs (Hendifar, Marchevsky, & Tuli, 2017). Carcinoids are easily treatable by surgical removal whereas SCLC is one of the most aggressive cancer forms and is associated with a short survival time after diagnosis (Rossi, Tay, Chiramel, Prelaj, & Califano, 2018).

There is limited understanding of the role of PNECs in each of these carcinomas in comparison to chronic diseases. Comprehensive sequencing has revealed a high mutation rate of p53 and Rb in SCLCs (George et al., 2015). In mice, inactivation of p53, Rb and/or Pten in PNECs leads to PNEC hyperplasia and SCLC lesions (Song et al., 2012). In 25% of human SCLC patients, loss-of-function mutation in Notch pathway genes was identified. Furthermore, activation of Notch signaling in the preclinical mouse model of SCLC led to the reduction of tumor load and increased survival (George et al., 2015). However, a recent study demonstrated that Notch signaling can play either a tumor suppressive role in repressing SCLC progress or an oncogenic role in promoting chemoresistant Notch-active non-neuroendocrine cells in SCLCs (Lim et al., 2017). Another study highlighted the heterogeneity in human SCLCs based on mosaic expression and function of *ASCL1* and *NEUROD1* (Borromeo et al., 2016). It was also demonstrated that MYC drives NEUROD1-high and ASCL1-low subtypes of SCLCs, which is sensitive to Aurora Kinase inhibition (Mollaoglu et al., 2017). Better understanding of the heterogeneity and drug sensitivity will lead to more effective treatment regimen.

9. Concluding remarks

PNECs, overlooked for many years due to their rarity, are emerging as a critical and multi-faceted player in lung function and pathogenesis. The current challenge is to define their *in vivo* impact, the specificity of their responses to stimuli and the heterogeneity within the population. Based on the frequent documentation of PNEC pathology in a plethora of lung diseases, it is critical to understand their roles in disease, and whether they present as novel targets for more effective treatment strategies.

References

Adriaensen, D., Timmermans, J. P., Brouns, I., Berthoud, H. R., Neuhuber, W. L., & Scheuermann, D. W. (1998). Pulmonary intraepithelial vagal nodose afferent nerve terminals are confined to neuroepithelial bodies: An anterograde tracing and confocal microscopy study in adult rats. *Cell and Tissue Research*, *293*, 395–405.

Aguayo, S. M., Kane, M. A., King, T. E., Jr., Schwarz, M. I., Grauer, L., & Miller, Y. E. (1989). Increased levels of bombesin-like peptides in the lower respiratory tract of asymptomatic cigarette smokers. *The Journal of Clinical Investigation*, *84*, 1105–1113.

Aguayo, S. M., Miller, Y. E., Waldron, J. A., Jr., Bogin, R. M., Sunday, M. E., Staton, G. W., Jr., et al. (1992). Brief report: Idiopathic diffuse hyperplasia of pulmonary neuroendocrine cells and airways disease. *The New England Journal of Medicine*, *327*, 1285–1288.

Aita, K., Doi, M., Tanno, K., Oikawa, H., Myo-Thaik, O., Ohashi, N., et al. (2000). Pulmonary neuroendocrine cell distribution in sudden infant death syndrome. *Legal Medicine (Tokyo, Japan)*, *2*, 134–142.

Alshehri, M., Cutz, E., Banzhoff, A., & Canny, G. (1997). Hyperplasia of pulmonary neuroendocrine cells in a case of childhood pulmonary emphysema. *Chest*, *112*, 553–556.

Archer, S. L., Reeve, H. L., Michelakis, E., Puttagunta, L., Waite, R., Nelson, D. P., et al. (1999). O_2 sensing is preserved in mice lacking the gp91 phox subunit of NADPH oxidase. *Proceedings of the National Academy of Sciences of the United States of America*, *96*, 7944–7949.

Asabe, K., Tsuji, K., Handa, N., Kajiwara, M., & Suita, S. (1999). Immunohistochemical distribution of bombesin-positive pulmonary neuroendocrine cells in a congenital diaphragmatic hernia. *Surgery Today*, *29*, 407–412.

Boers, J. E., den Brok, J. L., Koudstaal, J., Arends, J. W., & Thunnissen, F. B. (1996). Number and proliferation of neuroendocrine cells in normal human airway epithelium. *American Journal of Respiratory and Critical Care Medicine*, *154*, 758–763.

Borromeo, M. D., Savage, T. K., Kollipara, R. K., He, M., Augustyn, A., Osborne, J. K., et al. (2016). ASCL1 and NEUROD1 reveal heterogeneity in pulmonary neuroendocrine tumors and regulate distinct genetic programs. *Cell Reports*, *16*, 1259–1272.

Branchfield, K., Nantie, L., Verheyden, J. M., Sui, P., Wienhold, M. D., & Sun, X. (2016). Pulmonary neuroendocrine cells function as airway sensors to control lung immune response. *Science*, *351*, 707–710.

Brouns, I., Adriaensen, D., Burnstock, G., & Timmermans, J. P. (2000). Intraepithelial vagal sensory nerve terminals in rat pulmonary neuroepithelial bodies express P2X(3) receptors. *American Journal of Respiratory Cell and Molecular Biology*, *23*, 52–61.

Brouns, I., Van Genechten, J., Hayashi, H., Gajda, M., Gomi, T., Burnstock, G., et al. (2003). Dual sensory innervation of pulmonary neuroepithelial bodies. *American Journal of Respiratory Cell and Molecular Biology*, *28*, 275–285.

Brouns, I., Van Genechten, J., Scheuermann, D. W., Timmermans, J. P., & Adriaensen, D. (2002). Neuroepithelial bodies: A morphologic substrate for the link between neuronal nitric oxide and sensitivity to airway hypoxia? *The Journal of Comparative Neurology*, *449*, 343–354.

Burleson, M. L., Mercer, S. E., & Wilk-Blaszczak, M. A. (2006). Isolation and characterization of putative O_2 chemoreceptor cells from the gills of channel catfish (*Ictalurus punctatus*). *Brain Research*, *1092*, 100–107.
Carmichael, M. G., & Zacher, L. L. (2005). The demonstration of pulmonary neuroendocrine cell hyperplasia with tumorlets in a patient with chronic cough and a history of multiple medical problems. *Military Medicine*, *170*, 439–441.
Cattaneo, M. G., Codignola, A., Vicentini, L. M., Clementi, F., & Sher, E. (1993). Nicotine stimulates a serotonergic autocrine loop in human small-cell lung carcinoma. *Cancer Research*, *53*, 5566–5568.
Chang, R. B., Strochlic, D. E., Williams, E. K., Umans, B. D., & Liberles, S. D. (2015). Vagal sensory neuron subtypes that differentially control breathing. *Cell*, *161*, 622–633.
Chatterjee, K., Kamimoto, J. J., Dunn, A., Mittadodla, E., & Joshi, M. (2016). A case of DIPNECH presenting as usual interstitial pneumonia. *Pneumonologia i Alergologia Polska*, *84*, 174–177.
Chen, H., Thiagalingam, A., Chopra, H., Borges, M. W., Feder, J. N., Nelkin, B. D., et al. (1997). Conservation of the Drosophila lateral inhibition pathway in human lung cancer: A hairy-related protein (HES-1) directly represses achaete-scute homolog-1 expression. *Proceedings of the National Academy of Sciences of the United States of America*, *94*, 5355–5360.
Cho, T., Chan, W., & Cutz, E. (1989). Distribution and frequency of neuro-epithelial bodies in post-natal rabbit lung: Quantitative study with monoclonal antibody against serotonin. *Cell and Tissue Research*, *255*, 353–362.
Codignola, A., Tarroni, P., Cattaneo, M. G., Vicentini, L. M., Clementi, F., & Sher, E. (1994). Serotonin release and cell proliferation are under the control of alpha-bungarotoxin-sensitive nicotinic receptors in small-cell lung carcinoma cell lines. *FEBS Letters*, *342*, 286–290.
Cullen, A., Van Marter, L. J., Allred, E. N., Moore, M., Parad, R. B., & Sunday, M. E. (2002). Urine bombesin-like peptide elevation precedes clinical evidence of bronchopulmonary dysplasia. *American Journal of Respiratory and Critical Care Medicine*, *165*, 1093–1097.
Cutz, E., Fu, X. W., & Yeger, H. (2004). Methods to study neuroepithelial bodies as airway oxygen sensors. *Methods in Enzymology*, *381*, 26–40.
Cutz, E., Ma, T. K., Perrin, D. G., Moore, A. M., & Becker, L. E. (1997). Peripheral chemoreceptors in congenital central hypoventilation syndrome. *American Journal of Respiratory and Critical Care Medicine*, *155*, 358–363.
Cutz, E., Pan, J., Yeger, H., Domnik, N. J., & Fisher, J. T. (2013). Recent advances and contraversies on the role of pulmonary neuroepithelial bodies as airway sensors. *Seminars in Cell & Developmental Biology*, *24*, 40–50.
Cutz, E., Perrin, D. G., Pan, J., Haas, E. A., & Krous, H. F. (2007). Pulmonary neuroendocrine cells and neuroepithelial bodies in sudden infant death syndrome: Potential markers of airway chemoreceptor dysfunction. *Pediatric and Developmental Pathology*, *10*, 106–116.
Cutz, E., Speirs, V., Yeger, H., Newman, C., Wang, D., & Perrin, D. G. (1993). Cell biology of pulmonary neuroepithelial bodies—Validation of an in vitro model. I. Effects of hypoxia and Ca^{2+} ionophore on serotonin content and exocytosis of dense core vesicles. *The Anatomical Record*, *236*, 41–52.
Cutz, E., Yeger, H., & Pan, J. (2007). Pulmonary neuroendocrine cell system in pediatric lung disease-recent advances. *Pediatric and Developmental Pathology*, *10*, 419–435.
Davidson, L. M., & Berkelhamer, S. K. (2017). Bronchopulmonary dysplasia: Chronic lung disease of infancy and long-term pulmonary outcomes. *Journal of Clinical Medicine*, *6*, 4.
Deterding, R. R., Pye, C., Fan, L. L., & Langston, C. (2005). Persistent tachypnea of infancy is associated with neuroendocrine cell hyperplasia. *Pediatric Pulmonology*, *40*, 157–165.

Deutsch, G. H., Young, L. R., Deterding, R. R., Fan, L. L., Dell, S. D., Bean, J. A., et al. (2007). Diffuse lung disease in young children: Application of a novel classification scheme. *American Journal of Respiratory and Critical Care Medicine, 176*, 1120–1128.
Ebina, M., Hoyt, R. F., Jr., McNelly, N. A., Sorokin, S. P., & Linnoila, R. I. (1997). Effects of hydrogen and bicarbonate ions on endocrine cells in fetal rat lung organ cultures. *The American Journal of Physiology, 272*, L178–L186.
Emanuel, R. L., Torday, J. S., Mu, Q., Asokananthan, N., Sikorski, K. A., & Sunday, M. E. (1999). Bombesin-like peptides and receptors in normal fetal baboon lung: Roles in lung growth and maturation. *The American Journal of Physiology, 277*, L1003–L1017.
Engelhardt, J. F. (2001). Stem cell niches in the mouse airway. *American Journal of Respiratory Cell and Molecular Biology, 24*, 649–652.
Feyrter, F. (1954). Argyrophilia of bright cell system in bronchial tree in man. *Zeitschrift für Mikroskopisch-Anatomische Forschung, 61*, 73–81.
Fisseler-Eckhoff, A., & Demes, M. (2012). Neuroendocrine tumors of the lung. *Cancers, 4*, 777–798.
Fu, X. W., Nurse, C. A., & Cutz, E. (2004). Expression of functional purinergic receptors in pulmonary neuroepithelial bodies and their role in hypoxia chemotransmission. *Biological Chemistry, 385*, 275–284.
Fu, X. W., Nurse, C. A., Wong, V., & Cutz, E. (2002). Hypoxia-induced secretion of serotonin from intact pulmonary neuroepithelial bodies in neonatal rabbit. *The Journal of Physiology, 539*, 503–510.
Fu, X. W., Wang, D., Nurse, C. A., Dinauer, M. C., & Cutz, E. (2000). NADPH oxidase is an O_2 sensor in airway chemoreceptors: Evidence from K^+ current modulation in wild-type and oxidase-deficient mice. *Proceedings of the National Academy of Sciences of the United States of America, 97*, 4374–4379.
Gaxiola, A., Varon, J., & Valladolid, G. (2009). Congenital diaphragmatic hernia: An overview of the etiology and current management. *Acta Paediatrica, 98*, 621–627.
Gazdar, A. F., Savage, T. K., Johnson, J. E., Berns, A., Sage, J., Linnoila, R. I., et al. (2015). The comparative pathology of genetically engineered mouse models for neuroendocrine carcinomas of the lung. *Journal of Thoracic Oncology: Official Publication of the International Association for the Study of Lung Cancer, 10*, 553–564.
George, J., Lim, J. S., Jang, S. J., Cun, Y., Ozretic, L., Kong, G., et al. (2015). Comprehensive genomic profiles of small cell lung cancer. *Nature, 524*, 47–53.
Gillan, J. E., Curran, C., O'Reilly, E., Cahalane, S. F., & Unwin, A. R. (1989). Abnormal patterns of pulmonary neuroendocrine cells in victims of sudden infant death syndrome. *Pediatrics, 84*, 828–834.
Gillan, J. E., & Cutz, E. (1993). Abnormal pulmonary bombesin immunoreactive cells in Wilson-Mikity syndrome (pulmonary dysmaturity) and bronchopulmonary dysplasia. *Pediatric Pathology, 13*, 165–180.
Gorshtein, A., Gross, D. J., Barak, D., Strenov, Y., Refaeli, Y., Shimon, I., et al. (2012). Diffuse idiopathic pulmonary neuroendocrine cell hyperplasia and the associated lung neuroendocrine tumors: Clinical experience with a rare entity. *Cancer, 118*, 612–619.
Gu, X., Karp, P. H., Brody, S. L., Pierce, R. A., Welsh, M. J., Holtzman, M. J., et al. (2014). Chemosensory functions for pulmonary neuroendocrine cells. *American Journal of Respiratory Cell and Molecular Biology, 50*, 637–646.
Guha, A., Vasconcelos, M., Cai, Y., Yoneda, M., Hinds, A., Qian, J., et al. (2012). Neuroepithelial body microenvironment is a niche for a distinct subset of Clara-like precursors in the developing airways. *Proceedings of the National Academy of Sciences of the United States of America, 109*, 12592–12597.
He, L., Chen, J., Dinger, B., Sanders, K., Sundar, K., Hoidal, J., et al. (2002). Characteristics of carotid body chemosensitivity in NADPH oxidase-deficient mice. *American Journal of Physiology. Cell Physiology, 282*, C27–C33.

Heath, D., Yacoub, M., Gosney, J. R., Madden, B., Caslin, A. W., & Smith, P. (1990). Pulmonary endocrine cells in hypertensive pulmonary vascular disease. *Histopathology, 16*, 21–28.

Hendifar, A. E., Marchevsky, A. M., & Tuli, R. (2017). Neuroendocrine tumors of the lung: Current challenges and advances in the diagnosis and management of well-differentiated disease. *Journal of Thoracic Oncology: Official Publication of the International Association for the Study of Lung Cancer, 12*, 425–436.

Hockman, D., Burns, A. J., Schlosser, G., Gates, K. P., Jevans, B., Mongera, A., et al. (2017). Evolution of the hypoxia-sensitive cells involved in amniote respiratory reflexes. *eLife, 6*, e21231.

Hoepker, A., Seear, M., Petrocheilou, A., Hayes, D., Jr., Nair, A., Deodhar, J., et al. (2008). Wilson-Mikity syndrome: Updated diagnostic criteria based on nine cases and a review of the literature. *Pediatric Pulmonology, 43*, 1004–1012.

Hwang, J. S., & Rehan, V. K. (2018). Recent advances in bronchopulmonary dysplasia: Pathophysiology, prevention, and treatment. *Lung, 196*, 129–138.

Ijsselstijn, H., Gaillard, J. L., de Jongste, J. C., Tibboel, D., & Cutz, E. (1997). Abnormal expression of pulmonary bombesin-like peptide immunostaining cells in infants with congenital diaphragmatic hernia. *Pediatric Research, 42*, 715–720.

IJsselstijn, H., Perrin, D. G., de Jongste, J. C., Cutz, E., & Tibboel, D. (1995). Pulmonary neuroendocrine cells in neonatal rats with congenital diaphragmatic hernia. *Journal of Pediatric Surgery, 30*, 413–415.

Ito, T., Ogura, T., Ogawa, N., Udaka, N., Hayashi, H., Inayama, Y., et al. (2002). Modulation of pulmonary neuroendocrine cells in idiopathic interstitial pneumonia. *Histology and Histopathology, 17*, 1121–1127.

Ito, T., Udaka, N., Yazawa, T., Okudela, K., Hayashi, H., Sudo, T., et al. (2000). Basic helix-loop-helix transcription factors regulate the neuroendocrine differentiation of fetal mouse pulmonary epithelium. *Development, 127*, 3913–3921.

Jiramethee, N., Erasmus, D., Nogee, L., & Khoor, A. (2017). Pulmonary neuroendocrine cell hyperplasja associated with surfactant protein C gene mutation. *Case Reports in Pulmonology, 2017*, 9541419.

Johnson, D. E., Anderson, W. R., & Burke, B. A. (1993). Pulmonary neuroendocrine cells in pediatric lung disease: Alterations in airway structure in infants with bronchopulmonary dysplasia. *The Anatomical Record, 236*, 115–119. 72–3; discussion 20-1.

Johnson, D. E., Kulik, T. J., Lock, J. E., Elde, R. P., & Thompson, T. R. (1985). Bombesin-, calcitonin-, and serotonin-immunoreactive pulmonary neuroendocrine cells in acute and chronic neonatal lung disease. *Pediatric Pulmonology, 1*, S13–S20.

Johnson, D. E., Lock, J. E., Elde, R. P., & Thompson, T. R. (1982). Pulmonary neuroendocrine cells in hyaline membrane disease and bronchopulmonary dysplasia. *Pediatric Research, 16*, 446–454.

Johnson, D. E., & Wobken, J. D. (1987). Calcitonin gene-related peptide immunoreactivity in airway epithelial cells of the human fetus and infant. *Cell and Tissue Research, 250*, 579–583.

Johnson, D. E., Wobken, J. D., & Landrum, B. G. (1988). Changes in bombesin, calcitonin, and serotonin immunoreactive pulmonary neuroendocrine cells in cystic fibrosis and after prolonged mechanical ventilation. *The American Review of Respiratory Disease, 137*, 123–131.

Jonz, M. G., Fearon, I. M., & Nurse, C. A. (2004). Neuroepithelial oxygen chemoreceptors of the zebrafish gill. *The Journal of Physiology, 560*, 737–752.

Jull, B. A., Plummer, H. K., 3rd, & Schuller, H. M. (2001). Nicotinic receptor-mediated activation by the tobacco-specific nitrosamine NNK of a Raf-1/MAP kinase pathway, resulting in phosphorylation of c-myc in human small cell lung carcinoma cells and pulmonary neuroendocrine cells. *Journal of Cancer Research and Clinical Oncology, 127*, 707–717.

Kantarci, S., & Donahoe, P. K. (2007). Congenital diaphragmatic hernia (CDH) etiology as revealed by pathway genetics. *American Journal of Medical Genetics. Part C, Seminars in Medical Genetics, 145C*, 217–226.

Kardon, G., Ackerman, K. G., McCulley, D. J., Shen, Y., Wynn, J., Shang, L., et al. (2017). Congenital diaphragmatic hernias: From genes to mechanisms to therapies. *Disease Models & Mechanisms, 10*, 955–970.

Kazemian, P., Stephenson, R., Yeger, H., & Cutz, E. (2001). Respiratory control in neonatal mice with NADPH oxidase deficiency. *Respiration Physiology, 126*, 89–101.

Kerby, G. S., Wagner, B. D., Popler, J., Hay, T. C., Kopecky, C., Wilcox, S. L., et al. (2013). Abnormal infant pulmonary function in young children with neuroendocrine cell hyperplasia of infancy. *Pediatric Pulmonology, 48*, 1008–1015.

King, K. A., Torday, J. S., & Sunday, M. E. (1995). Bombesin and [Leu8]phyllolitorin promote fetal mouse lung branching morphogenesis via a receptor-mediated mechanism. *Proceedings of the National Academy of Sciences of the United States of America, 92*, 4357–4361.

Koliakos, E., Thomopoulos, T., Abbassi, Z., Duc, C., & Christodoulou, M. (2017). Diffuse idiopathic pulmonary neuroendocrine cell hyperplasia: A case report and review of the literature. *The American Journal of Case Reports, 18*, 975–979.

Kuo, C. S., & Krasnow, M. A. (2015). Formation of a neurosensory organ by epithelial cell slithering. *Cell, 163*, 394–405.

Larson, S. D., Schelegle, E. S., Hyde, D. M., & Plopper, C. G. (2003). The three-dimensional distribution of nerves along the entire intrapulmonary airway tree of the adult rat and the anatomical relationship between nerves and neuroepithelial bodies. *American Journal of Respiratory Cell and Molecular Biology, 28*, 592–599.

Lauweryns, J. M., & Cokelaere, M. (1973). Hypoxia-sensitive neuro-epithelial bodies. Intrapulmonary secretory neuroreceptors, modulated by the CNS. *Zeitschrift für Zellforschung und Mikroskopische Anatomie (Vienna, Austria: 1948), 145*, 521–540.

Lauweryns, J. M., Cokelaere, M., Deleersynder, M., & Liebens, M. (1977). Intrapulmonary neuro-epithelial bodies in newborn rabbits. Influence of hypoxia, hyperoxia, hypercapnia, nicotine, reserpine, L-DOPA and 5-HTP. *Cell and Tissue Research, 182*, 425–440.

Lauweryns, J. M., Cokelaere, M., Lerut, T., & Theunynck, P. (1978). Cross-circulation studies on the influence of hypoxia and hypoxaemia on neuro-epithelial bodies in young rabbits. *Cell and Tissue Research, 193*, 373–386.

Lauweryns, J. M., Cokelaere, J., & Theunynck, P. (1973). Serotonin producing neuroepithelial bodies in rabbit respiratory mucosa. *Science, 180*, 410–413.

Lauweryns, J. M., Van Lommel, A. T., & Dom, R. J. (1985). Innervation of rabbit intrapulmonary neuroepithelial bodies. Quantitative and qualitative ultrastructural study after vagotomy. *Journal of the Neurological Sciences, 67*, 81–92.

Lembrechts, R., Brouns, I., Schnorbusch, K., Pintelon, I., Timmermans, J. P., & Adriaensen, D. (2012). Neuroepithelial bodies as mechanotransducers in the intrapulmonary airway epithelium: Involvement of TRPC5. *American Journal of Respiratory Cell and Molecular Biology, 47*, 315–323.

Lim, J. S., Ibaseta, A., Fischer, M. M., Cancilla, B., O'Young, G., Cristea, S., et al. (2017). Intratumoural heterogeneity generated by Notch signalling promotes small-cell lung cancer. *Nature, 545*, 360–364.

Linnoila, R. I. (2006). Functional facets of the pulmonary neuroendocrine system. *Laboratory Investigation, 86*, 425–444.

Linnoila, R. I., Zhao, B., DeMayo, J. L., Nelkin, B. D., Baylin, S. B., DeMayo, F. J., et al. (2000). Constitutive achaete-scute homologue-1 promotes airway dysplasia and lung neuroendocrine tumors in transgenic mice. *Cancer Research, 60*, 4005–4009.

Livermore, S., Zhou, Y., Pan, J., Yeger, H., Nurse, C. A., & Cutz, E. (2015). Pulmonary neuroepithelial bodies are polymodal airway sensors: Evidence for CO_2/H^+ sensing. *American Journal of Physiology. Lung Cellular and Molecular Physiology, 308*, L807–L815.

Mollaoglu, G., Guthrie, M. R., Bohm, S., Bragelmann, J., Can, I., Ballieu, P. M., et al. (2017). MYC drives progression of small cell lung cancer to a variant neuroendocrine subtype with vulnerability to Aurora kinase inhibition. *Cancer Cell, 31*, 270–285.

Morimoto, M., Nishinakamura, R., Saga, Y., & Kopan, R. (2012). Different assemblies of Notch receptors coordinate the distribution of the major bronchial Clara, ciliated and neuroendocrine cells. *Development, 139*, 4365–4373.

Nevel, R. J., Garnett, E. T., Schaudies, D. A., & Young, L. R. (2018). Growth trajectories and oxygen use in neuroendocrine cell hyperplasia of infancy. *Pediatric Pulmonology, 53*, 656–663.

Nevel, R. J., Garnett, E. T., Worrell, J. A., Morton, R. L., Nogee, L. M., Blackwell, T. S., et al. (2016). Persistent lung disease in adults with NKX2.1 mutation and familial neuroendocrine cell hyperplasia of infancy. *Annals of the American Thoracic Society, 13*, 1299–1304.

Noguchi, M., Sumiyama, K., & Morimoto, M. (2015). Directed migration of pulmonary neuroendocrine cells toward airway branches organizes the stereotypic location of neuroepithelial bodies. *Cell Reports, 13*, 2679–2686.

Nonomura, K., Woo, S. H., Chang, R. B., Gillich, A., Qiu, Z., Francisco, A. G., et al. (2017). Piezo2 senses airway stretch and mediates lung inflation-induced apnoea. *Nature, 541*, 176–181.

Ochs, M., Nyengaard, J. R., Jung, A., Knudsen, L., Voigt, M., Wahlers, T., et al. (2004). The number of alveoli in the human lung. *American Journal of Respiratory and Critical Care Medicine, 169*, 120–124.

Pan, J., Copland, I., Post, M., Yeger, H., & Cutz, E. (2006). Mechanical stretch-induced serotonin release from pulmonary neuroendocrine cells: Implications for lung development. *American Journal of Physiology. Lung Cellular and Molecular Physiology, 290*, L185–L193.

Pan, J., Yeger, H., & Cutz, E. (2004). Innervation of pulmonary neuroendocrine cells and neuroepithelial bodies in developing rabbit lung. *The Journal of Histochemistry and Cytochemistry: Official Journal of the Histochemistry Society, 52*, 379–389.

Park, K. S., Liang, M. C., Raiser, D. M., Zamponi, R., Roach, R. R., Curtis, S. J., et al. (2011). Characterization of the cell of origin for small cell lung cancer. *Cell Cycle, 10*, 2806–2815.

Perrin, D. G., McDonald, T. J., & Cutz, E. (1991). Hyperplasia of bombesin-immunoreactive pulmonary neuroendocrine cells and neuroepithelial bodies in sudden infant death syndrome. *Pediatric Pathology, 11*, 431–447.

Plummer, H. K., 3rd, Sheppard, B. J., & Schuller, H. M. (2000). Interaction of tobacco-specific toxicants with nicotinic cholinergic regulation of fetal pulmonary neuroendocrine cells: Implications for pediatric lung disease. *Experimental Lung Research, 26*, 121–135.

Post, L. C., Ternet, M., & Hogan, B. L. (2000). Notch/Delta expression in the developing mouse lung. *Mechanisms of Development, 98*, 95–98.

Qin, Z., Lewis, J. E., & Perry, S. F. (2010). Zebrafish (*Danio rerio*) gill neuroepithelial cells are sensitive chemoreceptors for environmental CO_2. *The Journal of Physiology, 588*, 861–872.

Reyes, L. J., Majo, J., Perich, D., & Morell, F. (2007). Neuroendocrine cell hyperplasia as an unusual form of interstitial lung disease. *Respiratory Medicine, 101*, 1840–1843.

Reynolds, S. D., Giangreco, A., Power, J. H., & Stripp, B. R. (2000). Neuroepithelial bodies of pulmonary airways serve as a reservoir of progenitor cells capable of epithelial regeneration. *The American Journal of Pathology, 156*, 269–278.

Rossi, A., Tay, R., Chiramel, J., Prelaj, A., & Califano, R. (2018). Current and future therapeutic approaches for the treatment of small cell lung cancer. *Expert Review of Anticancer Therapy, 18*, 473–486.

Saad, A. G., Heffelfinger, S., & Stanek, J. (2003). Amniotic sac infection syndrome features fetal lung neuroendocrine cell hyperfunction. *Pediatric and Developmental Pathology: The Official Journal of the Society for Pediatric Pathology and the Paediatric Pathology Society, 6*, 484–494.

Santos, M., Nogueira-Silva, C., Baptista, M. J., Soares-Fernandes, J., Moura, R. S., & Correia-Pinto, J. (2007). Pulmonary epithelial cell differentiation in the nitrofen-induced congenital diaphragmatic hernia. *Journal of Pediatric Surgery, 42*, 1231–1237.

Sasai, Y., Kageyama, R., Tagawa, Y., Shigemoto, R., & Nakanishi, S. (1992). Two mammalian helix-loop-helix factors structurally related to Drosophila hairy and Enhancer of split. *Genes & Development, 6*, 2620–2634.

Schuller, H. M. (1994). Carbon dioxide potentiates the mitogenic effects of nicotine and its carcinogenic derivative, NNK, in normal and neoplastic neuroendocrine lung cells via stimulation of autocrine and protein kinase C-dependent mitogenic pathways. *Neurotoxicology, 15*, 877–886.

Schuller, H. M., Nylen, E., Park, P., & Becker, K. L. (1990). Nicotine, acetylcholine and bombesin are trophic growth factors in neuroendocrine cell lines derived from experimental hamster lung tumors. *Life Sciences, 47*, 571–578.

Shan, L., Aster, J. C., Sklar, J., & Sunday, M. E. (2007). Notch-1 regulates pulmonary neuroendocrine cell differentiation in cell lines and in transgenic mice. *American Journal of Physiology. Lung Cellular and Molecular Physiology, 292*, L500–L509.

Sheppard, B. J., Williams, M., Plummer, H. K., & Schuller, H. M. (2000). Activation of voltage-operated Ca^{2+}-channels in human small cell lung carcinoma by the tobacco-specific nitrosamine 4-(methylnitrosamino)-1-(3-pyridyl)-1-butanone. *International Journal of Oncology, 16*, 513–518.

Song, H., Yao, E., Lin, C., Gacayan, R., Chen, M. H., & Chuang, P. T. (2012). Functional characterization of pulmonary neuroendocrine cells in lung development, injury, and tumorigenesis. *Proceedings of the National Academy of Sciences of the United States of America, 109*, 17531–17536.

Subramaniam, M., Sugiyama, K., Coy, D. H., Kong, Y., Miller, Y. E., Weller, P. F., et al. (2003). Bombesin-like peptides and mast cell responses: Relevance to bronchopulmonary dysplasia? *American Journal of Respiratory and Critical Care Medicine, 168*, 601–611.

Sui, P., Wiesner, D. L., Xu, J., Zhang, Y., Lee, J., Van Dyken, S., et al. (2018). Pulmonary neuroendocrine cells amplify allergic asthma responses. *Science*.

Sunday, M. E., Hua, J., Dai, H. B., Nusrat, A., & Torday, J. S. (1990). Bombesin increases fetal lung growth and maturation in utero and in organ culture. *American Journal of Respiratory Cell and Molecular Biology, 3*, 199–205.

Sutherland, K. D., Proost, N., Brouns, I., Adriaensen, D., Song, J. Y., & Berns, A. (2011). Cell of origin of small cell lung cancer: Inactivation of Trp53 and Rb1 in distinct cell types of adult mouse lung. *Cancer Cell, 19*, 754–764.

Thompson, R. J., Farragher, S. M., Cutz, E., & Nurse, C. A. (2002). Developmental regulation of O(2) sensing in neonatal adrenal chromaffin cells from wild-type and NADPH-oxidase-deficient mice. *Pflügers Archiv, 444*, 539–548.

Tsao, P. N., Vasconcelos, M., Izvolsky, K. I., Qian, J., Lu, J., & Cardoso, W. V. (2009). Notch signaling controls the balance of ciliated and secretory cell fates in developing airways. *Development, 136*, 2297–2307.

Van Lommel, A., & Lauweryns, J. M. (1997). Postnatal development of the pulmonary neuroepithelial bodies in various animal species. *Journal of the Autonomic Nervous System, 65*, 17–24.

Van Lommel, A., Lauweryns, J. M., & Berthoud, H. R. (1998). Pulmonary neuroepithelial bodies are innervated by vagal afferent nerves: An investigation with in vivo anterograde DiI tracing and confocal microscopy. *Anatomy and Embryology, 197*, 325–330.

Verckist, L., Lembrechts, R., Thys, S., Pintelon, I., Timmermans, J. P., Brouns, I., et al. (2017). Selective gene expression analysis of the neuroepithelial body microenvironment in postnatal lungs with special interest for potential stem cell characteristics. *Respiratory Research*, *18*, 87.

Wang, D., Yeger, H., & Cutz, E. (1996). Expression of gastrin-releasing peptide receptor gene in developing lung. *American Journal of Respiratory Cell and Molecular Biology*, *14*, 409–416.

Wiebe, B. M., & Laursen, H. (1995). Human lung volume, alveolar surface area, and capillary length. *Microscopy Research and Technique*, *32*, 255–262.

Yao, E., Lin, C., Wu, Q., Zhang, K., Song, H., & Chuang, P. T. (2018). Notch signaling controls transdifferentiation of pulmonary neuroendocrine cells in response to lung injury. *Stem Cells*, *36*, 377–391.

Young, L. R., Brody, A. S., Inge, T. H., Acton, J. D., Bokulic, R. E., Langston, C., et al. (2011). Neuroendocrine cell distribution and frequency distinguish neuroendocrine cell hyperplasia of infancy from other pulmonary disorders. *Chest*, *139*, 1060–1071.

Youngson, C., Nurse, C., Yeger, H., & Cutz, E. (1993). Oxygen sensing in airway chemoreceptors. *Nature*, *365*, 153–155.

CHAPTER FOUR

Molecular regulation of mammalian hepatic architecture

Stacey S. Huppert[a,b,1], **Makiko Iwafuchi-Doi**[b,c]
[a]Division of Gastroenterology, Hepatology & Nutrition, Cincinnati Children's Hospital Medical Center, Cincinnati, OH, United States
[b]Department of Pediatrics, University of Cincinnati College of Medicine, Cincinnati, OH, United States
[c]Division of Developmental Biology, Cincinnati Children's Hospital Medical Center, Cincinnati, OH, United States
[1]Corresponding author: e-mail address: stacey.huppert@cchmc.org

Contents

Abstract

The essential liver exocrine and endocrine functions require a precise spatial arrangement of the hepatic lobule consisting of the central vein, portal vein, hepatic artery, intrahepatic bile duct system, and hepatocyte zonation. This allows blood to be carried through the liver parenchyma sampled by all hepatocytes and bile produced by the hepatocytes to be carried out of the liver through the intrahepatic bile duct system composed of cholangiocytes. The molecular orchestration of multiple signaling pathways and epigenetic factors is required to set up lineage restriction of the bipotential hepatoblast progenitor into the hepatocyte and cholangiocyte cell lineages, and to further refine cell fate heterogeneity within each cell lineage reflected in the functional heterogeneity of hepatocytes and cholangiocytes. In addition to the complex molecular regulation, there is a complicated morphogenetic choreography observed in building the refined hepatic epithelial architecture. Given the multifaceted molecular and cellular

Current Topics in Developmental Biology, Volume 132
ISSN 0070-2153
https://doi.org/10.1016/bs.ctdb.2018.12.003

regulation, it is not surprising that impairment of any of these processes can result in acute and chronic hepatobiliary diseases. To enlighten the development of potential molecular and cellular targets for therapeutic options, an understanding of how the intricate hepatic molecular and cellular interactions are regulated is imperative. Here, we review the signaling pathways and epigenetic factors regulating hepatic cell lineages, fates, and epithelial architecture.

1. Hepatic specification

1.1 Establishment of hepatic competence by pioneer factors

Multipotent progenitors establish competence to differentiate into specific lineages but not others. Embryonic foregut endoderm cells have not expressed liver-specific genes yet, but the isolated foregut progenitors retain the competence to respond to liver inductive signals (Gualdi et al., 1996). This result proposes that the molecular nature of competence must be cell intrinsic, such as the combined pattern of transcription factors and chromatin states. The transcription factors Forkhead Box A1 (FoxA1) and FoxA2 are expressed in foregut endoderm cells, and conditional deletion of both genes in foregut endoderm fails to induce hepatic genes, such as *albumin* (*Alb*) and *transthyretin*, in response to fibroblast growth factor 2 (Fgf2) signaling (Lee, Friedman, Fulmer, & Kaestner, 2005). This study provides genetic evidence that FoxA provides the hepatic competence to respond to the inductive cue. The *Alb* enhancer is pre-occupied by FoxA as well as Gata Binding Protein (Gata) factors in foregut endoderm cell chromatin before the onset of *Alb* expression (Bossard & Zaret, 1998; Gualdi et al., 1996). FoxA and Gata belong to a special class of transcription factors known as "pioneer factors" that have unique ability to engage closed and silent chromatin converting it to an open and permissive chromatin state (Fig. 1) (Iwafuchi-Doi & Zaret, 2016). A genome-wide study using mouse embryonic stem cell (ESC)-derived endoderm and early hepatic cells further support that FoxA genes function as pioneer factors in hepatic differentiation. FoxA pre-occupies promoters and enhancers of silent hepatic genes before their differentiation is directed to hepatic cells (Xu et al., 2012). Taken together, pioneer factors FoxA and Gata first engage silent chromatin in foregut endoderm to impart the competence for hepatic fate. Given that different pioneer factors also work in other developmental contexts

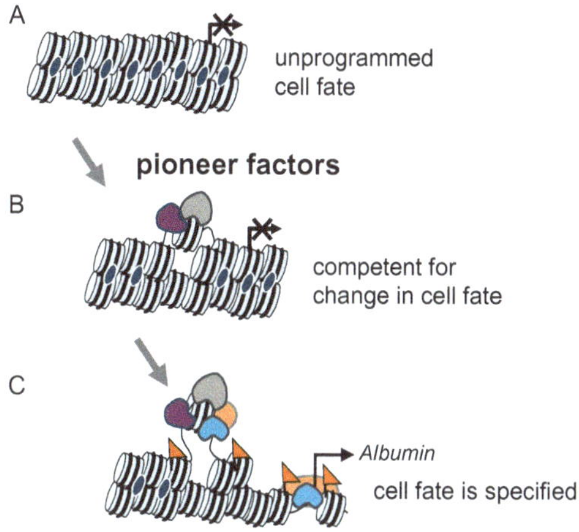

Fig. 1 Initial chromatin targeting of pioneer factor and subsequent events. (A) Closed chromatin configuration in either a low signal or repressed state. (B) Pioneer factors (*gray shape*) locally open gene regulatory regions in chromatin, referred to as a "poised" state. This enables cooperative binding with other transcription factors (*purple shape*) and establishes competence to respond to signals. (C) An active chromatin state is established (active histone marks, *orange triangles*). Additional transcription factors (*orange and blue shapes*) are recruited promoting cooperative and stable binding.

(Iwafuchi-Doi & Zaret, 2014), the studies of hepatic competence provide general principles by which multipotent progenitors gain the competence to differentiate into specific lineages but not others.

1.2 The onset of hepatic morphogenesis

The definitive endoderm is established during gastrulation and invaginates at the anterior end to generate the foregut, which ultimately gives rise to liver, pancreas, lung, and thyroid (Fig. 2). At mouse embryonic day 8.0 of gestation (E8.0), liver progenitors arise from paired lateral domains in the ventral foregut as well as from a small population in the ventral midline endodermal lip (Fig. 2) (Angelo, Guerrero-Zayas, & Tremblay, 2012; Tremblay & Zaret, 2005). At E8.5, morphogenetic movements help close off the foregut, where the paired lateral domains and ventral midline endodermal lip cells merge to create the hepatic endoderm, composed of cells known as hepatoblasts. Shortly after hepatic specification (E8.5–E9.0), the hepatic endoderm thickens as the cells transition from a columnar epithelium to a pseudostratified epithelium (Bort, Signore, Tremblay, Martinez Barbera, & Zaret, 2006). Between E9.0 and E9.5, hepatoblasts delaminate

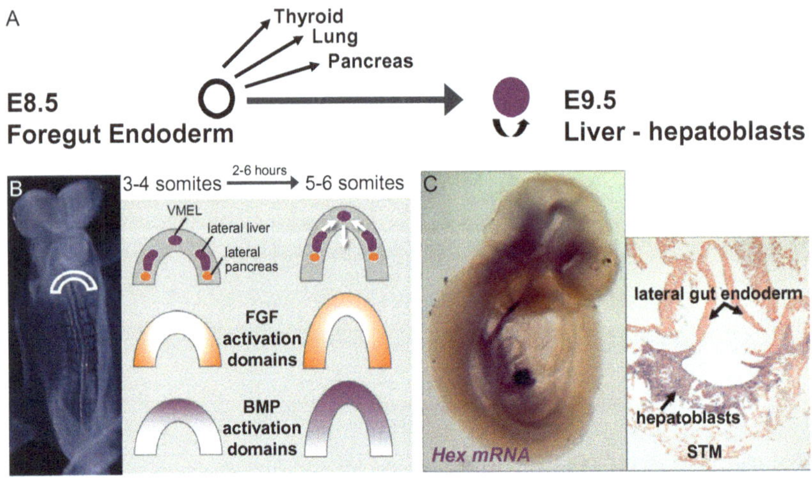

Fig. 2 Specification of the foregut endoderm. (A) During embryogenesis, the thyroid, lung, liver, and ventral pancreas arise by the budding of diverticula from the ventral foregut endoderm that contains multipotent progenitor cells. (B) Hepatic and pancreatic organogenesis requires complex and temporally precise FGF and BMP signaling. At the 3–4 somite stage of mouse development, FGF target genes are activated in lateral endoderm, whereas BMP targets are activated in the ventral midline endodermal lip (VMEL). Therefore, hepatic domains initially reside in a region of high BMP activity at the ventral midline endodermal lip and high FGF activity at the lateral endoderm. (C) Once specified, the *Hex*-positive hepatoblasts then delaminate from the epithelium and migrate into the septum transversum (STM).

from the epithelium and migrate into the adjacent septum transversum mesenchyme to form the nascent liver bud (Fig. 2) (Bort et al., 2006) (reviewed in Ober & Lemaigre, 2018).

1.3 Hepatic specification by signals from adjacent mesoderm tissues

Hepatic fate is specified by combinatorial signaling through a progressive series of reciprocal tissue interactions between the endoderm epithelium and nearby mesoderm during morphogenetic movements of foregut closure (Fig. 2). At early somite stages, Fgf from the cardiac mesoderm and bone morphogenetic protein (Bmp) from septum transversum mesenchyme cells coordinately specify the hepatic fate in the ventral foregut endoderm and suppress the pancreas program (Deutsch, Jung, Zheng, Lora, & Zaret, 2001; Jung, Zheng, Goldfarb, & Zaret, 1999; Rossi, Dunn, Hogan, & Zaret, 2001). As a downstream regulator of Fgf signaling, mitogen-activated protein kinase (Mapk) pathway is activated first in the lateral hepatic progenitors, and then in the ventral midline endodermal lip progenitors about 1 h later (Calmont et al., 2006). In contrast, downstream of Bmp signaling, mothers against decapentaplegic *Drosophila* homolog (Smad)1, 5, 8 are phosphorylated first in the ventral midline endodermal lip progenitors and then in the lateral hepatic progenitors about 1 h later (Wandzioch & Zaret, 2009). The effect of the different order of Fgf and Bmp signaling on gene expression and cellular function remains to be determined.

After foregut closure, the hepatoblasts proliferate and migrate into the surrounding mesenchyme, interacting to form a bud that becomes vascularized. Loss-of-function studies in mice have demonstrated that kinase insert domain receptor (Kdr) which encodes the vascular endothelial growth factor receptor 2 (Vegfr2) is required for hepatic outgrowth, but not hepatic specification (Matsumoto, Yoshitomi, Rossant, & Zaret, 2001). Because of the early embryonic lethality observed in homozygous Kdr mutants, an embryonic explant culture system was used to assess the expansion of the albumin-positive hepatic cells. The liver bud was isolated at E9.5 and put into culture for 72 h. The outgrowth of the hepatic endoderm was specifically affected in homozygous Kdr mutants, in contrast to the growth of the surrounding fibroblast cells or the initial expression of early liver genes in the endoderm. Additionally, the hepatic outgrowth observed in the explant cultures, induced by the endogenous endothelial cells in the wild-type and heterozygous explants, suggests that an intact vasculature and embryo are not necessary for hepatic outgrowth (Matsumoto et al., 2001).

With the advent of single-cell RNA sequencing and the ability to generate hepatic endoderm from human-induced pluripotent stem cells (iPSCs), a more thorough evaluation can be performed to determine the influence mesenchymal and endothelial cells impart on hepatic differentiation. Single-cell RNA sequencing of three-dimensional liver bud organoids reconstituted from human iPSC-derived hepatic endoderm, mesenchymal stem cells and endothelial cells established that the transcriptome states of the cells comprising the liver bud organoids more closely resemble the single-cell transcriptomes of human fetal hepatic cells isolated from samples at gestation weeks 10.5 and 17.5 than human adult liver (Camp et al., 2017; Takebe et al., 2015). *In silico* receptor-ligand pairing of the single-cell transcriptomes identified potential inter-lineage signaling mechanisms between specific cell lineages by means of complementary receptor-ligand pairs. The pairings implicated signaling pathways such as tumor necrosis factor (Tnf), Fgf, Janus tyrosine kinase/signal transducer, and activator of transcription (Jak/Stat), nuclear factor kappa-light-chain-enhancer of activated B cells (Nf-κB), hypoxia-inducible factor (Hif), and Vegf suggesting extensive crosstalk between hepatic, endothelial, and mesenchymal lineages (Camp et al., 2017). Using chemical inhibitors to investigate whether any of these signaling pathways influence the ratio of hepatic endoderm to endothelial cell composition of the liver bud organoids identified Nf-κB, Fgf, insulin-like growth factor (Igf), Jak, and Vegf as important (Camp et al., 2017). Focusing on the hepatic endoderm and endothelial cell interaction, chemical Kdr/Vegfr2 inhibitor treatment of human liver bud organoids or mouse ESC-derived hepatic endoderm co-cultured with endothelial cells resulted in impaired endothelial sprouting and hepatic differentiation (Camp et al., 2017; Han et al., 2018), independent of liver bud organoid self-condensation driven by mesenchymal cells (Camp et al., 2017; Takebe et al., 2015). As expected, since the majority of hepatic endoderm does not express Kdr/Vegfr2, the transcriptome of cultures consisting solely of human iPSC-derived hepatic endoderm treated with chemical Kdr/Vegfr2 inhibitors were not changed (Camp et al., 2017; Goldman et al., 2013). Because cultures consisting of only human iPSC-derived hepatic endoderm do express some mature hepatocyte markers, but their overall transcriptome is more similar to mouse E8.5–E10 liver, the conclusion that the endoderm-endothelial interaction is not required for initial hepatic specification is further supported (Camp et al., 2017).

The molecular regulation required for the *in vivo* process of forming a compact liver bud is unknown. However, *in vitro* studies strongly point to a requirement for mesenchymal cells. Using the liver bud organoid system, all possible combinations of human iPSC-derived hepatic endoderm, mesenchymal stem cells, and endothelial cells were tested for the critical cell type driving the dynamic collective movement of cells into a liver bud organoid (Takebe et al., 2015). The required cell lineage was identified as the mesenchymal stem cell component. Cell-cell contact appears to be required because conditioned medium generated by mesenchymal stem cells was not sufficient to drive collective cell movement and liver bud organoid formation (Takebe et al., 2015). Live imaging studies imply that the self-condensation dynamic is not based on active cell migration but rather on cell contraction through stress fibers, similar to mechanical contraction of viscoelastic body physical models (Shinozawa, Yoshikawa, & Takebe, 2016). During embryonic gastrulation, inward displacement of cell-cell junctions is driven by myosin II, a motor protein responsible cell contraction (Bertet, Sulak, & Lecuit, 2004; Pouille, Ahmadi, Brunet, & Farge, 2009; Shindo & Wallingford, 2014). Similarly, collective movement of cells to form the liver bud organoid requires myosin II. Self-condensation of liver bud organoids is antagonized by the treatment with blebbistatin, a myosin II ATPase inhibitor (Takebe et al., 2015). In total, these results suggest that mesenchymal stem cell-based traction force produced by the actomyosin cytoskeleton is central to the directed and drastic movements of cells observed in the formation of the liver bud organoid. Many *in vivo* experiments are needed to confirm whether these cell interactions are required for liver bud formation in a living organism.

The liver bud expansion intrinsically requires the divergent homeobox genes hematopoietically expressed homeobox (Hex) and prospero homeobox protein 1 (Prox1). *Hex* is the earliest gene known to be expressed and required for differentiation of hepatocytes (Fig. 2) (Keng et al., 2000; Martinez Barbera et al., 2000). At E9.5, Hex homozygous mutants form a ventral endoderm thickening, but no outgrowth of the specified hepatoblasts is observed. Similarly, Prox1 homozygous mutants exhibit a smaller liver and absence of hepatoblast migration due to an increase of E-cadherin expression and a failure to breakdown the laminin-rich membrane surrounding the liver bud. There are currently no known specific molecular markers associated with the function of initial endoderm thickening (Costa, Kalinichenko, Holterman, & Wang, 2003).

1.4 Chromatin basis of fate choice for liver *versus* pancreas

As discussed above, pioneer factors can open chromatin and establish permissive states for gene activation, but are alternative genes repressed? Small cell number chromatin immunoprecipitation (ChIP) studies using ventral foregut endoderm cells from mouse embryos identified a different "pre-pattern" of the chromatin states at hepatic *versus* pancreatic regulatory elements (Xu et al., 2011). In the ventral foregut, the pancreatic regulatory elements contained active (diacetylated histone 3 lysine 9 and 14 [H3K9acK14ac]) and repressed (trimethylated histone 3 lysine 27 [H3K27me3]) histone modifications, but the liver regulatory elements were devoid of these marks. During hepatic specification, the liver regulatory elements gain H3 acetylation, but regulatory elements of pancreatic genes retain the "bivalent" chromatin state in hepatoblasts. When enhancer of zeste homolog 2 (Ezh2), a key methyltransferase for H3K27me3 modification and catalytic subunit of polycomb repressive complex, was deleted in mouse foregut endoderm cells, pancreas specification was promoted, and the liver bud became smaller (Xu et al., 2011). Thus, the repressed chromatin state can modulate the liver *versus* pancreas fate choice by suppressing the pancreas lineage. More comprehensive, genome-wide studies reveal dynamic H3K27me3 patterns during endodermal lineages progression (van Arensbergen et al., 2010; Xie et al., 2013; Xu et al., 2014). It is important to understand how stage- and lineage-specific gene loss or gain of H3K27me3 pre-patterns a specific lineage in multipotent progenitor cells.

2. Hepatic architecture

The hepatic lobule is an approximate hexagonal or pentagonal shape with a portal vein at each vertex and a central vein branch in the middle (Fig. 3) (Desmet, 2011). The portal vein (PV), hepatic artery (HA), and intrahepatic bile duct (IHBD) branches are arranged in close association at the vertices in structures known as portal triads. The hierarchical, branched structures of the PV, HA, and IHBD bifurcate approximately 17–20 times in human forming smaller and smaller branches (Crawford, 2002).

Hepatocytes are arranged in cords to fill the space between the central vein (CV) and PV. Collectively, hepatocytes perform a wide variety of tasks, including metabolic, detoxification, synthetic, and immunologic (Bhatia, Underhill, Zaret, & Fox, 2014). These functions are segregated between different zonal subpopulations of hepatocytes enabling various metabolic pathways to run in parallel (Fig. 3). Zone 1 is the periportal zone (closest to PV), zone 2 is the intermediate zone, and zone 3 is the pericentral zone (closest to CV). Zone 1 hepatocytes specialize in gluconeogenesis, fatty acid oxidation, urea synthesis, and cholesterol synthesis while zone 3 hepatocytes specialize in glycolysis, lipogenesis, ketogenesis, glutamine synthesis, and bile acid synthesis (Kietzmann, 2017; Torre, Perret, & Colnot, 2011).

The assembly of hepatocytes as a polarized epithelium is critical for them to perform their exocrine and endocrine functions. Extending along the hepatocyte basal side are the sinusoids. Sinusoids are capillary-like structures that connect and permit blood flow from the PV and HA to the CV. Extending along the hepatocyte apical side are the bile canalicular channels. Canaliculi form a connected network extending throughout the parenchymal epithelium, allowing passive transport of bile produced by the hepatocytes (Gissen & Arias, 2015; Keppler, 2017). The multidrug resistance P-glycoprotein (Mdr1 or ATP binding cassette subfamily B member 1, ABCB1) encodes a large transmembrane efflux pump that decorates the

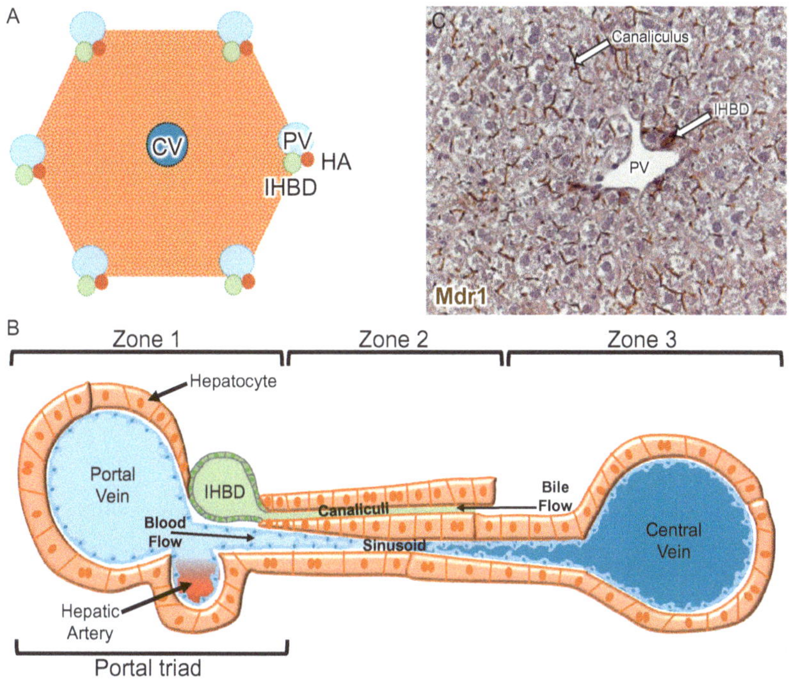

Fig. 3 Hepatic architecture. (A) Schematic of the approximate hexagonal or pentagonal hepatic lobule. (B) One radius of the approximate hexagonal or pentagonal hepatic lobule. Hepatocyte cords run along the radius of the lobule between portal veins (PV) and central veins (CV), and are arranged into three zones: zone 1 near the PV, zone 2, and zone 3 near the CV. Bile is secreted from hepatocytes into the canalicular channels and transported to the intrahepatic bile ducts (IHBD). The sinusoidal capillaries carry oxygenated blood from the PV and hepatic artery (HA) past the hepatocytes to the CV. (C) Localization of multidrug resistance (Mdr1) protein encoded by ATP binding cassette subfamily B member 1 (ABCB1) on the apical hepatocyte canalicular membrane and apical surface of cholangiocytes comprising IHBDs in mouse liver.

apical membrane of hepatocytes (i.e., canalicular membrane) and cholangiocytes (i.e., bile duct epithelial cells) (Fig. 3) (Boyer, 2013; Fickert & Wagner, 2017). Bile flows from the canaliculi to the canals of Hering that are conduits lined partly by hepatocytes and partly by cholangiocytes (Fig. 3). The canals of Hering are eventually connected to ductules, the smallest ramification of the IHBD system (Roskams et al., 2004). All sizes of ducts are composed entirely of cholangiocytes that establish an apicobasal epithelial polarity and form the connected IHBD system. The small peripheral IHBDs merge into fewer, larger ducts, until finally a single hilar duct carries the bile out of the liver and transports it into the

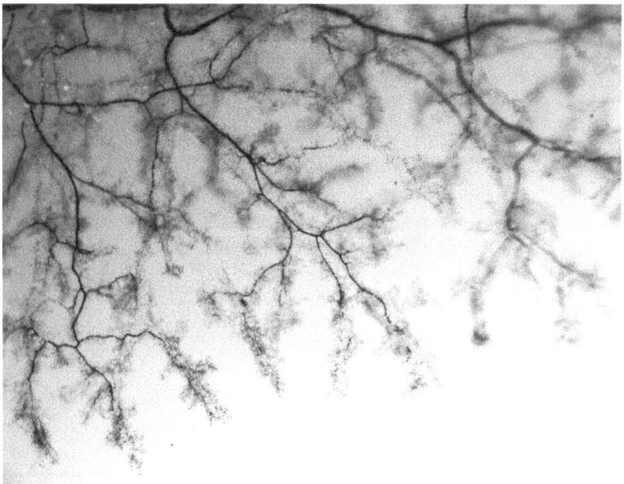

Fig. 4 Hierarchical branching architecture of the intrahepatic bile duct (IHBD) system. Retrograde ink injections into the left lobe of the mouse liver IHBD system. Liver lobe cleared with benzyl benzoate and benzyl alcohol (BABB) solution to visualize hierarchical structure.

gallbladder for storage and ultimately into the intestine to aid in digestion. The IHBD system relies on its intricate three-dimensional structure to access all of the hepatocytes and effectively clear bile out of the liver (Fig. 4).

3. Hepatocyte cell fate decision

3.1 Molecular regulation of hepatocyte differentiation

Hepatocyte-specific transcription factors are already expressed in bipotential hepatoblasts whether they are entering the hepatocyte or cholangiocyte lineage transcriptional program (Gerard, Tys, & Lemaigre, 2017). During mid-gestation until after birth, specified hepatocytes undergo a process of postnatal differentiation where they adopt the physiological functions and morphology associated with the adult liver. During the last weeks of gestation, inhibition of glycolytic enzymes is coupled with the rise in gluconeogenic enzyme levels, reflecting maturation of the liver from a primarily glycolytic role in the first trimesters to a gluconeogenic role before birth (Devi, Habeebullah, & Gupta, 1992b). Additionally, there is an age-dependent reduction in hepatocyte membrane fluidity with liver maturation. The decreased hepatocyte membrane fluidity is suggested to be due to a decrease of lipid content, an increase in plasma membrane cholesterol, and a progressive reduction in the lipid:protein ratio throughout the prenatal period and in adult human liver (Devi, Gupta, & Habeebullah, 1992a). Finally, hepatocyte maturation is assessed by the expression and activity of phase-I drug metabolizing enzymes, including cytochrome P450 (CYPs), to modify drugs and environmental toxins. The absence of most phase-I enzyme expression at birth is thought to be responsible for the substantial pharmacokinetic differences and toxicity between newborns and adults (Hart, Cui, Klaassen, & Zhong, 2009; Peng et al., 2013, 2012; Sadler et al., 2016).

3.1.1 Models for postnatal hepatocyte differentiation

During the process of human and mouse postnatal differentiation, the expression of hepatic specific genes is regulated by a progressively complex transcriptional network deemed the six key master regulators or liver-enriched factors: hepatocyte nuclear factor 4a (Hnf4a), Hnf3b/FoxA2, Hnf1a, Hnf6/Onecut 1(OC1), Hnf1b, and liver receptor homolog 1 (LRH1)/nuclear receptor subfamily 5 group A member 2 (Nr5a2) *in mouse* and HNF4A, HNF3B/FOXA2, HNF1A, HNF6/OC1, CAMP-responsive element binding protein 1 (CREB1), and upstream transcription factor 1 (USF1) *in human* (Kyrmizi et al., 2006; Odom et al., 2006, 2004; White, Brestelli, Kaestner, & Greenbaum, 2005). It is unclear how master regulators, expressed in both embryonic hepatoblasts and adult hepatocytes, coordinate the gene expression changes that are necessary for hepatocyte zonal architecture and mature hepatocyte cell function.

Published studies have identified a dichotomy in the severity of phenotype and number of differentially expressed hepatic genes that is dependent upon whether key master regulators such as Hnf4a and FoxA2 are deleted at embryonic *versus* adult stages of liver development (Battle et al., 2006; Bochkis et al., 2008; Hayhurst, Lee, Lambert, Ward, & Gonzalez, 2001; Kyrmizi et al., 2006; Lee et al., 2005; Li, Schug, Tuteja, White, & Kaestner, 2011; Parviz et al., 2003; Sund et al., 2000). Two explanations have been provided for why embryonic deletion causes more severe phenotypes and greater changes in gene expression (Fig. 5). First, the complex transcription factor promoter occupancies of the master regulators at adult postnatal ages provides a redundancy to maintain postnatal hepatocyte gene expression, stabilizing the transcription factor network and potentially demonstrating synergistic interdependence (Fig. 5) (Hayhurst et al., 2001; Sund et al., 2000). This is based on protein-DNA interactions (i.e., chromatin immunoprecipitation followed by PCR of enhancer regions or sequencing)

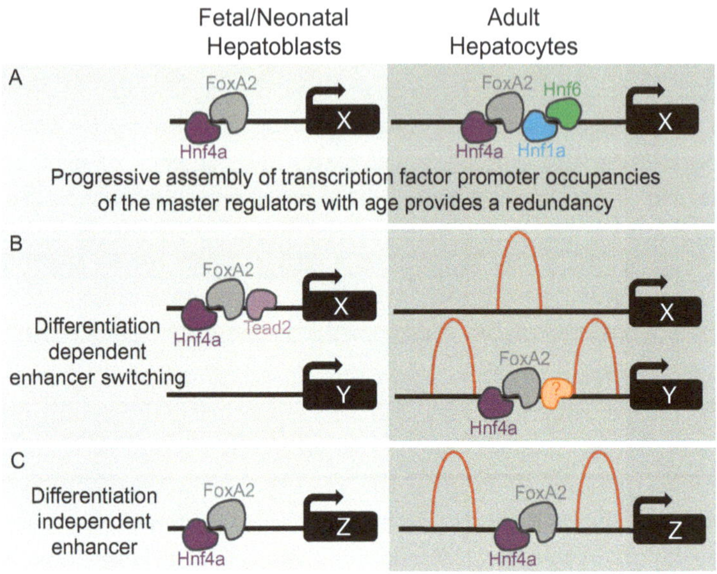

Fig. 5 Mechanistic models of hepatocyte differentiation. (A) Schematic demonstrating progressive assembly of transcription complexes. Fetal/neonatal hepatoblasts have less complicated transcription factor complexes associated at enhancers of the same genetic locus compared to adult hepatocytes. (B) Schematic of differentiation-dependent enhancers. Example demonstrates that the enhancer at genetic locus X is occupied in fetal/neonatal hepatoblasts but not occupied in adult hepatocytes, and the enhancer at genetic locus Y is not occupied in fetal/neonatal hepatoblasts but is occupied in adult hepatocytes. (C) Schematic of a differentiation-independent enhancer. Example demonstrates that the enhancer at genetic locus Z is similarly occupied in both fetal/neonatal hepatoblasts and adult hepatocytes. (B and C) Representation of H3K4me1 average enrichment profiles of binding (*red peaks*) at differentiation-dependent and -independent enhancers in adult hepatocytes. Distinct binding patterns (bimodal and monomodal distribution) are present, depending on whether FoxA2 and Hnf4a are bound to enhancers.

demonstrating that the number and complexity of transcription factor interactions on enhancer regions increase as hepatocyte maturation proceeds. There is a correlative rise in the expression of the key master regulators and each was found to occupy the gene regulatory regions of each other increasing occupation with age (Kyrmizi et al., 2006; Odom et al., 2006).

Second, different gene targets and/or enhancers are regulated by the similar transcription factor complexes in embryonic *versus* adult time points (Fig. 5) (Alder et al., 2014). Hnf4a and FoxA2 interact with thousands of enhancer regions in a differentiation-dependent manner during hepatic development. In fact, 55% of Hnf4a and 60% of FoxA2 binding sites are uniquely occupied in either the embryonic hepatoblasts or adult hepatocytes, suggesting that these sites are regulated in a differentiation-dependent manner (Alder et al., 2014). Differentiation-dependent enhancer switching would enable key master regulators to perform distinct roles in both hepatoblasts and hepatocytes. This switch of DNA-binding sites is distinct from situations in which a transcription factor binds only high-affinity sites in one cell type. A motif-finding algorithm oPOSSUM was used to identify hepatoblast-enriched motifs bound by Hnf4a and FoxA2 (Alder et al., 2014). This analysis identified known DNA-binding motifs for key master regulators such as Hnf1a and Hnf1b validating the approach of motif analysis as these master regulators are known to be bound within the same peaks as Hnf4a and FoxA2. Surprisingly, this analysis also identified a motif for TEA domain transcription factor (Tead) whose transcriptional activity is dependent on interactions with yes-associated protein 1 (Yap1), a downstream component of the Hippo signaling pathway (Alder et al., 2014; Wu, Liu, Zheng, Dong, & Pan, 2008; Zhang et al., 2008). To determine whether the presence of Tead2 and Yap1 can influence the recruitment of Hnf4a and FoxA2 to differentiation-dependent enhancers, a Yap1 inducible transgenic mouse model was expressed in the liver in response to doxycycline (Alder et al., 2014; Dong et al., 2007). No changes of Hnf4a and FoxA2 were observed at differentiation-independent enhancers, those found occupied in both hepatoblasts and hepatocytes. However, detection of Hnf4a and FoxA2 occupancy was decreased at adult hepatocyte-specific differentiation-dependent enhancers upon ectopic expression of Yap1 (Alder et al., 2014). Hippo signaling is a critical regulator of liver size (Camargo et al., 2007; Dong et al., 2007). Overexpression of Yap1 results in an increase of liver size by fourfold (Camargo et al., 2007; Dong et al., 2007). Hippo pathway activity results in phosphorylation and inactivation of the transcriptional coactivator Yap1 (Oka, Mazack, & Sudol, 2008; Zhao et al., 2007), and *in vivo* inactivation of Hippo signaling is sufficient to dedifferentiate adult hepatocytes into cells resembling hepatoblast characteristics, expressing both hepatocyte and cholangiocyte markers (Yimlamai et al., 2014). Taken together, Hippo signaling may affect hepatocyte differentiation through regulating the pool of nuclear Yap1 and thereby influence a temporal switch of DNA sites bound by Hnf4a and FoxA2 at differentiation-dependent enhancers *via* direct interaction with Yap1 or indirectly activating a repressor that masks adult hepatocyte enhancer regions in hepatoblasts.

3.1.2 Potential epigenetic regulators of enhancer switching

If enhancer switching is true at a more global level *versus* a few specific targets that recruit additional cofactors during hepatocyte postnatal differentiation, like Tead2 during hepatoblast stages (Alder et al., 2014), then it begs the question—how is occupation of transcriptional factors such as Hnf4a and FoxA2 at different targets regulated given that most of the master regulators are expressed in both embryonic and adult liver? Generally, enhancers exist in a primed state that is characterized by the presence of monomethylated histone 3 lysine 4 (H3K4me1) prior to acetylated histone 3 lysine 27 (H3K27ac) and enhancer activation (Calo & Wysocka, 2013). Indeed, differentiation-dependent enhancers in mouse and human ESCs are pre-marked by H3K4me1 (Creyghton et al., 2010; Rada-Iglesias et al., 2011; Zentner, Tesar, & Scacheri, 2011). Although H3K4me1 is promoted to be linked to enhancer activity, its presence at an enhancer is probably more accurately defined as providing a window of opportunity for activation, facilitating nucleosomal accessibility, and/or pioneer factor binding (Iwafuchi-Doi & Zaret, 2014). Adult liver ChIP sequencing data reveal that differentiation-independent (embryonic and adult) and -dependent (adult only) enhancer regions bound by Hnf4a and/or FoxA2 have the highest levels of H3K4me1 at each side of the central transcription factor binding site position (Fig. 5), exhibiting a bimodal distribution pattern of H3K4me1 at active enhancer regions (Alder et al., 2014; Hoffman et al., 2010). In contrast, differentiation-dependent embryonic sites that are not bound by Hnf4a and/or FoxA2 in adult liver show the highest levels of H3K4me1 in the central transcription factor binding site position (Fig. 5), exhibiting a monomodal distribution pattern of H3K4me1 at inactive enhancer regions (Alder et al., 2014; Hoffman et al., 2010). Conversely, enrichment patterns of the repressive histone modifications, H3K27me3, and trimethylated histone 3 lysine 9 (H3K9me3), are not predominately found at differentiation-dependent embryonic sites in adult liver as might be expected if this type of mechanism is used to exclude binding of transcription factors *via* chromatin compaction (Alder et al., 2014). In fact, only 3% H3K27me3 and 7% H3K9me3 of the differentiation-dependent embryonic enhancers were enriched in adult liver. The polycomb repressive complex protein Ezh2 expression is the highest in hepatoblasts at E9.5 and decreases after E13.5 with age, suggesting that Ezh2 downregulation is required for liver maturation, contrasting functional results have been obtained from *in vivo* loss of function and *ex vivo* knockdown studies

(Aoki et al., 2010; Koike et al., 2014). *In vivo* inducible deletion of Ezh2 in a non-lineage-specific manner starting at E10.5 and analyzed at E18.5 by global gene expression analysis demonstrated that progression of differentiation into functional hepatocytes is impaired by loss of Ezh2 (Koike et al., 2014). In contrast, *ex vivo* short hairpin RNA-mediated knockdown of Ezh2 from isolated E14.5 hepatoblasts promoted differentiation into hepatocytes based on increased expression of key master regulators or liver-enriched factors (Aoki et al., 2010). Further studies are required to determine whether non-hepatic lineage cells influence the phenotype of Ezh2 loss or Ezh2 loss at discrete time windows impacts the phenotype and transcriptional landscape as hepatoblasts differentiate into hepatocytes. Nonetheless, analysis of publicly available DNase I hypersensitivity data confirms that 60% of the differentiation-dependent embryonic sites remain accessible in adult liver (Alder et al., 2014). Therefore, differentiation-dependent enhancer switching is not simply explained by polycomb-mediated chromatin compaction.

3.1.3 Potential role of alternative histones

In addition to transcriptional networks and covalent modification on histones, incorporation of histone variants can regulate gene transcription. H2A.Z, a highly conserved variant of the histone H2A, is known to be associated with nucleosomes adjacent to the transcription start sites and linked to dynamic changes in gene expression (Bargaje et al., 2012). The differentiation of embryonic stem cells to endoderm/hepatic progenitors is regulated by FoxA2 binding to nucleosomal DNA on H2A.Z-containing nucleosomes, followed by FoxA2 and H2A.Z recruitment nucleosome disassembly complexes to enable nucleosome depletion and cell differentiation (Li et al., 2012), implicating chromatin remodeling as a regulator of hepatic lineage-specific gene regulation during embryonic stem cell differentiation. The *in vivo* data supporting incorporation of histone variants remain to be generated for hepatoblast cell fate decisions and postnatal differentiation of hepatocyte and cholangiocyte identities.

3.2 Molecular regulation of hepatocyte zonation

Hepatic metabolic zonation and further delineation of differentiated hepatocyte functional subsets begin in the first weeks after birth (Jungermann & Katz, 1989). Single-cell RNA sequencing of mouse hepatocytes highlights that half of the detected transcripts are not randomly expressed (3496 of 7277 genes) (Halpern et al., 2017). This emphasizes that liver zonation is a highly regulated process. However, only 25% (884 of 3496 genes) of all zonated genes are regulated by Wnt/beta-catenin (Benhamouche et al., 2006; Halpern et al., 2017; Sekine, Lan, Bedolli, Feng, & Hebrok, 2006), and an additional 9% (298 genes) are suggested to be regulated by hypoxia, Ras-dependent signaling, and pituitary hormones (Halpern et al., 2017). Therefore, the regulation of many zonally localized hepatocyte genes remains to be uncovered (Fig. 6).

3.2.1 Wnt/beta-catenin signaling pathway

As stated above, the main signaling pathway involved in setting up and maintaining hepatocyte zonation is canonical Wnt/beta-catenin. To activate canonical Wnt signaling, Wnt ligands bind to frizzled cell surface receptors and co-receptors, low-density lipoprotein receptor-related protein 5 (LRP5) and LRP6 (Pinson, Brennan, Monkley, Avery, & Skarnes, 2000; Tamai et al., 2000), blocking the beta-catenin destruction complex thereby allowing beta-catenin to accumulate and be transported into the nucleus. In the nucleus, beta-catenin acts as a coactivator for the transcription of Wnt

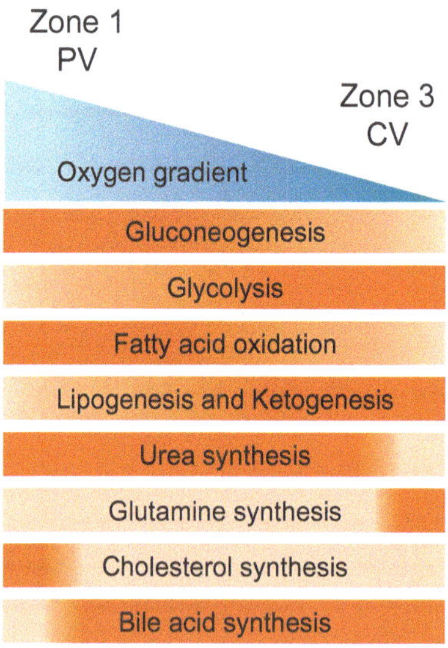

Fig. 6 Zonal hepatocyte functions. *Dark orange color* indicates more activity for hepatocyte functions.

target genes by binding to transcription factors from the T-cell factor (TCF) and lymphoid enhancer factor (Lef) family. In the absence of Wnt ligands, frizzled and LRP receptors are inactive, and the destruction complex including adenomatous polyposis coli (Apc) and Axin act as a scaffold, recruiting newly synthesized cytoplasmic beta-catenin for phosphorylation and targeted ubiquitin-mediated proteasomal degradation (Kretzschmar & Clevers, 2017). The R-spondins and their leucine-rich repeat containing G protein-coupled receptors, (LGR)4 and LGR5 (de Lau et al., 2011; Glinka et al., 2011; Ruffner et al., 2012), act as agonists by opposing the degradation of beta-catenin. Active R-spondin-Lgr complexes bind and inactivate the transmembrane E3 ubiquitin ligases zinc and ring finger 3 (Znrf3) and ring finger protein 43 (Rnf43) permitting prolonged stabilization of beta-catenin (Hao et al., 2012; Koo et al., 2012).

In vivo beta-catenin stabilization is both necessary and sufficient for expression of the pericentral enzyme glutamine synthetase in mouse liver (Fig. 6). Inducible loss of beta-catenin in hepatocytes results in all hepatocytes portraying a more periportal phenotype and lack of zone 3 hepatocytes expressing glutamine synthetase (Gougelet et al., 2014). Additional support comes from liver-specific deletion of the Wnt co-receptors (Lrp5 and 6) and hepatocyte-specific deletion of the Wnt agonists (Lgr4 and 5), resulting in the lack of liver zonation and failure of hepatocytes to express zone 3 metabolic genes (Planas-Paz et al., 2016; Yang et al., 2014). In contrast, stabilization of beta-catenin due to liver-specific deletion of Apc, a negative regulator of beta-catenin stabilization, induces a zone 3 hepatocyte phenotype and glutamine synthetase expression throughout the hepatic lobule (Benhamouche et al., 2006).

How Wnt activity is restricted or activated spatially in pericentral or zone 3 hepatocytes are unclear. The expression of Wnt ligands has been found in several hepatic cell types: hepatocytes, cholangiocytes, endothelial cells, liver sinusoidal endothelial cells (i.e., LSECs), stellate cells, and liver resident macrophages (i.e., Kupffer cells) (Zeng et al., 2007). Cell lineage-specific deletion of Wntless, a putative G protein-coupled receptor that transports all Wnts intracellularly for secretion, was used to determine which hepatic cell type is responsible for providing Wnt ligands for establishing and maintaining hepatic zonation (Yang et al., 2014). The results indicate that hepatocytes, cholangiocytes, and macrophages are not the source of Wnt ligands for beta-catenin activation and hepatic zonation. The loss of Wntless from endothelial cells using Tie2-Cre results in embryonic lethality, preventing assessment of proper hepatic zonation. However, screening all 19 Wnts by *in situ* hybridization identified that central vein (CV) endothelial cells specifically express Wnt2 and Wnt9b (Wang, Zhao, Fish, Logan, & Nusse, 2015). Finally, using VE-cadherin-CreERT2 to conditionally induce deletion of Wntless in endothelial cells resulted in loss of zone 3 pericentral expression of glutamine synthetase (Wang et al., 2015). Therefore, these data support a model where CV endothelial cells provide the source of Wnt ligands to activate the Wnt pathway in the pericentral zone 3.

3.2.2 Hedgehog signaling pathway

Another signaling pathway found to be important for maintaining liver zonation is hedgehog (Hh). To activate canonical Hh signaling, the ligands Sonic, Indian, and Desert hedgehog (Shh, Ihh, and Dhh) interact with the patched (Ptch1 and Ptch2) receptors, removing their inhibition of the co-receptor smoothened (Smo). Active Smo then triggers the nuclear localization and activation of the glioma-associated oncogene transcription factors (Gli1, 2, and 3) by preventing the conversion of Gli2 and Gli3 into transcriptional repressors (Petrov, Wierbowski, & Salic, 2017). *In vivo* conditional mouse hepatocyte-specific deletion of Smo at 8 weeks of age results in steatosis within 5 weeks (Matz-Soja et al., 2016). Interestingly, the lipid accumulation was observed in zone 1 hepatocytes near the PV. In correlation with the lipid accumulation in zone 1, upon Smo hepatocyte deletion, upregulation of the lipogenic transcription factors sterol regulatory element binding protein (Srebp) and peroxisome proliferator activator receptor (Ppar) as well as the enzyme fatty acid synthase (Fasn) is observed in zone 1. Normally these lipogenic genes are expressed in zone 3 to perform their liponeogenic function of generating fat by converting carbohydrates into fatty acids (Fig. 6). However, the ectopic expression of these lipogenic genes potentially drives ectopic liponeogenesis in zone 1 where hepatocytes do not possess the ability to use the generated lipid. This idea is mechanistically feasible, as Srebp1 does have binding sites located at −7000 and −500 bp from the transcriptional start of the *Fasn* gene (Amemiya-Kudo et al., 2002; Morishita, Mochizuki, & Goda, 2014), and overexpression of Srebp1c is sufficient to upregulate *Fasn* gene expression in hepatocytes (Dentin et al., 2004). In confirmation of a Hh role in hepatocyte zonation, knockdown of Gli transcription factors in hepatocyte cultures results in similar phenotypic characteristics as observed with Smo deletion (Matz-Soja et al., 2016). Importantly, the changes in lipid metabolic and liver zonation in mouse hepatocytes deficient for Smo are independent of changes in cholesterol biosynthesis, glycogen content, and glycolysis, which suggests a specific role for Hh signaling as these processes encompass general and zonal hepatocyte functions.

3.2.3 Other key signaling molecules involved in hepatocyte zonation

Two additional key molecules mediate hepatic zonation. Hnf4a loss-of-function results in liver with a normal pericentral zone 3 expression of glutamine synthetase, but glutamine synthetase is expanded to be expressed in zone 1 periportal hepatocytes (Stanulovic et al., 2007). Data support a model where Hnf4a and beta-catenin compete for binding to the DNA-binding cofactor transcription factor (Tcf) (Gougelet et al., 2014). Thereby, Hnf4a/Tcf drives zone 1 gene expression and beta-catenin/Tcf drives zone 3 gene expression. Thus, Hnf4a opposes Wnt signaling to promote the expression of zone 1 periportal hepatocyte genes and inhibit the expression of zone 3 pericentral hepatocyte genes. Glucagon loss-of-function mice also appear to have normal pericentral zone 3 expression of glutamine synthetase, but glutamine synthetase is expanded to be expressed in zone 2 and a gradient toward zone 1 periportal hepatocytes (Cheng et al., 2018). Glucagon is secreted from the pancreatic alpha-cells and transported to the liver through the PV to the CV in order to support glucose homeostasis by stimulating hepatic gluconeogenesis in zone 1 hepatocytes. Interestingly, glucagon infusion appears to modulate expression of many Wnt/beta-catenin target genes. The effects are most pronounced in zone 1 hepatocytes where 28% of all genes are activated by glucagon and inhibited by Wnt/beta-catenin (Cheng et al., 2018). Therefore, the counter concentration gradients of glucagon and Wnt ligands may maintain the liver metabolic zonation.

Finally, liver zonation is influenced by oxygen and reactive oxygen species (ROS) however certain genes and functions within the different hepatocyte zones diverge in their level of sensitivity to oxygen modulation. Genes involved in glucose and drug metabolism are more readily influenced by blood flow and oxygen tension, while genes involved in ammonia detoxification and glutamine synthesis have a more stable and defined expression pattern in the face of oxygen manipulations (Kietzmann, 2017). While it is clear that oxygen can influence hepatocyte zonation, an *in vivo* mechanism through which the regulation occurs and the importance of oxygen pressure during developmental zone establishment and homeostasis remains unknown. Data support the idea that Wnt/beta-catenin and the hedgehog pathway can be modulated by hypoxia and hypoxia-inducible transcription factors (Hif). *In vitro* studies using colon cancer cell lines demonstrated that hypoxic conditions reduced levels of Apc *via* a Hif1a-dependent mechanism (Newton, Kenneth, Appleton, Nathke, & Rocha, 2010). Hif1a represses the *Apc* gene by binding to a hypoxia responsive element within the *Apc* promoter. In contrast, Apc can mediate repression of Hif1a, but simultaneously requires low levels of beta-catenin and nuclear factor κB (Newton et al., 2010). Further, data support, at minimum, a parallel activity of hedgehog and Hif to induce the expression of Fasn. Hypoxic induction in adult mice can induce a rapid hedgehog response observed through expression of the ligand Shh and evidence of activity through expression of Ptch1 in multiple organs including the liver (Bijlsma et al., 2009). Also, Srebp1 expression is upregulated and has the potential to subsequently bind to the promoter of Fasn activating its expression as similarly demonstrated for hedgehog signaling (Bijlsma et al., 2009). Therefore, the dynamic interplay between oxygen levels and signaling pathways may influence liver zonation and contribute to modulations of hepatic physiological function.

3.2.4 Hepatocyte zonation is dynamic

It is important to note that hepatocyte zonation is dynamic. The liver performs its exocrine and endocrine functions by responding to nutrition levels, drugs, environmental toxin intake, and levels of hormone and other blood borne factors that result in changes of gene expression patterns. Therefore, gradients of morphogenesis, such as Wnt, Hh, and hormones, in concert with oxygen, induce and restrict gene expression in different subsets of hepatocytes located in different zones of the hepatic lobule to segregate hepatocyte function. Given the innate ability of the liver to regenerate, it is this flexibility that allows hepatocytes of different zonal origins to effectively evade and adeptly adapt and replace damaged hepatocytes (Planas-Paz et al., 2016).

4. Cholangiocyte cell fate decision and intrahepatic bile duct morphogenesis

4.1 Intrahepatic bile duct formation

Mammalian organs with a branched, multicellular epithelial network, such as the lung, kidney, pancreas, and salivary gland form tubes by processes involving wrapping of an epithelial sheet, branching *via* subdivision or proliferation from an existing cellular compartment, or cavitation of a cylindrical cluster of cells (Baer, Chanut-Delalande, & Affolter, 2009). In contrast to these organs, the liver forms a branched IHBD network through a multistep process including specification of cholangiocytes and subsequent morphogenesis of the specified cholangiocytes into a tube (Lemaigre, 2010; Ober & Lemaigre, 2018; Tanimizu & Mitaka, 2017). This alternative process of tubulogenesis could theoretically bestow the potential of the liver to continually generate a connected biliary system coordinated with a normal enlarging parenchymal mass and enable the unique ability of the liver to regenerate after injury (Desmet, 2011). A contrast is observed in organs where the process of tubulogenesis uses terminally differentiated cells such as alveoli, glomeruli, and beta cells from which to build tubes with very little potential for regeneration.

4.1.1 Cholangiocyte specification

IHBD architectural formation is a highly complex and regulated process that occurs in a coordinated fashion along the portal vein (PV) network to form a connected IHBD branched network that intimately follows the PV branched network (Fig. 7). The first step in IHBD formation is cholangiocyte specification where hepatoblasts enter into the cholangiocyte transcriptional program detected by expression of specific cytokeratins

Fig. 7 Three-dimensional association of the portal vein (PV) system (*white*) and the intrahepatic bile duct (IHBD) system (*black*). Retrograde ink injections into the left lobe of the mouse liver PV and IHBD system. Liver lobe cleared with benzyl benzoate and benzyl alcohol (BABB) solution to visualize hierarchical structure.

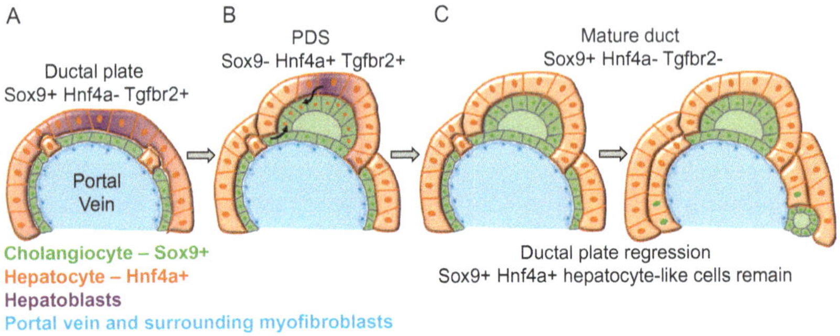

Fig. 8 Schematic of temporal cholangiocyte specification and morphogenesis process. (A) Hepatoblasts that are in close association with the portal vein (PV) myofibroblasts begin to express Sox9 and enter into the cholangiocyte transcriptional program in addition to repressing the hepatocyte program. These tightly associated Sox9+ cells form a temporary structure termed the ductal plate as they appear to form a cover surrounding the PV. (B) Primitive ductal structures (PDS) or luminal structures surrounded by asymmetrical gene expressing cells begin to form. These are luminal spaces are surrounded by Sox9+ Hnf4a− Tgfbr2− cells on the PV side and Sox9− Hnf4a+ Tgfbr2+ cells on the parenchymal side. The origin of these cells is uncertain. They may arise from the parenchymal hepatoblast-like cells or cells contributing from the initially formed ductal plate (*black arrows*). (C) Symmetrical ductal structures or mature ducts are formed of Sox9+ Hnf4a− Tgfbr2− cells encircling the luminal structure. The remaining ductal plate cells that are not incorporated into an intrahepatic bile duct (IHBD) regress back to Sox9+ Hnf4a+ hepatocytes-like cells.

(e.g., CK19) and the transcription factor Sry-related HMG box 9 (Sox9) (Fig. 8). Two-dimensional section analysis of human liver samples along with examination of experimental genetic mouse models exhibiting abnormal IHBD development have provided critical information regarding IHBD formation (Desmet, 1992; Raynaud et al., 2011; Terada, 2017; Van Eyken et al., 1988; Vestentoft et al., 2011). The cells that contribute to the IHBD structure are a subpopulation of the bipotential hepatoblasts in close proximity to the myofibroblasts surrounding PVs. Around E11–E14 in mice and 7–10 weeks of gestation in humans, a subpopulation of hepatoblasts forms a temporary structure appearing as a band of potential cholangiocytes, termed the ductal plate, encircling the PVs (Antoniou et al., 2009; Terada, 2017; Van Eyken et al., 1988; Vestentoft et al., 2011).

Hepatoblasts entering into the cholangiocyte program not only induce cholangiocyte-specific genes and repress hepatocyte genes but also undergo growth arrest during the specification process (Carpentier et al., 2011). There is evidence that growth arrest can initiate cholangiocyte specification, but it is not sufficient in all scenarios of genetic induced growth arrest.

For example, loss of Prox1 or T-Box transcription factor 3 (Tbx3) in mice results in growth arrest and increased commitment of hepatoblasts to the cholangiocyte program (Ludtke, Christoffels, Petry, & Kispert, 2009; Seth et al., 2014; Suzuki, Sekiya, Buscher, Izpisua Belmonte, & Taniguchi, 2008). The implicated regulator of growth arrest is p19ARF due to its over-expression in the absence of Tbx3 and known role as a tumor suppressor (Suzuki et al., 2008). In contrast, absence of the polycomb repressive complex protein Ezh2 leads to an increase of cyclin-dependent kinase inhibitor (Cdkn2A)/p19ARF and reduced hepatoblast proliferation, but without inducing the cholangiocyte program even though hepatocyte differentiation was inhibited (Koike et al., 2014). Consequently, there is a genetic context dependency that plays into the balance between proliferation and growth arrest, as well as initiation and repression of specific cell lineage transcriptional programs ultimately regulating the hepatoblast biopotential cell fate decision.

4.1.2 Transient primitive ductal structure as part of tubulogenesis

Tubulogenesis begins around E15–E17 in mice and 11–15 gestational weeks in humans (Antoniou et al., 2009; Terada, 2017; Van Eyken et al., 1988; Vestentoft et al., 2011). This morphogenesis step delineates the structure surrounding a forming lumen as either a primitive ductal structure or a mature duct (Fig. 8). Transient asymmetrical primitive ductal structures are observed in mice and humans (Raynaud et al., 2011). The primitive ductal structure is composed of two distinct cell types as distinguished by the presence or absence of marker expression (Sox9, Hnf4a, and transforming growth factor beta receptor type 2 [Tgfbr2]) compared to a mature duct. The primitive ductal structure is asymmetrical; cells on the PV side of the lumen express the marker Sox9, compared to cells on the parenchymal side that express Hnf4a and Tgfbr2. A mature duct is symmetrical, composed of cells expressing Sox9 (Antoniou et al., 2009; Raynaud et al., 2011). Additionally, the PV side of primitive ductal structure displays higher levels of E-cadherin expression and is in contact with laminin presumably produced from the PV myofibroblasts. Upon symmetrical mature duct formation, the lumen displays equal levels of E-cadherin expression and is surrounded by extracellular matrix and myofibroblasts (Antoniou et al., 2009). Detailed analysis using immunostaining suggests that there is a radial progression of differentiation—mature cells on the PV side of the lumen promote differentiation of the neighboring less mature cells on the parenchymal side of the lumen (Antoniou et al., 2009; Zong et al., 2009). However, it remains unclear if cells from the ductal plate move and contribute to the less mature cells of the forming lumen or more hepatoblasts are recruited to contribute to the lumen (Fig. 8). Given the expression of Hnf4a in the less mature cells on the parenchymal side of the primitive ductal structure, the more plausible explanation is that hepatoblasts are recruited to contribute to the forming lumen. Nevertheless, sandwich cultures of hepatoblasts suggest that monolayers of progenitors can fold up to form tubular structures as observed in a wrapping process of tubulogenesis (Tanimizu, Miyajima, & Mostov, 2009).

Finally, the ductal plate cells that remain unincorporated into an IHBD then either undergo apoptosis (Terada & Nakanuma, 1995) or turn off some of the cholangiocyte markers and contribute to periportal hepatocytes or the canal of Hering (Carpentier et al., 2011). Lineage tracing Sox9-positive cells in mouse strongly suggests that the unincorporated cholangiocytes transdifferentiate into hepatocytes without apoptosis or proliferation (Fig. 8) (Carpentier et al., 2011). If the unincorporated ductal plate cells do not receive or are unresponsive to the proper signals, they may contribute to ductal plate malformations as observed in patients with autosomal recessive polycystic kidney disease (ARPKD), congenital hepatic fibrosis, autosomal dominant polycystic kidney disease (ADPKD), Caroli's disease, Caroli's syndrome, and von Meyenburg complexes (Desmet, 1992; Huppert, 2011; Raynaud et al., 2011). Thus, there is a high level of coordination that must regulate sequential tubulogenesis and regression of the ductal plates along PVs within the three-dimensional space of the liver to connect the entire IHBD system to the extrahepatic ductal system. This indicates that careful orchestration of signals between epithelial and mesenchymal cells is required to guide IHBD formation (Gerard et al., 2017; Lemaigre, 2010).

4.1.3 Three-dimensional view of tubulogenesis

Due to the fact that the liver forms the IHBD system after the hepatic mass has already begun to expand and the lobes of the liver have formed, it is difficult to visualize and record the morphogenesis. Very interesting biology has been revealed using technical advancements to describe rendered IHBD three-dimensional models through quantitative measures and to visualize the IHBD three-dimensional architecture through retrograde ink injections in combination with tissue clearing (Figs. 4 and 7) (Takashima, Terada, Kawabata, & Suzuki, 2015; Tanimizu et al., 2016).

To map cholangiocyte specification and morphogenesis at a high resolution in three-dimension, serial sections of mouse mid-gestational to adult liver samples were immunostained with CK19 and digital three-dimensional reconstruction of images was performed. Using quantitative morphometric analyses, length, number of branch points, and distance from the PV were analyzed. Information garnered from the high resolution three-dimensional maps suggests that clusters of CK19-positive cholangiocytes are specified individually or in clusters rather than a cell layer as deduced from two-dimensional analyses (Takashima et al., 2015). As IHBD development proceeds, the clusters of CK19-positive cholangiocytes are further increased and form dense networks encircling the PV (Takashima et al., 2015). Thus, the ductal plate structure assumed to be a cell layer of specified cholangiocytes surrounding the PV may instead be a three-dimensional dense plexus-like network of cholangiocytes (Fig. 9). Finally, the network of CK19-positive cells surrounding the PV decreases with age as specific segments of the network become larger diameter CK19-positive tubes extending parallel to the PV, but at a greater distance away from the PV than when the CK19-positive tubes were initially formed (Takashima et al., 2015).

To provide a global view of IHBD formation, a retrograde injection of carbon ink into the luminal space from the common bile duct was performed at time points between mid-gestation and 1 week after birth followed by clearing the liver tissue to expose the ink filled IHBD structure (Kaneko, Kamimoto, Miyajima, & Itoh, 2015; Tanimizu et al., 2016). Ink fills the continuous luminal space of IHBDs but does not leak into the PV or

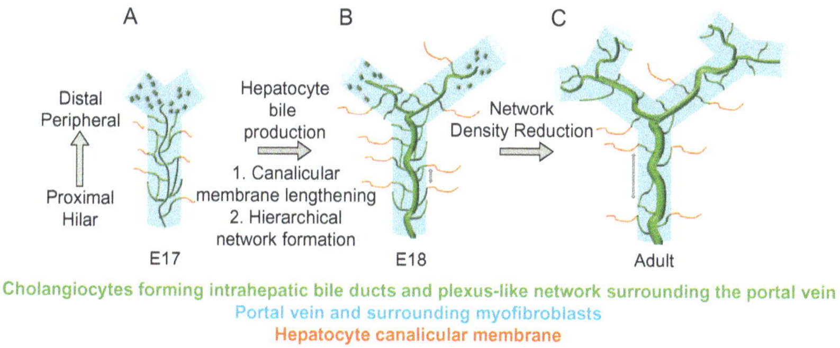

Fig. 9 Three-dimensional model of intrahepatic bile duct formation (IHBD) formation. (A) At the beginning of IHBD development, cholangiocytes are specified in the region adjacent to the portal vein (PV) system and are quickly incorporated into a dense homogeneous network that is communicating with the extrahepatic bile duct. (B) Upon hepatocyte bile production, secretion, and canalicular membrane lengthening, the homogenous network begins to reorganize into a hierarchical network between mouse E17 and E18. (C) As the liver parenchyma expands, new IHBDs are generated peripherally and the distance between the fine network structures surrounding the PV increases.

the hepatic parenchyma (Kaneko et al., 2015). Immunostaining used in combination with the ink injection detects the correlation between cholangiocyte specification and communicating luminal IHBD structures. The main findings are consistent with three-dimensional reconstruction, however a greater appreciation for the process of cholangiocyte morphogenesis and IHBD formation is revealed. Remodeling of specified cholangiocytes into IHBDs starts at the larger PVs at the hilar/proximal regions and based on serial sections and three-dimensional visualization moves toward the peripheral/distal region of liver following the PV system. Therefore, all steps of biliary morphogenesis progress in a hilum-to-periphery direction, allowing several stages of IHBD formation to be analyzed within a single liver during hepatogenesis (Fig. 9) (Antoniou et al., 2009; Tanimizu et al., 2016; Van Eyken et al., 1988).

Newly committed cholangiocytes are specified peripherally and are quickly incorporated into a plexus-like network of communicating luminal structures encircling the PV (Fig. 9). In mice, the homogenous communicating network begins to be rearranged into a hierarchical network between E17 and E18. The timing of the hierarchical arrangement correlates with lengthening of the bile canalicular network, hepatocyte excretion of bilirubin into bile, and presence of bile in the intestine (Tanimizu et al., 2016). Using the multidrug resistance-associated protein 2 (Mrp2) inhibitor benzbromarone to block the unidirectional efflux of bile from the hepatocyte canalicular membrane (thereby decreasing the flow of bile between mouse E16 and E18), the structural rearrangement and formation of the hierarchical network of IHBDs were inhibited (Colombo, Armstrong, Duan, & Rioux, 2012; Tanimizu et al., 2016). Additionally, co-localization between liver kinase B1 (Lkb1 encoded by Stk11), a serine/threonine kinase, and E-cadherin has been described at adherens junctions of kidney and intestinal polarized epithelial cell lines in culture (Sebbagh, Santoni, Hall, Borg, & Schwartz, 2009). Lkb1 is an evolutionary conserved serine/threonine protein kinase implicated in a wide range of cellular functions including inhibition of cellular proliferation, regulation of cellular polarity, and metabolism (Salvi & DeMali, 2018). Loss of Lkb1 at mid-gestation in hepatoblasts results in jaundice, loss of protein localization at the bile canalicular membrane, and inability of specified cholangiocytes to undergo formation of the IHBD hierarchical network (Woods et al., 2011). Lkb1-deficient cholangiocytes are able to generate primitive ductal structure but fail to form mature symmetrical ducts (Just et al., 2015). These results indicate that formation of a hepatocyte canalicular membrane and bile flow drives the structural transition of IHBDs from the homogenous tubular plexus-like network into the mature hierarchical network (Fig. 9). This model proposes that any communicating luminal duct that is part of the homogeneous network and connected to a bile canaliculus has the potential to receive more bile secreted from hepatocytes compared to the other communicating lumena and thereby its luminal space may enlarge. Whether the luminal enlargement occurs through localized proliferation of cholangiocytes, movement and incorporation of nearby cholangiocytes, or stretching and enlargement of cholangiocyte cell size remains to be experimentally determined.

After birth, a homogenous network surrounding the PV is still visible in the liver periphery of mice until 1 week of age (Tanimizu et al., 2016). The incomplete IHBD architecture at birth, also exists in humans where the IHBD system is still forming during the first years of life (Van Eyken et al., 1988). Therefore, the process by which the IHBD system forms allows for progressive assembly of a communicating IHBD architecture coincident with the enlargement of the liver during normal growth in childhood. The final IHBD hierarchical architecture consists of large ducts running along PVs and small channels forming a mesh-like network around PVs. The IHBD intricate structure is formed by a very complicated specification and movement of cells that somehow know their place in the three-dimensional space to form connections to drain bile out of the liver. Therefore, the model for IHBD formation can be refined as follows: (1) cholangiocyte specification initially occurs as discontinuous clusters of cells surrounding the PV, (2) specified cholangiocytes rapidly undergo morphogenesis to form a homogeneous communicating luminal network of small ducts, and (3) the homogeneous IHBD network is rearranged into a hierarchical IHBD system coincident with hepatocyte bile production and secretion.

4.2 Molecular regulation of intrahepatic bile duct formation

Cholangiocyte specification and tubulogenesis require regulated input from various signaling pathways (Fig. 10). Specification of cholangiocytes and the process of tubulogenesis to form IHBDs take place in the vicinity of the PV, suggesting that local cues drive differentiation of hepatoblasts toward cholangiocytes and control the formation of primitive ductal structures. Notch and Tgfb are the two main signaling pathways known to supply localized ligand expression and receptor activation within the PV region.

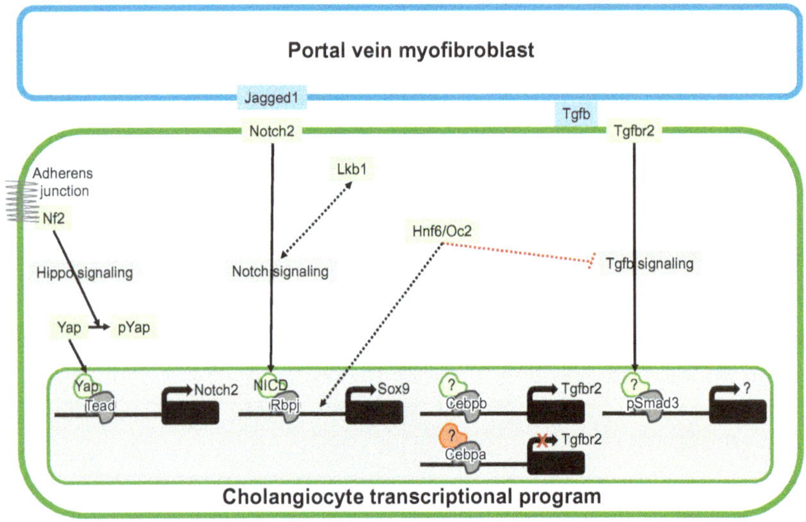

Fig. 10 Mechanistic model of cholangiocyte differentiation. Basic schematic of the regulatory nodes and pathways involved in setting up the cholangiocyte transcriptional program. *Green outlined cell* represents cholangiocyte and *blue outlined cell* represents portal vein (PV) myofibroblasts. Evidence of functional requirement indicated by *solid lines* and unknown level of interaction is indicated by *dotted lines*. Unidentified binding partners/coactivators or targets during cholangiocyte differentiation indicated by "?". *Double-headed arrow* indicates mutual cross regulation between Lkb1 and Notch signaling.

4.2.1 Notch signaling pathway

In humans, mutations in both Jagged1 (Jag1) and Notch2, a Notch pathway ligand and receptor, respectively, cause Alagille syndrome (ALGS1 and ALGS2) (Li et al., 1997; McDaniell et al., 2006; Oda et al., 1997). ALGS is an autosomal dominant disorder that is primarily characterized by neonatal jaundice, cholestasis, and paucity of intrahepatic bile ducts (IHBDs), and patients with ALGS also may have characteristic appearance of facial structure along with abnormalities of heart, eye, skeleton, vasculature, kidney, and pancreas (Emerick et al., 1999). Based on the ALGS hepatic phenotype, the primary pathway that has been implicated in IHBD formation is Notch signaling. In general, Notch ligands and receptors are both membrane-bound and participate in juxtacrine signaling between adjacent cells. Ligand binding triggers activation of the Notch receptor through a series of proteolytic events, culminating in a gamma-secretase, presenilin-dependent, proteolytic release of the Notch intracellular domain (NICD) from the membrane. NICD translocates to the nucleus where it associates with recombination signal binding protein for immunoglobulin kappa J region (Rbpj) to activate transcription of downstream targets. Therefore, the Notch signaling pathway is likely the main determinant that positions the proximal spatial localization of the PV and IHBD systems.

During cholangiocyte specification, myofibroblasts surrounding the PV express the Notch ligand Jag1 and most hepatic cells express Notch2 (Hofmann et al., 2010; Kodama, Hijikata, Kageyama, Shimotohno, & Chiba, 2004; Tanimizu & Miyajima, 2004; Zong et al., 2009). Importantly, cell lineage requirement studies have confirmed that Jag1 is required in the PV myofibroblasts, but not the endothelium (Hofmann et al., 2010) or hepatic epithelium (Loomes et al., 2007). Homozygous deletion of Jag1 in smooth muscle protein 22 alpha (SM22a)-positive cells, including PV myofibroblasts, leads to defective bile duct development resulting in jaundice, liver failure, and paucity of IHBDs. CK19-positive cells are detected surrounding the PV, but they are unable to undergo efficient tubulogenesis. Notably, initial cholangiocyte specification based on CK19 expression and the formation of the first layer of ductal plate appears to be independent of the PV myofibroblast supplied Jag1. However, Jag1-dependent signaling from the PV myofibroblast is essential for tubulogenesis and expansion of the mesenchymal cell population (Hofmann et al., 2010), all supporting a model whereby Jag1 interacts with Notch receptors expressed on bipotential hepatoblasts to activate the Notch pathway and regulate cholangiocyte specification and/or morphogenesis.

However, Jag1 expression is not only detected in the PV myofibroblasts but also in differentiating cholangiocytes (Zong et al., 2009), suggesting that Jag1-Notch activation within the differentiating cholangiocytes may contribute to cholangiocyte specification and/or tubulogenesis. Newly differentiated Jag1-positive cholangiocytes may activate the Notch receptors in adjacent cells. Therefore, hepatoblasts positioned between Jag1-positive cholangiocytes and near PV myofibroblasts would receive stronger induction signals than those outside the homogeneous IHBD network. Despite the observed expression data, initial *in vivo* studies investigating the requirement of Jagged1 in hepatoblasts suggests that Jag1 is not required in the hepatic epithelium for IHBD formation (Loomes et al., 2007).

Work with cultured hepatoblasts indicates that constitutive Notch activation represses expression of hepatocyte genes but induces expression of biliary genes, although no morphological indications of tubular structures were formed (Tanimizu & Miyajima, 2004). *In vivo*, when the ligand-independent constitutively active intracellular domain of Notch1 or Notch2 (i.e., NICD1 or NICD2) is expressed in hepatoblasts or hepatocytes, either is sufficient to promote cholangiocyte specification (Jeliazkova et al., 2013; Sparks, Huppert, Brown, Washington, & Huppert, 2010; Tchorz et al., 2009; Zong et al., 2009). Notably, not only is NICD1 able to induce cholangiocyte specification but also cholangiocyte morphogenesis, increasing the three-dimensional density of communicating IHBDs as assessed by retrograde resin injections (Sparks et al., 2010; Sparks, Perrien, Huppert, Peterson, & Huppert, 2011).

Inhibiting all canonical Notch activity by *in vivo* hepatoblast-specific deletion of Rbpj lead to a visible decrease of ductal plate cells and specified cholangiocytes (Sparks et al., 2010; Zong et al., 2009). Alternatively, using DAPT, a gamma-secretase inhibitor, to block all canonical Notch signaling between postnatal day 2 and postnatal day 6 in mice (when clusters of cholangiocytes are being specified and undergo tubulogenesis to form a communicating IHBD network), a deficiency of specified cholangiocytes to participate in generating the communicating network of IHBDs results (Tanimizu et al., 2016). Although, in the presence of DAPT, cholangiocytes were able to establish the apicobasal polarity (Tanimizu et al., 2016). Additionally, homozygous deletion of Notch2 in hepatoblasts using a mid-gestational Cre driver containing mouse albumin regulatory elements and the alpha-fetoprotein enhancers (i.e., Alfp-Cre, Kellendonk, Opherk, Anlag, Schutz, & Tronche, 2000) results in an absence of embryonic cholangiocyte specification (Falix et al., 2014). This result clearly indicates that Notch2 is the fundamental hepatic Notch receptor involved in cholangiocyte specification.

Inhibiting the expression of individual Notch receptors using a mid-gestational Cre driver containing the promoter and an upstream enhancer of the rat albumin gene (i.e., Alb-Cre, Postic & Magnuson, 2000), demonstrated that Notch2, but not Notch1 homozygous deletion results in tubulogenesis defects, but not clear cholangiocyte specification defects with regard to cell number (Geisler et al., 2008; Lozier, McCright, & Gridley, 2008; Sparks et al., 2010). The resulting phenotype is persistent unremodeled CK-positive cholangiocytes. These findings suggest that a reduction in Notch activity results in precocious differentiation, not proliferation, of ductal plate cells that are unable to either receive cues for tubulogenesis or for elimination if they are not incorporated into an IHBD. Therefore, although these specified cholangiocytes appear normal based on current cholangiocyte marker expression, they clearly are unable to complete morphogenesis. All results indicate that the Notch-deficient cholangiocytes are not correctly specified. This phenotype is consistent with what is observed in Jag1 haploinsufficient mice and when Jag1 is deleted from the myofibroblasts using the SM22-Cre transgenic line (Hofmann et al., 2010; Thakurdas et al., 2016). Therefore, it remains unknown if the observed tubulogenesis defects are due to incorrect cholangiocyte specification or whether Notch directly regulates gene targets important for tubulogenesis in the context of IHBD formation.

Sox9, one of the first indications of cells entering the cholangiocyte transcriptional program is a Notch target gene (Zong et al., 2009). There are 10 consensus Rbpj DNA-binding sites in the Sox9 promoter. Two *cis*-regulatory DNA-binding sites close to the Sox9 are verified to be *bona fide* Rbpj sites by ChIP PCR (Zong et al., 2009), suggesting that a specific level of Notch activity may be necessary to induce the correct level of Sox9 expression for correct cholangiocyte specification. Additionally, crosstalk between Lkb1 and Notch signaling may regulate tubulogenesis. Phenotypically, the phenotypes between loss of Lkb1 and Notch pathway components are very similar with jaundice and defects in bile duct tubulogenesis. Similarities between liver transcriptomes were analyzed using microarrays comparing the differential genes lists between liver-specific Rbpj loss of function to controls and liver-specific Lkb1 loss of function compared to controls. Parallels were found in 20–25% of the differential gene lists. Ingenuity pathway analysis highlighted metabolic pathways as the main signatures driving the similarity between LKB1 and Notch gene datasets (Just et al., 2015). Whether these are the genes driving the similar tubulogenesis defect remain unknown, especially given the role of Lkb1 in hepatocyte canalicular membrane formation. However, epistatic studies *in vitro* and *in vivo* reveal mutual cross regulation between Lkb1 and Notch signaling without clear evidence that one is epistatic to the other (Just et al., 2015).

4.2.2 Tgfb signaling pathway

Upon Tgfb ligand binding to the extracellular domain of Tgfbr2 a conformational change is elicited, resulting in the phosphorylation and activation of Tgfbr1. Tgfbr1 phosphorylates Smad2 or Smad3 on two serine residues within their C-terminus enabling binding to Smad4 to form heteromeric Smad complexes that enter the nucleus and initiate gene transcription (Vander Ark, Cao, & Li, 2018). In the liver, the mRNA of both *Tgfb2* and *Tgfb3* ligands are predominantly expressed by PV myofibroblasts while *Tgfb1* is produced throughout the liver parenchyma (Antoniou et al., 2009).

Tgfb activity assessed using a transgenic mouse line harboring an activin/Tgfb reporter consisting of 12 Smad-consensus binding sites upstream of enhanced green fluorescent protein is present in a portal to central gradient at E12.5 (Clotman et al., 2005). Therefore, the predominant spatial activation of the Tgfb pathway in the region of the PV suggests a role for Tgfb in regulating hepatoblast competency to become cholangiocytes and/or inducing cholangiocyte specification.

The dynamic expression pattern of Tgfbr2 suggests that the Tgfb pathway also plays a role in tubulogenesis. Immunostaining of Tgfb2 and Tgfb3 demonstrates binding specifically to the parenchymal side of primitive ductal structure coincident with localization of Tgfbr2 on the parenchymal side (Antoniou et al., 2009). At the onset of cholangiocyte specification, Tgfbr2 is expressed in all cells of the ductal plate. However, as stated above, during primitive ductal structure formation Tgfbr2 is specifically localized on cells surrounding the forming lumen on the parenchymal side and is absent on the side nearest the PV. In mature ducts, all cholangiocytes are negative for Tgfbr2 expression (Antoniou et al., 2009). Although speculative, this model suggests that Tgfb successively induces cholangiocyte specification and formation of the transient ductal plate and primitive ductal structure, then is downregulated.

Several functional sufficiency studies support the premise that Tgfb signaling plays a role in cholangiocyte specification. The first used E12.5 fetal liver explants treated with either Tgfb ligand added to the media or a bead supplying Tgfb ligand demonstrated an increase of cholangiocyte marker mRNA and induction of a local gradient of cholangiocyte cytokeratin markers (Clotman et al., 2005). Second, treating an E14.5 hepatoblast cell line (i.e., BMEL) in culture with any of the three Tgfb ligands (Tgfb1, Tgfb2, or Tgfb3) reduces the mRNA expression of hepatocyte markers and increases the mRNA expression of cholangiocyte markers (Antoniou et al., 2009). Coincident with the increase of cholangiocyte markers is the decrease of Tgfbr2 expression, indicating that activation of the Tgfb pathway ultimately prevents further activation. Finally, using human ESC generated hepatoblasts, treatment with Tgfbr2, Tgfb1, or Tgfb2, but not Tgfb3 increases mRNA expression of cholangiocyte markers and decreases mRNA expression of hepatocyte markers *in vitro* (Takayama et al., 2014), supporting the ability of the Tgfb pathway to induce cholangiocyte specification in the context of functional Notch signaling.

A clear *in vivo* requirement for Tgfb signaling in the process of IHBD formation is lacking. *In vitro* knockdown of Tgfbr2 in human ESC generated hepatoblasts show a decrease in mRNA expression of cholangiocyte markers in response to exogenous stimulus for cholangiocyte differentiation (Takayama et al., 2014). Further, the presence of small molecule Tgfb inhibitors in an *in vitro* epithelial cyst forming assay, using a hepatoblast cell line (i.e., HPPL), reduced the number of formed cysts (Tanimizu, Miyajima, & Mostov, 2007), suggesting that Tgfb signaling is required for cholangiocyte specification and tubulogeneis. One *in vivo* study, treating wild-type pregnant female mice with an anti-Tgfb neutralizing antibody at E10.5 and analysis performed at E14.5 show a qualitative reduction in the cholangiocyte cytokeratin-positive cells around the PV (Clotman et al., 2005). However, deletion of Tgfbr2 specifically in hepatoblasts using a mid-gestational Cre driver (i.e., Alb-Cre, Postic & Magnuson, 2000) does not result in an IHBD phenotype (Schaub et al., 2018). Therefore, the Tgfb pathway may only be required temporally at the very beginning of cholangiocyte specification either in the process of setting up competency or direct regulation of cholangiocyte target genes or has a role in a non-hepatoblast cell lineage.

The dynamic expression pattern of Tgfbr2 and the potential temporal requirement of the Tgfb pathway imply a high level of regulation. Indeed, persistent Tgfbr2 expression is associated with delayed transition from primitive ductal structure to mature bile duct formation observed when Sox9 is deleted specifically from hepatoblasts at mid-gestation (Antoniou et al., 2009; Poncy et al., 2015). Part of the phenotypic delay of mature bile duct formation may also be attributed to persistent expression of CCAAT enhancer-binding protein alpha (Cebpa) on both the portal and parenchymal side of the primitive ductal structure in Sox9-deficient liver. Deletion of Cebpa specifically in hepatoblasts at mid-gestation results in preferential differentiation of cholangiocytes rather than hepatocytes (Yamasaki et al., 2006). Interestingly, Tgfbr2 is negatively and positively regulated by Cebpa and Cebpb, respectively (Takayama et al., 2014). ChIP experiments demonstrate that Cebpa and Cebpb are recruited to the same DNA-binding site in the Tgfbr2 promoter region in cells differentiated to the hepatocyte-like or cholangiocyte-like fate from human ESCs, respectively (Takayama et al., 2014). Further supporting the model that Cebpa and Cebpb oppositely regulates Tgfbr2 promoter activity, reporter assays of the Tgfbr2 promoter and hepatoblast differentiation assays all verify that forced Cebpb expression activates expression of Tgfbr2 and phenotypically induces cholangiocyte-like cells from human ESC generated hepatoblasts and isolated E13.5 mouse hepatoblasts (Takayama et al., 2014). Therefore, the ratio of Cebpb to Cebpa appears to govern the hepatoblast cell fate decision by negatively and positively regulating the expression of Tgfbr2.

The Onecut (OC) family of transcription factors also impacts Tgfb activity. Hnf6 (i.e., OC1) and OC2 are transcription factors that bind to DNA as monomers utilizing their cut- and homeo-DNA-binding domains. A critical threshold of Hnf6 and OC2 dosage has been proposed for regulating hepatoblast differentiation. Too little or too much can result in premature biliary differentiation where hepatoblasts express both hepatocyte and cholangiocyte markers (Lemaigre, 2009). In adult mice, Hnf6 and OC2 are detected in both hepatocytes and cholangiocytes and therefore levels of Hnf6 may remain critical in maintaining adult hepatic cell fates (Lemaigre, 2009). Loss of Hnf6 and/or OC2 results in ectopic Tgfb activity. Changes to the portal to central gradient of Tgfb activity is detected, presumably allowing formation of intermediate cells and formation of primitive ductules within the parenchyma instead of restricting the lineage commitment in the vicinity of the PV (Clotman et al., 2005, 2002). The change in the portal to central gradient of Tgfb activity may be instigated through the observed stimulation of Tgfbr2 expression by Hnf6 and/or OC2 direct transcriptional regulation or reduced expression of the Activin/Tgfb inhibitors follistatin and/or alpha2-macroglobulin (Clotman et al., 2005). However, no evidence of Hnf6 binding to the Tgfbr2 promoter and directly regulating transcription in ChIP assays has been found (Takayama et al., 2014). Hnf6 can bind to the Sox9 promoter, but whether this is in response to Tgfb signaling or is recruited independently is unknown (Alder et al., 2014; Laudadio et al., 2012). These findings allude to indirect regulation of Tgfb by the Onecut family of transcription factors and imply a positive and negative feedback loop involving the Cebp family of transcription factors.

4.2.3 Other pathways with Notch-dependent and -independent mechanisms of IHBD formation

Although Notch signaling is the main pathway involved in IHBD formation, the emergence of Notch-independent cholangiocytes has been demonstrated in experimental mouse models (Falix et al., 2014; Walter, Vanderpool, Cast, & Huppert, 2014). In these models, mice initially lack peripheral IHBDs at neonatal stages, leading to severe cholestasis, necrotic liver injury, and liver fibrosis (Falix et al., 2014; Vanderpool et al., 2012). However, *de novo* IHBD formation occurs in both mouse models after weaning (Falix et al., 2014; Walter et al., 2014), allowing reversal of cholestasis and normalization of liver histology by postnatal day 120 (Schaub et al., 2018; Walter et al., 2014). The ductular reactive structures that emerge in these mutant mice resemble those seen in human, chronically injured livers containing intermediate cells expressing both hepatocyte and cholangiocyte markers. Hepatocytes not only have the ability to self-duplicate but also have the remarkable capability of assuming cholangiocyte identity in injured rodent and human liver (Ernst, Spinner, Piccoli, Mauger, & Russo, 2007; Fabris et al., 2007; Fukuda et al., 2004; Limaye et al., 2008; Michalopoulos, Barua, & Bowen, 2005; Schaub et al., 2018; Sekiya & Suzuki, 2014; Tanimizu, Nishikawa, Ichinohe, Akiyama, & Mitaka, 2014; Tarlow et al., 2014; Yanger et al., 2013; Yimlamai et al., 2014). Because the Notch-independent cholangiocytes emerge in the PV region, the most likely candidate for an alternative signaling pathway is the Tgfb pathway. Hepatocyte lineage tracing confirms that the *de novo* formed IHBDs are hepatocyte-derived and require Tgfbr2 (Schaub et al., 2018). Hence, the Tgfb pathway can compensate for the loss of Notch in the process of hepatocyte-to-cholangiocyte transdifferentiation but is unable to compensate for loss of Notch during hepatoblast-to-cholangiocyte differentiation. The targets of Tgfbr2/pSmad3 leading to activation of the cholangiocyte transcriptional program in hepatocytes are unknown. Even given this ability of Tgfb signaling, Notch activation appears to be more efficient to induce hepatocyte-to-cholangiocyte transdifferentiation than Tgfb (Sparks et al., 2010; Zong et al., 2009).

One other pathway has demonstrated the clear ability to influence the cholangiocyte fate in a Notch-dependent mechanism. Increased Yap activity induces hepatoblast and hepatocyte conversion to a cholangiocyte-like fate although the potential of these cells to undergo morphogenesis and form *de novo* IHBDs has not been demonstrated (Yimlamai et al., 2014). Similarly, the loss of the upstream tumor suppressor of the Hippo pathway, Neurofibromin 2 (Nf2), leads to increased Yap activity and generation of hepatoblast conversion to a cholangiocyte-like fate (Zhang et al., 2010). These phenotypes are very reminiscent of the induced hepatoblast and hepatocyte conversion to cholangiocytes upon NICD1 and NICD2 hepatic expression in mice (Jeliazkova et al., 2013; Sparks et al., 2010; Tchorz et al., 2009; Zong et al., 2009). Based on ChIP PCR confirmation of TEAD4 binding to a Notch2 promoter region, the resulting phenotype of increased Yap activity is thought to be a Notch2-dependent mechanism with Notch2 being a direct transcriptional target of the Hippo pathway effector Yap (Yimlamai et al., 2014). Epistatic studies have confirmed this model by reducing the Notch signaling activity through hepatoblast-specific deletion of Notch2. The Nf2 phenotype of increased cholangiocyte differentiation is suppressed while rescuing the paucity of IHBDs observed by mid-gestational hepatoblast-specific deletion of Notch2 (Wu et al., 2017). These results indicate that Notch signaling is one of the downstream effectors of the Hippo signaling pathway in regulating IHBD development (Fig. 10).

5. Conclusions

Future studies investigating the signaling pathways and epigenetic factors required for lineage restriction of the bipotential hepatoblast into the hepatocyte and cholangiocyte lineages, and further refinement of cell fate heterogeneity within each cell lineage for essential functional diversity of the hepatic epithelium remain critical for enlightening the disease processes of acute and chronic hepatobiliary diseases as well as the development of potential molecular and cellular targets for therapeutic options. The liver is alone among solid organs in its ability to regenerate full mass and function following extensive acute injury or partial resection (Huppert & Campbell, 2016). However, the reason for functional failure of the liver to regenerate mass or replace specific hepatic fate heterogeneity in most cases is unknown. Mouse studies using genetic reporters to trace the hepatocyte or cholangiocyte lineage reveal no evidence of contribution from a reserve stem cell population for liver homeostasis or liver regeneration (Malato et al., 2011; Raven et al., 2017; Schaub et al., 2018; Schaub, Malato, Gormond, & Willenbring, 2014; Yimlamai et al., 2014). Instead, during the liver injury or regenerative process, cells originating from the hepatocyte or cholangiocyte cell lineage can be found in different states of cell identity, transitioning to and from the hepatocyte and cholangiocyte lineage (Raven et al., 2017; Schaub et al., 2018). This highlights the amazing plasticity of adult hepatic epithelial cells. Still, we do not know whether all the signaling pathways and epigenetic factors used during hepatogenesis are the same molecular factors exploited during regeneration. Nevertheless, knowledge of molecular and cellular hepatogenesis regulators gives us a solid starting point. Discovery of the molecular regulators that bestow the ability of hepatic cell fates to be dynamic in response to injury, nutrition levels, drugs, environmental toxin intake, and levels of hormone, and other blood borne factors are vital to lessen the need for liver transplantation and inform potential pluripotent stem cell-derived hepatocyte and cholangiocyte cell replacement therapies.

References

Alder, O., Cullum, R., Lee, S., Kan, A. C., Wei, W., Yi, Y., et al. (2014). Hippo signaling influences HNF4A and FOXA2 enhancer switching during hepatocyte differentiation. *Cell Reports*, *9*, 261–271.

Amemiya-Kudo, M., Shimano, H., Hasty, A. H., Yahagi, N., Yoshikawa, T., Matsuzaka, T., et al. (2002). Transcriptional activities of nuclear SREBP-1a, -1c, and -2 to different target promoters of lipogenic and cholesterogenic genes. *Journal of Lipid Research*, *43*, 1220–1235.

Angelo, J. R., Guerrero-Zayas, M. I., & Tremblay, K. D. (2012). A fate map of the murine pancreas buds reveals a multipotent ventral foregut organ progenitor. *PLoS One*, *7*, e40707.

Antoniou, A., Raynaud, P., Cordi, S., Zong, Y., Tronche, F., Stanger, B. Z., et al. (2009). Intrahepatic bile ducts develop according to a new mode of tubulogenesis regulated by the transcription factor SOX9. *Gastroenterology*, *136*, 2325–2333.

Aoki, R., Chiba, T., Miyagi, S., Negishi, M., Konuma, T., Taniguchi, H., et al. (2010). The polycomb group gene product Ezh2 regulates proliferation and differentiation of murine hepatic stem/progenitor cells. *Journal of Hepatology*, *52*, 854–863.

Baer, M. M., Chanut-Delalande, H., & Affolter, M. (2009). Cellular and molecular mechanisms underlying the formation of biological tubes. *Current Topics in Developmental Biology*, *89*, 137–162.

Bargaje, R., Alam, M. P., Patowary, A., Sarkar, M., Ali, T., Gupta, S., et al. (2012). Proximity of H2A.Z containing nucleosome to the transcription start site influences gene expression levels in the mammalian liver and brain. *Nucleic Acids Research*, *40*, 8965–8978.

Battle, M. A., Konopka, G., Parviz, F., Gaggl, A. L., Yang, C., Sladek, F. M., et al. (2006). Hepatocyte nuclear factor 4alpha orchestrates expression of cell adhesion proteins during the epithelial transformation of the developing liver. *Proceedings of the National Academy of Sciences of the United States of America*, *103*, 8419–8424.

Benhamouche, S., Decaens, T., Godard, C., Chambrey, R., Rickman, D. S., Moinard, C., et al. (2006). Apc tumor suppressor gene is the "zonation-keeper" of mouse liver. *Developmental Cell*, *10*, 759–770.

Bertet, C., Sulak, L., & Lecuit, T. (2004). Myosin-dependent junction remodelling controls planar cell intercalation and axis elongation. *Nature*, *429*, 667–671.

Bhatia, S. N., Underhill, G. H., Zaret, K. S., & Fox, I. J. (2014). Cell and tissue engineering for liver disease. *Science Translational Medicine*, *6*(245), 245sr2. https://doi.org/10.1126/scitranslmed.3005975.

Bijlsma, M. F., Groot, A. P., Oduro, J. P., Franken, R. J., Schoenmakers, S. H., Peppelenbosch, M. P., et al. (2009). Hypoxia induces a hedgehog response mediated by HIF-1alpha. *Journal of Cellular and Molecular Medicine*, *13*, 2053–2060.

Bochkis, I. M., Rubins, N. E., White, P., Furth, E. E., Friedman, J. R., & Kaestner, K. H. (2008). Hepatocyte-specific ablation of Foxa2 alters bile acid homeostasis and results in endoplasmic reticulum stress. *Nature Medicine*, *14*, 828–836.

Bort, R., Signore, M., Tremblay, K., Martinez Barbera, J. P., & Zaret, K. S. (2006). Hex homeobox gene controls the transition of the endoderm to a pseudostratified, cell emergent epithelium for liver bud development. *Developmental Biology*, *290*, 44–56.

Bossard, P., & Zaret, K. S. (1998). GATA transcription factors as potentiators of gut endoderm differentiation. *Development*, *125*, 4909–4917.

Boyer, J. L. (2013). Bile formation and secretion. *Comprehensive Physiology*, *3*, 1035–1078.

Calmont, A., Wandzioch, E., Tremblay, K. D., Minowada, G., Kaestner, K. H., Martin, G. R., et al. (2006). An FGF response pathway that mediates hepatic gene induction in embryonic endoderm cells. *Developmental Cell*, *11*, 339–348.

Calo, E., & Wysocka, J. (2013). Modification of enhancer chromatin: What, how, and why? *Molecular Cell*, *49*, 825–837.
Camargo, F. D., Gokhale, S., Johnnidis, J. B., Fu, D., Bell, G. W., Jaenisch, R., et al. (2007). YAP1 increases organ size and expands undifferentiated progenitor cells. *Current Biology*, *17*, 2054–2060.
Camp, J. G., Sekine, K., Gerber, T., Loeffler-Wirth, H., Binder, H., Gac, M., et al. (2017). Multilineage communication regulates human liver bud development from pluripotency. *Nature*, *546*, 533–538.
Carpentier, R., Suner, R. E., van Hul, N., Kopp, J. L., Beaudry, J. B., Cordi, S., et al. (2011). Embryonic ductal plate cells give rise to cholangiocytes, periportal hepatocytes, and adult liver progenitor cells. *Gastroenterology*, *141*, 1432–1438. e1431–1434.
Cheng, X., Kim, S. Y., Okamoto, H., Xin, Y., Yancopoulos, G. D., Murphy, A. J., et al. (2018). Glucagon contributes to liver zonation. *Proceedings of the National Academy of Sciences of the United States of America*, *115*, E4111–E4119.
Clotman, F., Jacquemin, P., Plumb-Rudewiez, N., Pierreux, C. E., Van der Smissen, P., Dietz, H. C., et al. (2005). Control of liver cell fate decision by a gradient of TGF beta signaling modulated by Onecut transcription factors. *Genes & Development*, *19*, 1849–1854.
Clotman, F., Lannoy, V. J., Reber, M., Cereghini, S., Cassiman, D., Jacquemin, P., et al. (2002). The Onecut transcription factor HNF6 is required for normal development of the biliary tract. *Development*, *129*, 1819–1828.
Colombo, F., Armstrong, C., Duan, J., & Rioux, N. (2012). A high throughput in vitro mrp2 assay to predict in vivo biliary excretion. *Xenobiotica*, *42*, 157–163.
Costa, R. H., Kalinichenko, V. V., Holterman, A. X., & Wang, X. (2003). Transcription factors in liver development, differentiation, and regeneration. *Hepatology*, *38*, 1331–1347.
Crawford, J. M. (2002). Development of the intrahepatic biliary tree. *Seminars in Liver Disease*, *22*, 213–226.
Creyghton, M. P., Cheng, A. W., Welstead, G. G., Kooistra, T., Carey, B. W., Steine, E. J., et al. (2010). Histone H3K27ac separates active from poised enhancers and predicts developmental state. *Proceedings of the National Academy of Sciences of the United States of America*, *107*, 21931–21936.
de Lau, W., Barker, N., Low, T. Y., Koo, B. K., Li, V. S., Teunissen, H., et al. (2011). Lgr5 homologues associate with Wnt receptors and mediate R-spondin signalling. *Nature*, *476*, 293–297.
Dentin, R., Pegorier, J. P., Benhamed, F., Foufelle, F., Ferre, P., Fauveau, V., et al. (2004). Hepatic glucokinase is required for the synergistic action of ChREBP and SREBP-1c on glycolytic and lipogenic gene expression. *The Journal of Biological Chemistry*, *279*, 20314–20326.
Desmet, V. J. (1992). Congenital diseases of intrahepatic bile ducts: Variations on the theme "ductal plate malformation" *Hepatology*, *16*, 1069–1083.
Desmet, V. J. (2011). Ductal plates in hepatic ductular reactions. Hypothesis and implications. II. Ontogenic liver growth in childhood. *Virchows Archiv*, *458*, 261–270.
Deutsch, G., Jung, J., Zheng, M., Lora, J., & Zaret, K. S. (2001). A bipotential precursor population for pancreas and liver within the embryonic endoderm. *Development*, *128*, 871–881.
Devi, B. G., Gupta, P. D., & Habeebullah, C. M. (1992). Changes in membrane fluidity during human liver development. *Biochemistry International*, *28*, 41–49.
Devi, B. G., Habeebullah, C. M., & Gupta, P. D. (1992). Glycogen metabolism during human liver development. *Biochemistry International*, *28*, 229–237.
Dong, J., Feldmann, G., Huang, J., Wu, S., Zhang, N., Comerford, S. A., et al. (2007). Elucidation of a universal size-control mechanism in Drosophila and mammals. *Cell*, *130*, 1120–1133.

Emerick, K. M., Rand, E. B., Goldmuntz, E., Krantz, I. D., Spinner, N. B., & Piccoli, D. A. (1999). Features of Alagille syndrome in 92 patients: Frequency and relation to prognosis. *Hepatology*, *29*, 822–829.

Ernst, L. M., Spinner, N. B., Piccoli, D. A., Mauger, J., & Russo, P. (2007). Interlobular bile duct loss in pediatric cholestatic disease is associated with aberrant cytokeratin 7 expression by hepatocytes. *Pediatric and Developmental Pathology*, *10*, 383–390.

Fabris, L., Cadamuro, M., Guido, M., Spirli, C., Fiorotto, R., Colledan, M., et al. (2007). Analysis of liver repair mechanisms in Alagille syndrome and biliary atresia reveals a role for notch signaling. *The American Journal of Pathology*, *171*, 641–653.

Falix, F. A., Weeda, V. B., Labruyere, W. T., Poncy, A., de Waart, D. R., Hakvoort, T. B., et al. (2014). Hepatic Notch2 deficiency leads to bile duct agenesis perinatally and secondary bile duct formation after weaning. *Developmental Biology*, *396*, 201–213.

Fickert, P., & Wagner, M. (2017). Biliary bile acids in hepatobiliary injury—What is the link? *Journal of Hepatology*, *67*, 619–631.

Fukuda, K., Sugihara, A., Nakasho, K., Tsujimura, T., Yamada, N., Okaya, A., et al. (2004). The origin of biliary ductular cells that appear in the spleen after transplantation of hepatocytes. *Cell Transplantation*, *13*, 27–33.

Geisler, F., Nagl, F., Mazur, P. K., Lee, M., Zimber-Strobl, U., Strobl, L. J., et al. (2008). Liver-specific inactivation of Notch2, but not Notch1, compromises intrahepatic bile duct development in mice. *Hepatology*, *48*, 607–616.

Gerard, C., Tys, J., & Lemaigre, F. P. (2017). Gene regulatory networks in differentiation and direct reprogramming of hepatic cells. *Seminars in Cell & Developmental Biology*, *66*, 43–50.

Gissen, P., & Arias, I. M. (2015). Structural and functional hepatocyte polarity and liver disease. *Journal of Hepatology*, *63*, 1023–1037.

Glinka, A., Dolde, C., Kirsch, N., Huang, Y. L., Kazanskaya, O., Ingelfinger, D., et al. (2011). LGR4 and LGR5 are R-spondin receptors mediating Wnt/beta-catenin and Wnt/PCP signalling. *EMBO Reports*, *12*, 1055–1061.

Goldman, O., Han, S., Sourisseau, M., Dziedzic, N., Hamou, W., Corneo, B., et al. (2013). KDR identifies a conserved human and murine hepatic progenitor and instructs early liver development. *Cell Stem Cell*, *12*, 748–760.

Gougelet, A., Torre, C., Vebèr, P., Sartor, C., Bachelot, L., Denechaud, P. D., et al. (2014). T-cell factor 4 and beta-catenin chromatin occupancies pattern zonal liver metabolism in mice. *Hepatology*, *59*, 2344–2357.

Gualdi, R., Bossard, P., Zheng, M., Hamada, Y., Coleman, J. R., & Zaret, K. S. (1996). Hepatic specification of the gut endoderm in vitro: Cell signaling and transcriptional control. *Genes & Development*, *10*, 1670–1682.

Halpern, K. B., Shenhav, R., Matcovitch-Natan, O., Toth, B., Lemze, D., Golan, M., et al. (2017). Single-cell spatial reconstruction reveals global division of labour in the mammalian liver. *Nature*, *542*, 352–356.

Han, S., Tan, C., Ding, J., Wang, J., Ma'ayan, A., & Gouon-Evans, V. (2018). Endothelial cells instruct liver specification of embryonic stem cell-derived endoderm through endothelial VEGFR2 signaling and endoderm epigenetic modifications. *Stem Cell Research*, *30*, 163–170.

Hao, H. X., Xie, Y., Zhang, Y., Charlat, O., Oster, E., Avello, M., et al. (2012). ZNRF3 promotes Wnt receptor turnover in an R-spondin-sensitive manner. *Nature*, *485*, 195–200.

Hart, S. N., Cui, Y., Klaassen, C. D., & Zhong, X. B. (2009). Three patterns of cytochrome P450 gene expression during liver maturation in mice. *Drug Metabolism and Disposition*, *37*, 116–121.

Hayhurst, G. P., Lee, Y. H., Lambert, G., Ward, J. M., & Gonzalez, F. J. (2001). Hepatocyte nuclear factor 4alpha (nuclear receptor 2A1) is essential for maintenance of hepatic gene expression and lipid homeostasis. *Molecular and Cellular Biology*, *21*, 1393–1403.

Hoffman, B. G., Robertson, G., Zavaglia, B., Beach, M., Cullum, R., Lee, S., et al. (2010). Locus co-occupancy, nucleosome positioning, and H3K4me1 regulate the functionality of FOXA2-, HNF4A-, and PDX1-bound loci in islets and liver. *Genome Research*, *20*, 1037–1051.

Hofmann, J. J., Zovein, A. C., Koh, H., Radtke, F., Weinmaster, G., & Iruela-Arispe, M. L. (2010). Jagged1 in the portal vein mesenchyme regulates intrahepatic bile duct development: Insights into Alagille syndrome. *Development*, *137*, 4061–4072.

Huppert, S. S. (2011). A new set of classifications for ductal plate malformations. *Hepatology*, *53*, 1795–1797.

Huppert, S. S., & Campbell, K. M. (2016). Emerging advancements in liver regeneration and organogenesis as tools for liver replacement. *Current Opinion in Organ Transplantation*, *21*, 581–587.

Iwafuchi-Doi, M., & Zaret, K. S. (2014). Pioneer transcription factors in cell reprogramming. *Genes & Development*, *28*, 2679–2692.

Iwafuchi-Doi, M., & Zaret, K. S. (2016). Cell fate control by pioneer transcription factors. *Development*, *143*, 1833–1837.

Jeliazkova, P., Jors, S., Lee, M., Zimber-Strobl, U., Ferrer, J., Schmid, R. M., et al. (2013). Canonical Notch2 signaling determines biliary cell fates of embryonic hepatoblasts and adult hepatocytes independent of Hes1. *Hepatology*, *57*, 2469–2479.

Jung, J., Zheng, M., Goldfarb, M., & Zaret, K. S. (1999). Initiation of mammalian liver development from endoderm by fibroblast growth factors. *Science*, *284*, 1998–2003.

Jungermann, K., & Katz, N. (1989). Functional specialization of different hepatocyte populations. *Physiological Reviews*, *69*, 708–764.

Just, P. A., Poncy, A., Charawi, S., Dahmani, R., Traore, M., Dumontet, T., et al. (2015). LKB1 and Notch pathways interact and control biliary morphogenesis. *PLoS One*, *10*, e0145400.

Kaneko, K., Kamimoto, K., Miyajima, A., & Itoh, T. (2015). Adaptive remodeling of the biliary architecture underlies liver homeostasis. *Hepatology*, *61*, 2056–2066.

Kellendonk, C., Opherk, C., Anlag, K., Schutz, G., & Tronche, F. (2000). Hepatocyte-specific expression of Cre recombinase. *Genesis*, *26*, 151–153.

Keng, V. W., Yagi, H., Ikawa, M., Nagano, T., Myint, Z., Yamada, K., et al. (2000). Homeobox gene Hex is essential for onset of mouse embryonic liver development and differentiation of the monocyte lineage. *Biochemical and Biophysical Research Communications*, *276*, 1155–1161.

Keppler, D. (2017). Progress in the molecular characterization of hepatobiliary transporters. *Digestive Diseases*, *35*, 197–202.

Kietzmann, T. (2017). Metabolic zonation of the liver: The oxygen gradient revisited. *Redox Biology*, *11*, 622–630.

Kodama, Y., Hijikata, M., Kageyama, R., Shimotohno, K., & Chiba, T. (2004). The role of notch signaling in the development of intrahepatic bile ducts. *Gastroenterology*, *127*, 1775–1786.

Koike, H., Ouchi, R., Ueno, Y., Nakata, S., Obana, Y., Sekine, K., et al. (2014). Polycomb group protein Ezh2 regulates hepatic progenitor cell proliferation and differentiation in murine embryonic liver. *PLoS One*, *9*, e104776.

Koo, B. K., Spit, M., Jordens, I., Low, T. Y., Stange, D. E., van de Wetering, M., et al. (2012). Tumour suppressor RNF43 is a stem-cell E3 ligase that induces endocytosis of Wnt receptors. *Nature*, *488*, 665–669.

Kretzschmar, K., & Clevers, H. (2017). Wnt/beta-catenin signaling in adult mammalian epithelial stem cells. *Developmental Biology*, *428*, 273–282.

Kyrmizi, I., Hatzis, P., Katrakili, N., Tronche, F., Gonzalez, F. J., & Talianidis, I. (2006). Plasticity and expanding complexity of the hepatic transcription factor network during liver development. *Genes & Development*, *20*, 2293–2305.

Laudadio, I., Manfroid, I., Achouri, Y., Schmidt, D., Wilson, M. D., Cordi, S., et al. (2012). A feedback loop between the liver-enriched transcription factor network and miR-122 controls hepatocyte differentiation. *Gastroenterology*, *142*, 119–129.

Lee, C. S., Friedman, J. R., Fulmer, J. T., & Kaestner, K. H. (2005). The initiation of liver development is dependent on Foxa transcription factors. *Nature*, *435*, 944–947.

Lemaigre, F. P. (2009). Mechanisms of liver development: Concepts for understanding liver disorders and design of novel therapies. *Gastroenterology*, *137*, 62–79.

Lemaigre, F. P. (2010). Molecular mechanisms of biliary development. *Progress in Molecular Biology and Translational Science*, *97*, 103–126.

Li, Z., Gadue, P., Chen, K., Jiao, Y., Tuteja, G., Schug, J., et al. (2012). Foxa2 and H2A.Z mediate nucleosome depletion during embryonic stem cell differentiation. *Cell*, *151*, 1608–1616.

Li, L., Krantz, I. D., Deng, Y., Genin, A., Banta, A. B., Collins, C. C., et al. (1997). Alagille syndrome is caused by mutations in human Jagged1, which encodes a ligand for Notch1. *Nature Genetics*, *16*, 243–251.

Li, Z., Schug, J., Tuteja, G., White, P., & Kaestner, K. H. (2011). The nucleosome map of the mammalian liver. *Nature Structural & Molecular Biology*, *18*, 742–746.

Limaye, P. B., Alarcon, G., Walls, A. L., Nalesnik, M. A., Michalopoulos, G. K., Demetris, A. J., et al. (2008). Expression of specific hepatocyte and cholangiocyte transcription factors in human liver disease and embryonic development. *Laboratory Investigation*, *88*, 865–872.

Loomes, K. M., Russo, P., Ryan, M., Nelson, A., Underkoffler, L., Glover, C., et al. (2007). Bile duct proliferation in liver-specific Jag1 conditional knockout mice: Effects of gene dosage. *Hepatology*, *45*, 323–330.

Lozier, J., McCright, B., & Gridley, T. (2008). Notch signaling regulates bile duct morphogenesis in mice. *PLoS One*, *3*, e1851.

Ludtke, T. H., Christoffels, V. M., Petry, M., & Kispert, A. (2009). Tbx3 promotes liver bud expansion during mouse development by suppression of cholangiocyte differentiation. *Hepatology*, *49*, 969–978.

Malato, Y., Naqvi, S., Schurmann, N., Ng, R., Wang, B., Zape, J., et al. (2011). Fate tracing of mature hepatocytes in mouse liver homeostasis and regeneration. *The Journal of Clinical Investigation*, *121*, 4850–4860.

Martinez Barbera, J. P., Clements, M., Thomas, P., Rodriguez, T., Meloy, D., Kioussis, D., et al. (2000). The homeobox gene Hex is required in definitive endodermal tissues for normal forebrain, liver and thyroid formation. *Development*, *127*, 2433–2445.

Matsumoto, K., Yoshitomi, H., Rossant, J., & Zaret, K. S. (2001). Liver organogenesis promoted by endothelial cells prior to vascular function. *Science*, *294*, 559–563.

Matz-Soja, M., Rennert, C., Schonefeld, K., Aleithe, S., Boettger, J., Schmidt-Heck, W., et al. (2016). Hedgehog signaling is a potent regulator of liver lipid metabolism and reveals a GLI-code associated with steatosis. *eLife*, *5*, e13308. https://doi.org/10.7554/eLife.13308.

McDaniell, R., Warthen, D. M., Sanchez-Lara, P. A., Pai, A., Krantz, I. D., Piccoli, D. A., et al. (2006). NOTCH2 mutations cause Alagille syndrome, a heterogeneous disorder of the notch signaling pathway. *American Journal of Human Genetics*, *79*, 169–173.

Michalopoulos, G. K., Barua, L., & Bowen, W. C. (2005). Transdifferentiation of rat hepatocytes into biliary cells after bile duct ligation and toxic biliary injury. *Hepatology*, *41*, 535–544.

Morishita, S., Mochizuki, K., & Goda, T. (2014). Bindings of ChREBP and SREBP1, and histone acetylation around the rat liver fatty acid synthase gene are associated with induction of the gene during the suckling-weaning transition. *Journal of Nutritional Science and Vitaminology (Tokyo)*, *60*, 94–100.

Newton, I. P., Kenneth, N. S., Appleton, P. L., Nathke, I., & Rocha, S. (2010). Adenomatous polyposis coli and hypoxia-inducible factor-1{alpha} have an antagonistic connection. *Molecular Biology of the Cell, 21*, 3630–3638.

Ober, E. A., & Lemaigre, F. P. (2018). Development of the liver: Insights into organ and tissue morphogenesis. *Journal of Hepatology, 68*, 1049–1062.

Oda, T., Elkahloun, A. G., Pike, B. L., Okajima, K., Krantz, I. D., Genin, A., et al. (1997). Mutations in the human Jagged1 gene are responsible for Alagille syndrome. *Nature Genetics, 16*, 235–242.

Odom, D. T., Dowell, R. D., Jacobsen, E. S., Nekludova, L., Rolfe, P. A., Danford, T. W., et al. (2006). Core transcriptional regulatory circuitry in human hepatocytes. *Molecular Systems Biology, 2*, 2006.0017.

Odom, D. T., Zizlsperger, N., Gordon, D. B., Bell, G. W., Rinaldi, N. J., Murray, H. L., et al. (2004). Control of pancreas and liver gene expression by HNF transcription factors. *Science, 303*, 1378–1381.

Oka, T., Mazack, V., & Sudol, M. (2008). Mst2 and Lats kinases regulate apoptotic function of Yes kinase-associated protein (YAP). *The Journal of Biological Chemistry, 283*, 27534–27546.

Parviz, F., Matullo, C., Garrison, W. D., Savatski, L., Adamson, J. W., Ning, G., et al. (2003). Hepatocyte nuclear factor 4alpha controls the development of a hepatic epithelium and liver morphogenesis. *Nature Genetics, 34*, 292–296.

Peng, L., Cui, J. Y., Yoo, B., Gunewardena, S. S., Lu, H., Klaassen, C. D., et al. (2013). RNA-sequencing quantification of hepatic ontogeny of phase-I enzymes in mice. *Drug Metabolism and Disposition, 41*, 2175–2186.

Peng, L., Yoo, B., Gunewardena, S. S., Lu, H., Klaassen, C. D., & Zhong, X. B. (2012). RNA sequencing reveals dynamic changes of mRNA abundance of cytochromes P450 and their alternative transcripts during mouse liver development. *Drug Metabolism and Disposition, 40*, 1198–1209.

Petrov, K., Wierbowski, B. M., & Salic, A. (2017). Sending and receiving hedgehog signals. *Annual Review of Cell and Developmental Biology, 33*, 145–168.

Pinson, K. I., Brennan, J., Monkley, S., Avery, B. J., & Skarnes, W. C. (2000). An LDL-receptor-related protein mediates Wnt signalling in mice. *Nature, 407*, 535–538.

Planas-Paz, L., Orsini, V., Boulter, L., Calabrese, D., Pikiolek, M., Nigsch, F., et al. (2016). The RSPO-LGR4/5-ZNRF3/RNF43 module controls liver zonation and size. *Nature Cell Biology, 18*, 467–479.

Poncy, A., Antoniou, A., Cordi, S., Pierreux, C. E., Jacquemin, P., & Lemaigre, F. P. (2015). Transcription factors SOX4 and SOX9 cooperatively control development of bile ducts. *Developmental Biology, 404*, 136–148.

Postic, C., & Magnuson, M. A. (2000). DNA excision in liver by an albumin-Cre transgene occurs progressively with age. *Genesis, 26*, 149–150.

Pouille, P. A., Ahmadi, P., Brunet, A. C., & Farge, E. (2009). Mechanical signals trigger myosin II redistribution and mesoderm invagination in Drosophila embryos. *Science Signaling, 2*, ra16.

Rada-Iglesias, A., Bajpai, R., Swigut, T., Brugmann, S. A., Flynn, R. A., & Wysocka, J. (2011). A unique chromatin signature uncovers early developmental enhancers in humans. *Nature, 470*, 279–283.

Raven, A., Lu, W. Y., Man, T. Y., Ferreira-Gonzalez, S., O'Duibhir, E., Dwyer, B. J., et al. (2017). Cholangiocytes act as facultative liver stem cells during impaired hepatocyte regeneration. *Nature, 547*, 350–354.

Raynaud, P., Tate, J., Callens, C., Cordi, S., Vandersmissen, P., Carpentier, R., et al. (2011). A classification of ductal plate malformations based on distinct pathogenic mechanisms of biliary dysmorphogenesis. *Hepatology, 53*, 1959–1966.

Roskams, T. A., Theise, N. D., Balabaud, C., Bhagat, G., Bhathal, P. S., Bioulac-Sage, P., et al. (2004). Nomenclature of the finer branches of the biliary tree: Canals, ductules, and ductular reactions in human livers. *Hepatology, 39*, 1739–1745.

Rossi, J. M., Dunn, N. R., Hogan, B. L., & Zaret, K. S. (2001). Distinct mesodermal signals, including BMPs from the septum transversum mesenchyme, are required in combination for hepatogenesis from the endoderm. *Genes & Development, 15*, 1998–2009.

Ruffner, H., Sprunger, J., Charlat, O., Leighton-Davies, J., Grosshans, B., Salathe, A., et al. (2012). R-Spondin potentiates Wnt/beta-catenin signaling through orphan receptors LGR4 and LGR5. *PLoS One, 7*, e40976.

Sadler, N. C., Nandhikonda, P., Webb-Robertson, B. J., Ansong, C., Anderson, L. N., Smith, J. N., et al. (2016). Hepatic cytochrome P450 activity, abundance, and expression throughout human development. *Drug Metabolism and Disposition, 44*, 984–991.

Salvi, A. M., & DeMali, K. A. (2018). Mechanisms linking mechanotransduction and cell metabolism. *Current Opinion in Cell Biology, 54*, 114–120.

Schaub, J. R., Huppert, K. A., Kurial, S. N. T., Hsu, B. Y., Cast, A. E., Donnelly, B., et al. (2018). De novo formation of the biliary system by TGFbeta-mediated hepatocyte transdifferentiation. *Nature, 557*, 247–251.

Schaub, J. R., Malato, Y., Gormond, C., & Willenbring, H. (2014). Evidence against a stem cell origin of new hepatocytes in a common mouse model of chronic liver injury. *Cell Reports, 8*, 933–939.

Sebbagh, M., Santoni, M. J., Hall, B., Borg, J. P., & Schwartz, M. A. (2009). Regulation of LKB1/STRAD localization and function by E-cadherin. *Current Biology, 19*, 37–42.

Sekine, S., Lan, B. Y., Bedolli, M., Feng, S., & Hebrok, M. (2006). Liver-specific loss of beta-catenin blocks glutamine synthesis pathway activity and cytochrome p450 expression in mice. *Hepatology, 43*, 817–825.

Sekiya, S., & Suzuki, A. (2014). Hepatocytes, rather than cholangiocytes, can be the major source of primitive ductules in the chronically injured mouse liver. *The American Journal of Pathology, 184*, 1468–1478.

Seth, A., Ye, J., Yu, N., Guez, F., Bedford, D. C., Neale, G. A., et al. (2014). Prox1 ablation in hepatic progenitors causes defective hepatocyte specification and increases biliary cell commitment. *Development, 141*, 538–547.

Shindo, A., & Wallingford, J. B. (2014). PCP and septins compartmentalize cortical actomyosin to direct collective cell movement. *Science, 343*, 649–652.

Shinozawa, T., Yoshikawa, H. Y., & Takebe, T. (2016). Reverse engineering liver buds through self-driven condensation and organization towards medical application. *Developmental Biology, 420*, 221–229.

Sparks, E. E., Huppert, K. A., Brown, M. A., Washington, M. K., & Huppert, S. S. (2010). Notch signaling regulates formation of the three-dimensional architecture of intrahepatic bile ducts in mice. *Hepatology, 51*, 1391–1400.

Sparks, E. E., Perrien, D. S., Huppert, K. A., Peterson, T. E., & Huppert, S. S. (2011). Defects in hepatic Notch signaling result in disruption of the communicating intrahepatic bile duct network in mice. *Disease Models & Mechanisms, 4*, 359–367.

Stanulovic, V. S., Kyrmizi, I., Kruithof-de Julio, M., Hoogenkamp, M., Vermeulen, J. L., Ruijter, J. M., et al. (2007). Hepatic HNF4alpha deficiency induces periportal expression of glutamine synthetase and other pericentral enzymes. *Hepatology, 45*, 433–444.

Sund, N. J., Ang, S. L., Sackett, S. D., Shen, W., Daigle, N., Magnuson, M. A., et al. (2000). Hepatocyte nuclear factor 3beta (Foxa2) is dispensable for maintaining the differentiated state of the adult hepatocyte. *Molecular and Cellular Biology, 20*, 5175–5183.

Suzuki, A., Sekiya, S., Buscher, D., Izpisua Belmonte, J. C., & Taniguchi, H. (2008). Tbx3 controls the fate of hepatic progenitor cells in liver development by suppressing p19ARF expression. *Development, 135*, 1589–1595.

Takashima, Y., Terada, M., Kawabata, M., & Suzuki, A. (2015). Dynamic three-dimensional morphogenesis of intrahepatic bile ducts in mouse liver development. *Hepatology, 61*, 1003–1011.

Takayama, K., Kawabata, K., Nagamoto, Y., Inamura, M., Ohashi, K., Okuno, H., et al. (2014). CCAAT/enhancer binding protein-mediated regulation of TGFbeta receptor 2 expression determines the hepatoblast fate decision. *Development, 141*, 91–100.

Takebe, T., Enomura, M., Yoshizawa, E., Kimura, M., Koike, H., Ueno, Y., et al. (2015). Vascularized and complex organ buds from diverse tissues via mesenchymal cell-driven condensation. *Cell Stem Cell, 16*, 556–565.

Tamai, K., Semenov, M., Kato, Y., Spokony, R., Liu, C., Katsuyama, Y., et al. (2000). LDL-receptor-related proteins in Wnt signal transduction. *Nature, 407*, 530–535.

Tanimizu, N., Kaneko, K., Itoh, T., Ichinohe, N., Ishii, M., Mizuguchi, T., et al. (2016). Intrahepatic bile ducts are developed through formation of homogeneous continuous luminal network and its dynamic rearrangement in mice. *Hepatology, 64*, 175–188.

Tanimizu, N., & Mitaka, T. (2017). Epithelial morphogenesis during liver development. *Cold Spring Harbor Perspectives in Biology, 9*, a027862. https://doi.org/10.1101/cshperspect.a027862.

Tanimizu, N., & Miyajima, A. (2004). Notch signaling controls hepatoblast differentiation by altering the expression of liver-enriched transcription factors. *Journal of Cell Science, 117*, 3165–3174.

Tanimizu, N., Miyajima, A., & Mostov, K. E. (2007). Liver progenitor cells develop cholangiocyte-type epithelial polarity in three-dimensional culture. *Molecular Biology of the Cell, 18*, 1472–1479.

Tanimizu, N., Miyajima, A., & Mostov, K. E. (2009). Liver progenitor cells fold up a cell monolayer into a double-layered structure during tubular morphogenesis. *Molecular Biology of the Cell, 20*, 2486–2494.

Tanimizu, N., Nishikawa, Y., Ichinohe, N., Akiyama, H., & Mitaka, T. (2014). Sry HMG box protein 9-positive (Sox9+) epithelial cell adhesion molecule-negative (EpCAM-) biphenotypic cells derived from hepatocytes are involved in mouse liver regeneration. *The Journal of Biological Chemistry, 289*, 7589–7598.

Tarlow, B. D., Pelz, C., Naugler, W. E., Wakefield, L., Wilson, E. M., Finegold, M. J., et al. (2014). Bipotential adult liver progenitors are derived from chronically injured mature hepatocytes. *Cell Stem Cell, 15*, 605–618.

Tchorz, J. S., Kinter, J., Muller, M., Tornillo, L., Heim, M. H., & Bettler, B. (2009). Notch2 signaling promotes biliary epithelial cell fate specification and tubulogenesis during bile duct development in mice. *Hepatology, 50*, 871–879.

Terada, T. (2017). Human ductal plate and its derivatives express antigens of cholangiocellular, hepatocellular, hepatic stellate/progenitor cell, stem cell, and neuroendocrine lineages, and proliferative antigens. *Experimental Biology and Medicine (Maywood, N.J.), 242*, 907–917.

Terada, T., & Nakanuma, Y. (1995). Detection of apoptosis and expression of apoptosis-related proteins during human intrahepatic bile duct development. *The American Journal of Pathology, 146*, 67–74.

Thakurdas, S. M., Lopez, M. F., Kakuda, S., Fernandez-Valdivia, R., Zarrin-Khameh, N., Haltiwanger, R. S., et al. (2016). Jagged1 heterozygosity in mice results in a congenital cholangiopathy which is reversed by concomitant deletion of one copy of Poglut1 (Rumi). *Hepatology, 63*, 550–565.

Torre, C., Perret, C., & Colnot, S. (2011). Transcription dynamics in a physiological process: Beta-catenin signaling directs liver metabolic zonation. *The International Journal of Biochemistry & Cell Biology, 43*, 271–278.

Tremblay, K. D., & Zaret, K. S. (2005). Distinct populations of endoderm cells converge to generate the embryonic liver bud and ventral foregut tissues. *Developmental Biology, 280*, 87–99.

van Arensbergen, J., Garcia-Hurtado, J., Moran, I., Maestro, M. A., Xu, X., Van de Casteele, M., et al. (2010). Derepression of polycomb targets during pancreatic organogenesis allows insulin-producing beta-cells to adopt a neural gene activity program. *Genome Research, 20,* 722–732.
Van Eyken, P., Sciot, R., Callea, F., Van der Steen, K., Moerman, P., & Desmet, V. J. (1988). The development of the intrahepatic bile ducts in man: A keratin-immunohistochemical study. *Hepatology, 8,* 1586–1595.
Vander Ark, A., Cao, J., & Li, X. (2018). TGF-beta receptors: In and beyond TGF-beta signaling. *Cellular Signalling, 52,* 112–120.
Vanderpool, C., Sparks, E. E., Huppert, K. A., Gannon, M., Means, A. L., & Huppert, S. S. (2012). Genetic interactions between hepatocyte nuclear factor-6 and Notch signaling regulate mouse intrahepatic bile duct development in vivo. *Hepatology, 55,* 233–243.
Vestentoft, P. S., Jelnes, P., Hopkinson, B. M., Vainer, B., Mollgard, K., Quistorff, B., et al. (2011). Three-dimensional reconstructions of intrahepatic bile duct tubulogenesis in human liver. *BMC Developmental Biology, 11,* 56.
Walter, T. J., Vanderpool, C., Cast, A. E., & Huppert, S. S. (2014). Intrahepatic bile duct regeneration in mice does not require Hnf6 or Notch signaling through Rbpj. *The American Journal of Pathology, 184,* 1479–1488.
Wandzioch, E., & Zaret, K. S. (2009). Dynamic signaling network for the specification of embryonic pancreas and liver progenitors. *Science, 324,* 1707–1710.
Wang, B., Zhao, L., Fish, M., Logan, C. Y., & Nusse, R. (2015). Self-renewing diploid Axin2(+) cells fuel homeostatic renewal of the liver. *Nature, 524,* 180–185.
White, P., Brestelli, J. E., Kaestner, K. H., & Greenbaum, L. E. (2005). Identification of transcriptional networks during liver regeneration. *The Journal of Biological Chemistry, 280,* 3715–3722.
Woods, A., Heslegrave, A. J., Muckett, P. J., Levene, A. P., Clements, M., Mobberley, M., et al. (2011). LKB1 is required for hepatic bile acid transport and canalicular membrane integrity in mice. *The Biochemical Journal, 434,* 49–60.
Wu, S., Liu, Y., Zheng, Y., Dong, J., & Pan, D. (2008). The TEAD/TEF family protein Scalloped mediates transcriptional output of the Hippo growth-regulatory pathway. *Developmental Cell, 14,* 388–398.
Wu, N., Nguyen, Q., Wan, Y., Zhou, T., Venter, J., Frampton, G. A., et al. (2017). The Hippo signaling functions through the Notch signaling to regulate intrahepatic bile duct development in mammals. *Laboratory Investigation, 97,* 843–853.
Xie, R., Everett, L. J., Lim, H. W., Patel, N. A., Schug, J., Kroon, E., et al. (2013). Dynamic chromatin remodeling mediated by polycomb proteins orchestrates pancreatic differentiation of human embryonic stem cells. *Cell Stem Cell, 12,* 224–237.
Xu, C. R., Cole, P. A., Meyers, D. J., Kormish, J., Dent, S., & Zaret, K. S. (2011). Chromatin "prepattern" and histone modifiers in a fate choice for liver and pancreas. *Science, 332,* 963–966.
Xu, C. R., Li, L. C., Donahue, G., Ying, L., Zhang, Y. W., Gadue, P., et al. (2014). Dynamics of genomic H3K27me3 domains and role of EZH2 during pancreatic endocrine specification. *The EMBO Journal, 33,* 2157–2170.
Xu, C., Lu, X., Chen, E. Z., He, Z., Uyunbilig, B., Li, G., et al. (2012). Genome-wide roles of Foxa2 in directing liver specification. *Journal of Molecular Cell Biology, 4,* 420–422.
Yamasaki, H., Sada, A., Iwata, T., Niwa, T., Tomizawa, M., Xanthopoulos, K. G., et al. (2006). Suppression of C/EBPalpha expression in periportal hepatoblasts may stimulate biliary cell differentiation through increased Hnf6 and Hnf1b expression. *Development, 133,* 4233–4243.
Yang, J., Mowry, L. E., Nejak-Bowen, K. N., Okabe, H., Diegel, C. R., Lang, R. A., et al. (2014). Beta-catenin signaling in murine liver zonation and regeneration: A Wnt-Wnt situation! *Hepatology, 60,* 964–976.

Yanger, K., Zong, Y., Maggs, L. R., Shapira, S. N., Maddipati, R., Aiello, N. M., et al. (2013). Robust cellular reprogramming occurs spontaneously during liver regeneration. *Genes & Development, 27*, 719–724.

Yimlamai, D., Christodoulou, C., Galli, G. G., Yanger, K., Pepe-Mooney, B., Gurung, B., et al. (2014). Hippo pathway activity influences liver cell fate. *Cell, 157*, 1324–1338.

Zeng, G., Awan, F., Otruba, W., Muller, P., Apte, U., Tan, X., et al. (2007). Wnt'er in liver: Expression of Wnt and frizzled genes in mouse. *Hepatology, 45*, 195–204.

Zentner, G. E., Tesar, P. J., & Scacheri, P. C. (2011). Epigenetic signatures distinguish multiple classes of enhancers with distinct cellular functions. *Genome Research, 21*, 1273–1283.

Zhang, N., Bai, H., David, K. K., Dong, J., Zheng, Y., Cai, J., et al. (2010). The Merlin/NF2 tumor suppressor functions through the YAP oncoprotein to regulate tissue homeostasis in mammals. *Developmental Cell, 19*, 27–38.

Zhang, L., Ren, F., Zhang, Q., Chen, Y., Wang, B., & Jiang, J. (2008). The TEAD/TEF family of transcription factor Scalloped mediates Hippo signaling in organ size control. *Developmental Cell, 14*, 377–387.

Zhao, B., Wei, X., Li, W., Udan, R. S., Yang, Q., Kim, J., et al. (2007). Inactivation of YAP oncoprotein by the Hippo pathway is involved in cell contact inhibition and tissue growth control. *Genes & Development, 21*, 2747–2761.

Zong, Y., Panikkar, A., Xu, J., Antoniou, A., Raynaud, P., Lemaigre, F., et al. (2009). Notch signaling controls liver development by regulating biliary differentiation. *Development, 136*, 1727–1739.

Further reading

Yanger, K., Knigin, D., Zong, Y., Maggs, L., Gu, G., Akiyama, H., et al. (2014). Adult hepatocytes are generated by self-duplication rather than stem cell differentiation. *Cell Stem Cell, 15*, 340–349.

CHAPTER FIVE

Connecting muscle development, birth defects, and evolution: An essential role for muscle connective tissue

Elizabeth M. Sefton, Gabrielle Kardon*
Department of Human Genetics, University of Utah, Salt Lake City, UT, United States
*Corresponding author: e-mail address: gkardon@genetics.utah.edu

Contents

Abstract

Skeletal muscle powers all movement of the vertebrate body and is distributed in multiple regions that have evolved distinct functions. Axial muscles are ancestral muscles essential for support and locomotion of the whole body. The evolution of the head was accompanied by development of cranial muscles essential for eye movement, feeding, vocalization, and facial expression. With the evolution of paired fins and limbs and their associated muscles, vertebrates gained increased locomotor agility, populated the land, and acquired fine motor skills. Finally, unique muscles with specialized functions have evolved in some groups, and the diaphragm which solely evolved in mammals to increase respiratory capacity is one such example. The function of all these muscles requires their integration with the other components of the musculoskeletal system: muscle connective tissue (MCT), tendons, bones as well as nerves and vasculature. MCT is muscle's closest anatomical and functional partner. Not only is MCT critical in the adult for muscle structure and function, but recently MCT in the embryo has been found to be crucial for muscle development. In this review, we examine the important role of the MCT in axial, head, limb, and diaphragm muscles for regulating normal muscle development, discuss how defects in MCT-muscle interactions during development underlie the etiology of a range of birth defects, and explore how changes in MCT development or communication with muscle may have led to the modification and acquisition of new muscles during vertebrate evolution.

Current Topics in Developmental Biology, Volume 132
ISSN 0070-2153
https://doi.org/10.1016/bs.ctdb.2018.12.004

1. Introduction

Skeletal muscle powers many critical vertebrate functions: locomotion, postural support, feeding, respiration, and communication. Accounting for 30–50% of vertebrate body mass (Romer & Parsons, 1986), skeletal muscle is distributed in multiple regions that have evolved distinct functions. A shared feature of chordates are the axial muscles, metameric muscles that stabilize and flex the notochord or vertebrae to enable support and locomotion of the body (Fig. 1; reviewed by Wotton, Schubert, & Dietrich, 2015). The evolution of the head in vertebrates has been accompanied by the

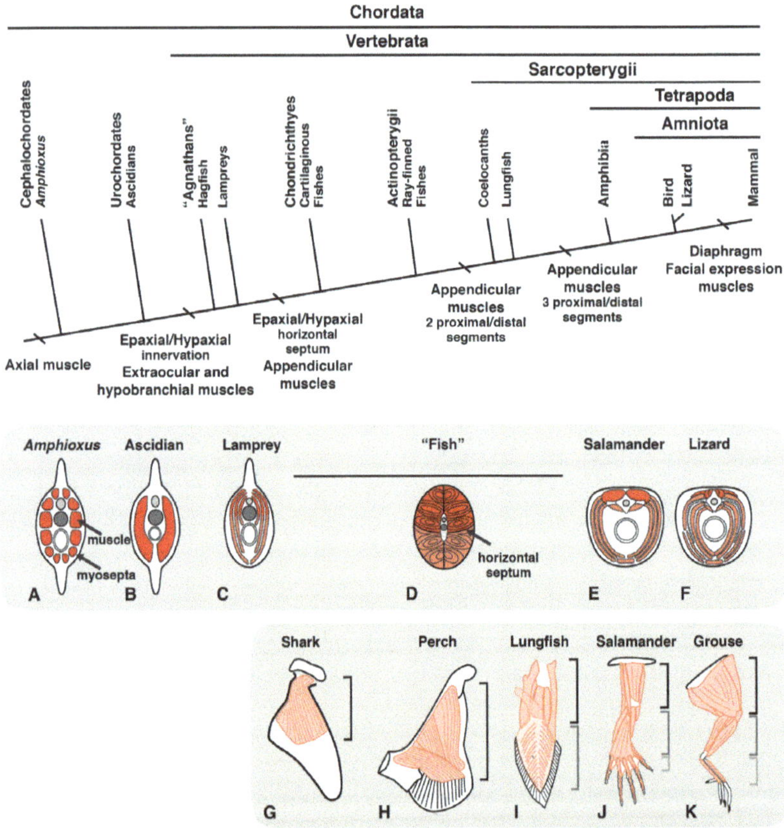

Fig. 1 During the course of chordate and vertebrate evolution, axial and limb muscles become increasingly complex. (A–B) Simple axial muscles in *Amphioxus* (A) and tail of ascidian free-swimming larva (B). (C) Lamprey myomeric axial muscle subdivided into epaxial muscle (red), innervated by dorsal roots, and hypaxial muscle (pink), innervated

development of muscles that allow for eye movement, feeding, vocalization, and facial expressions (Fig. 1). With the evolution of paired fins and limbs and their associated musculature, vertebrates have gained increased locomotor agility, populated the land, and been able to acquire fine motor skills, such as writing. In addition to axial, head, and limb muscles, different groups of vertebrates have evolved novel muscles that allow for specialized functions. Such novel muscles include the syringeal muscles which control song production in birds, facial muscles that allow for suckling in mammals, and the mammalian diaphragm that is essential for respiration and separating the thoracic and abdominal cavities. Development of all these different muscle groups has been crucial for the evolutionary success of vertebrates.

Muscle function requires its integration with the other components of the musculoskeletal system: bones, tendons, and muscle connective tissue as well as nerves and vasculature (Fig. 2A). The bones are the rigid part of the musculoskeletal system that support, and in response to muscle contraction, move the body. The tendons are the tensional links between the bones and muscles. The muscle connective tissue (MCT, sometimes called dense irregular connective tissue; Nassari, Duprez, & Fournier-Thibault, 2017) is the muscle's closest anatomical and functional partner (Fig. 2B). MCT is composed of three layers: the endomysium ensheathes individual myofibers, the perimysium

by ventral roots. (D). Myomeric axial muscle of Chondrichthyan, Actinopterygian, and Sarcopterygian fishes is subdivided into epaxial and hypaxial muscles innervated by dorsal and ventral roots, respectively, and separated by a horizontal septum. (E–F) Epaxial and hypaxial muscles of tetrapods, e.g., amphibians (E) and reptiles (F) are separated by vertebral transverse processes and subdivided into multiple muscles. (G) Simple shark ventral pectoral fin muscles extend from pectoral girdle to dermal rays. (H) Multiple ventral fin muscles of ray-finned perch extend pectoral girdle to bony rays. (I) Dorsal muscles of lungfish *Neoceratodus* pelvic fin are subdivided into two proximal-distal groups. (J) Dorsal muscles of salamander *Ambystoma* are subdivided into three proximal-distal groups. (K) Dorsal and ventral muscles of grouse *Dendragapus* are subdivided into three proximal-distal groups. *Panel (B and G) based on Walker, S. F. & Homberger, D. G. (1992).* Vertebrate dissection *(8th ed.). Fort Worth: Harcourt Brace College Publishers. Panel (C) based on Fetcho, J. R. (1987). A review of the organization and evolution of motoneurons innervating the axial musculature of vertebrates.* Brain Research, 434, *243–280. Panel (D–F) based on Romer, A. S., & Parsons, T.S. (1986).* The vertebrate body. *Fort Worth: Saunders College Publishing. Panel (H) based on Winterbottom, R. (1974). Descriptive synonymy of striated muscles of Teleostei.* Proceedings of the Academy of Natural Sciences of Philadelphia, 125, *225–317. Panel (I and J) after Diogo, R., Johnston, P., Molnar, J. L., & Esteve-Altava, B. (2016). Characteristic tetrapod musculoskeletal limb phenotype emerged more than 400 MYA in basal lobe-finned fishes.* Scientific Reports, 6, *37592. Panel (K) after Kardon, G. (1998). Muscle and tendon morphogenesis in the avian hind limb.* Development, 125, *4019–4032.*

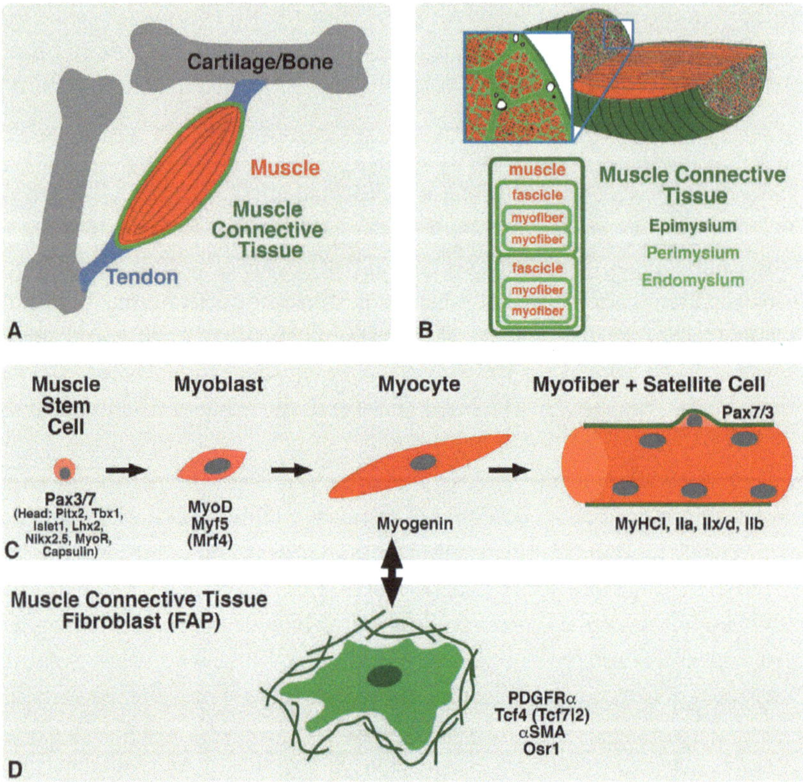

Fig. 2 Muscle and muscle connective tissue (MCT) structure, molecular markers, and development. (A–B) MCT surrounds myofibers, fascicles, and whole muscles to transmit contractile force of muscle to tendon and bone. (C) During myogenesis muscle progenitors/stem cells (which express Pax3/7 in axial and limb muscles and variety of transcription factors in the head) become committed MyoD/Myf5+ myoblasts, differentiate into Myogenin+ myocytes, which fuse into post-mitotic, multinucleate myofibers. (D) MCT fibroblasts (also known as FAPs in the adult) express Pdgfrα, Tcf7l22, and Osr1 and secrete ECM.

bundles myofibers into fascicles, and the epimysium surrounds anatomical muscles and links to the tendons at the myotendinous junctions (Sanes, 2004). All three MCT layers are composed primarily of extracellular matrix, largely collagens and proteoglycans, as well as the MCT fibroblasts that produce this matrix (Andrade & Brandan, 1991; Lipton, 1977; Sanes, 2004; Zou, Zhang, Sabatelli, Chu, & Bonnemann, 2008). The MCT is essential for maintaining muscle's structural integrity as well as transmitting muscle's contractile force to tendon and bone. In addition to MCT's important structural and functional roles in the adult, recently MCT has been found to have critical roles in regulating muscle development.

Muscle development occurs via two interrelated processes, myogenesis and muscle morphogenesis. Myogenesis is the cellular process by which muscle progenitors differentiate into multinucleate myofibers (Fig. 2C). The first step in myogenesis is the establishment of muscle progenitors. Axial and limb muscle progenitors originate from the somites, transient epithelial blocks of paraxial mesoderm located bilateral to the neural tube, while head muscle progenitors largely come from cranial pre-chordal and pharyngeal mesoderm, lateral to the developing brain (Bryson-Richardson & Currie, 2008; Buckingham, 2006; Michailovici, Eigler, & Tzahor, 2015; Murphy & Kardon, 2011; Wotton et al., 2015). Reflecting their different embryonic origins, the axial and limb muscle progenitors express the transcription factors Pax3 and Pax7, while the head progenitors express a heterogeneous array of transcription factors, including Pitx2, Tbx1, Islet1, Lhx2, Nkx2.5, MyoR and Capsulin (Michailovici et al., 2015; Murphy & Kardon, 2011; Relaix, Rocancourt, Mansouri, & Buckingham, 2005; Sambasivan, Kuratani, & Tajbakhsh, 2011; Tzahor, 2009). Subsequently, in all regions of the body muscle progenitors become committed myoblasts that express the myogenic regulatory transcription factors MyoD, Myf5 and/or Mrf4. Finally, myoblasts terminally differentiate into committed post-mitotic myocytes that express Myogenin, synthesize muscle-specific proteins (e.g., myosin heavy chain), fuse into multinucleate myofibers (Abmayr & Pavlath, 2012; Sampath, Sampath, & Millay, 2018), and mature and diversify into myofibers that differ in their contractile and metabolic properties (Schiaffino, Sandri, & Murgia, 2007). Concurrent with myogenesis is muscle morphogenesis, the process by which individual myofibers are assembled into defined, unique muscles (e.g., biceps) and these muscles are arranged into a stereotyped spatial array. This process involves the migration of muscle progenitors to particular regions, regulation of the number of myoblasts, selective differentiation of muscle progenitors into myofibers, elongation and increase in size of myofibers via selective myoblast fusion, and the integration of myofibers with MCT, tendons, and bones.

Considerably less in known about the development of the MCT (Fig. 2D). MCT fibroblasts are the cellular source of the MCT extracellular matrix, and they originate from multiple different embryonic sources. The MCT fibroblasts associated with the axial muscles likely derive from the somites (Nowicki, Takimoto, & Burke, 2003). However, the MCT fibroblasts associated with limb muscles arise from the lateral plate mesoderm, the mesoderm that will form the limb bud and give rise not only to the MCT, but the tendons, ligaments, and bones of the limb (Kardon, Harfe, & Tabin, 2003; Pearse et al., 2007). Finally, the MCT fibroblasts of head muscles originate from a third source, the neural crest cells (Le Lièvre & Le Douarin, 1975; Noden, 1983a; Olsson, Falck, Lopez, Cobb, & Hanken, 2001). Study of MCT fibroblasts had been hampered by the lack of molecular markers for these cells. Recently, several markers of embryonic and adult MCT fibroblasts have been identified, and these include Tcf4 (formally known as Tcf7l2), plate-derived growth factor receptor alpha (PDGRFRα), and alpha smooth muscle actin (αSMA) (Kardon et al., 2003; Mathew et al., 2011; Murphy, Lawson, Mathew, Hutcheson, & Kardon, 2011; Olson & Soriano, 2009; Tomasek, Gabbiani, Hinz, Chaponnier, & Brown, 2002). These proteins are expressed in MCT fibroblasts associated with muscles throughout the body. In addition, the transcription factor odd skipped-related 1 (Osr1) was recently found to identify a sub-population of embryonic limb MCT fibroblasts (Vallecillo-García et al., 2017). In the embryo Osr1 is expressed in limb MCT fibroblasts, with Osr1 and Tcf4 co-expressed in sub-populations of fibroblasts, and in the adult Osr1 is expressed in activated MCT fibroblasts following muscle injury (Stumm et al., 2018; Vallecillo-García et al., 2017). Interest in MCT fibroblasts in the adult has been accelerated by the finding of fluorescence-activated cell sorting strategies that allow MCT fibroblasts to be isolated from adult muscle and its associated MCT (Joe et al., 2010; Uezumi, Fukada, Yamamoto, Takeda, & Tsuchida, 2010). As these adult MCT fibroblasts can give rise to both fibroblasts and adipogenic cells, they have been called fibroadipogenic progenitors (FAPs; Joe et al., 2010; Uezumi et al., 2010). Since embryonic Osr1+ cells have been explicitly shown to give rise to FAPs (Vallecillo-García et al., 2017) and most adult MCT fibroblasts express Tcf4 and PDGFRα and activated FAPs express Osr1 (Murphy et al., 2011; Stumm et al., 2018; Uezumi et al., 2010), MCT fibroblasts and FAPs are likely the same cell population. However, molecular heterogeneity and sub-populations within these fibroblasts undoubtedly exist (also see discussions in Nassari et al., 2017; Wosczyna & Rando, 2018).

Development of skeletal muscle and its MCT is tightly linked and recent studies show that MCT fibroblasts are crucial regulators of muscle development throughout the vertebrate body. In this review we discuss and synthesize our current knowledge of the cellular and molecular mechanisms by which the MCT regulates myogenesis and muscle morphogenesis in four major types of vertebrate muscle: axial muscles, head muscles, limb muscles, and the mammalian-specific diaphragm. Defects in MCT development or its communication with neighboring muscle have also emerged as a source of congenital birth defects in head and limb muscles and the diaphragm. We review these studies and explore how studies of normal muscle-MCT development interactions and genetic mouse models of limb and diaphragm defects together enhance and deepen our understanding of normal development and the etiology of birth defects. Finally, we review new studies that suggest that changes in MCT development may have been important developmental innovations that led to the modification or acquisition of new muscles during vertebrate evolution.

2. Axial muscle

Locomotion of chordates and their vertebrate descendants requires the axial muscles. Axial muscles lie in close proximity to the dorsal notochord, vertebrae, and the ventral trunk regions and served as the main propulsive force prior to the evolution of limbs (Romer & Parsons, 1986). The ancestral axial muscle was likely similar to that of the Cephalochordate, *Amphioxus*, and the swimming larvae of Urochordates, both close relatives of vertebrates (Fig. 1A and B). In *Amphioxus* axial muscle extends from the head to the tip of the tail and consists of a series of V-shaped myomeres, separated by connective tissue myosepta (Holland, 1996). These myomeres function as a series of separate contractile units that laterally flex the notochord and enable undulatory swimming and stabilize the back. With the evolution of jawless vertebrates (lampreys and hagfish), the axial muscle became subdivided into epaxial and hypaxial muscles (Fig. 1C; Fetcho, 1987). The epaxial muscles are the intrinsic muscles of the back that are closely associated with the vertebral column and innervated by the dorsal rami of the spinal cord ventral roots, while the hypaxial muscles lie ventral to the vertebrae and are innervated by the ventral rami (Fetcho, 1987). This division of axial muscle into two separately innervated muscles increased the range of motion, allowing for both lateral undulations and dorsal-ventral flexion in the earliest vertebrates (Ahmed et al., 2017). Subsequently in jawed vertebrates, epaxial and hypaxial muscles became further functionally separated, as a horizontal septum of connective tissue developed between these muscles (Fig. 1D; reviewed in Wotton et al., 2015). While in sharks, some fishes, and salamanders the epaxial muscle is primarily composed of one muscle (the dorsalis trunci; Fig. 1E), in many reptiles and mammals the epaxial muscle is subdivided into three muscles: the transversospinalis, longissimus dorsi, and iliocostalis

(Fig. 1F; Romer & Parsons, 1986). Mammals further elaborate epaxial muscles into a complex and regionally varied series of muscles that are important for dynamic and static stabilization, spinal extension, and lateral bending (Fig. 1F; Goodrich, 1958; Romer & Parsons, 1986; Schilling & Carrier, 2010; Webster, Hudson, & Channon, 2014). Similarly, the hypaxial muscles (e.g., intercostal and abdominal muscles) became progressively more elaborate and regionally varied during the evolution of tetrapods (Fig. 1E and F; Romer & Parsons, 1986). Interestingly, there are both epaxial and hypaxial muscles that lose their segmental arrangement and fuse to form sheets of muscle that are critical for locomotion (Carrier, 1990; Romer & Parsons, 1986).

Epaxial and hypaxial trunk muscles develop from the somites (Fig. 3; Bryson-Richardson & Currie, 2008; Buckingham, 2006; Wotton et al., 2015). The somites mature to form a dorsal epithelial layer, the dermomyotome, which gives rise to the overlying dermis and, via multiple waves of migrating muscle progenitors, the underlying primary myotome (Fig. 3A; Gros, Scaal, & Marcelle, 2004; Hollway & Currie, 2005; Kalcheim, Cinnamon, & Kahane, 1999; Venters & Ordahl, 2002). The myotome is a metameric series of muscles transiently present in all amniote embryos. Ultimately, the dorsomedial lip of the dermomyotome and the medial myotome gives rise to the epaxial muscles, while the ventrolateral lip of the dermomyotome and the lateral myotome give rise to the hypaxial muscles (Fig. 3A–C; Cinnamon, Kahane, & Kalcheim, 1999; Denetclaw, Christ, & Ordahl, 1997; Gros et al., 2004; Pu et al., 2013). On a molecular level, the transcription factor, *Engrailed*, appears to be essential for establishing the boundary between the epaxial and hypaxial muscle domains (Ahmed et al., 2017; Wotton et al., 2015).

The somites are also the embryonic source for the cartilages, bones, and tendons of the axial musculoskeletal system. As the somites mature, they not only form the dorsal epithelial dermomyotome and myotome but a ventromedial sclerotome, a region of mesenchymal cells (Fig. 3A; reviewed in Scaal, 2016). The ventral sclerotome gives rise to the vertebral column and also the ribs (Scaal, 2016). In addition, in response to signals from the overlying myotome, the dorsal cranial and caudal edges of the sclerotome form the syndetome, the region that gives rise to the tendons of the axial muscles (Fig. 3A; Brent, Braun, & Tabin, 2005; Brent, Schweitzer, & Tabin, 2003; Brent & Tabin, 2004). Thus developmental interactions between the muscle cells in the myotome and the cartilage progenitors in the sclerotome lead to the induction of the tendon progenitors that are ultimately needed to connect muscle to bone. Furthermore, the close spatial

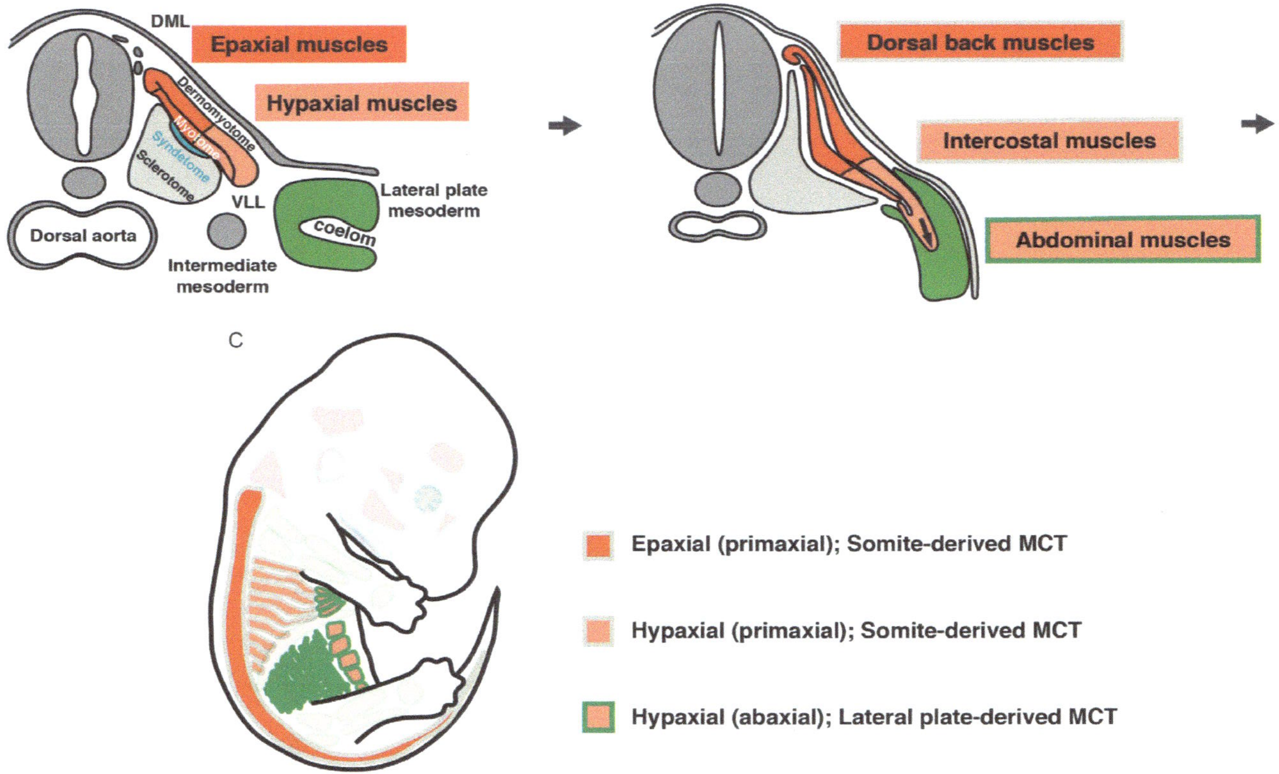

Fig. 3 Development of axial muscles in tetrapods. (A) Epaxial muscle derives from the dorsomedial region of the dermomyotome and myotome, while the hypaxial muscle derives from the ventrolateral region of the dermomyotome and myotome. DML, dorsomedial lip and VLL, ventrolateral lip. (B–C) Dorsomedial dermomyotome and myotome give rise to epaxial back muscles with MCT presumably derived from the somite (shown as red muscles outlined in gray). Hypaxial dermomyotome and myotome give rise to primaxial hypaxial muscles (e.g., intercostal muscles, pink muscles outlined in gray) with MCT presumably derived from the somite and also abaxial hypaxial muscles (e.g., abdominal muscles, pink muscles outlined in green) with MCT derived from lateral plate mesoderm (green).

relationship between the myotome, syndetome, and sclerotome prefigures and ensures the future linkage of muscle, tendon, and bone.

The developmental origin of the MCT of the axial muscles is complex and less well known. Lineage studies using quail-chick chimeras or using the *Prx1Cre* transgene in mice to follow the fate of lateral plate mesoderm have determined that the MCT of a subset of hypaxial trunk muscles (distal intercostals, pectoralis, and transverse, internal/external oblique, and rectus abdominal muscles) is derived from the lateral plate mesoderm (Fig. 3; Burke & Nowicki, 2003; Christ, Jacob, & Jacob, 1983; Durland, Sferlazzo, Logan, & Burke, 2008). This subset of hypaxial muscles with MCT derived from the lateral plate have been termed abaxial muscles (Burke & Nowicki, 2003). The MCT associated with the other hypaxial muscles (e.g., intercostals and longissimus dorsi) as well as the epaxial muscles are assumed to come from the somites and these muscles have been termed primaxial muscles (Burke & Nowicki, 2003). Thus there appears to be two different developmental sources for the MCT of the axial muscles: MCT of lateral plate origin that surrounds the abaxial muscles and MCT of somitic origin that surrounds the primaxial muscles (Nowicki et al., 2003). The boundary between the primaxial and abaxial muscles, with MCT of different developmental origins, has been termed the lateral somitic frontier. However, it should be noted that there is a dearth of formal evidence that the MCT of the primaxial muscles derives from the somites. Transplantation of quail somites into chick has shown that somitic cells can contribute to the MCT of some axial muscles (Saberi et al., 2017). In addition, analysis of a *ScxGFP* transgene that labels the syndetome and tendon cells also labels MCT fibroblasts in some epaxial muscles (Deries, Schweitzer, & Duxson, 2010). Together these data suggest that MCT fibroblasts of primaxial muscles may derive from the syndetome. However, a definitive test of the extent and particular location of somitic contribution to the MCT of primaxial muscles will require more extensive and detailed analyses of quail-chick chimeras or Cre-lineage experiments in mice in which particular regions of the somites are targeted.

The morphogenetic processes that transform the myotome and dermomyotome into the diverse array of epaxial and hypaxial muscles are just beginning to be elucidated. Via a detailed analysis of a developmental series of embryos labeled via whole mount immunofluorescence, Deries et al. (2010) have determined that the epaxial muscles develop gradually by the translocation, re-orientation, and elongation of the myotomal myocytes followed by cleavage of the myotomal masses. How the hypaxial muscles form has not been as fully detailed. Flank hypaxial muscles (both primaxial and abaxial) arise from epithelial extensions of the dermomyotome and the myotome (Christ et al., 1983; Evans, 2003). The primaxial hypaxial muscles undergo morphogenesis into individual muscles in the context of presumably somite-derived MCT and those muscles in the thoracic region develop in close association with the sclerotome forming the ribs (Fig. 3; Burke & Nowicki, 2003; Chevallier, 1975; Durland et al., 2008; Evans, 2003). The abaxial hypaxial muscles derive from the lateral-most regions of the dermomyotome that extend into the lateral plate mesoderm, and their morphogenesis occurs in the context of lateral plate-derived MCT (Fig. 3; Burke & Nowicki, 2003; Durland et al., 2008).

Not only does the MCT surround epaxial and hypaxial muscles as they develop, but it is a source of signals regulating muscle development. The morphogenesis of the dermomyotome and myotome into epaxial muscles appears to be tightly correlated with changes in the extracellular matrix (Deries et al., 2012). The dermomyotome and myotome are surrounded by a laminin matrix that is remodeled at distinct stages of myotome development. When the myotome forms the epaxial muscle masses, the laminin matrix disassembles and the fibronectin matrix increases in density around the myocytes. This correlation suggests that the fibronectin matrix may regulate the morphogenesis of the epaxial muscles (Deries et al., 2012). As MCT fibroblasts are intercalated among the myogenic cells of the myotome (Deries et al., 2010), this suggests that these fibroblasts may be the source of fibronectin and thus regulate epaxial muscle morphogenesis.

Several studies have provided evidence that the MCT fibroblasts regulate the morphogenesis of abaxial hypaxial muscles. The first evidence that MCT fibroblasts play a role in hypaxial muscle morphogenesis comes from quail-chick chimera studies (Nowicki & Burke, 2000). As with most axial tissues, axial muscle expresses *Hox* genes appropriate to their anterior-posterior location. Interestingly, the *Hox* gene expression of transplanted somites and their abaxial muscle derivatives gets re-specified to match the *Hox* genes of the surrounding lateral plate mesoderm, which will give rise to MCT (Nowicki & Burke, 2000). This suggests that the MCT in the lateral plate mesoderm may be an important regulator, via *Hox* expression, of abaxial muscle morphogenesis. Subsequently, two other studies have specifically examined the role of the lateral plate mesoderm in regulating morphogenesis of abaxial muscles. Genetic deletion of the transcription factor *Pitx2* leads to the almost complete absence of abdominal abaxial muscles (external and internal obliques and transverse and rectus abdominus) (Eng et al., 2012). As *Pitx2* is most strongly expressed in the lateral plate mesoderm when the muscle defects appear, the authors suggest that *Pitx2* in the lateral plate mesoderm, via its regulation of *Hox* genes, regulates abdominal muscle morphogenesis. A more direct test of the role of the lateral plate in muscle morphogenesis comes from an analysis of the development of the cutaneous maximus muscle, an abaxial muscle present in many mammals (but not humans) that functions to twitch the skin over the back (Pan et al., 2012). *Fat1* is an atypical Fat-like cadherin planar cell polarity molecule that is critical for development of the cutaneous maximus. While *Fat1* does have a cell-autonomous role controlling the polarity of muscle progenitor migration (Caruso et al., 2013), it also is essential in the surrounding lateral plate mesoderm for regulating spreading and differentiation of muscle progenitors (Helmbacher, 2018). Deletion of *Fat1* in the lateral plate mesoderm, via *Prx1Cre*, leads to hypoplasia of the ventral cutaneous maximus with disorganized myofibers and a reduction in the number of muscle progenitors (Helmbacher, 2018). Thus this study explicitly demonstrates that the lateral plate mesoderm (presumably the MCT fibroblasts) controls the morphogenesis of an abaxial hypaxial muscle.

In summary, during chordate and vertebrate evolution axial muscle becomes progressively more complex and allows for a greater range of motion. The muscle, tendon, and bone of the axial musculoskeletal system all derive from different regions of the somite and their close spatial relationship allows for their coordinated development. In contrast, the MCT of axial muscle arises from two different developmental sources: the somites and the lateral plate mesoderm. Similar to other regions of the embryo (see sections below), the MCT is likely a key source of patterning information that regulates axial muscle morphogenesis.

3. Cranial muscle

The evolution of the head has been a major innovation in vertebrates that concentrated the sense organs, neural tissues, and structures for feeding, breathing, and communication in the rostral region of the animal. Cranial muscles are essential for many of these functions, including feeding, breathing, talking and facial expressions (Edgeworth, 1935; Haas, 2001; Holliday & Witmer, 2007). Craniofacial muscles can be broadly divided into three groups: (1) extraocular muscles, which move and rotate the eye; (2) muscles developmentally associated with the pharyngeal arches (bilateral swellings on either side of the pharynx) that control the jaw, facial expressions, pharynx; and (3) tongue muscles (Fig. 4). Development of these

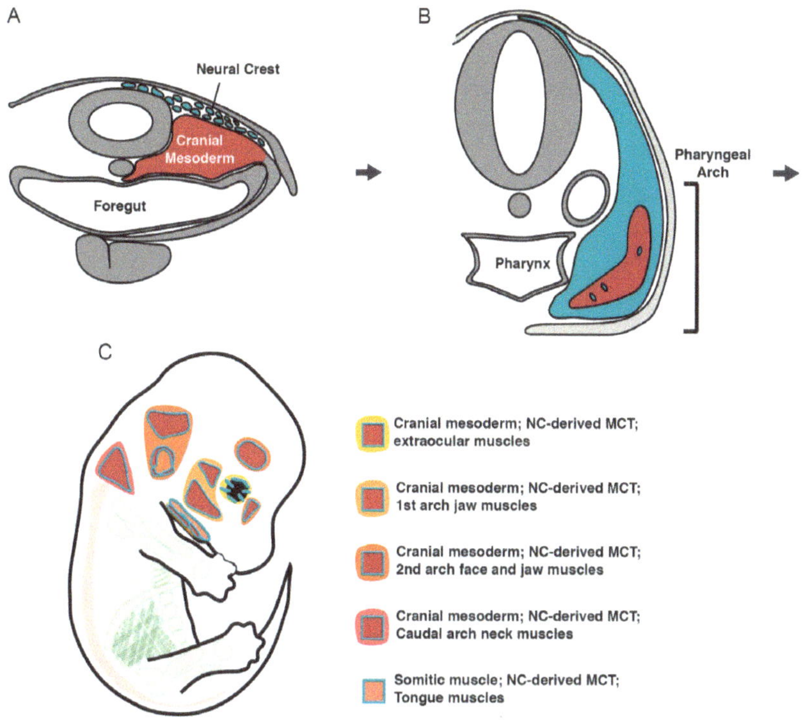

Fig. 4 Development of cranial muscle in tetrapods. (A) Transverse section through early head showing cranial neural crest migration and cranial mesoderm. (B) Transverse section through later developing head showing pharyngeal arch with a cranial mesodermal core surrounded by neural crest that has begun to infiltrate the mesoderm. (C) Groups of cranial muscles with their neural crest-derived MCT.

muscles involves the complex interplay of different embryonic tissues, and defects in these interactions are a source of cranial birth defects. Changes in development have also led to major evolutionary innovations in cranial muscles; for example, duplication of jaw muscles to increase the predation capabilities of Tetraodontoid fishes (Friel & Wainwright, 1997), evolution and diversification of facial expression muscles to allow increased modes of feeding and communication (Burrows, 2008), and the expansion of laryngeal muscles in humans to increase vocalization abilities (Diogo, Abdala, Lonergan, & Wood, 2008). Interactions between the MCT and muscle are critical for morphogenesis of cranial muscle and developmental changes in these interactions may be important for the etiology of birth defects and the evolutionary origin and modification of cranial muscles.

The head muscles and their MCT derive from several embryonic sources (Fig. 4). Most of the head muscles derive from the cranial mesoderm, mesoderm lying on either side of the neural tube and extending rostrally from the prosencephalon and caudally toward the somites. Unlike the paraxial mesoderm that segments into somites in the trunk, the cranial mesoderm is unsegmented, but does contain different regions that give rise to different cranial muscles (reviewed in Michailovici et al., 2015; Ziermann, Diogo, & Noden, 2018). The most rostral region (also termed the pre-chordal mesoderm) gives rise to the extraocular muscles. Pharyngeal mesoderm gives rise to the masticatory muscles (associated with pharyngeal arch 1), facial muscles (associated with pharyngeal arch 2), and the pharyngeal and laryngeal muscles (associated with more caudal arches). Interestingly, the pharyngeal cranial mesoderm region does not give rise exclusively to head muscles, but has also been found to give rise to major components of the heart (Lescroart et al., 2015, 2010). The tongue muscles derive from anterior somites, from which muscle progenitors migrate to the tongue region (reviewed in Diogo et al., 2015; Ziermann et al., 2018). The MCT comes from a third embryonic tissue, the neural crest. The neural crest is a migratory population derived from the dorsal neural tube and not only gives rise to the head's MCT, but also the tendons, facial bones, and many bones of the skull (Couly, Coltey, & Le Douarin, 1993; Noden, 1983b; Olsson et al., 2001; Piekarski, Gross, & Hanken, 2014). Unlike the true head muscles, the neck muscles and MCT represent a transitional region in which muscle derives from somites and cranial mesoderm and MCT derives from neural crest and the lateral plate mesoderm (Durland et al., 2008; Heude et al., 2018; Lescroart et al., 2015; Matsuoka et al., 2005; Piekarski & Olsson, 2007; Sefton, Bhullar, Mohaddes, & Hanken, 2016; Theis et al., 2010).

The development of the head muscles begins with the concurrent movement of muscle progenitors and neural crest cells and with the specification of muscle cells (Fig. 4). Muscle progenitors move into locations surrounding the eye, pharyngeal arches, and regions associated with the pharynx, larynx, and tongue. Concurrently, cranial neural crest cells collectively migrate in stereotyped streams to these same regions (Horigome et al., 1999; Landacre, 1921; Noden, 1975). In the pharyngeal region, muscle progenitors form the core of each arch, and neural crest cells circumscribe and infiltrate the muscle progenitors, leading to extensive mixing of neural crest cells and muscle progenitors (Grenier, Teillet, Grifone, Kelly, & Duprez, 2009; Hacker & Guthrie, 1998; Noden, 1983a; Trainor, Tan, & Tam, 1994). During this time, muscle progenitors begin the process of myogenesis and are specified into myoblasts. Unlike axial or limb muscles, Pax3 and Pax7 are not the initial key regulators of myogenesis, but instead distinct regulatory networks of multiple transcription factors including Pitx2, Tbx1, Islet1, Lhx2, Nkx2.5, MyoR and Capsulin are required, but then converge on the requirement of MyoD and Myf5 for myoblast specification and Myogenin for myocyte differentiation (reviewed by Diogo et al., 2015; Harel et al., 2012; Michailovici et al., 2015; Tzahor, 2009; Ziermann et al., 2018). The initial steps of myogenesis leading up to the specification of MyoD+ myoblasts are independent of signals from the neural crest (Tzahor et al., 2003).

The neural crest plays a key role in regulating the morphogenesis and pattern of cranial muscles (reviewed by Noden & Francis-West, 2006; Noden & Trainor, 2005; Ziermann et al., 2018). Initial evidence for the important role of the neural crest came from embryological transplantation experiments. When grafted in birds or amphibians to different anteroposterior levels, neural crest cells are able to dictate the pattern of muscles and skeletal structures that develop (Hörstadius & Sellman, 1946; Noden, 1983b). In contrast, prior to the emigration of cranial neural crest, all cranial mesoderm is equivalent in their potential to form different head muscles (Trainor et al., 1994; von Scheven, Alvares, Mootoosamy, & Dietrich, 2006). Extirpation of neural crest in axolotls reveals that the formation of head muscle with properly organized myofibers and origins and insertions depends on the neural crest (Ericsson, Cerny, Falck, & Olsson, 2004; Olsson et al., 2001). Experimental manipulations of chick embryos and analysis of mouse mutants further reveal that while neural crest regulates muscle morphogenesis, it is not essential for the initial steps of myogenesis. In the absence of neural crest, *MyoD+* and *Myf5+* myoblasts are specified (Rinon et al., 2007; Tzahor et al., 2003). However, neural crest via inhibition of WNT and BMP signaling regulates the proliferation, differentiation, and position of myoblasts and this, in turn, regulates the pattern of muscles formed (Rinon et al., 2007; Tzahor et al., 2003). Thus, initiation of myogenesis is independent of cranial neural crest, but subsequent patterning and proliferation rely on neural crest-derived signals.

Analyses of mouse mutants in which genes are deleted in neural crest have explicitly demonstrated that the neural crest regulates muscle morphogenesis in the head and also suggests that the MCT, which derives from the neural crest, may play a key role. First, analysis of the transcription factor *Pitx2*, which is required for development of extraocular muscles, has provided insights into neural crest function. Germline null mutations in *Pitx2* lead to the complete loss of extraocular muscles (Gage, Suh, & Camper, 1999; Kitamura et al., 1999; Lu, Pressman, Dyer, Johnson, & Martin, 1999), but the expression of *Pitx2* in both the cranial mesoderm and the neural crest (Gage, Rhoades, Prucka, & Hjalt, 2005) made it unclear in which tissue *Pitx2* was required. Subsequently, tissue-specific deletion of *Pitx2* determined that while *Pitx2* is required in cranial mesoderm cell-autonomously for survival of muscle primordia and activation of *Myf5* and *MyoD* in extraocular muscles (Zacharias, Lewandoski, Rudnicki, & Gage, 2011), *Pitx2* is also important in the neural crest (Evans & Gage, 2005). Neural crest specific deletion of *Pitx2* via *Wnt1Cre* demonstrated that *Pitx2* is required in neural crest to regulate the orientation of extraocular muscles relative to the optic cup. In a second series of experiments, the transcription factors *Dlx5/6* were shown to be critical in neural crest cells for development of the masseter muscle. By analyzing mice null for *Dlx5/6* or its regulator *Endra*, Heude et al. (2010) determined that *Dlx5/6* in non-skeletal neural crest cells is critical for formation of this jaw muscle. Furthermore, the expression of *Dx5/6* in neural crest-derived cells closely associated with the masseter suggested that *Dlx5/6+* MCT in particular may be critical for masseter morphogenesis. Finally, multiple studies have shown that signals from the neural crest are critical for several aspects of tongue muscle development (reviewed by Cobourne et al., 2018). The first step in tongue development is the migration of muscle progenitors following the migration of neural crest cells to the tongue bud (Han et al., 2012). Conditional deletion experiments have determined that cilia-dependent GLI processing in the neural crest is required for survival and migration of muscle progenitors into the bud (Millington et al., 2017). Once in the tongue bud, conditional mutagenesis experiments have established that reception of TGFβ signaling either through *Tgfβr2* or *Alk5* in the neural crest is essential for regulating proliferation and differentiation of tongue muscle cells (Han, Zhao, Li, Pelikan, & Chai, 2014; Hosokawa et al., 2010; Iwata, Suzuki, Pelikan, Ho, & Chai, 2013). The disorganized collagen and elastin fibers associated with the aberrant tongue muscle in *Tgfβr2* mutants further suggest that *Tgfβr2* is acting within the MCT to affect muscle morphogenesis (Iwata et al., 2013). Altogether these experiments demonstrate that the neural crest regulates the proliferation and differentiation of myoblasts and ultimately the pattern of head muscles that develop and suggests that signaling from the MCT may be key for this regulation.

Of babies born with birth defects, approximately one-third exhibit craniofacial defects (Gorlin, Cohen, & Blackburn, 1990) and usually these defects are attributed to problems in neural crest (Trainor, 2010). Several of these birth defects include aberrant cranial muscles and comparison of these human birth defects with mouse genetic experiments suggest that defects in signaling between neural crest and muscle progenitors may underlie the etiology of these defects. Human patients with hemifacial macrosomia have unilateral underdevelopment of facial structures that develop from pharyngeal arches 1 and 2 (Heude, Rivals, Couly, & Levi, 2011). In addition to underdevelopment of bones, they also exhibit poorly developed masticatory muscles. The finding of patients with masseter muscle defects but normal jaw bones (Heude et al., 2011), coupled with the mouse genetic data onthe development of the masseter (Heude et al., 2010), suggests that defects in the MCT associated with the masseter may be the cause of the defects in these patients. Another interesting example of cranial muscle defects is Axenfeld-Rieger syndrome, an autosomal dominant disorder characterized by defects in the eye (Lines et al., 2004; Meyer-Marcotty, Weisschuh, Dressler, Hartmann, & Stellzig-Eisenhauer, 2008; Semina et al., 1996), and although extraocular muscles are present, they may be smaller or have anomalous insertions (Bhate & Martin, 2012; Park, Kim, Heo, & Park, 2009). *PITX2* mutations are frequently associated with this syndrome and based on the mouse genetic analysis of the importance of *Pitx2* in the neural crest in patterning extraocular muscle insertions (Evans & Gage, 2005), it is possible that defects in the neural crest are responsible for this phenotype.

The head muscles have undergone substantial changes during vertebrate evolution and developmental modifications in neural crest are likely to be important drivers of these morphological changes. Comparative studies and interspecific transplantation experiments have demonstrated that species-specific muscle patterning information resides within the neural crest. Studies in amphibians provided the first evidence that the neural crest were the source of species-specific information in the head musculoskeleton; neural crest transplants from a toad into a newt resulted in a donor-specific cartilage as well as the induction of a host-derived, but toad-specific muscle (Wagner, 1959). Subsequent data largely come from the study of bird development. Quails and ducks have distinctly different jaw skeletal and muscle morphologies, associated with their different feeding habits (Tokita & Schneider, 2009). Transplantation of quail cranial neural crest into a duck host resulted in the production of quail-derived skeleton and MCT that transformed duck-derived muscle to acquire a quail-like shape and attachment sites (Tokita & Schneider, 2009). This establishes that, similar to amphibians, species-specific patterning information resides within the neural crest. Careful analysis of this transformation process revealed that the expression levels and location of the MCT marker, *Tcf4*, and tendon marker, *Scx*, reflected the donor neural crest program. Furthermore, analysis of cranial myogenesis and muscle morphogenesis showed that the neither muscle specification nor differentiation was changed by the presence of quail-derived neural crest, but the shape and attachment sites of muscle were transformed. This suggests that the neural crest-derived MCT and tendons determine the species-specific shape and attachment of head muscles. Subsequently, developmental studies of parrots, which have acquired two anatomically unique jaw muscles to allow for feeding on hard-shelled nuts and seeds, suggest that changes in the neural crest-derived MCT and tendon may also lead to the evolution of entirely new muscles (Tokita, Nakayama, Schneider, & Agata, 2013). Altogether these studies strongly implicate an important role for the neural crest in determining the species-specific pattern of cranial muscles and suggest that changes in neural crest development may be the underlying driver of the evolutionary modification and generation of new cranial muscles.

4. Limb muscle

The origin and diversification of paired appendages have played critical roles in vertebrate evolution (Fig. 1G–K). Muscle is a crucial component of fins and limbs, as the number, arrangement, and anatomy of muscles has been modified to enable activities such as swimming, running, flying, and manipulating tools. Tetrapod limbs contain over 40 distinct muscles and each muscle extends over one or two joints, attaching to bone via tendons that originate proximally and insert distally (Fig. 1J and K). Muscles are organized into three proximal-distal groups associated with the upper arm/thigh, forearm/shank, and hand/foot, and each of these groups has a dorsal subdivision that extends or ventral subdivision that flexes the limb (Fig. 5). Compared with other regions of the body, the morphogenesis of limb muscles has been the most intensively studied because of the accessibility of the limb to surgical manipulation in chick embryos, identification of genetic reagents in mice to manipulate components of the limb musculoskeleton, and the nonessential requirement of the limb for viability. These studies have revealed that an assemblage of transcription factors expressed in the MCT is critical for limb muscle morphogenesis. In addition, recent evidence suggests that mutations in these transcription factors are a cause of congenital limb abnormalities that include muscle defects and changes

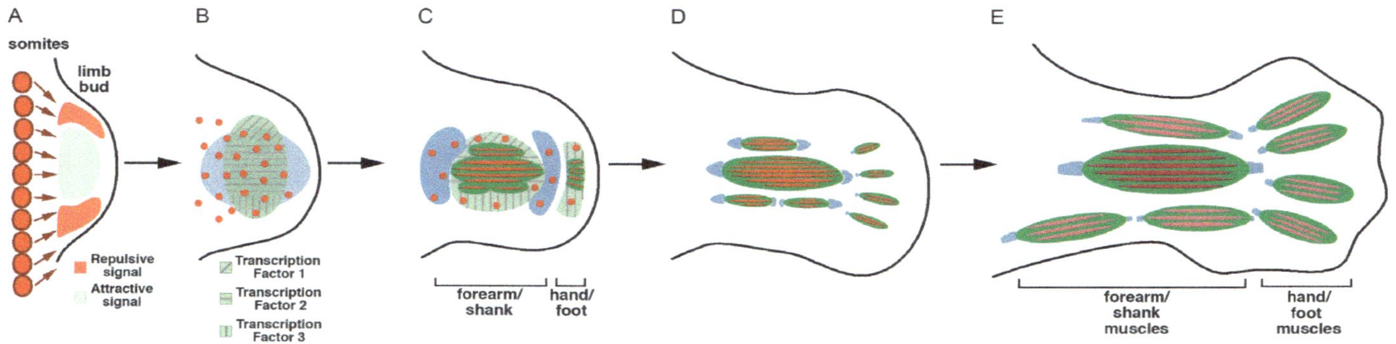

Fig. 5 Development of limb muscles in tetrapods. (A) Muscle progenitors (red) migrate from limb-level somites into the core of the limb, which contains attractive HGF and SDF signals, and avoids peripheral regions, which contains repulsive EphrinA5 signals. (B) Muscle progenitors populate MCT (green) regions that express an array of transcription factors (Tbx3–5, Tcf4, Osr1, Hoxa11/d11) and tendon (blue) regions. (C) Muscle progenitors that migrate into tendon regions do not differentiate, while those that migrate into MCT regions do differentiate. (D–E) In MCT regions muscle progenitors differentiate, and so the MCT pattern prefigures the future pattern of anatomical muscles. Tendons develop at the origin and insertion ends of forming muscles and their associated MCT.

in the expression of these factors can lead to the evolution of novel modifications of limb muscles.

The limb's muscle and MCT originate from two embryonic sources that are each accessible to surgical manipulation in chick and Cre-mediated manipulation in mouse (Fig. 5). Limb muscle derives from the ventrolateral dermomyotome of somites lying adjacent to the developing limb buds (Chevallier, Kieny, & Mauger, 1977; Christ, Jacob, & Jacob, 1977; Ordahl & Le Douarin, 1992). Quail-chick transplantations and chick viral lineage studies have provided a detailed map of which somites contribute to particular muscles in the fore and hind limbs (Beresford, 1983; Kardon, Campbell, & Tabin, 2002; Lance-Jones, 1988). Quail-chick studies have also determined that the lateral plate mesoderm proliferates to form the limb bud and the MCT, tendons, ligaments, and bones of the limb (Chevallier et al., 1977; Christ et al., 1977). In addition, more recent viral lineage studies have determined that the limb mesoderm gets progressively specified into the various cells types, with cartilage and perichondral cells coming from a common lineage and MCT and tendon deriving from another lineage (Pearse et al., 2007). The different embryonic sources of muscle and MCT have allowed each of these tissues to be individually genetically manipulated. $Pax3^{Cre}$ and $Pax7^{Cre}$ (and tamoxifen-inducible $Pax7^{CreERT2}$) have been used to lineage trace and manipulate muscle progenitors and their derivatives (Engleka et al., 2005; Keller, Hansen, Coffin, & Capecchi, 2004; Lepper, Partridge, & Fan, 2011; Murphy et al., 2011); $Pax3^{Cre}$ recombines in the earliest muscle progenitors, but also recombines in somitically-derived endothelial cells that populate the limb, while $Pax7^{Cre}$ and $Pax7^{CreERT2}$ recombine slightly later, but exclusively in muscle progenitors (Hutcheson, Zhao, Merrell, Haldar, & Kardon, 2009). The *Prx1Cre* transgene (and tamoxifen-inducible *Prx1CreERT2*) recombines in the lateral plate-derived mesoderm of the limb and so has been a useful reagent to manipulate the limb's MCT, tendons, and bones (Hasson et al., 2010; Logan et al., 2002).

The first step in the development of limb muscles is the emigration of muscle progenitors from the somites into the limb, and attractive and repulsive signals from the limb bud mesoderm are critical for this process (Fig. 5A and B). Two signaling molecules have been found to act as chemoattractants for muscle progenitors. The ligand Hepatocyte Growth Factor (HGF) is strongly expressed in the limb mesoderm, while the migrating muscle progenitors express the receptor tyrosine kinase MET (Adachi, Pascual-Anaya, Hirai, Higuchi, & Kuratani, 2018). Genetic loss-of-function experiments in mice and HGF bead experiments in chick demonstrate that HGF/MET signaling is necessary and sufficient to guide muscle progenitors into the limb (Bladt, Riethmacher, Isenmann, Aguzzi, & Birchmeier, 1995; Brand-Saberi, Müller, Wilting, Christ, & Birchmeier, 1996; Dietrich et al., 1999; Heymann, Koudrova, Arnold, Köster, & Braun, 1996). A second chemokine ligand Stromal Cell-Derived Factor 1 (SDF1 or Cxcl12) expressed in the limb bud mesoderm and its receptor Cxcr4 expressed in muscle progenitors also regulate muscle migration (Vasyutina et al., 2005). SDF1 bead experiments in chick and *Cxcr4* null mice show that SDF1/Cxcr4 signaling positively regulates the migration and survival of muscle progenitors into the limb. In addition to these positive regulators of migration, Ephrina5/Epha4 signaling negatively regulates limb muscle migration, as the ligand Ephrina5 expressed in the peripheral limb bud mesoderm repulses receptor Epha4-expressing muscle progenitors from populating these regions of the limb (Swartz, Eberhart, Pasquale, & Krull, 2001). Thus a combination of attractive HGF and SDF1 ligands expressed in the central limb and repulsive Ephrina5 ligand in the limb periphery determine where receptor Met+, Cxcr4+, Epha4+ muscle progenitors migrate to and ultimately reside in the limb (Fig. 5B). Although not yet explicitly tested, HGF and SDF1 are likely expressed in MCT fibroblasts or their precursors.

After entering the limb, muscle progenitors undergo a complex process of morphogenesis eventually resulting in the complex pattern of limb muscles. A detailed descriptive analysis of this process in developing chick hind limbs reveals that muscle progenitors initially form dorsal and ventral masses that gradually become subdivided into three dorsal and ventral muscle masses associated with the three proximal-distal limb regions and then these six masses subdivide into individual anatomical muscles (Fig. 5; Kardon, 1998). Early experiments in the chick suggested that the MCT may be an important regulator of this morphogenetic process. In particular, MCT forms in a muscle pre-pattern even in the absence of muscle, and MCT can organize even non-muscle cells to form muscle-like structures (Grim & Wachtler, 1991; Jacob & Christ, 1980; Lanser & Fallon, 1987). Subsequently, Tcf4 was identified as the first gene expressed in MCT of the limb in a muscle pre-pattern, independent of muscle (Kardon et al., 2003). Furthermore, virally expressed ectopic Tcf4 in chick limb non-muscle mesoderm led to induction of ectopic muscle and disruption of Tcf4 led to muscle mis-patterning (Kardon et al., 2003). These data suggested a model whereby the lateral plate-derived MCT establishes a pre-pattern in the limb that determines where muscle progenitors differentiate and ultimately the number and location of limb muscles (Fig. 5C–E).

Elucidation of the important role of MCT in regulating limb muscle morphogenesis has been facilitated by the identification in mouse of transcription factors that are expressed in the MCT and functionally important in establishing the pattern of limb muscles. Six transcription factors are expressed in different limb MCT regions. The T-box transcription factors Tbx5 and Tbx4 are expressed throughout the mesoderm of fore and hind limbs, respectively, including the MCT (Hasson, Del Buono, & Logan, 2007; Naiche & Papaioannou, 2007), while Tbx3 is expressed in the anterior and posterior MCT, posterior bones and a subset of bone eminences (eminences are the protusions on bones to which tendons attach; Colasanto et al., 2016). Hoxa11 is expressed in the MCT and also in the tendons and perichondrium of the forearm and shank regions of the fore and hind limbs (Swinehart, Schlientz, Quintanilla, Mortlock, & Wellik, 2013). The odd skipped-related 1, Osr1, transcription factor is expressed in multiple MCT regions of the fore and hind limb and also in the synovial joints (Stricker, Brieske, Haupt, & Mundlos, 2006; Vallecillo-García et al., 2017). Finally, analysis in mouse shows that Tcf4 is expressed in MCT throughout the limb (as well as some bones and bone eminences), but Tcf4 lineage studies show that it is most strongly expressed in MCT near the elbow and knee (Colasanto et al., 2016; Mathew et al., 2011). Genetic deletion of these genes leads to defects in the muscles in regions where these genes are expressed. Thus genetic deletion of *Tbx5* and *Tbx4* leads to mis-patterned fore and hind limb muscles, respectively (Hasson et al., 2010); compound mutants of *Hoxa11* and *Hoxd11* paralogue cause defects in forearm and shank muscles (Swinehart et al., 2013); mutations in *Tcf4* lead to muscle truncations near the knee (Mathew et al., 2011); and *Osr1* null mice have various muscle defects (Vallecillo-García et al., 2017). The phenotype of *Tbx3* mutants is most striking because two specific anterior muscles, the lateral triceps and brachialis, are absent in these mice (Colasanto et al., 2016). In addition to affecting the size and shape of muscles, Tcf4 also affects myofiber maturation and fiber type (Mathew et al., 2011). It is thought that all of these genes primarily function in the MCT, and at least in the case of Tbx3 and Tcf4 it has been demonstrated that these genes do not function cell-autonomously within muscle cells (via $Pax3^{Cre}$ deletion; Colasanto et al., 2016; Mathew et al., 2011). However, it should be noted that none of these transcription factors are exclusively expressed in nor have been specifically deleted just in the MCT and so may function in other tissues (e.g., tendons, bones, or bone eminences) to regulate muscle patterning.

The array of transcription factors expressed in the MCT regulates limb muscle morphogenesis through several mechanisms. Studies of Tbx3, Tbx5, and Osr1 examined when loss of these genes affected muscle development and all found that defects appeared by E12.5, 2 days before the pattern of limb muscles is established at E14.5 (Colasanto et al., 2016; Hasson et al., 2010; Vallecillo-García et al., 2017). Interestingly, while Osr1 and Tbx5 cell non-autonomously regulate the number and pattern of MyoD+ myoblasts at E12.5, Tbx3 does not regulate the number of myoblasts, but rather their differentiation into myofibers. There are at least three mechanisms by which these transcription factors may act within MCT fibroblasts to ultimately affect muscle morphogenesis. First, as found with Osr1, the transcription factors may regulate MCT fibroblast cell fate, and in the absence of *Osr1* lateral plate-derived limb mesodermal cells adopts a more chondrogenic or tendon-like fate (Vallecillo-García et al., 2017). Second, the transcription factors may regulate the number and/or pattern of MCT fibroblasts in particular limb regions, and in *Hoxa11/d11* mutants alterations in the pattern of Tcf4+ MCT fibroblasts are present (Swinehart et al., 2013). Finally, the transcription factors may regulate the MCT fibroblasts' secretion of signaling molecules or extracellular matrix that regulate the development of neighboring muscle cells; in *Tbx5* mutants MCT fibroblasts express less and more disorganized N-Cadherin and β-Catenin (Hasson et al., 2010) and OSR1 regulates a large number of secreted signals (e.g., SDF and Bmp4) and extracellular matrix proteins (e.g., Col6; Vallecillo-García et al., 2017).

Analysis of mutations of the transcription factors expressed in the MCT also provides insights into the relationships between the developing MCT and muscle with the developing tendons and bones. Analysis of *Hoxa11/d11*, *Tbx5*, and *Tbx3* mutants indicates that while muscle and MCT development are de-coupled from bone development, they are tightly linked with tendon and bone eminence development. In particular, in *Hoxa11/d11* compound mutants with one wild-type allele of either gene muscle and associated tendons in the forearm and shank are highly aberrant, but the bones appear normal (Swinehart et al., 2013). Careful analysis of the temporal requirements for *Tbx5* and *Tbx4* indicates that while these genes are required early (prior to E10.5–11.5) to regulate hind and fore limb bones, at later developmental times these genes coordinately regulate muscle and tendon morphogenesis without affecting bones (Hasson et al., 2010). However, a similar study of the temporal requirements of *Tbx3* indicates that while *Tbx3* is required early (prior to E9.25) for ulna and posterior digit development, the degree of development of the lateral triceps and brachialis muscles is tightly linked with the degree of development of their bone eminence attachment sites (Colasanto et al., 2016). Together these data, as well as lineage studies in the limb (Pearse et al., 2007), suggest that the long bones of the limb are specified and patterned early, while the MCT, tendons, and bone eminences are specified and undergo coordinated morphogenesis later in limb development.

In summary, an array of transcription factors are expressed in the lateral-plate derived mesoderm in different regions that are critical for patterning the musculoskeletal system. Some of these factors (e.g., Tbx3, 4, and 5) are expressed and required early in the limb (prior to E11.5 in mouse) for proper development of the long bones. All of the factors are expressed in the MCT as well as other lateral plate derivatives, such as the tendons and bone eminences. The expression of these transcription factors in the MCT is critical for regulating the morphogenesis of neighboring muscles and the expression of single or overlapping sets of transcription factors may function as a molecular address specifying particular muscles. The specification of two specific muscles by Tbx3 supports such a hypothesis (Colasanto et al., 2016). Mechanistically, these transcription factors act either by regulating the specification of MCT fibroblasts, the number and pattern of these fibroblasts, and/or the signaling or extracellular matrix molecules secreted by these fibroblasts. The MCT fibroblasts, in turn, determine the pattern of muscles that forms via their regulation of the number, survival, and/or differentiation of muscle progenitors. Finally, the coordinate morphogenesis of MCT, with muscle, tendon, and bone eminences, ensures that a functional musculoskeleton develops.

The MCT is not only important for regulating the normal development of limb muscles, but defects in MCT likely underlie the abnormal muscles found in congenital limb abnormalities. Mutations in *TBX3* and *TBX5* have been identified as the cause of two congenital limb syndromes, Ulnar-Mammary syndrome and Holt-Oram syndrome (Bamshad et al., 1997; Basson et al., 1997; Li et al., 1997). Individuals with Ulnar-Mammary syndrome exhibit a variety of limb phenotypes, ranging from the absence of the forearm and hand, to posterior bone defects, including loss of the ulna and/or posterior digits (Bamshad et al., 1999, 1997). Individuals with Holt-Oram syndrome are characterized by deformities in the forelimb, ranging from malformations of carpal bones, thumb abnormalities or phocomelia (Basson et al., 1997). In addition to the obvious bone defects, muscle defects are also associated with both syndromes. Individuals with Holt-Oram syndrome have been found to have hypoplastic pectoralis major, deltoid, and shoulder muscles (Newbury-Ecob, Leanage, Raeburn, & Young, 1996), and recently at least one individual with Ulnar-Mammary syndrome has completely lost the lateral triceps in one limb (Colasanto et al., 2016). Based on the genetic data from mouse mutants (Colasanto et al., 2016; Hasson et al., 2010), it is likely that loss of either *Tbx5* or *Tbx3* expression in the MCT is the cause of these muscle defects.

During vertebrate evolution fins and limbs have been subject to many evolutionary modifications, including changes in their musculature. One example of limb evolution is the striking variation in the feathering of feet in domestic pigeons, with some birds exhibiting many large feathers (Domyan et al., 2016). Intriguingly, the presence of feathered feet is also accompanied by changes in the leg musculature (Domyan et al., 2016). Underlying these morphological changes are genetic changes such that the presence of feathered feet and changed musculature results from a partial transformation of the hindlimb to a forelimb mediated by the decreased expression of the normally hindlimb-specific transcription factor *Pitx1* and the increased expression of the normally forelimb-specific *Tbx5*. Based on the data in mouse supporting a role of Tbx5 in MCT regulating forelimb muscle morphogenesis, it is likely that the increased expression of Tbx5 in the MCT during development of feathered footed pigeons is the cause of the morphological changes in their limb muscles. Potentially other changes in the genetic regulation of developing MCT underlie evolutionary changes in limb musculature.

5. Diaphragm

The diaphragm is a unique mammalian muscle that has been crucial for the evolutionary success of mammals. It is a domed muscle lying at the base of the thoracic cavity that is essential for the inspiratory phase of respiration and for separating the abdominal contents from the overlying thoracic cavity containing the heart and lungs (Merrell & Kardon, 2013). In humans, the common birth defect, Congenital Diaphragmatic Hernia (CDH), highlights the diaphragm's critical function. In CDH, the diaphragm fails to form properly and weaknesses in the diaphragm allow abdominal contents to herniate into the thoracic cavity and impede lung development. The resulting lung hypoplasia leads to 50% mortality and high morbidity and thus demonstrates the importance of proper diaphragm development (Kardon et al., 2017; Merrell & Kardon, 2013). The diaphragm is not only critical for humans, but also for all mammals. The evolution of mammals involved the appearance of tidally ventilated, alveolar lungs and increased aerobic requirements (Farmer, 2015; Perry, Similowski, Klein, & Codd, 2010). It is thought that the evolution of the diaphragm in mammals was essential for obtaining maximal lung function, as the diaphragm both opens the thoracic cavity during inspiration (allowing for influx of air) and keeps the abdominal contents caudal. Thus understanding the development of the diaphragm has important implications for human health and the overall evolutionary success of mammals. Strikingly, recent studies show that the MCT is critical for the development of the diaphragm, an important cellular source of CDH, and may play a key role in the evolution of the diaphragm.

During development the diaphragm's muscle and MCT originate from two different embryonic sources. Similar to axial and limb muscles, the diaphragm's muscle arises from the somites and recently was shown to specifically come from cervical-level somites (Sefton, Gallardo, & Kardon, 2018). The diaphragm's MCT comes from the pleuroperitoneal folds (PPFs), a lateral plate-derived pyramidal tissue located between the thoracic (pleural) and abdominal (peritoneal) cavities (Fig. 6A). Histological, immunofluorescent, and genetic lineage studies in mice reveal that the PPFs expand dorsally and ventrally to give rise to the diaphragm's MCT and central tendon (Fig. 6; Babiuk, Zhang, Clugston, Allan, & Greer, 2003; Merrell et al., 2015; Sefton et al., 2018). The muscle progenitors migrate from the cervical somites to the PPFs, and as the PPFs spread the muscle cells are carried throughout the nascent diaphragm and differentiate into the radially oriented myofibers of the costal diaphragm in close association with the MCT (Fig. 6; Merrell et al., 2015; Sefton et al., 2018).

Multiple studies demonstrate that the PPFs, and likely the PPF-derived MCT, control development of the diaphragm and are the cellular source of CDH. The normal morphogenetic expansion of the PPFs ahead of muscle (and also nerve and vasculature; Merrell et al., 2015; Sefton et al., 2018), as well as their expansion in the absence of muscle (Merrell et al., 2015) suggested that the PPFs regulate the morphogenesis of the diaphragm's muscle. Further studies conditionally deleting the CDH-associated genes, *Gata4*, *WT1*, and *β-catenin*, in specific embryonic tissues have now definitively shown that mutations in the PPFs (or their associated mesothelium) lead to defective diaphragm development and hernias (Carmona, Cañete, Cano, Ariza, & Muñoz-Chápuli, 2016; Merrell et al., 2015; Paris, Coles, & Ackerman, 2015) and so establish that PPFs regulate normal diaphragm development and are a cellular source of CDH. Detailed analysis of *Gata4* additionally revealed that the PPFs regulate the proliferation and survival of adjacent muscle progenitors, and because the PPF cells closest to the muscle are PPF-derived MCT fibroblasts, this suggests that these fibroblasts are the

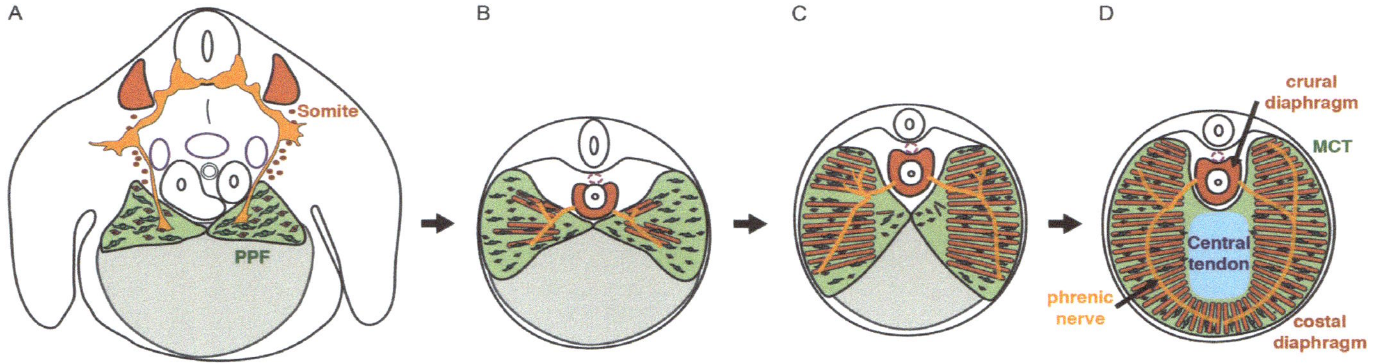

Fig. 6 Development of the mammalian diaphragm. (A) Muscle progenitors (red) migrate from cervical somites to the pleuroperitoneal folds (PPFs, green). The PPFs are also the target for the axons of the phrenic nerve (orange). (B–C) As PPFs spread dorsally and ventrally they carry the muscle progenitors, which differentiate into radially oriented myofibers. Outgrowth of phrenic nerves is also regulated by PPF expansion. (D) PPFs give rise to the MCT guiding expansion and differentiation of muscle and ultimately surrounding myofibers of the costal diaphragm.

critical regulators of the diaphragm's muscle (Merrell et al., 2015). Thus PPF-derived MCT fibroblasts guide normal diaphragm muscle development and defects in MCT development or MCT-muscle communication lead to CDH. What precisely is the molecular communication between the MCT fibroblasts and the muscle progenitors remains an open question.

How the diaphragm evolved at the origin of mammals during the Permian era is a major unanswered question (Farmer, 2015; Perry et al., 2010), but studies of diaphragm development in mouse suggest that the PPFs may be critical for diaphragm evolution. One way to elucidate the evolutionary origin of the diaphragm is to compare the developmental innovations present in mammals, but absent from birds and reptiles, which do not have a muscularized diaphragm. While cervical-level somites give rise to neck and limb muscles in birds, reptiles, and mammals, it is only in mammals that muscle progenitors migrate to the nascent diaphragm (Hirasawa, Fujimoto, & Kuratani, 2015; Hirasawa & Kuratani, 2013; Sefton et al., 2018). Thus recruitment of muscle progenitors from cervical-level somites to this region is a crucial event uniquely present in mammals. In mammals, the PPFs not only are the target of the migrating muscle progenitors, but they subsequently guide muscle morphogenesis (Babiuk et al., 2003; Hitachi et al., 2009; Merrell et al., 2015; Sefton et al., 2018). This suggests that the PPFs, which give rise to the diaphragm's MCT, may be the developmental innovation that allowed for the formation of a muscularized diaphragm. Critical to determine will be whether mammals are unique in their possession of PPFs or whether PPFs are present more broadly in amniotes, but only in mammals are the PPFs able to recruit muscle progenitors and guide their morphogenesis.

6. Conclusion

Muscles in different regions of the vertebrate body have evolved distinct anatomical structures and functions, but in all regions muscle connective tissue plays a central role in regulating muscle morphogenesis and determining the pattern of muscles that develop. The muscles in the trunk, head, limb, and diaphragm are composed of similar multinucleate post-mitotic myofibers, but they differ in the contractile and metabolic properties of their myofibers, the size and organization of myofibers, and the overall shape and attachment to tendons and bones. The generation of myofibers by the process of myogenesis is largely similar in axial, cranial, limb, and diaphragm muscles; following a heterogeneous network of genes regulating the development of muscle progenitors, the molecular and cellular processes of myoblast specification, differentiation, and fusion into myofibers are similar in all muscles (Fig. 2C). However, it is during the process of muscle morphogenesis that the distinctive pattern and anatomy of different muscles emerge, and MCT plays a crucial role in this developmental process. The MCT of the trunk, head, limbs, and diaphragm arises from different embryonic tissues—somites, neural crest, lateral plate mesoderm, and pleuroperitoneal folds—but nevertheless they function equivalently. In all regions, MCT does not regulate the initial development of muscle progenitors or their specification into myoblasts. Instead, MCT regulates several aspects of muscle morphogenesis. First, in the head, limbs, and diaphragm, MCT guides migration of muscle progenitors to target regions likely, as indicated by limb experiments (Bladt et al., 1995; Brand-Saberi et al., 1996; Dietrich et al., 1999; Heymann et al., 1996; Swartz, Nguyen, Mccarthy, & Eberhart, 2012; Vasyutina et al., 2005), by a combination of attractive (e.g., HGF, SDF), repulsive (e.g., Ephrin/Eph), and potentially permissive signals. Once muscle progenitors populate target regions, the MCT then locally promotes the proliferation, survival, and differentiation of neighboring myoblasts into myofibers. These localized regions of differentiated myofibers form the nucleus of anatomical muscles. Thus an attractive model is that the MCT serves as a mesodermal pre-pattern that determines the sites of myofiber differentiation and ultimately determines the number, location, size, and shape of muscles. Consistent with such a model, development of the MCT in the limb and diaphragm is independent of muscle development (Kardon et al., 2003; Merrell et al., 2015; Vallecillo-García et al., 2017). The data in the limb suggest that spatially complex and overlapping sets of transcription factors expressed in MCT fibroblasts in different regions may function as molecular addresses that specify particular muscles. The specification of two specific muscles by the transcription factor Tbx3 in the forelimb supports such a notion (Colasanto et al., 2016). How such genes expressed in MCT fibroblasts ultimately regulate muscle morphogenesis is unclear. These genes may regulate MCT cell fate (e.g., Osr1; Vallecillo-García et al., 2017), regulate the number and pattern of MCT fibroblasts, and/or the secretion by MCT fibroblasts of signaling molecules or extracellular matrix proteins (e.g., Hasson et al., 2010; Iwata et al., 2013) that regulate myoblast proliferation, survival, and/or differentiation. The experiments on MCT regulation of muscle morphogenesis also shed light on how the different components of the musculoskeletal system are assembled. Experiments in the limb and head (Colasanto et al., 2016; Hasson et al., 2010; Heude et al., 2010) suggest that specification of bones precedes and their development can be de-coupled from the rest of the musculoskeleton, while the development of the muscles, MCT, tendons, and bone eminences is tightly linked. The tight developmental linkage between muscles, MCT, tendons, and their bone attachment sites ensures that a functional musculoskeletal system develops.

While much progress has been made in our understanding of muscle morphogenesis, particularly with reference to limb muscles, many outstanding questions remain. First, the array of transcription factors in the limb suggests that overlapping sets of genes expressed in the MCT in different domains of the vertebrate body may be critical for specifying the pattern of muscles that develop. However, our current understanding of gene expression in MCT fibroblasts and their regulation of muscle patterning does not explain the complex array of axial, head, and limb muscles present. Second, our understanding of how MCT fibroblasts regulate the proliferation, survival, and differentiation of myoblasts is limited. For instance, the very first myofibers to differentiate develop in an orientation that prefigures the position of the future muscle (Kardon, 1998), and so this initial myofiber orientation determines the origin and insertion and therefore the function of the muscle that develops. Yet, we have little understanding of the molecular or cellular processes that determine where myofibers differentiate and their orientation. Finally, the coupled development of muscle, MCT, tendons, and bones is critical for development of a functional musculoskeleton. While signals from the MCT are critical for coupling muscle and MCT development, it is unclear how MCT, tendons, and bones are specified from a common embryonic source (e.g., somites in the trunk, neural crest in the head, or lateral plate in the limb) and then their development subsequently coordinated. In fact, one of the major impediments to studying MCT has been the lack of molecular markers and genetic reagents that are specific for MCT. To date, all MCT markers and reagents label MCT as well as at least one other component of the musculoskeleton, e.g., tendon or bone eminence, and this likely reflects their close lineage relationship. Finding unique molecular markers and creating more specific MCT Cre lines is a challenge for future research.

Birth defects in the development of the head, trunk, limbs, and diaphragm commonly arise in humans. Defects in development of the diaphragm muscle have long been recognized since CDH is both common and has major health consequences (Lally, 2016). In contrast, while birth defects in bone development have been readily recognized in the head and limbs, it is only more recently that abnormalities in cranial and limb muscle development have been identified (Colasanto et al., 2016; Heude et al., 2011; Newbury-Ecob et al., 1996). Strikingly, mouse models of CDH and cranial and limb birth defects suggest that MCT fibroblasts are an important cellular source for these muscle defects (Colasanto et al., 2016; Hasson et al., 2010; Heude et al., 2010; Merrell et al., 2015). Thus increasing our understanding of the normal role of MCT in regulating muscle morphogenesis will be essential for elucidating the molecular and cellular processes underlying the etiology of these birth defects.

Finally, the progressive modification of the form and function of different muscles and the acquisition of new muscles have been instrumental to the evolutionary success of vertebrates. How muscles have been modified or new muscles originate during evolution has been an outstanding question. Recent comparative studies in birds suggest that developmental modifications of the MCT have been key to evolutionary changes in their jaw musculature and have allowed birds to exploit new food sources (Tokita et al., 2013; Tokita & Schneider, 2009). Developmental innovations in the MCT may also underlie the evolution of novel muscles, such as the syringeal muscles in songbirds and the diaphragm in mammals. Future studies of the development of their MCT and muscle should provide insights into the mechanisms underlying the evolutionary origin of these important muscles.

In summary, the MCT is essential for regulating morphogenesis of muscles throughout the body; defects in MCT development have emerged as a source of congenital birth defects in head, limb, and diaphragm muscles; and developmental changes in MCT may have been key developmental innovations that enabled the evolution of the vertebrate musculature. Therefore the MCT serves as a critical link for understanding muscle development, birth defects, and evolution.

References

Abmayr, S. M., & Pavlath, G. K. (2012). Myoblast fusion: Lessons from flies and mice. *Development*, *139*, 641–656.

Adachi, N., Pascual-Anaya, J., Hirai, T., Higuchi, S., & Kuratani, S. (2018). Development of hypobranchial muscles with special reference to the evolution of the vertebrate neck. *Zoological Letters*, *4*, 5.

Ahmed, M. U., Maurya, A. K., Cheng, L., Jorge, E. C., Schubert, F. R., Maire, P., et al. (2017). Engrailed controls epaxial-hypaxial muscle innervation and the establishment of vertebrate three-dimensional mobility. *Developmental Biology*, *430*, 90–104.

Andrade, W., & Brandan, E. (1991). Isolation and characterization of rat skeletal muscle proteoglycan decorin and comparison with the human fibroblast decorin. *Comparative Biochemistry and Physiology. B*, *100*, 565–570.

Babiuk, R. P., Zhang, W., Clugston, R., Allan, D. W., & Greer, J. J. (2003). Embryological origins and development of the rat diaphragm. *The Journal of Comparative Neurology*, *455*, 477–487.

Bamshad, M., Le, T., Watkins, W. S., Dixon, M. E., Kramer, B. E., Roeder, A. D., et al. (1999). The spectrum of mutations in TBX3: Genotype/phenotype relationship in ulnar-mammary syndrome. *American Journal of Human Genetics*, *64*, 1550–1562.

Bamshad, M., Lin, R. C., Law, D. J., Watkins, W. S., Krakowiak, P. A., Moore, M. E., et al. (1997). Mutations in human TBX3 alter limb, apocrine and genital development in ulnar-mammary syndrome. *Nature Genetics*, *16*, 311–315.

Basson, C. T., Bachinsky, D. R., Lin, R. C., Levi, T., Elkins, J. A., Soults, J., et al. (1997). Mutations in human TBX5 cause limb and cardiac malformation in Holt-Oram syndrome. *Nature Genetics*, *15*, 30–35.

Beresford, B. (1983). Brachial muscles in the chick embryo: The fate of individual somites. *Journal of Embryology and Experimental Morphology*, 77, 99–116.

Bhate, M., & Martin, F. J. (2012). Unilateral inferior rectus hypoplasia in a child with Axenfeld-Rieger syndrome. *Journal of AAPOS*, *16*, 304–306.

Bladt, F., Riethmacher, D., Isenmann, S., Aguzzi, A., & Birchmeier, C. (1995). Essential role for the c-met receptor in the migration of myogenic precursor cells into the limb bud. *Nature*, *376*, 768–771.

Brand-Saberi, B., Müller, T. S., Wilting, J., Christ, B., & Birchmeier, C. (1996). Scatter factor/hepatocyte growth factor (SF/HGF) induces emigration of myogenic cells at interlimb level in vivo. *Developmental Biology*, *179*, 303–308.

Brent, A. E., Braun, T., & Tabin, C. J. (2005). Genetic analysis of interactions between the somitic muscle, cartilage and tendon cell lineages during mouse development. *Development*, *132*, 515–528.

Brent, A. E., Schweitzer, R., & Tabin, C. J. (2003). A somitic compartment of tendon progenitors. *Cell*, *113*, 235–248.

Brent, A. E., & Tabin, C. J. (2004). FGF acts directly on the somitic tendon progenitors through the Ets transcription factors Pea3 and Erm to regulate scleraxis expression. *Development*, *131*, 3885–3896.

Bryson-Richardson, R. J., & Currie, P. D. (2008). The genetics of vertebrate myogenesis. *Nature Reviews Genetics*, *9*, 632–646.

Buckingham, M. (2006). Myogenic progenitor cells and skeletal myogenesis in vertebrates. *Current Opinion in Genetics & Development*, *16*, 525–532.

Burke, A. C., & Nowicki, J. L. (2003). A new view of patterning domains in the vertebrate mesoderm. *Developmental Cell*, *4*, 159–165.

Burrows, A. M. (2008). The facial expression musculature in primates and its evolutionary significance. *BioEssays*, *30*, 212–225.

Carmona, R., Cañete, A., Cano, E., Ariza, L., & Muñoz-Chápuli, R. (2016). Conditional deletion of WT1 in the septum transversum mesenchyme causes congenital diaphragmatic hernia in mice. *eLife*, *19*, e16009.

Carrier, D. (1990). Activity of the hypaxial muscles during walking in the lizard Iguana iguana. *The Journal of Experimental Biology*, *152*, 453–470.

Caruso, N., Herberth, B., Bartoli, M., Puppo, F., Dumonceaux, J., Zimmermann, A., et al. (2013). Deregulation of the protocadherin hene FAT1 alters muscle shapes: Implications for the pathogenesis of facioscapulohumeral dystrophy. *PLoS Genetics*, *9*, e1003550.

Chevallier, A. (1975). Rôle du mésoderme somitique dans le développement de la cage thoracique de l'embryon d'oiseau. I. Origine du segment sternal et mécanismes de la différentiation des côtes. *Development*, *33*, 291–311.

Chevallier, A., Kieny, M., & Mauger, A. (1977). Limb-somite relationship: Origin of the limb musculature. *Journal of Embryology and Experimental Morphology*, *41*, 245–258.

Christ, B., Jacob, H. J., & Jacob, M. (1977). Experimental analysis of the origin of the wing musculature in avian embryos. *Anatomy and Embryology*, *150*, 171–186.

Christ, B., Jacob, M., & Jacob, H. J. (1983). On the origin and development of the ventrolateral abdominal muscles in the avian dmbryo—An experimental and ultrastructural-study. *Anatomy and Embryology, 166*, 87–101.

Cinnamon, Y., Kahane, N., & Kalcheim, C. (1999). Characterization of the early development of specific hypaxial muscles from the ventrolateral myotome. *Development, 126*, 4305–4315.

Cobourne, M. T., Iseki, S., Birjandi, A. A., Adel Al-Lami, H., Thauvin-Robinet, C., Xavier, G. M., et al. (2018). How to make a tongue: Cellular and molecular regulation of muscle and connective tissue formation during mammalian tongue development. *Seminars in Cell & Developmental Biology*, pii: S1084-9521(17)30147-7, https://www.ncbi.nlm.nih.gov/pubmed/29784581.

Colasanto, M. P., Eyal, S., Mohassel, P., Bamshad, M., Bonnemann, C. G., Zelzer, E., et al. (2016). Development of a subset of forelimb muscles and their attachment sites requires the ulnar-mammary syndrome gene Tbx3. *Disease Models & Mechanisms, 9*, 1257–1269.

Couly, G. F., Coltey, P. M., & Le Douarin, N. M. (1993). The triple origin of skull in higher vertebrates: A study in quail-chick chimeras. *Development, 117*, 409–429.

Denetclaw, W. F., Christ, B., & Ordahl, C. P. (1997). Location and growth of epaxial myotome precursor cells. *Development, 124*, 1601–1610.

Deries, M., Gonçalves, A. B., Vaz, R., Martins, G. G., Rodrigues, G., & Thorsteinsdóttir, S. (2012). Extracellular matrix remodeling accompanies axial muscle development and morphogenesis in the mouse. *Developmental Dynamics, 241*, 350–364.

Deries, M., Schweitzer, R., & Duxson, M. J. (2010). Developmental fate of the mammalian myotome. *Developmental Dynamics, 239*, 2898–2910.

Dietrich, S., Abou-Rebyeh, F., Brohmann, H., Bladt, F., Sonnenberg-Riethmacher, E., Yamaai, T., et al. (1999). The role of SF/HGF and c-Met in the development of skeletal muscle. *Development, 126*, 1621–1629.

Diogo, R., Abdala, V., Lonergan, N., & Wood, B. A. (2008). From fish to modern humans—Comparative anatomy, homologies and evolution of the head and neck musculature. *Journal of Anatomy, 213*, 391–424.

Diogo, R., Kelly, R. G., Christiaen, L., Levine, M., Ziermann, J. M., Molnar, J. L., et al. (2015). A new heart for a new head in vertebrate cardiopharyngeal evolution. *Nature, 520*, 466–473.

Domyan, E. T., Kronenberg, Z., Infante, C. R., Vickrey, A. I., Stringham, S. A., Bruders, R., et al. (2016). Molecular shifts in limb identity underlie development of feathered feet in two domestic avian species. *eLife, 5*, e12115.

Durland, J. L., Sferlazzo, M., Logan, M., & Burke, A. C. (2008). Visualizing the lateral somitic frontier in the Prx1Cre transgenic mouse. *Journal of Anatomy, 212*, 590–602.

Edgeworth, F. H. (1935). *The cranial muscles of vertebrates*. London: Cambridge University Press.

Eng, D., Ma, H.-Y., Xu, J., Shih, H. P., Gross, M. K., & Kiouss, C. (2012). Loss of abdominal muscle in Pitx2 mutants associated with altered axial specification of lateral plate mesoderm. *PLoS One*, 7, e42228.

Engleka, K. A., Gitler, A. D., Zhang, M., Zhou, D. D., High, F. A., & Epstein, J. A. (2005). Insertion of Cre into the Pax3 locus creates a new allele of Splotch and identifies unexpected Pax3 derivatives. *Developmental Biology, 280*, 396–406.

Ericsson, R., Cerny, R., Falck, P., & Olsson, L. (2004). Role of cranial neural crest cells in visceral arch muscle positioning and morphogenesis in the Mexican axolotl, *Ambystoma mexicanum*. *Developmental Dynamics, 231*, 237–247.

Evans, D. J. R. (2003). Contribution of somitic cells to the avian ribs. *Developmental Biology, 256*, 115–127.

Evans, A. L., & Gage, P. J. (2005). Expression of the homeobox gene Pitx2 in neural crest is required for optic stalk and ocular anterior segment development. *Human Molecular Genetics*, *14*, 3347–3359.

Farmer, C. G. (2015). The evolution of unidirectional pulmonary airflow. *Physiology (Bethesda)*, *30*, 260–272.

Fetcho, J. R. (1987). A review of the organization and evolution of motoneurons innervating the axial musculature of vertebrates. *Brain Research*, *434*, 243–280.

Friel, J. P., & Wainwright, P. C. (1997). A model system of structural duplication: Homologies of adductor mandibulae muscles in tetraodontiform fishes. *Systematic Biology*, *46*, 441–463.

Gage, P. J., Rhoades, W., Prucka, S. K., & Hjalt, T. (2005). Fate maps of neural crest and mesoderm in the mammalian eye. *Investigative Ophthalmology & Visual Science*, *46*, 4200–4208.

Gage, P. J., Suh, H. Y., & Camper, S. A. (1999). Dosage requirement of Pitx2 for development of multiple organs. *Development*, *126*, 4643–4651.

Goodrich, E. S. (1958). *Studies on the structure and development of vertebrates*. New York: Dover publications.

Gorlin, R. J., Cohen, M. M., & Blackburn, L. D. (1990). *Syndromes of the head and neck* (3rd ed.). New York: Oxford University Press.

Grenier, J., Teillet, M.-A., Grifone, R., Kelly, R. G., & Duprez, D. (2009). Relationship between neural crest cells and cranial mesoderm during head muscle development. *PLoS One*, *4*, e4381.

Grim, M., & Wachtler, F. (1991). Muscle morphogenesis in the absence of myogenic cells. *Anatomy and Embryology*, *183*, 67–70.

Gros, J., Scaal, M., & Marcelle, C. (2004). A two-step mechanism for myotome formation in chick. *Developmental Cell*, *6*, 875–882.

Haas, A. (2001). Mandibular arch musculature of anuran tadpoles, with comments on homologies of amphibian jaw muscles. *Journal of Morphology*, *247*, 1–33.

Hacker, A., & Guthrie, S. (1998). A distinct developmental programme for the cranial paraxial mesoderm in the chick embryo. *Development*, *125*, 3461–3472.

Han, A., Zhao, H., Li, J., Pelikan, R., & Chai, Y. (2014). ALK5-mediated transforming growth factor β signaling in neural crest cells controls craniofacial muscle development via tissue-tissue interactions. *Molecular and Cellular Biology*, *34*, 3120–3131.

Han, D., Zhao, H., Parada, C., Hacia, J. G., Bringas, P., & Chai, Y. (2012). A TGFβ-Smad4-Fgf6 signaling cascade controls myogenic differentiation and myoblast fusion during tongue development. *Development*, *139*, 1640–1650.

Harel, I., Maezawa, Y., Avraham, R., Rinon, A., Ma, H.-Y., Cross, J. W., et al. (2012). Pharyngeal mesoderm regulatory network controls cardiac and head muscle morphogenesis. *Proceedings of the National Academy of Sciences of the United States of America*, *109*, 18839–18844.

Hasson, P., Del Buono, J., & Logan, M. P. O. (2007). Tbx5 is dispensable for forelimb outgrowth. *Development*, *134*, 85–92.

Hasson, P., DeLaurier, A., Bennett, M., Grigorieva, E., Naiche, L. A., Papaioannou, V. E., et al. (2010). Tbx4 and tbx5 acting in connective tissue are required for limb muscle and tendon patterning. *Developmental Cell*, *18*, 148–156.

Helmbacher, F. (2018). Tissue-specific activities of the Fat1 cadherin cooperate to control neuromuscular morphogenesis. *PLoS Biology*, *16*.

Heude, E., Bouhali, K., Kurihara, Y., Kurihara, H., Couly, G., Janvier, P., et al. (2010). Jaw muscularization requires Dlx expression by cranial neural crest cells. *Proceedings of the National Academy of Sciences of the United States of America*, *107*, 11441–11446.

Heude, E., Rivals, I., Couly, G., & Levi, G. (2011). Masticatory muscle defects in hemifacial microsomia: A new embryological concept. *American Journal of Medical Genetics. Part A*, *155A*, 1991–1995.

Heude, E., Tesarova, M., Sefton, E. M., Jullian, E., Adachi, N., Grimaldi, A., et al. (2018). Unique morphogenetic signatures define mammalian neck muscles and associated connective tissues. *eLife*, *7*, e40179.
Heymann, S., Koudrova, M., Arnold, H., Köster, M., & Braun, T. (1996). Regulation and function of SF/HGF during migration of limb muscle precursor cells in chicken. *Developmental Biology*, *180*, 566–578.
Hirasawa, T., Fujimoto, S., & Kuratani, S. (2015). Expansion of the neck reconstituted the shoulder-diaphragm in amniote evolution. *Development, Growth & Differentiation*, *58*, 143–153.
Hirasawa, T., & Kuratani, S. (2013). A new scenario of the evolutionary derivation of the mammalian diaphragm from shoulder muscles. *Journal of Anatomy*, *222*, 504–517.
Hitachi, K., Kondow, A., Danno, H., Nishimura, Y., Okabayashi, K., & Asashima, M. (2009). Molecular analyses of Xenopus laevis Mesp-related genes. *Integrative Zoology*, *4*, 387–394.
Holland, L. Z. (1996). Muscle development in amphioxus: Morphology, biochemistry, and molecular biology. *Israel Journal of Zoology*, *42*(Suppl. 1), S235–S246.
Holliday, C. M., & Witmer, L. M. (2007). Archosaur adductor chamber evolution: Integration of musculoskeletal and topological criteria in jaw muscle homology. *Journal of Morphology*, *268*, 457–484.
Hollway, G., & Currie, P. (2005). Vertebrate myotome development. *Birth Defects Research. Part C, Embryo Today*, *75*, 172–179.
Horigome, N., Myojin, M., Ueki, T., Hirano, S., Aizawa, S., & Kuratani, S. (1999). Development of cephalic neural crest cells in embryos of Lampetra japonica, with special reference to the evolution of the jaw. *Developmental Biology*, *207*, 287–308.
Hörstadius, S., & Sellman, S. (1946). Experimentelle Untersuchungen uber die Determination des knorpeligen Kopfskelettes bei Urodelen. *Nova acta Regiae Societatis Scientiarum Upsaliensis Series IV*, *13*, 1–170.
Hosokawa, R., Oka, K., Yamaza, T., Iwata, J., Urata, M., Xu, X., et al. (2010). TGF-beta mediated FGF10 signaling in cranial neural crest cells controls development of myogenic progenitor cells through tissue-tissue interactions during tongue morphogenesis. *Developmental Biology*, *341*, 186–195.
Hutcheson, D. A., Zhao, J., Merrell, A., Haldar, M., & Kardon, G. (2009). Embryonic and fetal limb myogenic cells are derived from developmentally distinct progenitors and have different requirements for beta-catenin. *Genes & Development*, *23*, 997–1013.
Iwata, J., Suzuki, A., Pelikan, R. C., Ho, T. V., & Chai, Y. (2013). Noncanonical transforming growth factor beta (TGFbeta) signaling in cranial neural crest cells causes tongue muscle developmental defects. *The Journal of Biological Chemistry*, *288*, 29760–29770.
Jacob, H. J., & Christ, B. (1980). On the formation of muscle pattern in the chick limb. In H.-J. Merker, H. Nau, & D. Neubert (Eds.), *Teratology of the Limbs* (pp. 89–97). Berlin: Walter de Gruyter and Co.
Joe, A. W., Yi, L., Natarajan, A., Le Grand, F., So, L., Wang, J., et al. (2010). Muscle injury activates resident fibro/adipogenic progenitors that facilitate myogenesis. *Nature Cell Biology*, *12*, 153–163.
Kalcheim, C., Cinnamon, Y., & Kahane, N. (1999). Myotome formation: A multistage process. *Cell and Tissue Research*, *296*, 161–173.
Kardon, G. (1998). Muscle and tendon morphogenesis in the avian hind limb. *Development*, *125*, 4019–4032.
Kardon, G., Ackerman, K. G., McCulley, D. J., Shen, Y., Wynn, J., Shang, L., et al. (2017). Congenital diaphragmatic hernias: From genes to mechanisms to therapies. *Disease Models & Mechanisms*, *10*, 955–970.
Kardon, G., Campbell, J. K., & Tabin, C. J. (2002). Local extrinsic signals determine muscle and endothelial cell fate and patterning in the vertebrate limb. *Developmental Cell*, *3*, 533–545.

Kardon, G., Harfe, B. D., & Tabin, C. J. (2003). A Tcf4-positive mesodermal population provides a prepattern for vertebrate limb muscle patterning. *Developmental Cell*, *5*, 937–944.

Keller, C., Hansen, M. S., Coffin, C. M., & Capecchi, M. R. (2004). Pax3:Fkhr interferes with embryonic Pax3 and Pax7 function: Implications for alveolar rhabdomyosarcoma cell of origin. *Genes & Development*, *18*, 2608–2613.

Kitamura, K., Miura, H., Miyagawa-Tomita, S., Yanazawa, M., Katoh-Fukui, Y., Suzuki, R., et al. (1999). Mouse Pitx2 deficiency leads to anomalies of the ventral body wall, heart, extra- and periocular mesoderm and right pulmonary isomerism. *Development*, *126*, 5749–5758.

Lally, K. P. (2016). Congenital diaphragmatic hernia—The past 25 (or so) years. *Journal of Pediatric Surgery*, *51*, 695–698.

Lance-Jones, C. (1988). The somitic level of origin of embryonic chick hindlimb muscles. *Developmental Biology*, *126*, 394–407.

Landacre, F. (1921). The fate of the neural crest in the head of the urodeles. *Journal of Comparative Neurology*, *33*, 1–43.

Lanser, M. E., & Fallon, J. F. (1987). Development of wing-bud-derived muscles in normal and wingless chick embryos: A computer-assisted three-dimensional reconstruction study of muscle pattern formation in the absence of skeletal elements. *The Anatomical Record*, *217*, 61–78.

Le Lièvre, C. S., & Le Douarin, N. M. (1975). Mesenchymal derivatives of the neural crest: Analysis of chimaeric quail and chick embryos. *Journal of Embryology and Experimental Morphology*, *34*, 125–154.

Lepper, C., Partridge, T. A., & Fan, C. M. (2011). An absolute requirement for Pax7-positive satellite cells in acute injury-induced skeletal muscle regeneration. *Development*, *138*, 3639–3646.

Lescroart, F., Hamou, W., Francou, A., Théveniau-Ruissy, M., Kelly, R. G., & Buckingham, M. (2015). Clonal analysis reveals a common origin between nonsomite-derived neck muscles and heart myocardium. *Proceedings of the National Academy of Sciences of the United States of America*, *112*, 1446–1451.

Lescroart, F., Kelly, R. G., Le Garrec, J.-F., Nicolas, J.-F., Meilhac, S. M., & Buckingham, M. (2010). Clonal analysis reveals common lineage relationships between head muscles and second heart field derivatives in the mouse embryo. *Development*, *137*, 3269–3279.

Li, Q. Y., Newbury-Ecob, R. A., Terrett, J. A., Wilson, D. I., Curtis, A. R., Yi, C. H., et al. (1997). Holt-Oram syndrome is caused by mutations in TBX5, a member of the Brachyury (T) gene family. *Nature Genetics*, *15*, 21–29.

Lines, M. A., Kozlowski, K., Kulak, S. C., Allingham, R. R., Heon, E., Ritch, R., et al. (2004). Characterization and prevalence of PITX2 microdeletions and mutations in Axenfeld-Rieger malformations. *Investigative Ophthalmology & Visual Science*, *45*, 828.

Lipton, B. H. (1977). Collagen synthesis by normal and bromodeoxyuridine-modulated cells in myogenic culture. *Developmental Biology*, *61*, 153–165.

Logan, M., Martin, J. F., Nagy, A., Lobe, C., Olson, E. N., & Tabin, C. J. (2002). Expression of Cre recombinase in the developing mouse limb bud driven by a Prxl enhancer. *Genesis*, *33*, 77–80.

Lu, M. F., Pressman, C., Dyer, R., Johnson, R. L., & Martin, J. F. (1999). Function of Rieger syndrome gene in left-right asymmetry and craniofacial development. *Nature*, *401*, 276–278.

Mathew, S. J., Hansen, J. M., Merrell, A. J., Murphy, M. M., Lawson, J. A., Hutcheson, D. A., et al. (2011). Connective tissue fibroblasts and Tcf4 regulate myogenesis. *Development*, *138*, 371–384.

Matsuoka, T., Ahlberg, P. E., Kessaris, N., Iannarelli, P., Dennehy, U., Richardson, W. D., et al. (2005). Neural crest origins of the neck and shoulder. *Nature*, *436*, 347–355.

Merrell, A. J., Ellis, B. J., Fox, Z. D., Lawson, J. A., Weiss, J. A., & Kardon, G. (2015). Muscle connective tissue controls development of the diaphragm and is a source of congenital diaphragmatic hernias. *Nature Genetics*, *47*, 496–504.

Merrell, A. J., & Kardon, G. (2013). Development of the diaphragm—A skeletal muscle essential for mammalian respiration. *The FEBS Journal*, *280*, 4026–4035.

Meyer-Marcotty, P., Weisschuh, N., Dressler, P., Hartmann, J., & Stellzig-Eisenhauer, A. (2008). Morphology of the sella turcica in Axenfeld-Rieger syndrome with PITX2 mutation. *Journal of Oral Pathology & Medicine*, *37*, 504–510.

Michailovici, I., Eigler, T., & Tzahor, E. (2015). Craniofacial muscle development. *Current Topics in Developmental Biology*, *115*, 3–30.

Millington, G., Elliott, K. H., Chang, Y.-T., Chang, C.-F., Dlugosz, A., & Brugmann, S. A. (2017). Cilia-dependent GLI processing in neural crest cells is required for tongue development. *Developmental Biology*, *424*, 124–137.

Murphy, M., & Kardon, G. (2011). Origin of vertebrate limb muscle: The role of progenitor and myoblast populations. *Current Topics in Developmental Biology*, *96*, 1–32.

Murphy, M. M., Lawson, J. A., Mathew, S. J., Hutcheson, D. A., & Kardon, G. (2011). Satellite cells, connective tissue fibroblasts and their interactions are crucial for muscle regeneration. *Development*, *138*, 3625–3637.

Naiche, L. A., & Papaioannou, V. E. (2007). Tbx4 is not required for hindlimb identity or post-bud hindlimb outgrowth. *Development*, *134*, 93–103.

Nassari, S., Duprez, D., & Fournier-Thibault, C. (2017). Non-myogenic contribution to muscle development and homeostasis: The role of connective tissues. *Frontiers in Cell and Development Biology*, *5*, 22.

Newbury-Ecob, R. A., Leanage, R., Raeburn, J. A., & Young, I. D. (1996). Holt-Oram syndrome: A clinical genetic study. *Journal of Medical Genetics*, *33*, 300–307.

Noden, D. M. (1975). Analysis of migratory behavior of avian cephalic neural crest cells. *Developmental Biology*, *42*, 106–130.

Noden, D. M. (1983a). The embryonic origins of avian cephalic and cervical muscles and associated connective tissues. *The American Journal of Anatomy*, *168*, 257–276.

Noden, D. M. (1983b). The role of the neural crest in patterning of avian cranial skeletal, connective, and muscle tissues. *Developmental Biology*, *96*, 144–165.

Noden, D. M., & Francis-West, P. (2006). The differentiation and morphogenesis of craniofacial muscles. *Developmental Dynamics*, *235*, 1194–1218.

Noden, D. M., & Trainor, P. A. (2005). Relations and interactions between cranial mesoderm and neural crest populations. *Journal of Anatomy*, *207*, 575–601.

Nowicki, J., & Burke, A. (2000). Hox genes and morphological identity: Axial versus lateral patterning in the vertebrate mesoderm. *Development*, *127*, 4265–4275.

Nowicki, J. L., Takimoto, R., & Burke, A. C. (2003). The lateral somitic frontier: Dorso-ventral aspects of anterio-posterior regionalization in avian embryos. *Mechanisms of Development*, *120*, 227–240.

Olson, L. E., & Soriano, P. (2009). Increased PDGFRalpha activation disrupts connective tissue development and drives systemic fibrosis. *Developmental Cell*, *16*, 303–313.

Olsson, L., Falck, P., Lopez, K., Cobb, J., & Hanken, J. (2001). Cranial neural crest cells contribute to connective tissue in cranial muscles in the anuran amphibian, *Bombina orientalis*. *Developmental Biology*, *237*, 354–367.

Ordahl, C. P., & Le Douarin, N. M. (1992). Two myogenic lineages within the developing somite. *Development*, *114*, 339–353.

Pan, B., Grünewald, B., Nguyen, T., Farah, M., Polydefkis, M., McDonald, J., et al. (2012). The lateral thoracic nerve and the cutaneous maximus muscle—A novel in vivo model system for nerve degeneration and regeneration studies. *Experimental Neurology*, *236*, 6–18.

Paris, N. D., Coles, G. L., & Ackerman, K. G. (2015). *Wt1* and *β-catenin* cooperatively regulate diaphragm development in the mouse. *Developmental Biology*, *407*, 40–56.
Park, S. W., Kim, H. G., Heo, H., & Park, Y. G. (2009). Anomalous scleral insertion of superior oblique in Axenfeld-Rieger syndrome. *Korean Journal of Ophthalmology*, *23*, 62–64.
Pearse, R. V., 2nd, Scherz, P. J., Campbell, J. K., & Tabin, C. J. (2007). A cellular lineage analysis of the chick limb bud. *Developmental Biology*, *310*, 388–400.
Perry, S. F., Similowski, T., Klein, W., & Codd, J. R. (2010). The evolutionary origin of the mammalian diaphragm. *Respiratory Physiology & Neurobiology*, *171*, 1–16.
Piekarski, N., Gross, J. B., & Hanken, J. (2014). Evolutionary innovation and conservation in the embryonic derivation of the vertebrate skull. *Nature Communications*, *5*, 5661.
Piekarski, N., & Olsson, L. (2007). Muscular derivatives of the cranialmost somites revealed by long-term fate mapping in the Mexican axolotl (*Ambystoma mexicanum*). *Evolution & Development*, *9*, 566–578.
Pu, Q., Abduelmula, A., Masyuk, M., Theiss, C., Schwandulla, D., Hans, M., et al. (2013). The dermomyotome ventrolateral lip is essential for the hypaxial myotome formation. *BMC Developmental Biology*, *13*, 37.
Relaix, F., Rocancourt, D., Mansouri, A., & Buckingham, M. (2005). A Pax3/Pax7-dependent population of skeletal muscle progenitor cells. *Nature*, *435*, 948–953.
Rinon, A., Lazar, S., Marshall, H., Büchmann-Møller, S., Neufeld, A., Elhanany-Tamir, H., et al. (2007). Cranial neural crest cells regulate head muscle patterning and differentiation during vertebrate embryogenesis. *Development*, *134*, 3065–3075.
Romer, A. S., & Parsons, T. S. (1986). *The vertebrate body*. Fort Worth: Saunders College Publishing.
Saberi, M., Pu, Q., Valasek, P., Norizadeh-Abbariki, T., Patel, K., & Huang, R. (2017). The hypaxial origin of the epaxially located rhomboid muscles. *Annals of Anatomy*, *214*, 15–20.
Sambasivan, R., Kuratani, S., & Tajbakhsh, S. (2011). An eye on the head: The development and evolution of craniofacial muscles. *Development*, *138*, 2401–2415.
Sampath, S. C., Sampath, S. C., & Millay, D. P. (2018). Myoblast fusion confusion: The resolution begins. *Skeletal Muscle*, *8*, 3.
Sanes, J. (2004). The extracellular matrix. In A. Engel & C. Franzini-Armstrong (Eds.), *Myology* (pp. 471–487). New York: McGraw-Hill.
Scaal, M. (2016). Early development of the vertebral column. *Seminars in Cell & Developmental Biology*, *49*, 83–91.
Schiaffino, S., Sandri, M., & Murgia, M. (2007). Activity-dependent signaling pathways controlling muscle diversity and plasticity. *Physiology (Bethesda)*, *22*, 269–278.
Schilling, N., & Carrier, D. R. (2010). Function of the epaxial muscles in walking, trotting and galloping dogs: Implications for the evolution of epaxial muscle function in tetrapods. *The Journal of Experimental Biology*, *213*, 1490–1502.
Sefton, E. M., Bhullar, B., Mohaddes, Z., & Hanken, J. (2016). Evolution of the head-trunk interface in tetrapod vertebrates. *eLife*, *5*, e09972.
Sefton, E. M., Gallardo, M., & Kardon, G. (2018). Developmental origin and morphogenesis of the diaphragm, an essential mammalian muscle. *Developmental Biology*, *440*, 64–73.
Semina, E. V., Reiter, R., Leysens, N. J., Alward, W., Small, K. W., Datson, N. A., et al. (1996). Cloning and characterization of a novel bicoid-related homeobox transcription factor gene, RIEG, involved in Rieger syndrome. *Nature Genetics*, *14*, 392–399.
Stricker, S., Brieske, N., Haupt, J., & Mundlos, S. (2006). Comparative expression pattern of Odd-skipped related genes Osr1 and Osr2 in chick embryonic development. *Gene Expression Patterns*, *6*, 826–834.

Stumm, J., Vallecillo-Garcia, P., Vom Hofe-Schneider, S., Ollitrault, D., Schrewe, H., Economides, A. N., et al. (2018). Odd skipped-related 1 (Osr1) identifies muscle-interstitial fibro-adipogenic progenitors (FAPs) activated by acute injury. *Stem Cell Research*, *32*, 8–16.

Swartz, M. E., Eberhart, J., Pasquale, E. B., & Krull, C. E. (2001). EphA4/ephrin-A5 interactions in muscle precursor cell migration in the avian forelimb. *Development*, *128*, 4669–4680.

Swartz, M. E., Nguyen, V., Mccarthy, N. Q., & Eberhart, J. K. (2012). Hh signaling regulates patterning and morphogenesis of the pharyngeal arch-derived skeleton. *Developmental Biology*, *369*, 65–75.

Swinehart, I. T., Schlientz, A. J., Quintanilla, C. A., Mortlock, D. P., & Wellik, D. M. (2013). Hox11 genes are required for regional patterning and integration of muscle, tendon and bone. *Development*, *140*, 4574–4582.

Theis, S., Patel, K., Valasek, P., Otto, A., Pu, Q., Harel, I., et al. (2010). The occipital lateral plate mesoderm is a novel source for vertebrate neck musculature. *Development*, *137*, 2961–2971.

Tokita, M., Nakayama, T., Schneider, R. A., & Agata, K. (2013). Molecular and cellular changes associated with the evolution of novel jaw muscles in parrots. *Proceedings of the Biological Sciences*, *280*, 20122319.

Tokita, M., & Schneider, R. A. (2009). Developmental origins of species-specific muscle pattern. *Developmental Biology*, *331*, 311–325.

Tomasek, J. J., Gabbiani, G., Hinz, B., Chaponnier, C., & Brown, R. A. (2002). Myofibroblasts and mechano-regulation of connective tissue remodelling. *Nature Reviews. Molecular Cell Biology*, *3*, 349–363.

Trainor, P. A. (2010). Craniofacial birth defects: The role of neural crest cells in the etiology and pathogenesis of Treacher Collins syndrome and the potential for prevention. *American Journal of Medical Genetics. Part A*, *152A*, 2984–2994.

Trainor, P. A., Tan, S. S., & Tam, P. P. (1994). Cranial paraxial mesoderm: Regionalisation of cell fate and impact on craniofacial development in mouse embryos. *Development*, *120*, 2397–2408.

Tzahor, E. (2009). Heart and craniofacial muscle development: A new developmental theme of distinct myogenic fields. *Developmental Biology*, *327*, 273–279.

Tzahor, E., Kempf, H., Mootoosamy, R. C., Poon, A. C., Abzhanov, A., Tabin, C. J., et al. (2003). Antagonists of Wnt and BMP signaling promote the formation of vertebrate head muscle. *Genes & Development*, *17*, 3087–3099.

Uezumi, A., Fukada, S., Yamamoto, N., Takeda, S., & Tsuchida, K. (2010). Mesenchymal progenitors distinct from satellite cells contribute to ectopic fat cell formation in skeletal muscle. *Nature Cell Biology*, *12*, 143–152.

Vallecillo-García, P., Orgeur, M., Vom Hofe-Schneider, S., Stumm, J., Kappert, V., Ibrahim, D. M., et al. (2017). Odd skipped-related 1 identifies a population of embryonic fibro-adipogenic progenitors regulating myogenesis during limb development. *Nature Communications*, *8*, 1218.

Vasyutina, E., Stebler, J., Brand-Saberi, B., Schulz, S., Raz, E., & Birchmeier, C. (2005). CXCR4 and Gab1 cooperate to control the development of migrating muscle progenitor cells. *Genes & Development*, *19*, 2187–2198.

Venters, S. J., & Ordahl, C. P. (2002). Persistent myogenic capacity of the dermomyotome dorsomedial lip and restriction of myogenic competence. *Development*, *129*, 3873–3885.

von Scheven, G., Alvares, L., Mootoosamy, R., & Dietrich, S. (2006). Neural tube derived signals and Fgf8 act antagonistically to specify eye versus mandibular arch muscles. *Development*, *133*, 2731–2745.

Wagner, G. (1959). Untersuchungen an *Bombinator-Triton*-chimaeren: Das skelett larvaler *Triton*-köpfe mit *Bombinator*-Mesektoderm. *Wilhelm Roux' Archiv für Entwicklungsmechanik der Organismen*, *151*, 136–158.

Webster, E. L., Hudson, P. E., & Channon, S. B. (2014). Comparative functional anatomy of the epaxial musculature of dogs (*Canis familiaris*) bred for sprinting vs. fighting. *Journal of Anatomy*, *225*, 317–327.

Wosczyna, M. N., & Rando, T. A. (2018). A muscle stem cell support group: Coordinated cellular responses in muscle regeneration. *Developmental Cell*, *46*, 135–143.

Wotton, K. R., Schubert, F. R., & Dietrich, S. (2015). Hypaxial muscle: Controversial classification and controversial data? *Results and Problems in Cell Differentiation*, *56*, 25–48.

Zacharias, A. L., Lewandoski, M., Rudnicki, M. A., & Gage, P. J. (2011). *Pitx2* is an upstream activator of extraocular myogenesis and survival. *Developmental Biology*, *349*, 395–405.

Ziermann, J. M., Diogo, R., & Noden, D. M. (2018). Neural crest and the patterning of vertebrate craniofacial muscles. *Genesis*, *56*, e23097.

Zou, Y., Zhang, R.-Z., Sabatelli, P., Chu, M.-L., & Bonnemann, C. G. (2008). Muscle interstitial fibroblasts are the main source of collagen VI synthesis in skeletal muscle: Implications for congenital muscular dystrophy types Ullrich and Bethlem. *Journal of Neuropathology and Experimental Neurology*, *67*, 144–154.

CHAPTER SIX

Vascularizing organogenesis: Lessons from developmental biology and implications for regenerative medicine

Edward Daniel, Ondine Cleaver*
Department of Molecular Biology and Center for Regenerative Science and Medicine, University of Texas Southwestern Medical Center, Dallas, TX, United States
*Corresponding author: e-mail address: ondine.cleaver@utsouthwestern.edu

Contents

Abstract

Organogenesis requires tightly coordinated and patterned growth of numerous cell types to form a fully mature and vascularized organ. Endothelial cells (ECs) that line blood vessels develop alongside the growing organ, but only recently has their role in directing epithelial and stromal growth been appreciated. Endothelial, epithelial, and stromal cells in embryonic organs actively communicate with one another throughout development to ensure that the organ forms appropriately. What signals tell blood vessel progenitors where to go? How are tissues influenced by the vasculature that pervades it? In this chapter, we review the ways in which crosstalk between ECs and

Current Topics in Developmental Biology, Volume 132
ISSN 0070-2153
https://doi.org/10.1016/bs.ctdb.2018.12.012

epithelial or stromal cells during development leads to a fully patterned pancreas, lung, or kidney. ECs in all of these organs are necessary for proper epithelial and stromal growth, but how they direct this process is organ- and time-specific, highlighting the concept of dynamic EC heterogeneity. We end with a discussion on how understanding cell-cell crosstalk during development can be applied therapeutically through the generation of transplantable miniature organ-like tissues called "organoids." We will discuss the current state of organoid technology and highlight the major challenges in forming a properly patterned vascular network that will be critical in transforming them into a viable therapeutic option.

1. Introduction

Blood vessels are the conduits for life. Through them, organ systems across the body communicate to one another through the exchange of nutrients, gas, and signaling factors. Over the last several years, there has been a growing appreciation of the role blood vessels play in paracrine signaling within developing organs. The endothelial cells (ECs) lining blood vessels actively communicate with surrounding stromal (or mesenchymal) and epithelial tissue to ensure proper patterning of both the endothelium and other tissues. In fact, endothelial-derived signals are required for proper development and patterning in organogenesis. One of the main challenges facing vascular biologists today is understanding both what these signals are and how they regulate organ development.

Where do vascular progenitors come from? How do they become patterned to accommodate growing organs? Do they all have the same potential and plasticity? These are all questions that the field of vascular developmental biology is eager to answer. Some general principles have emerged. Among them, the concept that angiogenic factors are expressed by non-vascular cells to recruit and maintain forming blood vessels. In addition, it is now well accepted that blood vessels are not equivalent in all tissues, but rather they adapt to local needs and conform to distinct environments. What is less clear is how ECs in turn influence the tissues they perfuse. Are there identifiable non-nutritional signals? Most cells within tissues are understood to communicate with each other, but what the language of that conversation says, remains a significant mystery to decode. Experimental data, often contradictory and spatiotemporally dependent, remain a challenge to interpret.

Many recent efforts have been aimed at generating tissues or organs *in vitro* to replace those that are diseased or defective. Successful clinical application has been of limited success as only rudimental vascularization has been achieved. Whereas these efforts have focused extensively on improving epithelial development in these *ex vivo* organs, there has been limited success to promote proper vascularization. This is, in part, due to the lack in understanding how organs are normally vascularized during development. A better understanding of the coordinated growth of tissues during embryonic development, the interface of blood vessels with different cells within these tissues, and the patterning, remodeling, and heterogeneity of ECs that form these vessels will therefore be critical to develop fully vascularized and therapeutically viable tissues for transplant.

In this chapter, we will focus on three important aspects of blood vessel formation in organs: (1) how the vasculature remodels and organizes within developing tissues; (2) establishment of vascular heterogeneity, as vessels differentiate and display organ specificity and regionalization within tissues; and (3) ways in which the vasculature and surrounding tissues communicate to one another to direct development. We will examine these themes of vascular biology in the context of three vital organs, the pancreas, lung, and kidney with a special focus on the kidney as greater relevant literature is available (Fig. 1). Lastly, we will highlight the current efforts to apply the principles guiding development toward generating functional kidney tissue *ex vivo* and discuss challenges related to vascularization of these *ex vivo* tissues.

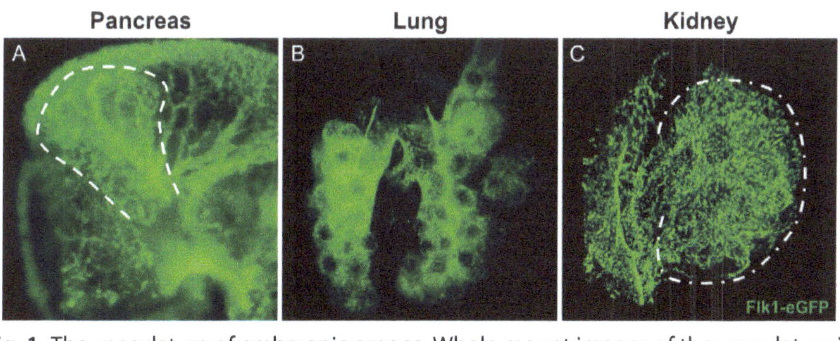

Fig. 1 The vasculature of embryonic organs. Whole mount images of the vasculature in E12.5 Flk1-eGFP^{+} pancreas (A, outline), lung (B), and kidney (C, outline). Flk1-eGFP is expressed exclusively in the endothelium throughout development.

2. EC development, heterogeneity, and the challenges we face

2.1 Vasculogenesis and angiogenesis

Blood vessels grow *via* two main mechanisms: *vasculogenesis* and *angiogenesis*. *Vasculogenesis* is the process by which vessels assemble *de novo* from individual endothelial precursors termed "angioblasts." These angioblasts coalesce to form a vascular cord, polarize, and open a lumen, differentiating into mature ECs to form a functional blood vessel (Azizoglu & Cleaver, 2016). Conversely, *angiogenesis* is the process by which new vessels grow from pre-existing vessels. ECs in these vessels respond to local cues, undergo morphologic changes, and sprout and generate a new lumenized vessel (Azizoglu & Cleaver, 2016). Whether the vessels in an organ form through a vasculogenic or angiogenic process is entirely organ-specific. For example, the early vessels of the yolk sac and dorsal aortae form through vasculogenesis, while the retinal vasculature forms primarily through an angiogenic process (Garcia & Larina, 2014; Siemerink, Klaassen, Van Noorden, & Schlingemann, 2013; Strilic et al., 2009). Some vascular beds, such as those in the lung, form through both angiogenesis and vasculogenesis (Herriges & Morrisey, 2014). Through these processes, vessels extend throughout the growing embryo, coalesce, and remodel until they yield a fully functional and hierarchical vascular network.

2.2 Endothelial heterogeneity

A fascinating aspect of blood vessels is how different they are in every tissue: some vessels contain small pores called fenestrae, some vessels form strict continuous barriers, and some form highly discontinuous lumens or sinusoids. In fact, at times it seems the only unifying characteristic is that they carry blood! The concept of endothelial heterogeneity is underscored by organ-to-organ variations in how blood vessels develop. ECs are highly adapted to their specific local environment. The fact that ECs display stark heterogeneity based on their local environment has been known for some time (Aird, 2003, 2007a, 2007b). On a gross scale, the vascular system can be broken down into three main components: high pressure arteries that transport blood away from the heart, low pressure veins that bring blood back to the heart, and capillaries where nutrient and gas exchange occurs. ECs in each of these regions possess unique characteristics that facilitate their functions. Arteries and veins can be distinguished from each other by histology and by gene expression profiles (Chi et al., 2003; Chong, Koo, Xu, Fu, & Cleaver, 2011; dela Paz & D'Amore, 2009). In fact, arteriovenous differentiation, the process by which some early ECs commit toward an arterial or venous fate, is one of the first ways in which ECs begin to demonstrate this heterogeneity. Arteriovenous differentiation is evident as early as E8.0 when the first major vessels, the dorsal aortae and vitelline veins, show signs of formation and transcriptional differences (Chong et al., 2011).

ECs that compose capillaries also display heterogeneous morphology and gene expression dependent on their resident tissue. Capillaries can be classified as continuous, fenestrated, or sinusoidal/discontinuous based on morphology and vessel permeability (Augustin & Koh, 2017). The permeability of a capillary bed depends on the needs of the organ. For example, the capillaries supplying the central nervous system (CNS) possess a continuous endothelium that contributes to the maintenance of the blood-brain barrier (BBB), a highly restrictive interface that tightly regulates the ions, small solutes, nutrients, drugs, water, and other molecules that can be transferred from the plasma into the CNS in order to protect the brain from potentially harmful substances (Zhao, Nelson, Betsholtz, & Zlokovic, 2015). On the other hand, the sinusoids within the liver allow for the free flow of solutes in the blood into the space of Disse adjacent to hepatocytes (Augustin & Koh, 2017). Recently, transcriptional profiling of perfused adult ECs from several different organs—the kidney, liver, lung, heart, testes, brain, muscle, spleen, and bone marrow—and from human fetal kidney, liver, lung, and heart brought new insights into the degree of EC heterogeneity (Marcu et al., 2018; Nolan et al., 2013). ECs from each organ displayed significantly different transcriptomes and functional capabilities. EC heterogeneity has also been observed not only between different endothelial beds, but also at the cellular level within the same vessel. For example, the ECs that line the aorta demonstrate a mosaic and dynamic pattern of vWF expression (Yuan et al., 2016). Similar mosaicism has also been observed in the developing pancreatic arteries as well (Azizoglu et al., 2016). Taken together, these data clearly demonstrate that ECs display tissue-specific heterogeneity.

Although ECs are understood to be heterogeneous, there are still several challenges facing the field. First, it is unclear what upstream regulators coordinate the transcriptional programs that define ECs in each region. Are these programs dependent on EC intrinsic cues, extrinsic cues from the surrounding microenvironment, or a mixture of both? Next, when is this heterogeneity established during organogenesis? As will be discussed below, EC differentiation—including morphogenesis, arteriovenous specification, and mature capillary formation—is entirely organ-dependent. For example, the first evidence of arteriovenous differentiation in the pancreas is a single vessel present at E11.5, when connexin40 (Cx40) expression is seen in vessels undergoing coalescence (Azizoglu et al., 2016). In the kidney, arterial development begins at E13.5, possibly due to differences in when the organs begin to develop, and emerges as a hierarchical tree instead of a single vessel (Daniel et al., 2018). These questions must be answered to understand how organs form blood vessels adapted to suit their needs.

2.3 Organotypic blood vessels: Challenges facing the field

Currently, we lack some basic tools necessary to study organotypic variations between ECs and answer these organ-specific questions about the vasculature. The current genetic tools available to study endothelial development rely on broadly expressed endothelial genes, namely, VE-Cadherin or Tie2. While powerful means to study EC biology, there are important limitations to these systems if we want to answer questions in an organ-specific manner. First, deletion of a gene in all ECs can cause off-target effects independent of the effect in a specific organ. Any gene that prevents ECs from forming a continuous, lumenized network will lead to vascular anomalies with far reaching effects and, possibly, early embryonic lethality (Ferrara et al., 1996; Xu et al., 2011). Additionally, transcriptional profiling of organ-specific ECs revealed that these genes, and other broadly expressed endothelial markers, vary in expression levels between organs (Nolan et al., 2013). Therefore, even the available pan-endothelial promoters used to make reporters or Cre lines expressed in "all" ECs may not be well suited to answer questions in all organs.

To overcome this challenge and generate both EC- and organ-specific mouse models, one method utilized two recombinase systems to specifically label coronary or brain ECs (Pu et al., 2018). In this method, a Dre recombinase was expressed in ECs using the Tie2 promoter. A second recombinase, a Cre-ERT2, was placed downstream of the Wt1 or Mfsd2a promoters, which are expressed in coronary or brain ECs, respectively, as well as other tissues. This Cre-ERT2 cassette contained Dre-specific Rox sites around the ERT2 cassette (CrexER). In ECs where Dre is expressed, the ERT2 domain will be eliminated and Cre will be able to translocate to the nucleus to undergo additional recombination events. Therefore, Cre-mediated excision will only occur in cell types that express both the Dre and Cre recombinases, providing a means to achieve both endothelial- and tissue-specific expression pattern. Although potentially useful in certain contexts, this approach would only be useful if lines that overlap in expression exist. In addition, it may also not prove suitable in tissues that lack a single tissue-specific promoter, like the kidney. It is also limited by the lack of suitable mouse lines with the CrexER transgene. Thus, the tools needed to answer many of the questions about organ-specific vascular biology need to be further developed.

3. Vascular morphogenesis in specific organ systems

3.1 Pancreas

3.1.1 Overview of pancreatic vascular development

The pancreas consists of two organs combined into one: an exocrine compartment that secretes digestive proteins into the duodenum and an endocrine compartment that releases hormones that regulate glucose homeostasis, including insulin and glucagon (Pan & Wright, 2011). Both of these lineages arise from the same progenitor population marked by *Pdx1* (Gu, Brown, & Melton, 2003). The first morphologic event of pancreatic development occurs with specification and thickening of $Pdx1^{+}$ pre-pancreatic epithelium on both dorsal and ventral sides of the foregut-midgut junction at E8.5–E9.0 (Villasenor, Chong, Henkemeyer, & Cleaver, 2010). The dorsal pancreatic epithelium lies in direct contact with the dorsal aortae, while the ventral epithelium is separated from the nearby vitelline veins by a small layer of mesenchymal cells (Fig. 2A) (Lammert, Cleaver, & Melton, 2001; Yoshitomi & Zaret, 2004). The surrounding mesenchyme also contains fine capillaries and isolated angioblasts (Azizoglu et al., 2016; Katsumoto & Kume, 2011; Yoshitomi & Zaret, 2004). In a process termed the "primary transition" (from approximately E9.0 to E12.0), the thickened pancreatic epithelium undergoes additional morphogenesis to form a stratified epithelial bud that will give rise to the exocrine and endocrine pancreas (Villasenor et al., 2010). At this stage, small capillaries, often in direct contact with the epithelium, can be identified in the surrounding mesenchyme around but not within the bud, forming a mesh-like network or plexus (Fig. 2B and C) (Azizoglu et al., 2016; Pierreux et al., 2010). The majority of these vessels at this stage are not perfused (Shah et al., 2011).

Signs of arteriovenous differentiation can be observed in the pancreatic bud as early as E10.5. As vessels expand and coalesce, they initiate expression of ephrin $B2^{+}$ or $EphB4^{+}$ vessels, marking arteries and veins, respectively. Within 1 day, a central artery, but no large caliber vein, can be seen,

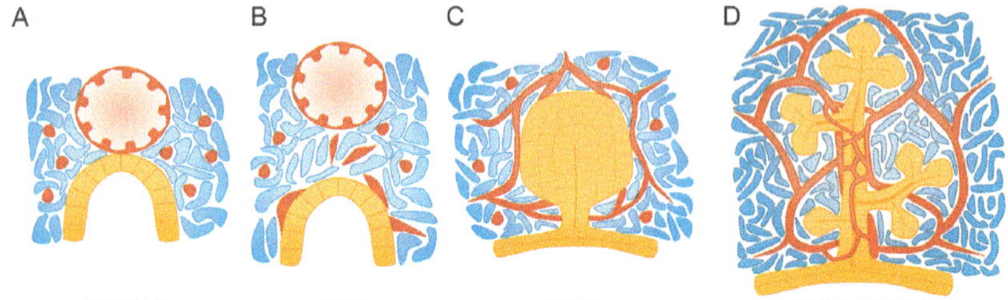

Fig. 2 Vascular development in the pancreas. (A) At E8.5–E9.0 when the pancreas initially forms, the dorsal pancreas (endodermal epithelium in yellow) lies in direct contact with the dorsal aorta (red). The dorsal aorta communicates directly with the early pancreas to regulate early pancreatic morphogenesis and to maintain Pdx1 expression. Surrounding the dorsal aorta is the trunk mesenchyme (blue), which also regulates pancreatic development. A fine network of capillaries as well as isolated angioblasts can be identified in this mesenchyme. (B) By E9.5, mesenchyme and capillaries intervene between the dorsal aorta and the dorsal pancreatic epithelium. (C) At E10.5, the dorsal pancreatic epithelium has thickened into a stratified bud. An endothelial plexus surrounds this bud making occasional direct contact with the epithelium, but not invading it. (D) After the secondary transition beginning at E12.5, ECs preferentially enrich around the trunk of the pancreatic epithelial tree (thicker branches) and are more sparse around the distal tips, or acini.

indicating that arterial development precedes venous development in the pancreas. Importantly, these early arteries, veins, and capillaries present around the epithelium arise through remodeling of the plexus cords present in the mesenchyme at E9.5 to eventually form a hierarchical tree (Azizoglu et al., 2016). Angiogenesis does not seem to play a significant a role in early pancreatic endothelial development, as pancreatic branches are largely seen to integrate between existing vessel loops rather than *via* formation of new sprouts. These studies in early pancreatic morphogenesis underscore how closely associated early vessels are with the growing pancreatic epithelium at early stages of organogenesis.

The next phase of pancreatic development beginning around E12.5 is called the "secondary transition." This period marks a dramatic change in epithelial growth and differentiation. During this time frame, the stratified epithelial bud begins to resolve into a single-layered epithelium and the epithelium experiences a burst of proliferation. The result is the emergence of two distinct epithelial regions: the distal tips containing acinar cells that secrete digestive enzymes and the trunk region where bipotential cells differentiate to form ductal and endocrine lineages (Pan & Wright, 2011). At this stage, ECs are more sparse around the distal acinar tips, while being more dense around the trunk region, coincident with increased expression of the powerful endothelial mitogen VEGF-A in the trunk (Fig. 2D) (Pierreux et al., 2010). This organization is maintained through the rest of development.

3.1.2 Endothelium regulates early pancreatic morphogenesis

During initial stages of pancreatic development, ECs promote formation of the early pancreatic bud. Pre-pancreatic endoderm explants form pancreatic-like structures only when co-cultured with the dorsal aorta or other EC-containing tissues (Lammert et al., 2001). Utilizing *Flk1*$^{-/-}$ mutant mouse embryos which lack all ECs to expand upon this observation, Yoshitomi and Zaret observed that early dorsal pancreatic endoderm still expressed Pdx1; however, the dorsal endoderm failed to expand into the early stratified bud while the ventral pancreatic endoderm remained unaffected (Yoshitomi & Zaret, 2004). These results indicate that ECs are required for pancreatic growth but dispensable for initial pancreatic induction (Yoshitomi & Zaret, 2004). These *Flk1*$^{-/-}$ mice also exhibited a loss of Ptf1a, an important driver of the pancreatic fate in the early endoderm (Kawaguchi et al., 2002), which was rescued when cultured with fragments of the dorsal aorta. Experiments performed in chick embryos suggest a reciprocal relationship between ECs and the early pancreas. Early pre-pancreatic endoderm secretes the chemokine Cxcl12, which binds to its cognate receptor Cxcr4 expressed on nearby angioblasts. These angioblasts, in turn, signal back to the endoderm to promote pancreatic development (Katsumoto & Kume, 2011). Together, these experiments demonstrate the promotion of early pancreatic development by vascular structures.

ECs also communicate to the pancreatic bud indirectly by supporting the surrounding mesenchyme. The mesenchyme is necessary for epithelial development from the onset of pancreatic morphogenesis partly through secreting Fgf10 (Bhushan et al., 2001; Landsman et al., 2011). In *Flk1*$^{-/-}$ embryos, the mesenchyme is nearly absent due to increased apoptosis (Jacquemin et al., 2006). Likewise, dorsal mesenchyme explants can only grow when cultured with the dorsal aorta suggesting that signals from the ECs themselves support mesenchymal growth. Adding Fgf10 to *Flk1*$^{-/-}$ pancreas explants rescued pancreatic growth (Jacquemin et al., 2006). Thus, ECs are not only important for early pancreatic epithelial development through direct endothelial-epithelial crosstalk, but they also communicate *via* the surrounding mesenchyme to support mesenchyme-epithelial crosstalk.

Systemic factors have also been shown to be important for early pancreas formation. *Cdh2*$^{-/-}$ mice, which are able to form the dorsal aortae but lack blood flow due to cell adhesion defects in the heart, display initiation of Pdx1 expression but failure of dorsal pancreas expansion. Dorsal pancreatic formation was rescued when Cdh2 activity was restored in the heart alone, suggesting that systemic factors may be regulating this phenotype (Edsbagge et al., 2005). Indeed, addition of Sphingosine-1-phosphate (S1P) to *Cdh2*$^{-/-}$ early pancreatic explants rescued dorsal pancreatic morphogenesis. S1P appears to signal through the endothelium as genetic deletion of the endothelial-specific S1P receptor $S1P_1$ also resulted in a dramatic reduction of both dorsal and ventral pancreas growth by E12.5 (Edsbagge et al., 2005; Sand et al., 2011). These data argue that ECs promote growth and differentiation of the early dorsal pancreas both directly *via* cell-cell communication between the epithelium and endothelium and indirectly *via* supplying systemic factors and supporting the mesenchyme after the initial specification.

3.1.3 The evolving role of ECs in pancreatic morphogenesis

By the secondary transition, ECs become more dense around the trunk and more sparse around the acinar tips. This vascular organization in the developing pancreas suggests that ECs may differentially regulate pancreatic morphogenesis depending on their location. Specifically, ECs may positively regulate trunk development—and the bipotential progenitors therein—while inhibiting acinar development. In support of this idea, overexpression of VEGF using the Ptf1a promoter (Pierreux et al., 2010) or a TET off system under the control of the Pdx1 promoter (Magenheim et al., 2011) both lead to hypervascularization of the pancreas including around the distal tips, decreased epithelial branching, and inhibition of acinar differentiation. Conversely, inhibition of VEGFR2 in explants decreases vascularization while increasing epithelial growth specifically at the tips but not at the trunk (Magenheim et al., 2011; Pierreux et al., 2010; Sand et al., 2011). These data strongly support the model in which ECs inhibit acinar differentiation during the secondary transition by keeping adjacent epithelial cells in a more immature or progenitor-like state. Therefore, ECs can be argued to serve two opposite roles in pancreatic development: (1) during initial pancreas formation and primary transition, ECs promote expansion and development of the early dorsal pancreatic bud and (2) when pancreatic organogenesis shifts to the secondary transition, ECs perform the opposite role and inhibit acinar differentiation.

3.1.4 Endothelium regulates endocrine fates

Based on the localization of ECs around the trunk, one could hypothesize that ECs promote the development of trunk cells into ductal and endocrine cell fates. Indeed, the precise role of ECs in directing endocrine fates appears to be time-dependent, mirroring the change in the role of ECs during epithelial development. ECs appear to promote endocrine differentiation early in pancreas development, but inhibit it at later stages. VEGF-A overexpression in Pdx1^{+} cells from the onset of pancreatic development and into perinatal stages leads to islet hyperplasia and ectopic insulin expression in the stomach, suggesting that ECs promote endocrine differentiation (Lammert et al., 2001). Likewise, *Flk1*$^{-/-}$ embryos fail to develop insulin^{+} or glucagon^{+} cells (Yoshitomi & Zaret, 2004). Conversely, deletion of VEGF-A from Pdx1^{+} cells during this same time frame led to a reduction in endocrine cell mass (Reinert et al., 2013). These results may be due to changes in oxygen levels. Perfused blood vessels lie in close proximity to glucagon^{+} cell clusters and pancreas explants grown in hypoxia demonstrate decreased endocrine differentiation suggesting that early endocrine differentiation relies on increased oxygen (Shah et al., 2011). Alternatively, inhibition of VEGFR2 in pancreas explants for a short period of time decreased endocrine differentiation, indicating that ECs themselves also play a role in promoting endocrine differentiation (Pierreux et al., 2010). As seen in early pancreatic epithelial morphogenesis, the endothelium promotes early endocrine development both directly and indirectly.

At later stages of pancreas development, ECs inhibit endocrine differentiation. Overexpression of VEGF exclusively in β-cells, which emerge starting at E11.5, led to disrupted islet morphology and decreased endocrine mass by E16.5 (Cai et al., 2012). In a separate series of experiments, ECs were actually shown to inhibit endocrine differentiation in contrast to the previous work (Magenheim et al., 2011). By overexpressing VEGF in Pdx1^{+} cells using the Tet system *in vivo* and inhibiting VEGFR2 *in vitro*, this study demonstrated that endocrine differentiation is inversely related to the amount of vascularization. Why these studies see disparate responses to hyper- and hypovascularization remain unclear. Potential reasons include differences in (1) the transgenic system utilized, (2) the explant culturing techniques, and (3) the small molecule VEGFR2 inhibitor. All of these variables could introduce unknown additional effects on the developing pancreatic epithelium and the progenitors therein, which ultimately manifest in the different phenotypes observed. The exact mechanism by which ECs influence endocrine differentiation—and how to reconcile these studies—remains to be elucidated.

3.2 Lung

3.2.1 Overview of pulmonary vascular development

The lung facilitates the exchange of O_2, CO_2, and other gases between the air and the blood. As such, endothelial-epithelial interaction is central for proper lung development and function. Lung development begins at E9.0 with specification of the early lung progenitors in the embryonic foregut, formation and splitting of the tracheoesophageal ridge, and initial branching events (Herriges & Morrisey, 2014). This early developmental timeframe is termed the "pseudoglandular stage" due to its resemblance to glandular organs.

During this early morphogenesis, a delicate vascular plexus surrounds the early lung as it courses throughout the mesenchyme (deMello, Sawyer, Galvin, & Reid, 1997; Parera et al., 2005; Schachtner, Wang, & Scott Baldwin, 2000). By E12.5, the lung progenitors differentiate into two main progenitor populations: one that will give rise to the distal airway spaces where gas exchange occurs and another that generates the proximal (or conductive) airways responsible for transporting gases between the distal airways and external environment. Likewise, the endothelium associated with either of these epithelial compartments is thought to arise from two separate sources. The endothelium of the large caliber pulmonary arteries and veins that form next to the proximal airways arise from $Wnt2^+$ cardiopulmonary progenitors that also give rise to the associated smooth muscle cells (Peng et al., 2013). On the other hand, ECs surrounding the distal airways arise from a VE-Cadherin$^+$ population of ECs present in the trunk at E8.5 before initial lung development (Herriges & Morrisey, 2014). It is striking that the lung endothelium arises from two separate endothelial populations and that they must coordinate to anastomose and form a fully functional endothelial network. How these two endothelial populations form a vascular network remains to be determined as the current data support both angiogenic and vasculogenic processes for both populations.

3.2.2 Distinct origins determine the role of the pulmonary vasculature

The distinct origin of the two lung EC populations and their relation with the associated airways raise the possibility that endothelial-epithelial crosstalk is region-specific. Early ECs have been shown to be dispensable for initial specification of lung progenitors and *vice versa* (Havrilak, Melton, & Shannon, 2017; Peng et al., 2013). Instead, ECs are necessary for proper maturation and branching of the embryonic lung. Studies that investigated how ECs regulate lung development have utilized an indirect approach of modulating VEGF-A levels in specific regions of the lung—which thereby affects EC patterning and function—and assessing changes in the pulmonary epithelium. Overexpression of VEGF-A in proximal epithelial cells (specifically Clara Cell Secretory Protein ($CCSP^+$) cells) during development or after birth led to disruption of the proximal airway morphology and dilation of the associated arteries with minimal to no involvement of the distal airway spaces (Akeson, Cameron, Le Cras, Whitsett, & Greenberg, 2005; Akeson et al., 2003; Le Cras et al., 2004). Likewise, overexpression of VEGF-A in distal epithelial cells (specifically Surfactant protein-C ($SP\text{-}C^+$) cells) during development led to disruption of the normal lung vasculature with dilated distal airway spaces (Akeson et al., 2003; Zeng, Wert, Federici, Peters, & Whitsett, 1998). Proximal airways appeared unaffected in these models. Interestingly, deletion of VEGF-A in these same distal epithelial cells also led to dilation of the distal airways, although this was due to impaired septal formation (Yamamoto et al., 2007). These data indicate that the role of VEGF-A—and, by extension, the role of ECs—during lung organogenesis is spatially defined.

3.2.3 Lung blood vessels influence lung development and maturation

Ex vivo lung explant models and three-dimensional analyses of *in vivo* lungs have demonstrated that ECs promote branching morphogenesis. The addition of VEGF-A in lung explant models increases epithelial branching morphogenesis through increased Bmp4 and SP-C expression (Del Moral et al., 2006). This effect seems to be specifically affecting the distal tips of the lung epithelial tree. Conversely, explants cultured with antisense oligonucleotides against VEGF-A or VEGFR2 demonstrate decreased epithelial branching (Del Moral et al., 2006; van Tuyl et al., 2005). When VEGF-A was blocked *in vivo* through increased production of a soluble VEGFR1 dummy receptor, the rate of epithelial branching did not change but, rather, the orientation of the branches was disrupted (Lazarus et al., 2011). All of these data show that ECs influence epithelial maturation by modulating epithelial morphogenesis during development.

ECs are also necessary for postnatal maturation of the lung epithelium, specifically the alveoli where gas exchange occurs. The majority of alveoli form in the immediate postnatal period in a process termed "alveolarization," which leads to a massive expansion in the surface area for gas exchange to occur (Warburton et al., 2010). Treatment of neonatal pups with anti-angiogenic agents or an inhibitor against VEGFR2 resulted in decreased postnatal lung growth, decreased endothelial arborization, and fewer but dilated alveoli, all indicating disrupted alveolar maturation (Jakkula et al., 2000). Likewise, genetic deletion of PECAM-1, an endothelial-specific gene that is necessary for angiogenesis (Cao et al., 2002; DeLisser et al., 1997; Newman & Newman, 2003), resulted in decreased alveolarization and thickened airway spaces in postnatal mice without affecting EC proliferation or apoptosis.

3.2.4 Pulmonary pericyte signals to lung epithelium

Recently, lung pericytes have also been implicated in directing alveolar maturation. Deletion of the Hippo pathway proteins Yap1 and Taz in PDGFR-β^+ lung pericytes using the PDGFR-β-CreERT2 inducible mouse line resulted in impaired alveorization leading to dilated airspaces, decreased capillary formation, and reduced pulmonary vasculature (Kato et al., 2018). These phenotypes were likely caused by a decrease in angiopoietin-1, a ligand of the EC-specific Tie2 gene and important modulator of EC activity (Dumont et al., 1994; Suri et al., 1996). This phenotype was recapitulated when angiopoietin-1 was deleted specifically in pericytes. These studies underscore that EC migration and angiogenesis, mediated by crosstalk between each cell population in the lung, must occur for full maturation of the alveolar sacs during the postnatal period when air exchange is necessary.

3.2.5 Lung regeneration: Vascular contributions

Unlike the other organs discussed in this chapter, the lung displays an incredible ability to regenerate in response to injury. Unilateral pneumectomy of the left lung induces compensatory expansion of mass and volume of the right lung (Cowan & Crystal, 1975). Alternatively, chemical injury with bleomycin or naphthalene damages distal AT2 (SP-C^+) or proximal non-ciliated bronchiolar cells ($CCSP^+$), respectively, and induces a set of resident lung stem cells called “bronchioalveolar stem cells” (BASCs) (defined by coexpression of SP-C and CCSP) to repair the damage (Aso, Yoneda, & Kikkawa, 1976; Kim et al., 2005; Plopper, Suverkropp, Morin, Nishio, & Buckpitt, 1992). In all of these models, ECs play a critical role in directing lung regeneration *via* expansion of epithelial stem cells. With bleomycin injury, BASCs secrete Bmp4, which activates NFATc1 signaling in ECs to produce Tsp-1 and promote epithelial repair (Lee et al., 2014). Left lung pneumectomy leads to an alternate response: synergistic activation of the VEGFR2 and FGFR1 and the Cxcr4–Cxcl12 (also called SDF-1) pathways in pulmonary ECs leads to secretion of MMP14, which promotes the expansion of epithelial progenitors through the release of cryptic EGF ligands (Ding et al., 2011; Rafii et al., 2015). The differences in the signaling pathways induced in these studies may be due to the differences in the models employed, whereas chemical injury induces a regenerative response to replace the damaged tissue, pneumectomy leads to hyperplasia of the remaining lung lobes and, thus, creates new alveolar spaces. Importantly, these multiple models of lung injury confirm that endothelial-epithelial crosstalk is indispensable for proper lung regeneration. The obvious question emerges as to whether these same pathways are also implicated during lung development or if they are injury-specific responses.

3.3 Kidney

3.3.1 *Blood vessels mirror nephron development*

The kidney is a highly patterned and organized organ that carries out a wide variety of vital functions. The primary role of the kidney is to filter the blood and concentrate the urine, thus eliminating wastes from the body and reclaiming important small molecules and water from the filtrate. As a part of these roles, the kidneys help regulate serum ion levels, osmolarity, and pH, adjusting as necessary to fit the needs of the individual at that moment. The kidneys also secrete hormones that regulate blood pressure (renin) and red blood cell production (erythropoietin). Most of these functions are carried out by structures called "nephrons," which consist of a glomerulus—the filter apparatus—at the proximal end followed by a single, highly-patterned tubule that eventually drains into the ureter. Each region of the tubule is adapted to perform specific cellular duties. For example, the proximal convoluted tubule is better suited for reclaiming ions and small molecules while the descending loop of Henle can only absorb water to concentrate the urine. Not only is the nephron epithelia patterned along its length, the kidney itself is patterned from the cortex to the medulla whereby different nephron segments and stromal populations are located exclusively in one region or the other.

To be able to carry out all of the diverse roles of the kidney efficiently, the renal vasculature must be highly adapted and regionalized. The path of blood flow through the kidney is the first level of endothelial organization. Blood flows from the (1) renal artery branching off of the abdominal aorta, to the (2) segmental arteries, (3) interlobar arteries, (4) afferent arterioles, (5) glomerular capillaries, (6) efferent arterioles, (7) peritubular capillaries in the cortex and vasa recta in the medulla that surround and perfuse the entire nephron tubule, (8) interlobar and segmental veins, and (9) the renal

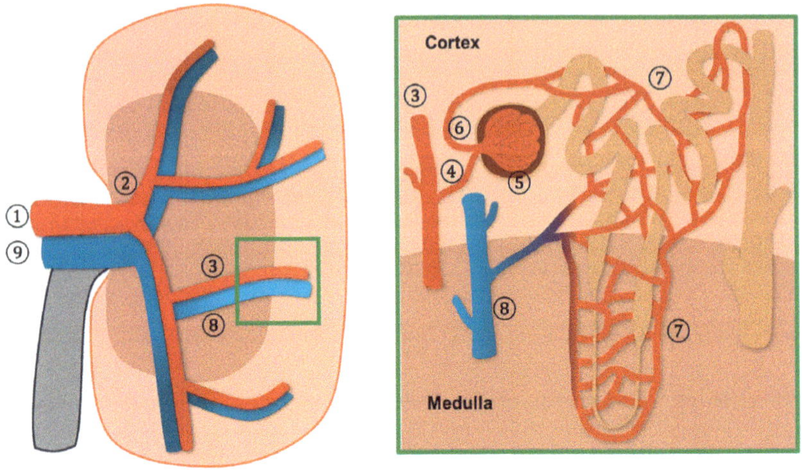

Fig. 3 The path of blood flow through the kidney. Blood flow through the kidney progresses through a well-established path on both the gross level and through each nephron (boxed area). Renal blood flow begins in the renal artery (1), which branches into the segmental arteries (2) and lobar arteries (3). Next, many afferent arterioles (4) branch off the lobar arteries, each feeding into capillaries within a single glomerulus (5). Blood exits the glomerulus through the efferent arteriole (6), which drains into the second set of fine vessels called the "peritubular capillaries" in the cortex (7, lighter area, top) and vasa recta in the medulla (7, darker area, bottom) that surround the nephron tubules. These capillaries then feed into lobar and segmental veins (8), which all eventually drain into the renal vein (9) and back into the systemic circulation.

vein that drains into the vena cava (Fig. 3). One of the most striking features of this organization is that the renal vasculature possesses two capillary beds in series: the glomerulus followed by peritubular capillaries and vasa recta. Additionally, ECs in each of these regions carry out regionalized exchanges and must be adapted to fit their role. For example, glomerular capillaries possess fenestrae to allow for filtration of solutes in the plasma. Although the peritubular capillaries and vasa recta represent a "single" endothelial group, the high degree of nephron epithelial patterning suggests that these capillaries may also be patterned along each region of the nephron. Which cues direct the formation of individual capillaries and vasa recta and how ECs specialize in these individual niches are currently unknown.

3.3.2 Kidney epithelium and endothelium grow and expand coordinately

Kidney development begins at E10.5 when the Wolffian duct buds into the adjacent metanephric mesenchyme. This mesenchyme contains both $Six2^{+}$

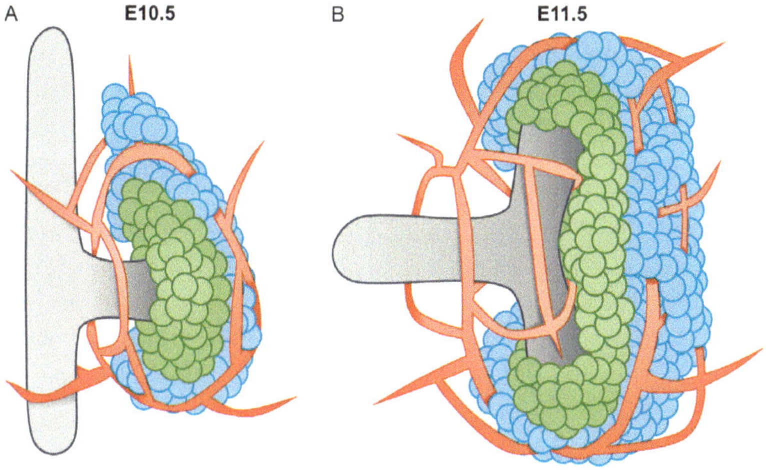

Fig. 4 Vascularization of the early metanephros. (A) At E10.5, the nephric duct (gray) or ureteric bud (UB) buds into the adjacent metanephric mesenchyme population, which consists of NPCs (green) and stromal progenitor cells (blue). The vasculature (red) exists as a plexus within the surrounding stroma, but does not penetrate the NPC clusters. (B) At E11.5 after initial UB branching, the vascular plexus expands along with the developing kidney. Note that the vasculature still does not invade the NPCs.

nephron progenitor cells (NPCs) and FoxD1^{+} stromal progenitor cells. At this stage, ECs can be identified surrounding (but not penetrating) the metanephric blastema (Fig. 4A) (Daniel et al., 2018; Munro, Hohenstein, & Davies, 2017; Robert, St John, & Abrahamson, 1998). Next, the Wolffian duct bud, called the "ureteric bud" (or UB), evaginates into the metanephric mesenchyme in response to inductive signals emanating from the NPCs (O'Brien & McMahon, 2014). At the same time, the vasculature surrounding the early metanephric mesenchyme forms a continuous vascular plexus both around and within the developing kidney (Fig. 4B). Through reciprocal signaling between the UB and NPCs, the UB undergoes extensive growth and branching while the NPCs cluster at distal UB tips to form "caps" and proliferate. This process is maintained until NPCs are exhausted at around P3 in mice (O'Brien & McMahon, 2014).

Recently, three-dimensional imaging of the distal tips revealed that ECs grow coordinately with the distal UB tips and NPC caps. ECs form cruciform structures around the bifurcation of the UB tips and circumscribe each NPC cap (Daniel et al., 2018; Munro, Hohenstein, & Davies, 2017). Notably, these ECs do not directly contact either the UB or the NPCs at first, as they are predominately surrounded by stromal cells with rare connections to the UB epithelium (Daniel et al., 2018). This organization suggests that the endothelium at distal UB tips is patterned, at least in part, by the stroma. Indeed, ablation of FoxD1$^+$ stromal cells leads to hypervascularization of the progenitor region and disruption of endothelial structure (Hum, Rymer, Schaefer, Bushnell, & Sims-Lucas, 2014), suggesting it acts to restrain EC proliferation. Interestingly, in both mutant and normal mice, the ECs rarely contact NPCs directly (Daniel et al., 2018; Hum et al., 2014; Munro, Hohenstein, & Davies, 2017). In fact, NPCs preferentially cluster away from ECs. This cellular organization may suggest that NPCs reside in a state of hypoxia. In agreement with this model, kidney explants cultured under hypoxia show an increase in NPC numbers compared to those cultured under normoxia (Rymer et al., 2014); however, the mechanisms that regulate this are unknown at this time.

3.3.3 New paradigms of the vasculature in nephrogenesis

Beginning around E12.5, some NPCs in the caps will undergo a mesenchymal to epithelial transition (MET) and begin nephrogenesis, the process by which new nephrons are formed. The first step of nephrogenesis is formation of a pretubular aggregate followed by the renal vesicle (RV) stage when a lumen forms. The RV undergoes extensive morphogenesis to form first a comma-shaped body, then an S-shaped body, and eventually a full nephron. As NPCs undergo nephrogenesis and new nephrons form, the vasculature integrates with the expanding nephric tree. Previous models of nephrogenesis suggested that individual angioblasts migrate into the cleft of the developing S-shaped body where presumptive podocytes reside (Eremina et al., 2003; Saxén & Sariola, 1987).

Early studies suggested nephrogenesis occurred without significant vascularization until the S-shaped body stage. This was likely due to methods available at the time, including limited markers and imaging techniques. Utilizing three-dimensional imaging, our lab showed that the endothelium remodels to become intimately associated with the developing nephron as early as the renal vesicle stage, forming a plexus around the structure (Daniel et al., 2018). The endothelium and epithelium grow coordinately: as the nephron progresses through the stages of nephrogenesis, the endothelial plexus surrounding it elaborates with it and becomes preferentially enriched along the portion fated to become the distal tubule. The ECs that invade the cleft of the S-shaped body are not individual cells but, rather, an angiogenic sprout from the plexus that began early during nephrogenesis. This model is supported by recent high-resolution live imaging data of embryonic kidney explant cultures that identified an angiogenic sprout invading the developing S-shaped body (Saarela et al., 2017). Therefore, models of endothelial involvement during nephrogenesis have evolved as methods have technically improved (Daniel et al., 2018).

Although ECs appear to organize in a specific manner during nephrogenesis, it remains to be determined if they are truly patterned at a transcriptional level along different regions of the developing nephron. The nephron itself is patterned as early as the renal vesicle stage: cells closest to the UB are fated to become the distal nephron, while those farther from the UB will become the podocytes and proximal tubule (Lindstrom et al., 2018). At the S-shaped body stage, additional discrete regions can be identified using different markers including separation of the proximal tubular and podocyte fates as well as additional intermediate populations between each segment. That the epithelium demonstrates clear signs of patterning during nephrogenesis raises several questions concerning the surrounding endothelium. First, is the endothelium patterned on a transcriptional level along the developing nephron? If so, at what stage is this heterogeneity established? Additionally, is there crosstalk between the endothelium and epithelium during nephrogenesis or is EC organization along the nephron entirely cell autonomous? Does EC-epithelial crosstalk play a role in establishing heterogeneity in either compartment? Answering these questions will shed new light on the pathways necessary for full nephron development.

3.3.4 Factors regulating glomerular development

The glomerulus is a highly patterned structure responsible for ultrafiltration of the blood. Each glomerulus consists of fenestrated ECs, highly specialized epithelial cells called "podocytes," and stromal cells called "mesangial cells." During development, ECs and podocytes both lay down several extracellular matrix (ECM) proteins to generate a fused basement membrane called the "glomerular basement membrane," or GBM, which is the filtering apparatus of the glomerulus (Fig. 5A) (Miner, 2012). Although not directly contributing to the GBM, mesangial cells support the activity of the other glomerular cell types in maintaining proper glomerular function.

Glomerular development requires coordinated crosstalk between the endothelium and developing podocytes. A key angiogenic factor responsible for establishing and maintaining ECs within the glomerulus is VEGF-A. VEGF-A is secreted by presumptive podocytes in the S-shaped bodies. VEGF-A then binds to its cognate receptor VEGFR2 on ECs in the surrounding plexus to induce them to invade the cleft of the S-shaped body

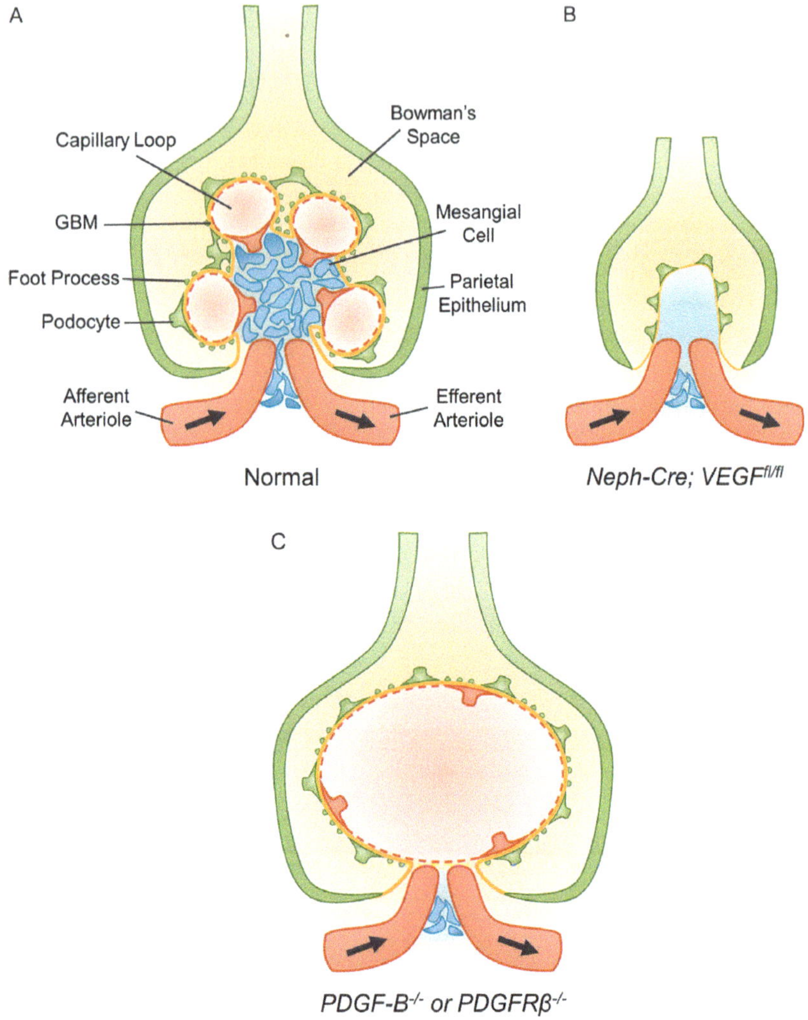

Fig. 5 Models of mature glomeruli in specific genetic mutants. (A) Healthy mature glomerulus. ECs (red) of the afferent arteriole (left red vessel) feed into multiple capillary loops, which drain into the efferent arteriole (right red vessel). These ECs interact with podocytes (green) to form the glomerular basement membrane or GBM (yellow). Mesangial cells (blue) are present around the ECs to support their activity. (B) VEGF deletion from podocytes results in shrunken glomeruli with a loss of ECs and mesangial cells. Although podocytes are present, they lack foot processes and the GBM is noticeably thinner. (C) Whole body deletion of PDGF-B or PDGFR-β results in loss of mesangial cells and formation of large capillary aneurysms instead of multiple capillary loops. Podocytes and the GBM in these mutants appear normal.

(Eremina et al., 2003). Podocyte-specific deletion of VEGF-A leads to hypovascularization of the glomerulus, loss of GBM fenestrations, and mesangial cell loss in a dose-dependent manner (Eremina et al., 2006, 2003) (Fig. 5B). Similar results were seen when newborn mice were administered a neutralizing antibody against $VEGF_{165}$ (Kitamoto, Tokunaga, & Tomita, 1997). Perhaps not unexpectedly, overexpression of VEGF-A also led to drastic alterations in glomerular function and morphology, indicating that VEGF-A expression must by tightly regulated to ensure proper glomerular development (Eremina et al., 2003; Sison et al., 2010). The VEGF-VEGFR2 axis between podocytes and glomerular ECs is not only necessary during development, but also for postnatal glomerular function. Whole-body deletion of VEGFR2 beginning 9 days after birth leads to pronounced glomerular damage highlighted by loss of glomerular ECs (Sison et al., 2010).

The ligand receptor pair PDGF and PDGFR also plays an integral role in glomerular formation. In general, interaction between PDGFs and PDGFRs is known to be important during angiogenesis and vessel maturation. PDGF-B, expressed by sprouting ECs, binds to PDGFR-β on vascular smooth muscle cells to recruit the smooth muscle cells to the developing vessel (Gerhardt & Betsholtz, 2003). The kidney follows a similar paradigm whereby PDGF-B is expressed by the renal endothelium and PDGFR-β is expressed in mesangial cells, vascular smooth muscle cells around the renal arterial tree, and stromal cells in the renal cortex (Lindahl et al., 1998). Germline deletion of either PDGF-B or PDGFR-β both result in aneurysms within glomeruli and a nearly complete absence of mesangial cells (Leveen et al., 1994; Soriano, 1994) (Fig. 5C). When PDGF-B is specifically deleted from the EC compartment, these results can be recapitulated although the phenotype is more variable and mild compared to the germline knockout, possibly due to incomplete deletion with available Cre lines used for conditional deletion (Bjarnegard et al., 2004). These results suggest that EC-derived PDGF-B is the predominant factor directing PDGF signaling and mesangial recruitment during glomerulogenesis. Notably, the GBM appeared normal in all mutants, indicating that the PDGF-PDGFR axis mediates the interaction between ECs and mesangial cells but not between ECs and podocytes.

Based on these data, one can build a model of the cell-cell signaling pathways that direct glomerular development and maturation. Initially, VEGF-VEGFR2 signaling is necessary to induce the EC plexus around the developing nephron to create an angiogenic sprout and invade the S-shaped body (Fig. 6A). Soon after this step, PDGF-B secreted by

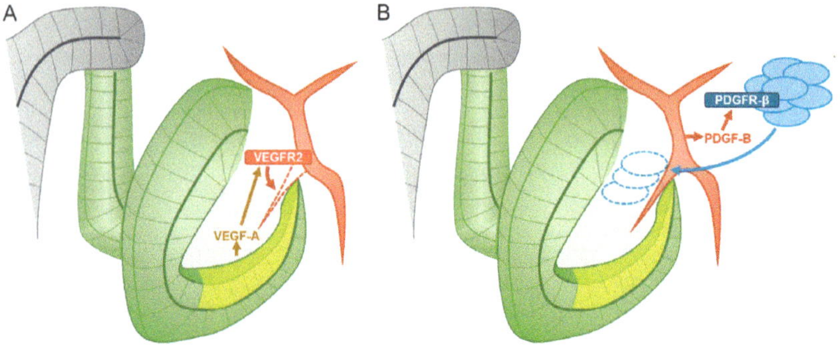

Fig. 6 Glomerulogenesis requires coordinated communication between podocytes, ECs, and mesangial cells. (A) During nephrogenesis, presumptive podocytes (yellow) within the S-shaped early nephron secrete VEGF-A. The VEGF-A ligand binds to VEGFR2 expressed on ECs (red), which are located in the plexus surrounding the developing nephron. This induces the ECs to undergo an angiogenic invasion of the developing nephron. (B) As ECs invade the nephron, they secrete PDGF-B. PDGF-B binds to PDGFR-β on presumptive mesangial cells (blue) to promote chemotactic attraction of these cells and induce them to migrate into the developing glomerular cleft. Note that part of the developing nephron is hidden to allow for easier visualization of this process.

these ECs binds to PDGFR-β on presumptive mesangial cells to recruit them into the glomerulus (Fig. 6B). This interaction continues during glomerulogenesis to generate multiple capillary loops. Meanwhile, ECs and podocytes communicate to lay down the ECM proteins necessary to form the GBM. Once the glomerulus has fully matured, ECs and podocytes require persistent communication through VEGF signaling in order to properly maintain the GBM and full glomerular function.

Several other signaling pathways have been implicated in directing glomerular formation through communication between ECs, podocytes, and mesangial cells (Mohamed & Sequeira-Lopez, 2018). These pathways include Notch (Alabi et al., 2016; Farber et al., 2018; Lin, Sequeira-Lopez, & Gomez, 2014; McCright et al., 2001), Bmps (Dunn et al., 1997; Ueda et al., 2008) Semaphorins (Bartlett et al., 2017; Reidy et al., 2009), and Angiopoietins and Tie (Jeansson et al., 2011) although direct functional analyses by specifically ablating these genes in the endothelium or mesangium have been relatively sparse (Table 1). This is due to, in part, the lack of useful genetic tools to answer these questions. Although several podocyte-specific Cre lines exist (Podocin-Cre, Nephrin-Cre) (Belteki et al., 2005; Eremina et al., 2003), such tools do not exist for either glomerular ECs or mesangial cells. Because several of the pathways implicated regulate many non-renal processes, whole-body or even vascular-specific deletion of these genes

Table 1 Table of pathways and genes affecting glomerular development.

Gene	Cell type expressed	Mutated allele	Phenotype					References
			ECs	Podocytes	Mesangial cells	GBM	Other	
VEGF-A	Podocytes	*Neph-Cre; $VEGF^{fl/fl}$*	Reduced or absent	N	A	Loss of EC fenestrations	Fatal within 18h	Eremina et al. (2006, 2003)
		Neph-Cre; $VEGF^{fl/+}$	Reduced or absent	Progressive loss	A	Loss of EC fenestrations		Eremina et al. (2003)
		Neph-Cre; $VEGF^{hypo/fl}$	Progressive loss	N	Progressive loss	Loss of EC fenestrations	Fatal within 3 weeks	Eremina et al. (2006)
		Neph-Cre; $VEGF_{164}$	Few, dilated capillary loops	N		Collapsed	Proteinurea	Eremina et al. (2003)
PDGFR-B	Mesangium	*$Pdgfr\text{-}\beta^{-/-}$*	Aneurysm	N	A	Normal		Soriano (1994) and Lindahl et al. (1998)
PDGF-B	ECs	*$Pdgfb^{-/-}$*	Aneurysm	N	A	Normal		Leveen et al. (1994) and Lindahl et al. (1998)
		Tie1-Cre; $Pdgfb^{fl/-}$	Aneurysm	N	A	Normal		Bjarnegard et al. (2004)
Notch1	ECs	*Tie2-Cre; $Notch1^{fl/fl}$*	Hypercellular					Alabi et al. (2016)
Notch2		*$Notch2^{hypo/hypo}$*	Absent or Aneurysm	Abnormal structure	A			McCright et al. (2001)

Continued

Table 1 Table of pathways and genes affecting glomerular development.—cont'd

Gene	Cell type expressed	Mutated allele	Phenotype					References
			ECs	Podocytes	Mesangial cells	GBM	Other	
RBP-J	Mesangium	*FoxD1-Cre; Rbp-J*$^{fl/fl}$	Aneurysm	N	A			Lin et al. (2014)
Adam10	ECs	*Tie2-Cre; Adam10*$^{fl/fl}$	Enlarged, hypercellular	N	Normal	EC fenestrations contain diaphragms		Farber et al. (2018)
Bmp4	Podocytes	*Bmp4*$^{+/-}$	Reduced or absent	N			Large glomerular cysts in Bowman's space	Ueda et al. (2008)
		Nephrin-Bmp4	Reduced or absent	N	Reduced or absent			Ueda et al. (2008)
Noggin	Podocytes	*Nephrin-Noggin*	Collapsed tufts, ECs in Bowman's capsule	N	Reduced or absent		Podocyte-like cells replace parietal epithelial cells	Ueda et al. (2008)
Nrp1	Mesangium/ ECs	*Pdgfrb-Cre; Nrp1*$^{fl/fl}$	Aneurysm		A	EC detachment	Proteinurea, glomerular sclerosis	Bartlett et al. (2017)

Sema3a	Podocytes	*Podocin-rtTA; tet-O-Sema3a*[a]	Progressive loss	Delayed differentiation		Loss of EC fenestrations, Immature foot processes.		Reidy et al. (2009)
		Sema3a$^{-/-}$	Hypercellular	N		Loss of EC fenestrations. Foot processes effaced.	Proteinurea	Reidy et al. (2009)
Angiopoietin-1	Podocytes	*ROSA-rtTA; tetO-Cre; Angpt1*$^{fl/fl}$[b]	Aneurysm	N	Reduced	EC detachment		Jeansson et al. (2011)

[a]Tet system induced with doxycycline from E12 to birth.
[b]Tet system induced beginning at E10.5.
Multiple different pathways regulate glomerular development. Genetic manipulation of these pathways can affect ECs, podocytes, mesangial cells, the GBM, or some combination of all four. This table summarizes the phenotypes that manifest when some of the genes and pathways implicated in glomerular development are mutated. A, absent; N, normal. Empty cell indicates that the phenotype was not assessed in that study.

can cause several off-target effects, limiting the analyses that can be performed on the kidney. Full understanding of molecular machinery necessary for glomerular formation and function will require development of these tools.

3.3.5 Renal arterial specification and development

The onset of AV specification marks an important initial step in EC differentiation throughout the embryo and each of its tissues. The first evidence of arterial specification in the kidney is relatively late, at E13.5, when vessels expressing the arterial markers Cx40 or Nrp1 can first be identified (Daniel et al., 2018; Munro, Hohenstein, Coate, & Davies, 2017). Arterial vessels display a predictable branching pattern. First, two large arteries branch off of renal artery extending in the cranial-caudal axis. Next, these arteries split to form three to four additional branches each. Unlike the arteries in the developing pancreas (Azizoglu et al., 2016) and lung (Lazarus et al., 2011) that follow and parallel growing epithelial branches, arteries in the kidney do not follow the branching UB tree but rather course along the cortex of the kidney avoiding the medulla altogether (Daniel et al., 2018). Although the onset and anatomy of the renal arteries has been established, how these vessels form remains controversial. Some have proposed a model in which arterial ECs form through angiogenic sprouts from the aorta (Stolz & Sims-Lucas, 2015), while others suggest arteries form through vasculogenic remodeling of the existing renal vasculature prior to E13.5 (Sequeira Lopez & Gomez, 2011). Similar coordination of angiogenic and vasculogenic mechanisms has been proposed in other organs, including the lung (Parera et al., 2005; Schachtner et al., 2000). Careful lineage-tracing experiments will need to be performed to fully determine how the renal arterial tree forms.

The known cues that regulate renal arterial development arise primarily from the stromal and perivascular cells supporting the arterial ECs. Stromal, perivascular, and mesangial cells in the kidney derive from a FoxD1^{+} stromal progenitor population present at the onset of metanephric development (Guillaume, Bressan, & Herzlinger, 2009; Hatini, Huh, Herzlinger, Soares, & Lai, 1996; Humphreys et al., 2010). Genetic deletion of *FoxD1* or mosaic ablation of FoxD1^{+} cells both lead to drastically altered renal artery formation with arterial vessels perfusing from the outer cortex to the glomeruli instead of from the hilum (Sequeira-Lopez et al., 2015). Whereas all perivascular cells derive from FoxD1^{+} cells in the controls, kidneys with mosaic ablation of FoxD1^{+} cells demonstrated recruitment of perivascular cells that came from a non FoxD1^{+} source indicating that there may be compensatory mechanisms to direct arterial development. The downstream mechanism of how FoxD1 regulates arterial development remains unknown, although it could be attributed at least in part to PDGF signaling. Indeed, deletion of Pbx1, a TALE transcription factor expressed in the mural cells lining the renal arterial ECs, in $FoxD1^+$ cells alters normal arterial branching morphogenesis through dysregulation of PDGFR-β (Hurtado et al., 2015). Additionally, the Notch pathway, known to regulate arterial development in other contexts (Deng et al., 2013; Duarte et al., 2004), has also been implicated as a potential mechanism by which the stroma regulates arteriogenesis (Lin et al., 2014). Deletion of RBP-J, a common downstream transcription factor of the Notch pathway, in $FoxD1^+$ stromal cells results in kidneys with fewer arteries and arterioles that possess a thinner layer of surrounding smooth muscle cells (Lin et al., 2014). Whether other molecular pathways known to regulate arterial development in other contexts, including VEGF (Lanahan et al., 2010, 2013; Simons & Eichmann, 2015), also facilitate stromal-endothelial crosstalk during renal arterial formation remains to be elucidated.

There are several other mechanisms that can affect arterial specification and development that have yet to be explored in the kidney. In the skin, sensory nerves determine where arteries, but not veins, form through VEGF-mediated signaling pathways (Mukouyama, Shin, Britsch, Taniguchi, & Anderson, 2002). Although where and how nerves form remains to be determined in the kidney, these data suggest that neuronal and arterial development may be co-regulated in the kidney. Additionally, the number of studies identifying EC-intrinsic mechanisms that regulate renal arterial development are limited. One study identified $S1P_1$ as an intrinsic factor regulating renal arterial development. Deletion of $S1P_1$ from the vascular and hemogenic compartments using Scl-Cre resulted in disorganized and dilated renal arterials (Hu, Li, Gothert, Gomez, & Sequeira-Lopez, 2016). Mechanical factors, namely, blood flow, may also be important in kidney arterial development. Blood flow plays a critical role in EC maturation in multiple contexts including arterial-specification in the aorta (Chong et al., 2011; le Noble et al., 2004; Simons & Eichmann, 2015). The renal arterial tree matures in a radial manner from the hilum—closest to the origin of the renal artery—to the outer cortex (Daniel et al., 2018). This pattern suggests blood flow through these early arteries helps to mature those ECs; however, more direct methods to test the effect of blood flow on renal artery development need to be performed.

3.3.6 Patterning of the peritubular capillaries and vasa recta

For the remaining vascular compartments in the kidney—the peritubular capillaries and vasa recta—most work has examined the role of general vascular signaling pathways in regulating their formation. VEGFR2, which is expressed throughout all ECs during kidney development (Daniel et al., 2018), is only expressed in glomerular ECs and peritubular capillaries, but not in the vasa recta or arteries in the adult kidney (Dimke et al., 2015). Deletion of its ligand VEGF-A from $Pax8^{+}$ tubular epithelial cells specifically abrogates peritubular capillary formation without appreciably affecting the vasa recta. On the other hand, angiopoietins and Tie signaling pathways regulate vasa recta formation. Deletion of Tie2 or both Angiopoietin-1 and Angiopoietin-2 in late gestation lead to a decrease in vascular density in the renal medulla (Kenig-Kozlovsky et al., 2018). The decrease in vascular density is due to a loss of ascending vasa recta, while the descending vasa recta appeared normal. These results point to distinct molecular pathways that regulate EC formation and maintenance in region-dependent manners. Outside of the studies described above, surprisingly little has been done to understand the molecular pathways that regulate peritubular capillary and vasa recta formation.

4. "Therapeutic" development: Using development to guide regenerative medicine

Chronic kidney disease (CKD) affects approximately 10% of the American adult population and is characterized by a decrease in glomerular filtration as a result of a loss of functional nephrons (Oxburgh et al., 2017; Stevens, Levin, Kidney Disease: Improving Global Outcomes Chronic Kidney Disease Guideline Development Work Group Members, 2013). Because adult kidneys cannot generate new nephrons, nephron replacement strategies to treat CKD must rely on non-endogenous sources. Allogeneic kidney transplantation is an effective means to treat patients with CKD; however, the demand for kidney transplants far exceeds the supply of transplantable kidney with over 100,000 Americans remaining on kidney transplant waiting lists (Tullius & Rabb, 2018). Thus, an alternative source for functional nephrons is necessary to meet this demand. Regenerative medicine has the promise to address the gap by directly generating new kidney tissue *ex vivo*. Importantly, the replacement tissue must (1) be able to perform all normal renal functions, (2) be fully vascularized, and (3) be able to integrate with the host vasculature. This section will focus on the current status of regenerative therapies for transplantation and the challenges in developing a fully functional and vascularized *ex vivo* organ.

4.1 Using organoids to generate de novo replacement kidney tissue

Current efforts to create organs *ex vivo* have revolved around the use of organoids. Organoids are clusters of cells organized in three-dimensional culture that contain multiple organ-specific cell types and resemble the intended tissue by morphology, gene expression, or function (Lancaster & Knoblich, 2014). Organoids demonstrate a greater potential for regenerative therapies than directed differentiation of pluripotent stem cells into a single cell type because they more closely mimic normal developmental programs as they generate multiple organ cell types in parallel. To date, organoids have been generated for several different organs, including the kidney (Bantounas et al., 2018; Freedman et al., 2015; Morizane et al., 2015; Takasato et al., 2015), intestines (Spence et al., 2011; Workman et al., 2017), lung (Chen et al., 2017; Dye et al., 2015; Lee et al., 2014), cerebrum (Eiraku et al., 2008; Jo et al., 2016; Lancaster et al., 2017, 2013), retina (Eiraku et al., 2011; Nakano et al., 2012), liver (Takebe et al., 2013), and esophagus (Trisno et al., 2018; Zhang et al., 2018). Organoid development in all of these examples follows a common theme: a stem cell population, either pluripotent or tissue-specific multipotent, is differentiated into the target tissue by recapitulating normal developmental pathways *in vitro*. In kidney organoids, human pluripotent cells are differentiated first into intermediate mesoderm, the precursor population to the kidney, then into metanephric mesenchyme and ureteric bud populations (Takasato et al., 2015).

4.2 Technical challenges of kidney organoids

Although organoids have shown great promise for future novel tissue replacement therapies, there are still several limitations of this system (Fig. 7). First, transcriptional profiling of kidney organoids reveals that they are more similar to the fetal version of the tissue than to the adult counterpart (Combes et al., 2017). Therefore, additional steps are needed in order to generate an adult-like organoid. Next, these organoids are several orders of magnitude smaller than the intended organ and current culture techniques are not easily scalable to sizes necessary for transplantation. Currently, kidney organoids are formed at an air-liquid interface on floating membranes similar to embryonic kidney explant models (Morizane et al., 2015; Takasato et al., 2015). The size of the organoid is limited by the extent to which nutrients in

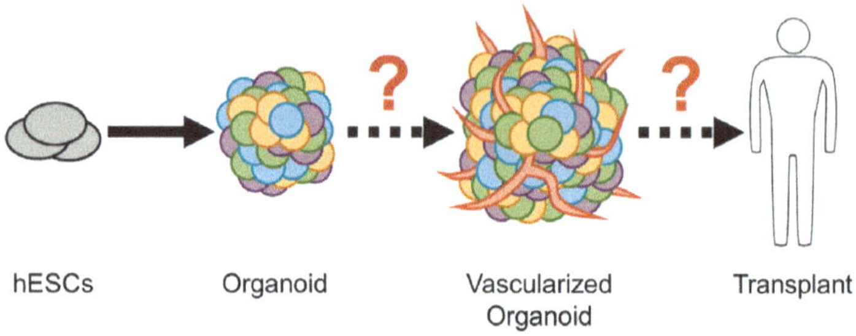

Fig. 7 Flow chart for regenerative therapies. The goal of regenerative therapies is to recapitulate development *ex vivo* to create replacement tissues. To date, many groups have been able to differentiate human embryonic stem cells (gray) into tissue-specific organoids consisting of multiple different lineages. However, these organoids remain relatively small and immature due to, in part, the inability to create a perfused vascular network that is necessary to support larger scale tissues. Are there feasible ways to vascularize organoids such that it integrates properly with the developing tissue? If and when vascularization is achieved, the next major challenge will be to identify ways to successfully transplant the organoids into a recipient. Can these organoids successfully integrate into a recipient's vasculature and carry out a normal organ's function? These are the critical big picture questions that must be answered if organoids can become a viable regenerative therapy.

the media can diffuse across the membrane. Overcoming the diffusion limit in order to support larger organoids will require a perfusable vascular network; however, vascularization of the organoids remains fairly limited. Although $CD31^+$ cells can be identified in many of these systems, vascularization is variable and often the handful of positive cells within an organoid that can be observed are progressively lost over time (Takasato et al., 2015; van den Berg et al., 2018). These results mirror those seen in embryonic kidney explant culture, whereby ECs present within the developing kidney are lost within several days of culture while the epithelial structures continue to grow (Robert et al., 1998). The ability of these ECs to form mature capillary structures *in vitro* is also unclear, as some groups have identified vascularized glomerular structures (Freedman et al., 2015; Takasato et al., 2015; van den Berg et al., 2018) while others do not (Bantounas et al., 2018; Morizane et al., 2015; Taguchi & Nishinakamura, 2017). Additionally, CD31 sometimes marks regions of the cell other than the plasma membrane and is present in cells that do not exhibit normal endothelial morphology or form lumenized vascular structures (O. Cleaver, unpublished data). These issues must be addressed in order for organoids to become a viable therapeutic option.

4.3 Organoid vascularization: Endogenous versus exogenous origins

The issue of vascularization in organoids may be solved through transplantation of these organoids into live hosts. The results of transplantations of the kidney explant model suggest that this will be a viable approach for organoids. Kidney explants transplanted in either the eye chamber, under the kidney capsule, or on the chorioallantoic membrane of quail embryos develop increased vascularization of the graft (Dekel et al., 2003; Hu et al., 2016; Hyink et al., 1996; Robert et al., 1998; Robert, St John, Hyink, & Abrahamson, 1996; Sariola, Ekblom, Lehtonen, & Saxen, 1983). The ECs found in the transplants are usually graft-derived with the exception of the vasculature of transplants placed under the newborn kidney capsule which arises from the host (Robert et al., 1996).

Recent work on transplantation of human organoids mirrors the encouraging results seen from transplantation of kidney explants. One group transplanted human-derived podocytes and proximal tubule cells with human umbilical vein endothelial cells (HUVECs) and mesenchymal stem cells (MSCs) under the kidney capsule (Sharmin et al., 2016). These transplants displayed vascularized glomeruli containing host-derived ECs. Likewise, transplantation of organoids grown from an established human kidney organoid protocol (Takasato et al., 2015) led to increased vascularization, improved epithelial maturation including vascularized glomerular structures, and perfusion of these organoids when transplanted under the kidney capsule (van den Berg et al., 2018) (Fig. 8). Time-matched non-transplanted organoid controls remained in an immature state and did not show signs of increased vascularization. Additionally, the vasculature within the organoids contained a mix of human- and mouse-derived ECs anastomosed together. These studies stand in contrast to the majority of the explant transplantations above in that the majority of the vasculature is host-derived rather than graft-derived. Notably, the explants that did manifest host-derived vasculature were transplanted under the kidney capsule where the organoids were also transplanted, suggesting that whether the transplant contains host-derived or graft-derived vasculature depends on transplant location. Both of these studies show that transplantation of immature human pluripotent stem cell-derived renal organoids promotes further development and maturation than what can be achieved through current *in vitro* systems alone.

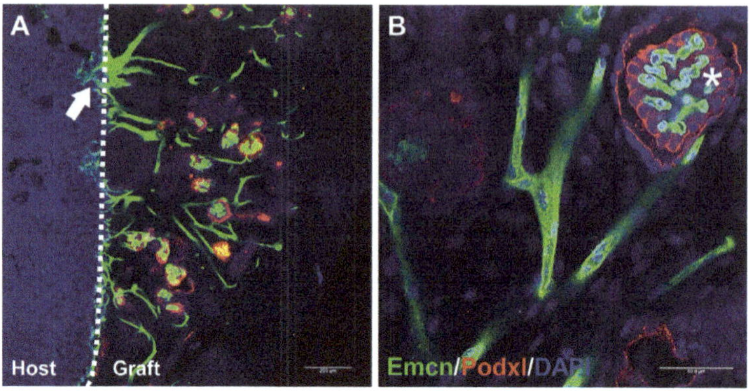

Fig. 8 Improved graft vascularization after engraftment of organoids. Human ESC-derived organoids were transplanted under the kidney capsule of recipient mice and grown for 3 weeks. (A) Grafted organoid exhibits extensive vascularization marked by Endomucin (Emcn$^+$, *green*), anastomosis with host vasculature (*arrow*), and several epithelial structures including glomerular-like structures marked by the podocyte marker Podocalyxin (Podxl$^+$, *red*). (B) High magnification image demonstrates connection between glomerular capillary tuft (*asterisk*) and vascular network in the graft. Scale bars = 200 μm (A), 50 μm (B).

4.3.1 Importance of hemodynamic flow for vascular stability

The results from these transplantation experiments indicate that developing kidney tissues, including the vasculature, require systemic factors in order to fully mature. Indeed, ECs present within organoids regress over time even though the organoids continuously produce VEGF even after nearly 2 months *in vitro* (van den Berg et al., 2018). A major difference between the transplanted and non-transplanted organoids is the presence of blood flow. Although flow is dispensable for EC activity prior to lumenogenesis, the mechanical stimuli created by flow are critical for formation and maintenance of lumens, arteriovenous specification (as mentioned previously), and EC homeostasis (Baeyens & Schwartz, 2016; Culver & Dickinson, 2010; Lucitti et al., 2007; Potter, Lao, Zeng, & Xu, 2014; Udan, Vadakkan, & Dickinson, 2013). Furthermore, the absence of flow in other contexts leads to vessel regression as seen in the current organoid models (Chen et al., 2012; Franco et al., 2015). Importantly, a fully perfused vascular network will be critical to scale these organoids to transplantable sizes.

This raises potential concerns with respect to tissue engineering: if flow within the vasculature is needed to generate fully mature vessels that can support the growth of the organoid, how can we generate a fully perfused vascular network within an organoid or transplantable organ *in vitro*? More simply, can any perfused vascular network be engineered and sustained *in vitro*?

4.3.2 Engineering blood vessels for transplantable organs

Several groups have tackled the latter question through the creation of microfluidic devices (Alonzo, Moya, Shirure, & George, 2015; Kolesky, Homan, Skylar-Scott, & Lewis, 2016; Shirure, Lezia, Tao, Alonzo, & George, 2017; Wang et al., 2016; Zheng et al., 2012). These devices contain perfusable channels that can be lined with ECs. There are two main approaches to generating these devices. One method involves the creation of pre-formed channels that are then lined with ECs by seeding purified EC populations through the same port where media flows (Kolesky et al., 2016; Zheng et al., 2012). ECs attach to the channel walls, proliferate, and form polarized tight junctions indicating EC maturation (Fig. 9A). However, these systems are limited due to the high number of cells needed to seed the lumens, the purity of the population needed to seed the channels, and

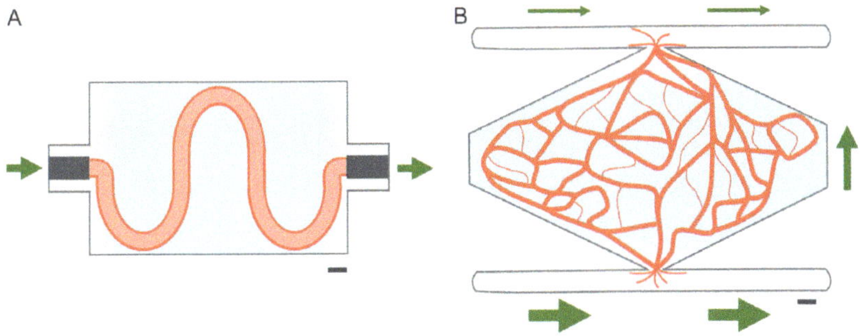

Fig. 9 Microfluidic devices as methods to generate perfused organoids. (A) An example of a three-dimensional printed endothelial tube as described in Kolesky et al. (2016). An initial cast of the tube is printed into the mold with an inlet and outlet. Next, an ECM matrix with or without cells (*blue*) is laid over the sacrificial cast and set. The inlet and outlets are attached to a microfluidic device to both flush out the original cast and ECs (*red*) are perfused through the now open channel. Flow in this system travels through the endothelial tube as marked by the *green arrows*. ECs will then adhere to the walls of the now empty tube and generate a patent vessel-like structure. Scale bar ≈ 100 μm. (B) An example of an engineered capillary network after capillaries have formed as described in Wang et al. (2016). ECs (*red*) and fibroblasts (not pictured) are seeded in a fibrin gel (*blue*) within a mold. This mold lies adjacent and connected to two parallel channels, one with higher rate of flow (*bottom, larger green arrows*) and one with a lower rate of flow (*top, small green arrows*). Because these channels are connected to each other through the mold, an additional direction of flow moves perpendicular to the channel orientation from the high-flow channel to the low-flow channel, perfusing the mold containing the ECs and fibroblasts (*green arrow on right*). Once seeded and subjected to this flow, ECs remodel, form cords, and open patent lumens that anastomose with the two channels allowing for flow through the newly generated endothelial lumens. Note the difference in vessel size between this system and three-dimensional printed channels shown in (A). Scale bar ≈ 100 μm.

the current limitation in creating smaller lumens (<100 μm in diameter) that is more appropriate for capillary flow.

Alternative approaches take advantage of the inherent ability of individual ECs to form a lumenized vascular network by themselves. One approach first seeds a chamber with ECs and perivascular cells, then generates fluid flow across the chamber (Alonzo et al., 2015; Shirure et al., 2017; Wang et al., 2016). In this system, there are two directions of flow: across the chamber that contains the ECs and parallel to this chamber (Fig. 9B). This organization mimics endogenous blood flow in which the flow moving parallel to the chamber represents larger caliber vessels and the flow across the chamber represents capillary flow. The main challenges with this approach are the inability to precisely control where and how lumens form and the lack of larger caliber vessels. To date, there have been no approaches that have attempted to create these vascularized networks in mixed populations of cells as seen in organoids.

4.3.3 The road ahead: Tackling challenges to create transplantable vascularized organs

The challenge of creating a purely *in vitro* perfused vascular network in an organoid may rely on a combination of these approaches. Initial perfusion of the organoid can follow the second design by first seeding organoids in microfluidic devices and stimulating the ECs within the organoid to form capillary-like vessels with patent lumens. These perfused organoids can then be fused to larger, pre-formed, EC-lined channels using techniques as described in the first approach, mimicking the larger caliber renal artery and vein. Therefore, the combination of these approaches may successfully generate a fully vascularized organoid containing different vascular compartments in the correct series as seen *in vivo*. Alternatively, organoids with perfusable vascular networks may be able to form when grown on other scaffolding methods, including porous silk scaffolds or decellularized kidneys (Subramanian et al., 2010; Uzarski et al., 2015). It remains to be seen if any of these approaches will be viable for organoid biology moving forward. Additionally, there may be other systemic factors important for organoid maturation as seen in a transplant model that are not completely recapitulated by the introduction of perfusion *in vitro*. Lastly, these devices may need to be very finely tuned to generate properly patterned and perfused nephrons. This includes laying down the proper ECM matrix, incorporating the right growth factors in the correct location on a cellular level, and ensuring that each cell meets its own metabolic requirements.

The field of organoid biology is still very young and many challenges stand in the way to transform this approach into viable therapies.

The combined effort of multiple disciplines, including developmental biology, bioengineering, and bioinformatics among others, will be absolutely necessary to move this field forward.

5. Concluding remarks

Organ development is intricately tied to and dependent on vascular development. The endothelium is not simply an inert tube but rather a highly active coordinator of organogenesis. Any attempt to recapitulate development to create transplantable organs must find ways to promote and establish organ-specific vasculature to ensure full functionality. Transcriptional analyses of different vascular beds both between organs and within the same organ have revealed stark heterogeneity between these different EC groups (Daniel et al., 2018; Nolan et al., 2013); however, the mechanistic analyses that clearly delineate the factors that determine these differences remain to be elucidated.

Studies that have investigated endothelial-epithelial crosstalk and endothelial patterning have mostly focused on well-established pathways that regulate endothelial activity, such as VEGF among others. Although these studies utilized the same general theme of inhibiting or over-activating these pathways in an organ-specific manner, the exact EC response to these perturbations was entirely location-dependent. These data demonstrate that, while these common signaling cascades are generally important for EC behavior in many organs, there are additional unknown modifiers that transform these common signaling pathways into an organ-specific signal. Determining EC heterogeneity, then, requires defining these unknown modifiers. Ultimately, identifying the mechanisms that determine EC heterogeneity, including their local character, signaling potential, and behaviors, will help propel research into organoid systems or other *in vitro* organ generation approaches and advance the translation of these approaches into viable regenerative therapies.

Acknowledgments

We thank Janet Rossant for the Flk1-eGFP mouse line as well as members of the Cleaver Lab, including Haley Barlow, Xiaowu Gu, Caitlin Braitsch, and Anne Ryan for discussions and critical reading of the manuscript. We also thank Anne Ryan, Alethia Villasenor, Ashwani Gupta, and Leif Oxburgh for figure contribution. This work was supported by National Institute of Diabetes and Digestive and Kidney Diseases including the ReBuilding the Kidney (RBK) consortium and DK106743, DK079862 to O.C.; and the National Institute of Heart, Lung and Blood HL113498 to O.C. Deposited in PMC for release after 12 months.

References

Aird, W. C. (2003). Endothelial cell heterogeneity. *Critical Care Medicine*, *31*, S221–S230.

Aird, W. C. (2007a). Phenotypic heterogeneity of the endothelium: I. Structure, function, and mechanisms. *Circulation Research*, *100*, 158–173.

Aird, W. C. (2007b). Phenotypic heterogeneity of the endothelium: II. Representative vascular beds. *Circulation Research*, *100*, 174–190.

Akeson, A. L., Cameron, J. E., Le Cras, T. D., Whitsett, J. A., & Greenberg, J. M. (2005). Vascular endothelial growth factor-A induces prenatal neovascularization and alters bronchial development in mice. *Pediatric Research*, *57*, 82–88.

Akeson, A. L., Greenberg, J. M., Cameron, J. E., Thompson, F. Y., Brooks, S. K., Wiginton, D., et al. (2003). Temporal and spatial regulation of VEGF-A controls vascular patterning in the embryonic lung. *Developmental Biology*, *264*, 443–455.

Alabi, R. O., Glomski, K., Haxaire, C., Weskamp, G., Monette, S., & Blobel, C. P. (2016). ADAM10-dependent signaling through Notch1 and Notch4 controls development of organ-specific vascular beds. *Circulation Research*, *119*, 519–531.

Alonzo, L. F., Moya, M. L., Shirure, V. S., & George, S. C. (2015). Microfluidic device to control interstitial flow-mediated homotypic and heterotypic cellular communication. *Lab on a Chip*, *15*, 3521–3529.

Aso, Y., Yoneda, K., & Kikkawa, Y. (1976). Morphologic and biochemical study of pulmonary changes induced by bleomycin in mice. *Laboratory Investigation*, *35*, 558–568.

Augustin, H. G., & Koh, G. Y. (2017). Organotypic vasculature: From descriptive heterogeneity to functional pathophysiology. *Science*, *357*.

Azizoglu, D. B., Chong, D. C., Villasenor, A., Magenheim, J., Barry, D. M., Lee, S., et al. (2016). Vascular development in the vertebrate pancreas. *Developmental Biology*, *420*, 67–78.

Azizoglu, D. B., & Cleaver, O. (2016). Blood vessel crosstalk during organogenesis-focus on pancreas and endothelial cells. *Wiley Interdisciplinary Reviews: Developmental Biology*, *5*, 598–617.

Baeyens, N., & Schwartz, M. A. (2016). Biomechanics of vascular mechanosensation and remodeling. *Molecular Biology of the Cell*, *27*, 7–11.

Bantounas, I., Ranjzad, P., Tengku, F., Silajdzic, E., Forster, D., Asselin, M. C., et al. (2018). Generation of functioning nephrons by implanting human pluripotent stem cell-derived kidney progenitors. *Stem Cell Reports*, *10*, 766–779.

Bartlett, C. S., Scott, R. P., Carota, I. A., Wnuk, M. L., Kanwar, Y. S., Miner, J. H., et al. (2017). Glomerular mesangial cell recruitment and function require the co-receptor neuropilin-1. *American Journal of Physiology. Renal Physiology*, *313*, F1232–F1242.

Belteki, G., Haigh, J., Kabacs, N., Haigh, K., Sison, K., Costantini, F., et al. (2005). Conditional and inducible transgene expression in mice through the combinatorial use of Cre-mediated recombination and tetracycline induction. *Nucleic Acids Research*, *33*, e51.

Bhushan, A., Itoh, N., Kato, S., Thiery, J. P., Czernichow, P., Bellusci, S., et al. (2001). Fgf10 is essential for maintaining the proliferative capacity of epithelial progenitor cells during early pancreatic organogenesis. *Development*, *128*, 5109–5117.

Bjarnegard, M., Enge, M., Norlin, J., Gustafsdottir, S., Fredriksson, S., Abramsson, A., et al. (2004). Endothelium-specific ablation of PDGFB leads to pericyte loss and glomerular, cardiac and placental abnormalities. *Development*, *131*, 1847–1857.

Cai, Q., Brissova, M., Reinert, R. B., Pan, F. C., Brahmachary, P., Jeansson, M., et al. (2012). Enhanced expression of VEGF-A in beta cells increases endothelial cell number but impairs islet morphogenesis and beta cell proliferation. *Developmental Biology*, *367*, 40–54.

Cao, G., O'Brien, C. D., Zhou, Z., Sanders, S. M., Greenbaum, J. N., Makrigiannakis, A., et al. (2002). Involvement of human PECAM-1 in angiogenesis and in vitro endothelial cell migration. *American Journal of Physiology. Cell Physiology, 282*, C1181–C1190.
Chen, Y. W., Huang, S. X., de Carvalho, A., Ho, S. H., Islam, M. N., Volpi, S., et al. (2017). A three-dimensional model of human lung development and disease from pluripotent stem cells. *Nature Cell Biology, 19*, 542–549.
Chen, Q., Jiang, L., Li, C., Hu, D., Bu, J. W., Cai, D., et al. (2012). Haemodynamics-driven developmental pruning of brain vasculature in zebrafish. *PLoS Biology, 10*, e1001374.
Chi, J. T., Chang, H. Y., Haraldsen, G., Jahnsen, F. L., Troyanskaya, O. G., Chang, D. S., et al. (2003). Endothelial cell diversity revealed by global expression profiling. *Proceedings of the National Academy of Sciences of the United States of America, 100*, 10623–10628.
Chong, D. C., Koo, Y., Xu, K., Fu, S., & Cleaver, O. (2011). Stepwise arteriovenous fate acquisition during mammalian vasculogenesis. *Developmental Dynamics, 240*, 2153–2165.
Combes, A. N., Phipson, B., Zappia, L., Lawlor, K., Er, P. X., Oshlack, A., et al. (2017). High throughput single cell RNA-seq of developing mouse kidney and human kidney organoids reveals a roadmap for recreating the kidney. *bioRxiv*.
Cowan, M. J., & Crystal, R. G. (1975). Lung growth after unilateral pneumonectomy: Quantitation of collagen synthesis and content. *The American Review of Respiratory Disease, 111*, 267–277.
Culver, J. C., & Dickinson, M. E. (2010). The effects of hemodynamic force on embryonic development. *Microcirculation, 17*, 164–178.
Daniel, E., Azizoglu, D. B., Ryan, A. R., Walji, T. A., Chaney, C. P., Sutton, G. I., et al. (2018). Spatiotemporal heterogeneity and patterning of developing renal blood vessels. *Angiogenesis, 21*, 617–634.
Dekel, B., Burakova, T., Arditti, F. D., Reich-Zeliger, S., Milstein, O., Aviel-Ronen, S., et al. (2003). Human and porcine early kidney precursors as a new source for transplantation. *Nature Medicine, 9*, 53–60.
Del Moral, P. M., Sala, F. G., Tefft, D., Shi, W., Keshet, E., Bellusci, S., et al. (2006). VEGF-A signaling through Flk-1 is a critical facilitator of early embryonic lung epithelial to endothelial crosstalk and branching morphogenesis. *Developmental Biology, 290*, 177–188.
dela Paz, N. G., & D'Amore, P. A. (2009). Arterial versus venous endothelial cells. *Cell and Tissue Research, 335*, 5–16.
DeLisser, H. M., Christofidou-Solomidou, M., Strieter, R. M., Burdick, M. D., Robinson, C. S., Wexler, R. S., et al. (1997). Involvement of endothelial PECAM-1/CD31 in angiogenesis. *The American Journal of Pathology, 151*, 671–677.
deMello, D. E., Sawyer, D., Galvin, N., & Reid, L. M. (1997). Early fetal development of lung vasculature. *American Journal of Respiratory Cell and Molecular Biology, 16*, 568–581.
Deng, Y., Larrivee, B., Zhuang, Z. W., Atri, D., Moraes, F., Prahst, C., et al. (2013). Endothelial RAF1/ERK activation regulates arterial morphogenesis. *Blood, 121*, 3988–3996. S3981–3989.
Dimke, H., Sparks, M. A., Thomson, B. R., Frische, S., Coffman, T. M., & Quaggin, S. E. (2015). Tubulovascular cross-talk by vascular endothelial growth factor a maintains peritubular microvasculature in kidney. *Journal of the American Society of Nephrology, 26*, 1027–1038.
Ding, B. S., Nolan, D. J., Guo, P., Babazadeh, A. O., Cao, Z., Rosenwaks, Z., et al. (2011). Endothelial-derived angiocrine signals induce and sustain regenerative lung alveolarization. *Cell, 147*, 539–553.
Duarte, A., Hirashima, M., Benedito, R., Trindade, A., Diniz, P., Bekman, E., et al. (2004). Dosage-sensitive requirement for mouse Dll4 in artery development. *Genes & Development, 18*, 2474–2478.
Dumont, D. J., Gradwohl, G., Fong, G. H., Puri, M. C., Gertsenstein, M., Auerbach, A., et al. (1994). Dominant-negative and targeted null mutations in the endothelial receptor

tyrosine kinase, tek, reveal a critical role in vasculogenesis of the embryo. *Genes & Development, 8*, 1897–1909.

Dunn, N. R., Winnier, G. E., Hargett, L. K., Schrick, J. J., Fogo, A. B., & Hogan, B. L. (1997). Haploinsufficient phenotypes in Bmp4 heterozygous null mice and modification by mutations in Gli3 and Alx4. *Developmental Biology, 188*, 235–247.

Dye, B. R., Hill, D. R., Ferguson, M. A., Tsai, Y. H., Nagy, M. S., Dyal, R., et al. (2015). In vitro generation of human pluripotent stem cell derived lung organoids. *Elife, 4*.

Edsbagge, J., Johansson, J. K., Esni, F., Luo, Y., Radice, G. L., & Semb, H. (2005). Vascular function and sphingosine-1-phosphate regulate development of the dorsal pancreatic mesenchyme. *Development, 132*, 1085–1092.

Eiraku, M., Takata, N., Ishibashi, H., Kawada, M., Sakakura, E., Okuda, S., et al. (2011). Self-organizing optic-cup morphogenesis in three-dimensional culture. *Nature, 472*, 51–56.

Eiraku, M., Watanabe, K., Matsuo-Takasaki, M., Kawada, M., Yonemura, S., Matsumura, M., et al. (2008). Self-organized formation of polarized cortical tissues from ESCs and its active manipulation by extrinsic signals. *Cell Stem Cell, 3*, 519–532.

Eremina, V., Cui, S., Gerber, H., Ferrara, N., Haigh, J., Nagy, A., et al. (2006). Vascular endothelial growth factor a signaling in the podocyte-endothelial compartment is required for mesangial cell migration and survival. *Journal of the American Society of Nephrology, 17*, 724–735.

Eremina, V., Sood, M., Haigh, J., Nagy, A., Lajoie, G., Ferrara, N., et al. (2003). Glomerular-specific alterations of VEGF-A expression lead to distinct congenital and acquired renal diseases. *Journal of Clinical Investigation, 111*, 707–716.

Farber, G., Hurtado, R., Loh, S., Monette, S., Mtui, J., Kopan, R., et al. (2018). Glomerular endothelial cell maturation depends on ADAM10, a key regulator of Notch signaling. *Angiogenesis, 21*, 335–347.

Ferrara, N., Carver-Moore, K., Chen, H., Dowd, M., Lu, L., O'Shea, K. S., et al. (1996). Heterozygous embryonic lethality induced by targeted inactivation of the VEGF gene. *Nature, 380*, 439–442.

Franco, C. A., Jones, M. L., Bernabeu, M. O., Geudens, I., Mathivet, T., Rosa, A., et al. (2015). Dynamic endothelial cell rearrangements drive developmental vessel regression. *PLoS Biology, 13*, e1002125.

Freedman, B. S., Brooks, C. R., Lam, A. Q., Fu, H., Morizane, R., Agrawal, V., et al. (2015). Modelling kidney disease with CRISPR-mutant kidney organoids derived from human pluripotent epiblast spheroids. *Nature Communications, 6*, 8715.

Garcia, M. D., & Larina, I. V. (2014). Vascular development and hemodynamic force in the mouse yolk sac. *Frontiers in Physiology, 5*, 308.

Gerhardt, H., & Betsholtz, C. (2003). Endothelial-pericyte interactions in angiogenesis. *Cell and Tissue Research, 314*, 15–23.

Gu, G., Brown, J. R., & Melton, D. A. (2003). Direct lineage tracing reveals the ontogeny of pancreatic cell fates during mouse embryogenesis. *Mechanisms of Development, 120*, 35–43.

Guillaume, R., Bressan, M., & Herzlinger, D. (2009). Paraxial mesoderm contributes stromal cells to the developing kidney. *Developmental Biology, 329*, 169–175.

Hatini, V., Huh, S. O., Herzlinger, D., Soares, V. C., & Lai, E. (1996). Essential role of stromal mesenchyme in kidney morphogenesis revealed by targeted disruption of Winged Helix transcription factor BF-2. *Genes & Development, 10*, 1467–1478.

Havrilak, J. A., Melton, K. R., & Shannon, J. M. (2017). Endothelial cells are not required for specification of respiratory progenitors. *Developmental Biology, 427*, 93–105.

Herriges, M., & Morrisey, E. E. (2014). Lung development: Orchestrating the generation and regeneration of a complex organ. *Development, 141*, 502–513.

Hu, Y., Li, M., Gothert, J. R., Gomez, R. A., & Sequeira-Lopez, M. L. (2016). Hemovascular progenitors in the kidney require sphingosine-1-phosphate receptor 1 for vascular development. *Journal of the American Society of Nephrology, 27*, 1984–1995.

Hum, S., Rymer, C., Schaefer, C., Bushnell, D., & Sims-Lucas, S. (2014). Ablation of the renal stroma defines its critical role in nephron progenitor and vasculature patterning. *PLoS One*, *9*, e88400.

Humphreys, B. D., Lin, S. L., Kobayashi, A., Hudson, T. E., Nowlin, B. T., Bonventre, J. V., et al. (2010). Fate tracing reveals the pericyte and not epithelial origin of myofibroblasts in kidney fibrosis. *The American Journal of Pathology*, *176*, 85–97.

Hurtado, R., Zewdu, R., Mtui, J., Liang, C., Aho, R., Kurylo, C., et al. (2015). Pbx1-dependent control of VMC differentiation kinetics underlies gross renal vascular patterning. *Development*, *142*, 2653–2664.

Hyink, D. P., Tucker, D. C., St John, P. L., Leardkamolkarn, V., Accavitti, M. A., Abrass, C. K., et al. (1996). Endogenous origin of glomerular endothelial and mesangial cells in grafts of embryonic kidneys. *The American Journal of Physiology*, *270*, F886–F899.

Jacquemin, P., Yoshitomi, H., Kashima, Y., Rousseau, G. G., Lemaigre, F. P., & Zaret, K. S. (2006). An endothelial-mesenchymal relay pathway regulates early phases of pancreas development. *Developmental Biology*, *290*, 189–199.

Jakkula, M., Le Cras, T. D., Gebb, S., Hirth, K. P., Tuder, R. M., Voelkel, N. F., et al. (2000). Inhibition of angiogenesis decreases alveolarization in the developing rat lung. *American Journal of Physiology. Lung Cellular and Molecular Physiology*, *279*, L600–L607.

Jeansson, M., Gawlik, A., Anderson, G., Li, C., Kerjaschki, D., Henkelman, M., et al. (2011). Angiopoietin-1 is essential in mouse vasculature during development and in response to injury. *The Journal of Clinical Investigation*, *121*, 2278–2289.

Jo, J., Xiao, Y., Sun, A. X., Cukuroglu, E., Tran, H. D., Goke, J., et al. (2016). Midbrain-like organoids from human pluripotent stem cells contain functional dopaminergic and neuromelanin-producing neurons. *Cell Stem Cell*, *19*, 248–257.

Kato, K., Dieguez-Hurtado, R., Park, D. Y., Hong, S. P., Kato-Azuma, S., Adams, S., et al. (2018). Pulmonary pericytes regulate lung morphogenesis. *Nature Communications*, *9*, 2448.

Katsumoto, K., & Kume, S. (2011). Endoderm and mesoderm reciprocal signaling mediated by CXCL12 and CXCR4 regulates the migration of angioblasts and establishes the pancreatic fate. *Development*, *138*, 1947–1955.

Kawaguchi, Y., Cooper, B., Gannon, M., Ray, M., MacDonald, R. J., & Wright, C. V. (2002). The role of the transcriptional regulator Ptf1a in converting intestinal to pancreatic progenitors. *Nature Genetics*, *32*, 128–134.

Kenig-Kozlovsky, Y., Scott, R. P., Onay, T., Carota, I. A., Thomson, B. R., Gil, H. J., et al. (2018). Ascending vasa recta are angiopoietin/Tie2-dependent lymphatic-like vessels. *Journal of the American Society of Nephrology*, *29*, 1097–1107.

Kim, C. F., Jackson, E. L., Woolfenden, A. E., Lawrence, S., Babar, I., Vogel, S., et al. (2005). Identification of bronchioalveolar stem cells in normal lung and lung cancer. *Cell*, *121*, 823–835.

Kitamoto, Y., Tokunaga, H., & Tomita, K. (1997). Vascular endothelial growth factor is an essential molecule for mouse kidney development: Glomerulogenesis and nephrogenesis. *The Journal of Clinical Investigation*, *99*, 2351–2357.

Kolesky, D. B., Homan, K. A., Skylar-Scott, M. A., & Lewis, J. A. (2016). Three-dimensional bioprinting of thick vascularized tissues. *Proceedings of the National Academy of Sciences of the United States of America*, *113*, 3179–3184.

Lammert, E., Cleaver, O., & Melton, D. (2001). Induction of pancreatic differentiation by signals from blood vessels. *Science*, *294*, 564–567.

Lanahan, A. A., Hermans, K., Claes, F., Kerley-Hamilton, J. S., Zhuang, Z. W., Giordano, F. J., et al. (2010). VEGF receptor 2 endocytic trafficking regulates arterial morphogenesis. *Developmental Cell*, *18*, 713–724.

Lanahan, A., Zhang, X., Fantin, A., Zhuang, Z., Rivera-Molina, F., Speichinger, K., et al. (2013). The neuropilin 1 cytoplasmic domain is required for VEGF-A-dependent arteriogenesis. *Developmental Cell*, *25*, 156–168.

Lancaster, M. A., Corsini, N. S., Wolfinger, S., Gustafson, E. H., Phillips, A. W., Burkard, T. R., et al. (2017). Guided self-organization and cortical plate formation in human brain organoids. *Nature Biotechnology*, *35*, 659–666.

Lancaster, M. A., & Knoblich, J. A. (2014). Organogenesis in a dish: Modeling development and disease using organoid technologies. *Science*, *345*, 1247125.

Lancaster, M. A., Renner, M., Martin, C. A., Wenzel, D., Bicknell, L. S., Hurles, M. E., et al. (2013). Cerebral organoids model human brain development and microcephaly. *Nature*, *501*, 373–379.

Landsman, L., Nijagal, A., Whitchurch, T. J., Vanderlaan, R. L., Zimmer, W. E., Mackenzie, T. C., et al. (2011). Pancreatic mesenchyme regulates epithelial organogenesis throughout development. *PLoS Biology*, *9*, e1001143.

Lazarus, A., Del-Moral, P. M., Ilovich, O., Mishani, E., Warburton, D., & Keshet, E. (2011). A perfusion-independent role of blood vessels in determining branching stereotypy of lung airways. *Development*, *138*, 2359–2368.

Le Cras, T. D., Spitzmiller, R. E., Albertine, K. H., Greenberg, J. M., Whitsett, J. A., & Akeson, A. L. (2004). VEGF causes pulmonary hemorrhage, hemosiderosis, and air space enlargement in neonatal mice. *American Journal of Physiology. Lung Cellular and Molecular Physiology*, *287*, L134–L142.

le Noble, F., Moyon, D., Pardanaud, L., Yuan, L., Djonov, V., Matthijsen, R., et al. (2004). Flow regulates arterial-venous differentiation in the chick embryo yolk sac. *Development*, *131*, 361–375.

Lee, J. H., Bhang, D. H., Beede, A., Huang, T. L., Stripp, B. R., Bloch, K. D., et al. (2014). Lung stem cell differentiation in mice directed by endothelial cells via a BMP4-NFATc1-thrombospondin-1 axis. *Cell*, *156*, 440–455.

Leveen, P., Pekny, M., Gebre-Medhin, S., Swolin, B., Larsson, E., & Betsholtz, C. (1994). Mice deficient for PDGF B show renal, cardiovascular, and hematological abnormalities. *Genes & Development*, *8*, 1875–1887.

Lin, E. E., Sequeira-Lopez, M. L., & Gomez, R. A. (2014). RBP-J in FOXD1+ renal stromal progenitors is crucial for the proper development and assembly of the kidney vasculature and glomerular mesangial cells. *American Journal of Physiology. Renal Physiology*, *306*, F249–F258.

Lindahl, P., Hellstrom, M., Kalen, M., Karlsson, L., Pekny, M., Pekna, M., et al. (1998). Paracrine PDGF-B/PDGF-Rbeta signaling controls mesangial cell development in kidney glomeruli. *Development*, *125*, 3313–3322.

Lindstrom, N. O., Tran, T., Guo, J., Rutledge, E., Parvez, R. K., Thornton, M. E., et al. (2018). Conserved and divergent molecular and anatomic features of human and mouse nephron patterning. *Journal of the American Society of Nephrology*, *29*, 825–840.

Lucitti, J. L., Jones, E. A., Huang, C., Chen, J., Fraser, S. E., & Dickinson, M. E. (2007). Vascular remodeling of the mouse yolk sac requires hemodynamic force. *Development*, *134*, 3317–3326.

Magenheim, J., Ilovich, O., Lazarus, A., Klochendler, A., Ziv, O., Werman, R., et al. (2011). Blood vessels restrain pancreas branching, differentiation and growth. *Development*, *138*, 4743–4752.

Marcu, R., Choi, Y. J., Xue, J., Fortin, C. L., Wang, Y., Nagao, R. J., et al. (2018). Human organ-specific endothelial cell heterogeneity. *iScience*, *4*, 20–35.

McCright, B., Gao, X., Shen, L., Lozier, J., Lan, Y., Maguire, M., et al. (2001). Defects in development of the kidney, heart and eye vasculature in mice homozygous for a hypomorphic notch2 mutation. *Development*, *128*, 491–502.

Miner, J. H. (2012). The glomerular basement membrane. *Experimental Cell Research*, *318*, 973–978.

Mohamed, T., & Sequeira-Lopez, M. L. S. (2018). Development of the renal vasculature. *Seminars in Cell & Developmental Biology*.

Morizane, R., Lam, A. Q., Freedman, B. S., Kishi, S., Valerius, M. T., & Bonventre, J. V. (2015). Nephron organoids derived from human pluripotent stem cells model kidney development and injury. *Nature Biotechnology*, *33*, 1193–1200.

Mukouyama, Y. S., Shin, D., Britsch, S., Taniguchi, M., & Anderson, D. J. (2002). Sensory nerves determine the pattern of arterial differentiation and blood vessel branching in the skin. *Cell*, *109*, 693–705.

Munro, D. A. D., Hohenstein, P., Coate, T. M., & Davies, J. A. (2017). Refuting the hypothesis that semaphorin-3f/neuropilin-2 exclude blood vessels from the cap mesenchyme in the developing kidney. *Developmental Dynamics*, *246*, 1047–1056.

Munro, D. A. D., Hohenstein, P., & Davies, J. A. (2017). Cycles of vascular plexus formation within the nephrogenic zone of the developing mouse kidney. *Scientific Reports*, *7*, 3273.

Nakano, T., Ando, S., Takata, N., Kawada, M., Muguruma, K., Sekiguchi, K., et al. (2012). Self-formation of optic cups and storable stratified neural retina from human ESCs. *Cell Stem Cell*, *10*, 771–785.

Newman, P. J., & Newman, D. K. (2003). Signal transduction pathways mediated by PECAM-1: New roles for an old molecule in platelet and vascular cell biology. *Arteriosclerosis, Thrombosis, and Vascular Biology*, *23*, 953–964.

Nolan, D. J., Ginsberg, M., Israely, E., Palikuqi, B., Poulos, M. G., James, D., et al. (2013). Molecular signatures of tissue-specific microvascular endothelial cell heterogeneity in organ maintenance and regeneration. *Developmental Cell*, *26*, 204–219.

O'Brien, L. L., & McMahon, A. P. (2014). Induction and patterning of the metanephric nephron. *Seminars in Cell & Developmental Biology*, *36*, 31–38.

Oxburgh, L., Carroll, T. J., Cleaver, O., Gossett, D. R., Hoshizaki, D. K., Hubbell, J. A., et al. (2017). (Re)building a kidney. *Journal of the American Society of Nephrology*, *28*, 1370–1378.

Pan, F. C., & Wright, C. (2011). Pancreas organogenesis: From bud to plexus to gland. *Developmental Dynamics*, *240*, 530–565.

Parera, M. C., van Dooren, M., van Kempen, M., de Krijger, R., Grosveld, F., Tibboel, D., et al. (2005). Distal angiogenesis: A new concept for lung vascular morphogenesis. *American Journal of Physiology. Lung Cellular and Molecular Physiology*, *288*, L141–L149.

Peng, T., Tian, Y., Boogerd, C. J., Lu, M. M., Kadzik, R. S., Stewart, K. M., et al. (2013). Coordination of heart and lung co-development by a multipotent cardiopulmonary progenitor. *Nature*, *500*, 589–592.

Pierreux, C. E., Cordi, S., Hick, A. C., Achouri, Y., Ruiz de Almodovar, C., Prevot, P. P., et al. (2010). Epithelial: Endothelial cross-talk regulates exocrine differentiation in developing pancreas. *Developmental Biology*, *347*, 216–227.

Plopper, C. G., Suverkropp, C., Morin, D., Nishio, S., & Buckpitt, A. (1992). Relationship of cytochrome P-450 activity to clara cell cytotoxicity. I. Histopathologic comparison of the respiratory tract of mice, rats and hamsters after parenteral administration of naphthalene. *The Journal of Pharmacology and Experimental Therapeutics*, *261*, 353–363.

Potter, C. M., Lao, K. H., Zeng, L., & Xu, Q. (2014). Role of biomechanical forces in stem cell vascular lineage differentiation. *Arteriosclerosis, Thrombosis, and Vascular Biology*, *34*, 2184–2190.

Pu, W., He, L., Han, X., Tian, X., Li, Y., Zhang, H., et al. (2018). Genetic targeting of organ-specific blood vessels. *Circulation Research*, *123*, 86–99.

Rafii, S., Cao, Z., Lis, R., Siempos, I. I., Chavez, D., Shido, K., et al. (2015). Platelet-derived SDF-1 primes the pulmonary capillary vascular niche to drive lung alveolar regeneration. *Nature Cell Biology*, *17*, 123–136.

Reidy, K. J., Villegas, G., Teichman, J., Veron, D., Shen, W., Jimenez, J., et al. (2009). Semaphorin3a regulates endothelial cell number and podocyte differentiation during glomerular development. *Development*, *136*, 3979–3989.

Reinert, R. B., Brissova, M., Shostak, A., Pan, F. C., Poffenberger, G., Cai, Q., et al. (2013). Vascular endothelial growth factor-a and islet vascularization are necessary in developing, but not adult, pancreatic islets. *Diabetes*, *62*, 4154–4164.

Robert, B., St John, P. L., & Abrahamson, D. R. (1998). Direct visualization of renal vascular morphogenesis in Flk1 heterozygous mutant mice. *The American Journal of Physiology*, *275*, F164–F172.

Robert, B., St John, P. L., Hyink, D. P., & Abrahamson, D. R. (1996). Evidence that embryonic kidney cells expressing flk-1 are intrinsic, vasculogenic angioblasts. *The American Journal of Physiology*, *271*, F744–F753.

Rymer, C., Paredes, J., Halt, K., Schaefer, C., Wiersch, J., Zhang, G., et al. (2014). Renal blood flow and oxygenation drive nephron progenitor differentiation. *American Journal of Physiology. Renal Physiology*, *307*, F337–F345.

Saarela, U., Akram, S. U., Desgrange, A., Rak-Raszewska, A., Shan, J., Cereghini, S., et al. (2017). Novel fixed z-direction (FiZD) kidney primordia and an organoid culture system for time-lapse confocal imaging. *Development*, *144*, 1113–1117.

Sand, F. W., Hornblad, A., Johansson, J. K., Loren, C., Edsbagge, J., Stahlberg, A., et al. (2011). Growth-limiting role of endothelial cells in endoderm development. *Developmental Biology*, *352*, 267–277.

Sariola, H., Ekblom, P., Lehtonen, E., & Saxen, L. (1983). Differentiation and vascularization of the metanephric kidney grafted on the chorioallantoic membrane. *Developmental Biology*, *96*, 427–435.

Saxén, L., & Sariola, H. (1987). Early organogenesis of the kidney. *Pediatric Nephrology*, *1*, 385–392.

Schachtner, S. K., Wang, Y., & Scott Baldwin, H. (2000). Qualitative and quantitative analysis of embryonic pulmonary vessel formation. *American Journal of Respiratory Cell and Molecular Biology*, *22*, 157–165.

Sequeira Lopez, M. L., & Gomez, R. A. (2011). Development of the renal arterioles. *Journal of the American Society of Nephrology*, *22*, 2156–2165.

Sequeira-Lopez, M. L., Lin, E. E., Li, M., Hu, Y., Sigmund, C. D., & Gomez, R. A. (2015). The earliest metanephric arteriolar progenitors and their role in kidney vascular development. *American Journal of Physiology. Regulatory, Integrative and Comparative Physiology*, *308*, R138–R149.

Shah, S. R., Esni, F., Jakub, A., Paredes, J., Lath, N., Malek, M., et al. (2011). Embryonic mouse blood flow and oxygen correlate with early pancreatic differentiation. *Developmental Biology*, *349*, 342–349.

Sharmin, S., Taguchi, A., Kaku, Y., Yoshimura, Y., Ohmori, T., Sakuma, T., et al. (2016). Human induced pluripotent stem cell-derived podocytes mature into vascularized glomeruli upon experimental transplantation. *Journal of the American Society of Nephrology*, *27*, 1778–1791.

Shirure, V. S., Lezia, A., Tao, A., Alonzo, L. F., & George, S. C. (2017). Low levels of physiological interstitial flow eliminate morphogen gradients and guide angiogenesis. *Angiogenesis*, *20*, 493–504.

Siemerink, M. J., Klaassen, I., Van Noorden, C. J., & Schlingemann, R. O. (2013). Endothelial tip cells in ocular angiogenesis: Potential target for anti-angiogenesis therapy. *The Journal of Histochemistry and Cytochemistry*, *61*, 101–115.

Simons, M., & Eichmann, A. (2015). Molecular controls of arterial morphogenesis. *Circulation Research*, *116*, 1712–1724.

Sison, K., Eremina, V., Baelde, H., Min, W., Hirashima, M., Fantus, I. G., et al. (2010). Glomerular structure and function require paracrine, not autocrine, VEGF-VEGFR-2 signaling. *Journal of the American Society of Nephrology*, *21*, 1691–1701.

Soriano, P. (1994). Abnormal kidney development and hematological disorders in PDGF beta-receptor mutant mice. *Genes & Development*, *8*, 1888–1896.

Spence, J. R., Mayhew, C. N., Rankin, S. A., Kuhar, M. F., Vallance, J. E., Tolle, K., et al. (2011). Directed differentiation of human pluripotent stem cells into intestinal tissue in vitro. *Nature, 470*, 105–109.

Stevens, P. E., Levin, A., & Kidney Disease: Improving Global Outcomes Chronic Kidney Disease Guideline Development Work Group Members. (2013). Evaluation and management of chronic kidney disease: Synopsis of the kidney disease: Improving global outcomes 2012 clinical practice guideline. *Annals of Internal Medicine, 158*, 825–830.

Stolz, D. B., & Sims-Lucas, S. (2015). Unwrapping the origins and roles of the renal endothelium. *Pediatric Nephrology, 30*, 865–872.

Strilic, B., Kucera, T., Eglinger, J., Hughes, M. R., McNagny, K. M., Tsukita, S., et al. (2009). The molecular basis of vascular lumen formation in the developing mouse aorta. *Developmental Cell, 17*, 505–515.

Subramanian, B., Rudym, D., Cannizzaro, C., Perrone, R., Zhou, J., & Kaplan, D. L. (2010). Tissue-engineered three-dimensional in vitro models for normal and diseased kidney. *Tissue Engineering. Part A, 16*, 2821–2831.

Suri, C., Jones, P. F., Patan, S., Bartunkova, S., Maisonpierre, P. C., Davis, S., et al. (1996). Requisite role of angiopoietin-1, a ligand for the TIE2 receptor, during embryonic angiogenesis. *Cell, 87*, 1171–1180.

Taguchi, A., & Nishinakamura, R. (2017). Higher-order kidney organogenesis from pluripotent stem cells. *Cell Stem Cell, 21*, 730-746.e736.

Takasato, M., Er, P. X., Chiu, H. S., Maier, B., Baillie, G. J., Ferguson, C., et al. (2015). Kidney organoids from human iPS cells contain multiple lineages and model human nephrogenesis. *Nature, 526*, 564–568.

Takebe, T., Sekine, K., Enomura, M., Koike, H., Kimura, M., Ogaeri, T., et al. (2013). Vascularized and functional human liver from an iPSC-derived organ bud transplant. *Nature, 499*, 481–484.

Trisno, S. L., Philo, K. E. D., McCracken, K. W., Cata, E. M., Ruiz-Torres, S., Rankin, S. A., et al. (2018). Esophageal organoids from human pluripotent stem cells delineate Sox2 functions during esophageal specification. *Cell Stem Cell, 23*, 501–515.

Tullius, S. G., & Rabb, H. (2018). Improving the supply and quality of deceased-donor organs for transplantation. *The New England Journal of Medicine, 378*, 1920–1929.

Udan, R. S., Vadakkan, T. J., & Dickinson, M. E. (2013). Dynamic responses of endothelial cells to changes in blood flow during vascular remodeling of the mouse yolk sac. *Development, 140*, 4041–4050.

Ueda, H., Miyazaki, Y., Matsusaka, T., Utsunomiya, Y., Kawamura, T., Hosoya, T., et al. (2008). Bmp in podocytes is essential for normal glomerular capillary formation. *Journal of the American Society of Nephrology, 19*, 685–694.

Uzarski, J. S., Bijonowski, B. M., Wang, B., Ward, H. H., Wandinger-Ness, A., Miller, W. M., et al. (2015). Dual-purpose bioreactors to monitor noninvasive physical and biochemical markers of kidney and liver scaffold recellularization. *Tissue Engineering. Part C, Methods, 21*, 1032–1043.

van den Berg, C. W., Ritsma, L., Avramut, M. C., Wiersma, L. E., van den Berg, B. M., Leuning, D. G., et al. (2018). Renal subcapsular transplantation of PSC-derived kidney organoids induces neo-vasculogenesis and significant glomerular and tubular maturation in vivo. *Stem Cell Reports, 10*, 751–765.

van Tuyl, M., Liu, J., Wang, J., Kuliszewski, M., Tibboel, D., & Post, M. (2005). Role of oxygen and vascular development in epithelial branching morphogenesis of the developing mouse lung. *American Journal of Physiology. Lung Cellular and Molecular Physiology, 288*, L167–L178.

Villasenor, A., Chong, D. C., Henkemeyer, M., & Cleaver, O. (2010). Epithelial dynamics of pancreatic branching morphogenesis. *Development, 137*, 4295–4305.

Wang, X., Phan, D. T., Sobrino, A., George, S. C., Hughes, C. C., & Lee, A. P. (2016). Engineering anastomosis between living capillary networks and endothelial cell-lined microfluidic channels. *Lab on a Chip*, *16*, 282–290.

Warburton, D., El-Hashash, A., Carraro, G., Tiozzo, C., Sala, F., Rogers, O., et al. (2010). Lung Organogenesis. *Current Topics in Developmental Biology*, *90*, 73–158.

Workman, M. J., Mahe, M. M., Trisno, S., Poling, H. M., Watson, C. L., Sundaram, N., et al. (2017). Engineered human pluripotent-stem-cell-derived intestinal tissues with a functional enteric nervous system. *Nature Medicine*, *23*, 49–59.

Xu, K., Sacharidou, A., Fu, S., Chong, D. C., Skaug, B., Chen, Z. J., et al. (2011). Blood vessel tubulogenesis requires Rasip1 regulation of GTPase signaling. *Developmental Cell*, *20*, 526–539.

Yamamoto, H., Yun, E. J., Gerber, H. P., Ferrara, N., Whitsett, J. A., & Vu, T. H. (2007). Epithelial-vascular cross talk mediated by VEGF-A and HGF signaling directs primary septae formation during distal lung morphogenesis. *Developmental Biology*, *308*, 44–53.

Yoshitomi, H., & Zaret, K. S. (2004). Endothelial cell interactions initiate dorsal pancreas development by selectively inducing the transcription factor Ptf1a. *Development*, *131*, 807–817.

Yuan, L., Chan, G. C., Beeler, D., Janes, L., Spokes, K. C., Dharaneeswaran, H., et al. (2016). A role of stochastic phenotype switching in generating mosaic endothelial cell heterogeneity. *Nature Communications*, *7*, 10160.

Zeng, X., Wert, S. E., Federici, R., Peters, K. G., & Whitsett, J. A. (1998). VEGF enhances pulmonary vasculogenesis and disrupts lung morphogenesis in vivo. *Developmental Dynamics*, *211*, 215–227.

Zhao, Z., Nelson, A. R., Betsholtz, C., & Zlokovic, B. V. (2015). Establishment and dysfunction of the blood-brain barrier. *Cell*, *163*, 1064–1078.

Zheng, Y., Chen, J., Craven, M., Choi, N. W., Totorica, S., Diaz-Santana, A., et al. (2012). In vitro microvessels for the study of angiogenesis and thrombosis. *Proceedings of the National Academy of Sciences of the United States of America*, *109*, 9342–9347.

Zhang, Y., Yang, Y., Jiang, M., Huang, S. X., Zhang, W., Al Alam, D., et al. (2018). 3D modeling of esophageal development using human PSC-derived basal progenitors reveals a critical role for notch signaling. *Cell Stem Cell*, *23*, 516–529.

CHAPTER SEVEN

Pancreas organogenesis: The interplay between surrounding microenvironment(s) and epithelium-intrinsic factors

Corinna Cozzitorto[a], Francesca M. Spagnoli[a,b,*]
[a]Max Delbrück Center for Molecular Medicine (MDC), Berlin, Germany
[b]Centre for Stem Cell and Regenerative Medicine, King's College London, London, United Kingdom
*Corresponding author: e-mail addresses: francesca.spagnoli@mdc-berlin.de; francesca.spagnoli@kcl.ac.uk

Contents

Abstract

During embryonic development, pancreatic epithelial cells engage in concomitant morphogenetic and fate specification events that will give rise to the final organ architecture and functions. Cues from the surrounding microenvironment are known to influence the behavior of epithelial progenitors and orchestrate these concomitant events throughout pancreas development. Nevertheless, the composition of the pancreatic microenvironment remains elusive; also, the interplay between components of the surrounding microenvironment and the epithelium is poorly characterized.

We present here a comprehensive overview of the pancreatic microenvironment and what is known regarding distinct cell types, signaling molecules, ECM, that constitute it. We focus on the molecular circuits governing cell-cell interactions, which are at

Current Topics in Developmental Biology, Volume 132
ISSN 0070-2153
https://doi.org/10.1016/bs.ctdb.2018.12.005

play in the developing pancreas, controlling pancreatic progenitor proliferation, morphogenesis, and differentiation. Finally, open questions and implication of future research in this field are discussed in the context of pancreatic diseases, such as diabetes and cancer, as well as therapeutic approaches for these diseases.

The adult pancreas is a composite organ responsible for two vital functions in our body: production of digestive enzymes and glucose metabolism (Slack, 1995). The exocrine portion of the pancreas is composed of acinar and ductal cells, which produce and collect, respectively, the enzymes that facilitate nutrient absorption in the gut. The acinar cells secrete important enzymes for the catalysis of proteins, lipids and carbohydrates. The bicarbonate- and mucin-secreting ductal cells form a complex ductal network allowing the release of the pancreatic secretions into the duodenum. The endocrine compartment is organized into the islets of Langerhans and is responsible for the production and secretion of glucose modulating hormones in the systemic circulation. The most abundant cell type in human and mouse islets are the insulin-producing β-cells ($\geq$80%); the remaining fraction includes glucagon-producing α-cells, somatostatin-producing δ-cells and pancreatic polypeptide-producing PP cells (Pan & Wright, 2011; Slack, 1995). Islet-secreted hormones play a crucial role in the regulation of glucose homeostasis (Romer & Sussel, 2015).

The vital functions controlled by the pancreas are reflected in the morbidity and mortality associated with pancreatic diseases, such as diabetes and pancreatic cancer (Ellis, Ramzy, & Kieffer, 2017; Feig et al., 2012; Pociot & Lernmark, 2016; Zheng, Ley, & Hu, 2018). Diabetes is generally characterized by hyperglycemia due to loss or dysfunction of the endocrine β-cells (Pociot & Lernmark, 2016). The challenge to curing diabetes relies on replacing the lost or damaged insulin-producing β-cells in diabetic patients (Sneddon et al., 2018). On the other hand, the exocrine compartment is the target of pancreatitis and pancreatic ductal adenocarcinoma (PDAC), which is a devastating malignancy with an extremely poor prognosis and unmet medical need (Feig et al., 2012; Sinha & Leach, 2016). Work from the past two decades clearly showed how much the understanding of the mechanisms underlying pancreas formation has helped elucidating the pathogenesis of pancreatic diseases and developing therapeutic strategies for these diseases. Hence, remarkable advances in generating a sustainable source of functional human stem cell-derived insulin-producing cells for cell transplantation have been made thanks to the identification of developmental factors, which are released from the pancreatic microenvironment and regulate expansion and differentiation of pancreatic cells (Sneddon et al., 2018). Moreover, aberrant activation of embryonic signaling pathways is frequent in PDAC, making developmental regulators therapeutically attractive (Biankin et al., 2012; Roy et al., 2016).

In this review, we focus on the pancreatic microenvironment and its different components with a particular emphasis on their interplay with the epithelium during pancreas organogenesis. We draw mainly from findings in the mouse model, but that are conserved in other model organisms, including in humans. We discuss the implications of these findings in the context of pancreatic diseases and therapeutic approaches.

1. Overview of embryonic pancreas development

Many excellent and comprehensive reviews have been written on the transcriptional control underlying pancreas development (Arda, Benitez, & Kim, 2013; Jørgensen et al., 2006; Larsen & Grapin-Botton, 2017; Pan & Wright, 2011; Shih, Wang, & Sander, 2013; Slack, 1995). Here, we provide only a brief overview of the key morphogenetic events and transcription factors (TF) important for establishing pancreatic cell identity.

The pancreas, together with the liver, biliary system, lung, thymus, thyroid and epithelial cells lining the respiratory and digestive systems, originates from the endoderm germ layer (Slack, 1995; Spagnoli, 2007; Zorn & Wells, 2007). Pancreatic fate specification occurs in the mouse embryo at approximately embryonic day (E)8.5–8.75, when the expression of the homeodomain TF *Pancreatic and duodenal homeobox 1* (*Pdx1*) starts in two distinct domains of the anterior region of the gut endoderm (*a.k.a.* foregut): the ventral and dorsal *anlagen* (Slack, 1995; Spagnoli, 2007; Zorn & Wells, 2007).

Commitment of endodermal cells to a pancreatic fate is a multistep process, involving a continuous crosstalk between the endoderm and surrounding tissues that eventually lead to the activation of a pancreatic gene expression program (Gittes, 2009; Spagnoli, 2007). At E9.5–10.5 in the mouse, mesenchymal cells condense around the two pancreatic buds, which consist of a pseudostratified epithelium positive for Pdx1 (Gittes, 2009; Landsman et al., 2011). Shortly after, at E11.5 pancreatic morphogenesis starts, allowing the pancreas to acquire the characteristic tree-like epithelial structure composed of a network of well-formed branches and connected ducts, necessary to the organ to fulfill its exocrine functions (Larsen & Grapin-Botton, 2017; Shih et al., 2013). Importantly, branching morphogenesis has been shown to be tightly coupled, both in a mechanistic and temporal manner, with growth and differentiation of the pancreatic cells (Larsen & Grapin-Botton, 2017; Villasenor et al., 2012). The first step of this complex process is the acquisition of an apico-basal polarity from the pancreatic epithelial cells, which is dependent on the Rho GTPase Cdc42 (Kesavan et al., 2009). Subsequently, some of these cells undergo apical constriction to form rosette structures with microlumens at the center (Hick et al., 2009; Villasenor et al., 2012). Microlumen formation and subsequent epithelial remodeling further depend on Eph/ephrin signaling (Villasenor et al., 2012). Moreover, actin cytoskeleton remodeling underlies apical constriction in pancreatic progenitors; this is regulated by Rho and the RhoGAP Stard13, whose disruption leads to reduced epithelial branching and associated pancreatic growth defects in the mouse (Petzold, Naumann, & Spagnoli, 2013). Around E12.5, the fusion of the microlumina allows the establishment of primary branches, which subsequently undergo a combination of lateral branching and tip-splitting to generate a complex tubular plexus (Hick et al., 2009; Pan & Wright, 2011; Puri & Hebrok, 2007; Villasenor et al., 2012). This coincides with a gradual spatial segregation of the multipotent pancreatic progenitor cells (MPC) and establishment of distinct pancreatic lineages. MPCs initially co-express a set of pancreatic TFs, including *Pdx1*, *Pancreatic-specific Transcription Factor 1a* (*Ptf1a*), *SRY-box 9* (*Sox9*), and *Hepatocyte nuclear factor 1β* (*Hnf1β*) (Ahlgren, Pfaff, Jessell, Edlund, & Edlund, 1997; Arda et al., 2013; Haumaitre et al., 2005; Kawaguchi et al., 2002; Offield et al., 1996; Seymour et al., 2007). After E12.5, pancreatic epithelial cells organize themselves in two different domains, called "tip" and "trunk" domains: the tip domain is marked by the expression of *Ptf1a*, *Gata4*, *c-myc* and *Cpa1* and gives rise to acinar cells (Pan & Wright, 2011; Zhou et al., 2007), whereas the trunk region contains bipotent cells that are committed to become ductal and endocrine cells and express *Nkx6 homeobox 1* (*Nkx6.1*), *Nkx6.2*, *Sox9*, *Hnf1β*, *Gata6* and *Hairy and enhancer of split* (*Hes1*) (Haumaitre et al., 2005; Henseleit et al., 2005; Jensen et al., 2000; Sander et al., 2000; Seymour et al., 2007; Solar et al., 2009). The mechanisms responsible for this early regionalization and underlying signaling pathways remain unknown. In other developing organ systems, focal sources of morphogens in the surrounding mesenchyme were shown to establish distinct gene expression domains (Hogan, 1999). If and how the pancreatic mesenchyme establishes "niche-like" environments responsible for distinct differentiation program and tissue-regionalization remain open questions.

Subsequently, following gut rotation, the ventral bud moves distally and fuses with the dorsal bud to form a single nascent pancreas (Jørgensen et al., 2006; Slack, 1995). This coincides with the major wave of pancreatic cell differentiation in a process known as "secondary transition" (Larsen & Grapin-Botton, 2017; Pan & Wright, 2011; Shih et al., 2013). Acinar cells at the tips of the branches begin expressing digestive enzymes, such as *Elastase* and *Amylase* (Pan & Wright, 2011). Within the trunk compartment, bipotent progenitor cells are directed into endocrine fate by induction of *Neurogenin 3* (*Ngn3*), whereas those that do not activate *Ngn3* will differentiate along the ductal compartment pathway (Gouzi, Kim, Katsumoto, Johansson, & Grapin-Botton, 2011; Magenheim, Klein, et al., 2011; Solar et al., 2009). Once *Ngn3* transcription has initiated, it triggers exit from the cell cycle and delamination from the trunk into the surrounding mesenchyme where the endocrine progenitors will differentiate into the five different endocrine cell types (Gouzi et al., 2011; Larsen & Grapin-Botton, 2017; Miyatsuka, Kosaka, Kim, & German, 2011). How the activation or repression of *Ngn3* in ductal cells is mechanistically achieved is still under active investigation (Bankaitis, Bechard, & Wright, 2015; Larsen & Grapin-Botton, 2017; Löf-Öhlin et al., 2017). The level of Notch signaling has been demonstrated to exert antagonistic effects on *Neurog3* transcription. The Notch target and effector *Hes1* has been shown to repress *Ngn3* transcription and accelerate Ngn3 protein degradation (Jensen et al., 2000; Larsen & Grapin-Botton, 2017; Qu et al., 2013); but, how different trunk cells acquire different Notch activation levels is still unknown. *Ngn3* activates a set of TFs, which, in turn, control further differentiation of endocrine progenitor cells into the five endocrine cell types, such as the LIM homeobox transcription factor *Islet1* (*Isl1*), *Nkx2.2*, *Insulinoma-associated 1* (*Insm1*), *Rfx6* and *Neurogenic differentiation 1* (*NeuroD1*) (Ahlgren et al., 1997; Mastracci, Anderson, Papizan, & Sussel, 2013; Osipovich et al., 2014; Romer & Sussel, 2015).

Of note, the majority of these TFs play multiple roles in pancreatic development as well as in the adult organ, implying that the same TF can exert different functions depending on the cellular context. For instance, TFs that are required for β-cell differentiation, such as Pdx1 and Nkx6.1, must be actively maintained throughout the cell's lifetime; indeed, their loss triggers β-cell de-differentiation or switch of identity (Gao et al., 2014; Romer & Sussel, 2015; Taylor, Liu, & Sander, 2013). Importantly, the same events as well as transcriptional silencing of these TFs have recently emerged to be associated with β-cell dysfunction and Type 2 diabetes (T2D) in humans (reviewed in Romer & Sussel, 2015), stressing the relevance of further elucidating the islet-cell specific transcriptional networks.

2. The pancreatic microenvironment

During embryonic development, the pancreatic microenvironment is composed of multiple cell types, including mesenchymal cells, endothelial cells, neural crest-derived and immune cells, as well as structural molecules, which are part of the extracellular matrix (ECM) (Azizoglu & Cleaver, 2016; Gittes, 2009; Hisaoka, Haratake, & Hashimoto, 1993; Landsman et al., 2011; Nekrep, Wang, Miyatsuka, & German, 2008; Pierreux et al., 2010; Shih, Panlasigui, Cirulli, & Sander, 2016) (Fig. 1). Most of these components may exert distinct roles at subsequent steps of pancreatic development, influencing fate specification, growth, morphogenesis and/or differentiation. External inputs from the pancreatic microenvironment,

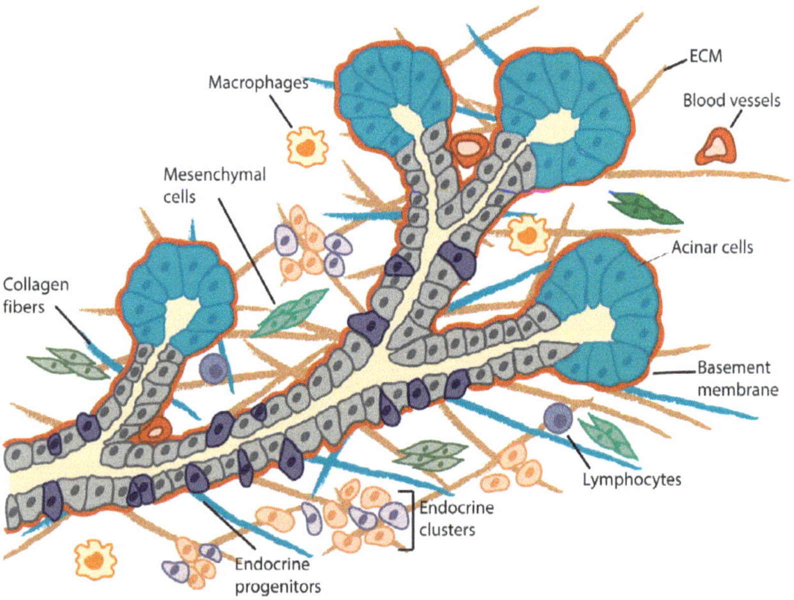

Fig. 1 The pancreatic microenvironment during embryonic development. Schematic representation of the developing pancreatic epithelium and surrounding microenvironment. This is composed by multiple cell populations, such as fibroblast-like mesenchymal, endothelial, immune and neural crest-derived cells, and ECM components. In blue, tip pancreatic cells; in gray duct cells; in purple, delaminating endocrine progenitors.

including signaling effectors and cell interactions, and their impact(s) on pancreatic epithelial cells at different developmental stages are discussed in the following sections.

2.1 Early tissue interactions underlying pancreatic fate specification

Studies in different vertebrate species have shown that during gastrulation instructive signals from the adjacent germ layers, including Fibroblast Growth Factor (FGF), Retinoic Acid (RA), Bone Morphogenetic Protein (BMP), render the endoderm competent to receive subsequent pro-pancreatic signals signaling effectors (Spagnoli, 2007; Zorn & Wells, 2007). After gastrulation, once the foregut endoderm has completed its ventral closure, the prospective pancreatic endoderm is exposed to a series of tissue interactions, which are required for its fate specification (Pan & Wright, 2011; Shih et al., 2013; Spagnoli, 2007). Interestingly, because the organ arises from two separate endoderm domains, different tissue interactions are established during early development. Specifically, the dorsal bud is in contact with the notochord and dorsal aorta, while the ventral buds with the lateral plate mesoderm, cardiac mesoderm and vitelline veins (Deutsch, Jung, Zheng, Lóra, & Zaret, 2001; Gittes, 2009; Hebrok, Kim, & Melton, 1998; Kim, Hebrok, & Melton, 1997; Kumar, Jordan, Melton, & Grapin-Botton, 2003; Lammert, Cleaver, & Melton, 2001; Yoshitomi & Zaret, 2004). Despite these contrasting local niches and some transcriptional heterogeneity revealed by bulk and single-cell RNASeq (Li et al., 2018; Rodríguez-Seguel et al., 2013), both pancreatic buds possess the ability to differentiate into all pancreatic cell types, suggesting that different developmental cues can be used to promote pancreatic fate.

The notochord lies next to the dorsal midline endoderm and remains in contact with the dorsal pancreatic endoderm until E8 in the mouse embryo, when it is displaced by the fusion of the paired dorsal aortas (Lammert et al., 2001; Slack, 1995). Seminal studies in the chick showed that the notochord is required and sufficient for pancreatic fate induction (Hebrok et al., 1998; Kim et al., 1997). Removal of the notochord in explanted chick embryos prevents dorsal pancreas formation, due to an ectopic expression of *Sonic Hedgehog* (*Shh*) in the pancreatic region (Hebrok et al., 1998; Kim et al., 1997). Conversely, grafting an ectopic notochord adjacent to the ventral foregut in a chick embryo results into suppression of endodermal *Shh* together with induction of pancreas formation. Additionally, these studies discovered that Activin-βB and FGF2 are the morphogens released by the chick notochord and responsible for repression of *Shh* expression in the pre-pancreatic region, allowing the initiation of pancreatic gene expression (Hebrok et al., 1998; Kim et al., 1997). Consistently, mouse mutants for ActRIIA and ActRIIB activin receptors exhibit pancreatic hypoplasia and defects in foregut patterning (Kim et al., 2000). Finally, gain-of-function studies of the Shh pathway in the mouse further established its role as repressor of pancreatic fate. Specifically, inactivation of two Hedgehog inhibitors, the receptor *Patched 1* (*Ptch1*) and *Hedgehog-interacting protein* (*Hhip*), leads to increased Hedgehog signaling and disruption of pancreas formation (Kawahira et al., 2003; Kawahira, Scheel, Smith, German, & Hebrok, 2005). Also, ectopic expression of *Shh* in the pancreatic endoderm induces partial conversion of the endoderm into duodenal fate and change in identity of the surrounding pancreatic mesenchyme (Apelqvist, Ahlgren, & Edlund, 1997).

Subsequently, around E9 in the mouse embryo, the paired dorsal aortae start to fuse at the midline and become interposed between the notochord and the dorsal region of the foregut endoderm (Azizoglu & Cleaver, 2016; Lammert et al., 2001). The newly established contact with endothelial cells is important for proper dorsal pancreas formation (Lammert et al., 2001; Yoshitomi & Zaret, 2004). In vitro tissue recombination experiments showed that aortic endothelial cells induce *Pdx1* and *insulin* expression in the adjacent dorsal endoderm, together with formation of structures resembling pancreatic tissue (Lammert et al., 2001). However, in vivo studies in mouse embryos null for *Flk1* [encoding Vascular Endothelial Growth Factor receptor (VEGFR) 2], which lack all endothelial cells, indicated that aortic endothelial cells are not required for the induction of *Pdx1* in the dorsal foregut, but for the later *Ptf1a* expression and expansion of the dorsal bud (Jacquemin et al., 2006; Yoshitomi & Zaret, 2004). Moreover, stimuli produced by the aorta are essential for the survival of the dorsal mesenchyme, which in turn influences dorsal pancreas development (Jacquemin et al., 2006; Yoshitomi & Zaret, 2004).

Retinoic acid (RA) also exerts a conserved role in pancreatic fate induction of the dorsal endoderm but not of the ventral bud (Arregi et al., 2016; Chen et al., 2004; Kumar et al., 2003; Martín et al., 2005; Stafford & Prince, 2002) (see Section 2.2).

The ventral region of the foregut is exposed to a different set of interactions, including with the lateral plate mesoderm and its derivatives, the cardiac mesoderm and septum transversum mesenchyme as well as to the vitelline veins (Azizoglu & Cleaver, 2016; Deutsch et al., 2001; Gualdi et al., 1996; Lammert et al., 2001). These unique interactions are primarily important for the lineage restriction between liver and pancreas cell fate within the ventral foregut endoderm (Deutsch et al., 2001; Gualdi et al., 1996; Wandzioch & Zaret, 2009). The ventral foregut endoderm contains a putative bipotent hepato-pancreatic progenitor population, which by default is fated to adopt a pancreatic fate (Angelo, Guerrero-Zayas, & Tremblay, 2012; Deutsch et al., 2001; Rodríguez-Seguel et al., 2013; Tremblay & Zaret, 2005). Mouse foregut explants cultured in vitro activate pancreatic gene expression, while exposure to FGF- and BMP-mediated signalings from the cardiac mesoderm induces hepatic fate (Deutsch et al., 2001; Gualdi et al., 1996; Rossi, Dunn, Hogan, & Zaret, 2001). More specifically, BMP inhibition has been suggested to be necessary for inducing pancreatic cell fate, whereas both BMP and FGF signaling are required for hepatic cell fate acquisition within the ventral foregut in a conserved fashion across species (Chung, Shin, & Stainier, 2008; Deutsch et al., 2001; Rossi et al., 2001; Spagnoli & Brivanlou, 2008). Zaret and colleagues discovered that the TF *Hematopoietically expressed homeobox gene* (*Hex*) helps to define the position of the prospective ventral pancreatic endoderm with respect to the cardiac mesoderm in the mouse (Bort, Martinez-Barbera, Beddington, & Zaret, 2004). In the absence of *Hex*, ventral pancreatic progenitors fail to proliferate beyond the cardiac mesoderm to escape its hepatic-inducing signals and be specified (Bort et al., 2004). The establishment of proper levels of Wnt/β-catenin signaling and its temporal sequential activation in the anterior endoderm are also known to be essential for foregut identity and organ formation (Li et al., 2008; Ober, Verkade, Field, & Stainier, 2006; Rodríguez-Seguel et al., 2013). Recent observations identified the non-canonical Wnt pathway as a developmental regulator of the liver *versus* pancreas fate decision, promoting pancreatic fate in the endoderm of *Xenopus* embryos and mouse ES cultures (Rodríguez-Seguel et al., 2013).

The Hh pathway needs to be repressed during ventral pancreas fate specification, like in the dorsal bud, however, this must occur *via* a notochord-independent mechanism (Gittes, 2009; Hebrok, Kim, Jacques, McMahon, & Melton, 2000). Recent observations from Sussel and colleagues showed that *Shh* expression is upregulated in the pancreatic endoderm upon *Gata4/6* gene ablation, which results in arrested pancreatic development and cell fate switch of ventral pancreatic cells into intestinal fate and dorsal pancreas into stomach (Xuan & Sussel, 2016). This suggests a repressive interaction between Shh and the Gata4/6 TFs, representing a unifying mechanism for dorsal and ventral pancreas fate specification (Xuan & Sussel, 2016).

2.2 The pancreatic mesenchyme

Mesenchymal cells derived from the splanchnic mesoderm condense around the evaginating pancreatic epithelium at E10 in the mouse (Gittes, 2009; Landsman et al., 2011). Seminal studies in the 1960s based on ex vivo tissue recombination experiments demonstrated the absolute requirement for pancreatic mesenchyme in epithelial growth and morphogenesis (Golosow & Grobstein, 1962; Wessells & Cohen, 1967). Subsequent ex vivo organ culture studies also revealed a critical role for the mesenchyme in pancreatic cell differentiation (Attali et al., 2007). Specifically, "naked epithelium" cultured in the absence of pancreatic mesenchyme sustains islet growth, while the presence of the pancreatic mesenchyme is required for induction of acinar fate (Miralles, Czernichow, & Scharfmann, 1998). Only recently, Landsman et al. provided the first in vivo genetic evidence for the requirement of the pancreatic mesenchyme throughout pancreas organogenesis in the mouse (Landsman et al., 2011). Combined use of Cre-lox technology with diphtheria toxin (DT)-induced apoptosis enabled tissue-specific depletion of pancreatic mesenchymal cells at different developmental stages in vivo (Landsman et al., 2011). Mesenchymal cells ablation at the onset of pancreas development results into pancreas agenesis, whereas depletion of mesenchymal cells at later stages leads to pancreatic hypoplasia and impaired epithelial morphogenesis, due to reduced epithelial proliferation (Landsman et al., 2011).

Even though the relevance of the mesenchyme during pancreatic organogenesis is well established, its embryonic origin is still elusive. Recent DiI fate mapping in mouse embryos cultured ex vivo suggests a common origin of dorsal and ventral pancreatic mesenchyme, both arising from a portion of the coelomic mesothelium (Angelo & Tremblay, 2018). Yet, a common origin of the pancreatic mesenchyme is at odds with the differences reported so far between the transcriptional programs of dorsal and ventral mesenchyme(s). For instance, the TF *Isl1* is expressed in the dorsal mesenchyme, but not in the ventral one (Ahlgren et al., 1997). Consistently, *Isl1* gene deletion in the mouse leads to failure of dorsal mesenchyme development, which in turn affects dorsal Pdx1+ endoderm outgrowth and differentiation (Ahlgren et al., 1997). Similarly, *N-cadherin* (encoded by *Cdh2*) is expressed at high levels in the dorsal pancreatic mesenchyme starting from E9.5 and, subsequently, in the endocrine progenitors at E12.5, once delamination has started (Esni, Johansson, Radice, & Semb, 2001). Genetic ablation of *Cdh2* results into dorsal pancreas agenesis due to impaired dorsal pancreatic mesenchyme survival (Esni et al., 2001). Despite the similarities in expression pattern and mutant phenotypes, Isl1 and N-cadherin seem to function independently, since *Isl1* expression is unchanged in N-cadherin-deficient mesenchyme (Esni et al., 2001). Future genetic lineage studies are required to further define the origin of the pancreatic mesenchyme and clarify if ventral and dorsal mesenchyme(s) share a common origin; the Dil-labeling results do not exclude the possibility of an additional mesenchymal source, as previously reported in other organs (Kfoury & Scadden, 2015; Lee et al., 2017; Zepp et al., 2017).

The dorsal pancreatic mesenchyme not only is crucial for the development of the dorsal pancreatic epithelium, but it also contributes to spleen formation (Asayesh et al., 2006; Hecksher-Sørensen et al., 2004; Koss et al., 2012). Notably, the TFs *Nkx3.2* and *Nkx2.5*, which initially mark dorsal pancreatic mesenchymal cells, become restricted to the spleen at E12.5, and mice mutants for these genes are asplenic (Asayesh et al., 2006; Koss et al., 2012). Specifically, in the absence of *Nkx3.2* the splenic mesenchyme fails to condense and does not segregate from the pancreas, resulting in metaplastic transformation of the pancreatic epithelium and mesenchyme into intestinal-like tissue, possibly due to ectopic *Shh*-expression (Asayesh et al., 2006; Hecksher-Sørensen et al., 2004). Interestingly, rare human congenital disorders present concurrent splenic defects (asplenia or polyspenia) and dorsal pancreatic aplasia/hypoplasia is likely to be explained by the common origin of the splenic and pancreatic mesenchyme (Drut, Drut, Gilbert-Barness, & Reynolds, 1993; Hadar, Gadoth, Herskovitz, & Heifetz, 1991; Herman & Siegel, 1991; Kim, Shin, Kim, Bae, & Park, 2009; Low, Williams, & Chaganti, 2011; Sener & Alper, 1994; Sriplung, 1991; Wainwright & Nelson, 1993).

2.3 Mesenchymal signaling molecules

To date, several signaling molecules have been identified to be released by pancreatic mesenchymal cells and participate in the epithelial-mesenchymal crosstalk during pancreatic development. Often these signaling molecules exert distinct roles during development, influencing either the expansion of epithelial progenitors, their differentiation or both. A selection of the major signaling pathways involved in the pancreatic epithelial-mesenchymal crosstalk is reviewed in the following section.

2.3.1 FGF signaling pathway

In the mouse embryo from E9.5 onward, different members of the *Fgf* family are expressed in the pancreatic mesenchyme, including *Fgf1*, *Fgf7* and *Fgf10*, whereas the *Fgf receptor 2b* (*Fgfr2b*) and *Fgfr3* are expressed in the pancreatic epithelium (Arnaud-Dabernat et al., 2007; Bhushan et al., 2001; Miralles, Czernichow, Ozaki, Itoh, & Scharfmann, 1999; Ye, Duvillié, & Scharfmann, 2005). FGF10 is among the best characterized signaling molecules in the pancreatic mesenchyme and its function has been extensively studied both ex vivo in pancreatic explants as well as in vivo in the mouse. *Fgf10−/−* embryos display severe pancreatic hypoplasia both dorsally and ventrally, due to failure of proliferation of specified Pdx1+ progenitors (Bhushan et al., 2001). Consistently, the pancreas in *Fgfr2b−/−* embryos is smaller than in wild-type littermates and pancreatic ductal branching as well as duct cell proliferation are significantly reduced, even though to a lesser extent than in the absence of FGF10 (Pulkkinen, Spencer-Dene, Dickson, & Otonkoski, 2003). Further investigation showed that *Fgf10* is required for the maintenance of *Pdx1* expression and induction of *Ptf1a* in MPCs (Jacquemin et al., 2006). A more recent study identified a relationship between FGF10 and another the pancreatic TF, Sox9, which is required for progenitor proliferation and survival (Seymour et al., 2012). Specifically, mesenchymal-derived FGF10, transduced through the FGFR2B receptor, sustains *Sox9* expression in the epithelium, which in turn is required for *Fgfr2b* expression and, therefore, FGF10 response (Seymour et al., 2012). Mechanistically, this process involves a SOX9/FGF10/FGFR2B feed-forward loop in the maintenance of pancreatic identity, and its perturbation results into loss of pancreatic cell identity and conversion into liver fate (Seymour et al., 2012). Interestingly, such a functional relationship between FGF10 and Sox9 is conserved in zebrafish embryos, where combined action of FGF10 and Sox9 is essential for the morphogenesis of the hepatopancreatic ducts and prevents pancreatic, hepatic and intestinal cell plasticity (Dong et al., 2007; Manfroid et al., 2007).

Unlike the other FGFR-driven signaling, FGFR3 has been reported to inhibit the expansion of the pancreatic epithelium (Arnaud-Dabernat et al., 2007). Likely, these differences in activities reflect temporal changes in expression of the signaling components and the dynamic nature of cellular competence in response to extracellular signals.

2.3.2 TGFβ/BMP signaling pathway

The TGFβ ligands, *Tgfβ1*, *Tgfβ2* and *Tgfβ3*, are expressed in the mouse pancreatic epithelium starting from E12.5, becoming subsequently restricted to the acinar cells (Crisera et al., 2000; Gittes, 2009). At the same embryonic stage, the TGFβ receptors, *Tgfbr1* and *Tgfbr2*, show an expression pattern similar to their ligands, but localize to ductal cells as development proceeds (Gittes, 2009; Tulachan et al., 2007). In addition, *Tgfbr1* is also expressed in the pancreatic mesenchyme at early stages (Tulachan et al., 2007). Multiple studies supported a role for TGFβ in the process of delamination of endocrine progenitors from the trunk compartment into the surrounding mesenchyme and consequent coalescence into islets (Miralles, Battelino, Czernichow, & Scharfmann, 1998; Sanvito et al., 1995; Tulachan et al., 2007). In particular, TGFβ1 has been shown to control the activity of Matrix metalloproteinase-2 (MMP-2) and islet morphogenesis in vitro (Miralles, Battelino, et al., 1998).

Activins and their receptors are important in the early foregut endoderm for pancreatic specification (see above), but also for endocrine differentiation at later stages (Maldonado et al., 2000; Verschueren et al., 1995). In particular, the ligands Activins A and B localize to the developing endocrine cells, being enriched in the glucagon-expressing α-cells (Maldonado et al., 2000). *Activin receptor type IIB* (*Acvr2b*) and *Activin receptor type IIA* (*Acvr2a*) mutant mice display pancreatic hypoplasia with reduced number of endocrine cells, which in turn lead to hypoplastic islets (Kim et al., 2000). Moreover, ex vivo treatment of mouse pancreatic explants with recombinant Activin results in reduced branching morphogenesis and an increased number of insulin-expressing cells (Demeterco, Beattie, Dib, Lopez, & Hayek, 2000; Ritvos et al., 1995). Consistently, when pancreatic explants are treated with Follistatin, a well-known Activin antagonist, cells undergo enhanced exocrine differentiation at the expenses of the endocrine cell lineage (Miralles, Czernichow, et al., 1998).

BMP molecules are also expressed in the developing pancreas and have been suggested to regulate the development of both pancreatic epithelium and mesenchyme (Ahnfelt-Rønne, Ravassard, Pardanaud-Glavieux, Scharfmann, & Serup, 2010; Dichmann, Miller, Jensen, Heller, & Serup, 2003; Goulley, Dahl, Baeza, Mishina, & Edlund, 2007). Bmp signaling is repeatedly used in this developmental process, exerting different, sometimes even opposite, actions. At early developmental stages, BMP is antagonistic to pancreatic fate and acts as a pro-hepatic factor, as seen in multiple vertebrate species, while few hours later it has opposite effects, promoting adoption of pancreatic fate in mouse foregut explants (Chung et al., 2008; Deutsch et al., 2001; Rossi et al., 2001; Spagnoli & Brivanlou, 2008; Wandzioch & Zaret, 2009). Additionally, conditional deletion of the BMP receptor 1A (ALK3) in insulin-positive cells identified a role for BMP in glucose-stimulated insulin secretion in the mouse adult pancreas (Goulley et al., 2007). Similarly, ectopic expression of BMP inhibitors, such as *Noggin* and *Smad6*, under the *Pdx1* promoter results into glucose intolerance and diabetes, but normal pancreatic morphology (Goulley et al., 2007).

Likewise, in mouse or human ES cultures differentiating along the pancreatic cell lineage BMP exerts versatile functions. While inhibition of BMP is important for inducing pancreatic fate specification from pluripotent ESCs, the late exposure of induced pancreatic progenitors to BMP inhibitors helps to ensure precise temporal activation of NEUROG3, favoring the generation of mono-hormonal insulin-secreting cells *versus* polyhormonal cells (D'Amour et al., 2006; Kroon et al., 2008; Mfopou, Chen, Mateizel, Sermon, & Bouwens, 2010; Russ et al., 2015).

The intracellular mediators of BMP and TGFβ signaling are the SMAD molecules (Massagué, Seoane, & Wotton, 2005). SMAD1, SMAD5 and SMAD8 are activated by BMP and, specifically, inhibited by SMAD6, whereas SMAD2 and SMAD3 are downstream of TGFβ isoforms and Activin. Beside a short window of time at E8.5, when phosphorylated (active) (p)-Smad1,5,8 has been reported in the foregut endoderm (Wandzioch & Zaret, 2009), BMP-mediated Smad activation seems entirely restricted to the mesenchyme of both chicken and mouse pancreas at subsequent stages (Ahnfelt-Rønne et al., 2010). Consistently, ex vivo studies in mouse explants indicate that BMP signaling is primarily required in the mesenchyme and that the effects on the epithelium might be indirect (Ahnfelt-Rønne et al., 2010).

Together these findings underscore once more the importance of time in signaling events and changes of cellular competence in response to extracellular signals. It is also unclear how BMPs mediate such a wide variety of responses. It may be, for instance, a result of diversity in the components of signal transduction or crosstalk with other pathways. Possibly, better in vivo models for fine-tuned spatio-temporal control and precise manipulation of signaling pathways at key developmental-equivalent stages are required to answer these questions.

2.3.3 Notch signaling pathway

Notch receptors and ligands are expressed in the developing pancreas starting from E9.5 in the mouse pancreatic epithelium, with the exception of Notch 3 and Notch 4, which are in mesenchymal and endothelial cells, respectively (Lammert, Brown, & Melton, 2000).

At early stages, Notch signaling is necessary to maintain proliferation of the undifferentiated pancreatic progenitors and prevent endocrine and exocrine differentiation, acting in a conserved fashion across species (Ahnfelt-Rønne et al., 2007; Apelqvist et al., 1999; Jensen et al., 2000; Murtaugh, Stanger, Kwan, & Melton, 2003). Null mutant mice for either Notch ligand genes, such as *Dll1*, or for Notch target genes, like *Rbpj-k* and *Hes1*, display pancreatic hypoplasia due to accelerated and increased endocrine differentiation (Apelqvist et al., 1999; Jensen et al., 2000). Moreover, constitutive expression of an active intracellular domain of *Notch1* under the control of the *Pdx1* promoter results into failure of endocrine and exocrine differentiation, which is accompanied by expansion of the pancreatic progenitor number (Murtaugh et al., 2003).

An interplay between Notch and FGF signaling has been reported, whereby FGF10 would maintain active Notch signaling to control epithelial cell proliferation and differentiation (Hart, Papadopoulou, & Edlund, 2003; Norgaard, Jensen, & Jensen, 2003). Specifically, Notch signaling controls proliferation of pancreatic progenitors downstream of FGF10 through Hes1 transcriptional control of *Cdkn1c* (Georgia, Soliz, Li, Zhang, & Bhushan, 2006; Miralles, Lamotte, Couton, & Joshi, 2006; Norgaard et al., 2003). In contrast, endocrine and exocrine cells are induced within the ductal epithelium *via* Notch signaling by lateral inhibition (Ahnfelt-Rønne et al., 2012; Apelqvist et al., 1999; Murtaugh et al., 2003). The mechanism through which Notch controls endocrine differentiation involves a Hes1-dependent suppression of *Ngn3* expression, whereas its modulation of acinar differentiation depends on the inhibition of PTF1a (Ahnfelt-Rønne et al., 2012; Esni et al., 2004; Murtaugh et al., 2003).

2.3.4 Hedgehog signaling pathway

Hh signaling needs to be tightly-regulated throughout pancreas development. Pancreas fate specification requires suppression of Hh signaling activation in the foregut endoderm (see above) (Hebrok et al., 2000; Kim et al., 1997). Subsequently in development, disrupted Hh signaling affects the development of both endocrine and exocrine cells and impairs β-cell function and mass (Hebrok et al., 2000; Kawahira et al., 2003, 2005). Efforts have been made to differentiate between Hh signaling contribution to pancreas epithelium and mesenchyme. Manipulation of the pathway in epithelial and β-cells showed a cell-autonomous role of Hh signaling in regulating endocrine mass and β-cell function (Thomas, Lee, Rastalsky, & Habener, 2001; Thomas, Rastalsky, Lee, & Habener, 2000). More recent work from Landsman and colleagues has provided direct evidence for non-autonomous roles of the Hh pathway in pancreatic development (Hibsher, Epshtein, Oren, & Landsman, 2016). Specifically, by deleting the receptor *Ptch1* in the pancreatic mesenchyme, Shh signaling activity is increased, due to the abrogation of the inhibitory function of Ptch1 on Smoothened. Such deregulated Hh signaling in mesenchymal cells leads to hyperplasia of the pancreatic mesenchyme along with disrupted epithelial growth and islet cellular composition (Hibsher et al., 2016). How the Hh pathway signals in the mesenchyme is unknown. It is unclear whether the epithelial defects are consequences of the mesenchymal hyperplasia, which might hamper proper mesenchymal-epithelial interactions. Finally, Hh-dependent mesenchymal cues that regulate pancreatic growth and differentiation are yet to be identified.

2.3.5 Retinoids signaling pathway

Vitamin A-derived RA, retinoid binding proteins and RA receptors are expressed in both pancreatic epithelium and mesenchyme during pancreas development (Arregi et al., 2016; Martín et al., 2005; Molotkov, Molotkova, & Duester, 2005; Stafford & Prince, 2002; Tulachan et al., 2003). Specifically, the enzymes RDH10 and RALDH2, responsible for the production of RA, are present in the lateral plate mesoderm at E8.25, before *Pdx1* induction, and later in the pancreatic mesenchyme (Arregi et al., 2016; Martín et al., 2005; Molotkov et al., 2005; Stafford & Prince, 2002; Tulachan et al., 2003).

In vertebrate species, including zebrafish, *Xenopus* and mouse, mesoderm-derived RA signaling induces pancreatic specification within the endoderm. Deletion of *Rdh10* or *Raldh2* and expression of dominant-negative RA receptors result into impaired pancreatic specification within the dorsal foregut endoderm (Arregi et al., 2016; Chen et al., 2004; Martín et al., 2005; Molotkov et al., 2005; Stafford & Prince, 2002; Tulachan et al., 2003). Later in the mouse embryo, RA has been shown to control endocrine and ductal differentiation and influence ductal *versus* acinar cell fate decision (Arregi et al., 2016). Specifically, endoderm specific-*Rdh10* deletion results into pancreatic hypoplasia with impaired endocrine differentiation observed at later stages (Arregi et al., 2016). Consistently, treatment of embryonic pancreatic explants with RA promotes ductal cell differentiation with premature formation of endocrine clusters at the expense of exocrine differentiation and branching morphogenesis (Shen et al., 2007; Tulachan et al., 2003). Interestingly, in *Rdh10* and *Raldh2* null mutant embryos the dorsal pancreatic mesenchyme is also disrupted, suggesting a further role of RA signaling in formation and patterning of mesenchymal cells (Arregi et al., 2016; Martín et al., 2005). Yet, specific mesenchymal contribution of the RA signaling has not been investigated and the distinction between cell-autonomous and non-cell-autonomous contributions needs to be addressed.

2.3.6 Epidermal growth factor (EGF) signaling pathway

EGF receptors (EGFR) and EGF ligands are present in both epithelium and mesenchyme of the developing pancreas and the pathway has a role in pancreatic morphogenesis and differentiation (Gittes, 2009; Kritzik et al., 2000). *Egfr* KO mouse embryos display impaired branching morphogenesis, due to proliferation defects, and delayed endocrine differentiation with perturbed islet formation (Miettinen et al., 2000). Moreover, overexpression of a dominant-negative form of *Egfr* under the *Pdx1*-promoter results into impaired postnatal β-cell proliferation and diabetes (Miettinen et al., 2000, 2006). In addition to *Egfr*, the *Erbb4* receptor is expressed in pancreatic epithelial cells at E12.5 and becomes restricted to the ductal epithelium by E16.5 (Kritzik et al., 2000). This receptor has been suggested to play a role in δ-cell development, since treatment of pancreatic explants with an antibody neutralizing Neuregulin 4, the Erbb4 ligand, specifically blocked δ-cell expansion (Huotari et al., 2002).

Recent work by Semb and colleagues has started shedding light into how EGFR signaling coordinates epithelial morphogenesis and cell differentiation in the pancreas (Löf-Öhlin et al., 2017). They identified an interesting and conserved mechanism whereby EGF/EGFR signaling controls β-cell differentiation *via* modulation of apical polarity in endocrine progenitor cells. Moreover, they reported sequential temporal activation of the EGFR by distinct EGF ligands during pancreas development (Löf-Öhlin et al., 2017). Specifically, during the primary transition (E9.5–E12) EGF ligand counteracts polarization of pancreatic epithelial progenitors, while later Betacellulin (BTC) ligand appears to induce β-cell commitment by reducing the apical domain size during the secondary transition. Apical domain constriction results into reduced Notch signaling, increased *Ngn3* expression and β-cell differentiation (Löf-Öhlin et al., 2017).

2.3.7 Hepatocyte growth factor (HGF) signaling pathway

Hgf and its receptor *Met* are expressed in the pancreatic mesenchyme and epithelium, respectively (Garcia-Ocaña et al., 2000; Otonkoski et al., 1996). HGF signaling was shown to induce β-cell formation both in vitro and in vivo, including in human fetal pancreas cultures

(Garcia-Ocaña et al., 2000; Otonkoski et al., 1996). Moreover, transgenic mice expressing *Hgf* under the rat *insulin* promoter display an increase in islet number with enhanced insulin content (Garcia-Ocaña et al., 2000). Consistently, deletion of *Met* in β-cells results in mice that exhibit glucose intolerance with reduced insulin secretion and islet size (Dai, Huh, Thorgeirsson, & Liu, 2005; Roccisana et al., 2005).

Interestingly, recent findings in zebrafish suggest a role for HGF-Met signaling in pancreas morphogenesis through control of directional migration of pancreatic progenitors (Anderson et al., 2013). If a similar mechanism contributes to pancreatic branching morphogenesis in mammals remains to be tested.

2.3.8 Wingless/integrated (Wnt)signaling pathway

In the developing pancreas, Wnt signaling plays multiple roles that are spatially and temporally restricted. An early role for Wnt signaling in pancreatic fate specification has been described in *Xenopus* and mouse embryos (see above) (McLin, Rankin, & Zorn, 2007; Rodríguez-Seguel et al., 2013). In the frog, canonical Wnt signaling needs to be repressed in the anterior endoderm in order to maintain foregut identity and allow pancreas and liver development (McLin et al., 2007). Once the foregut territory has been established, the non-canonical Wnt signaling regulates the pancreas *versus* liver cell fate decision in both *Xenopus* and mouse (Rodríguez-Seguel et al., 2013). In particular, non-canonical Wnt family members, such as *Wnt5a*, display a differential expression pattern between pancreatic and hepatic progenitors within the foregut endoderm (Rodríguez-Seguel et al., 2013). Consistently, Wnt5A treatment of *Xenopus* endodermal cells and mouse ESC (mESC) cells induces *Pdx1* expression (Rodríguez-Seguel et al., 2013).

Subsequently, several Wnt ligands localize in the mouse pancreatic mesenchyme, whereas Frizzled receptors (Fzd) and secreted Frizzled-Related Proteins (SFRP) are present in both the epithelium and mesenchyme (Heller et al., 2002). Mice overexpressing *Wnt1* and *Wnt5a* under the control of the *Pdx1* promoter display pancreatic agenesis and pancreatic hypoplasia, respectively, which is in line with its early role in foregut specification (Heller et al., 2002). Similarly, constitutive activation of β-catenin at early stages of pancreatic development leads to dramatic pancreatic hypoplasia (Heiser, Lau, Taketo, Herrera, & Hebrok, 2006). By contrast, ablation of *β-catenin* in *Pdx1*-expressing pancreatic progenitors reduces epithelial proliferation and, specifically, prevents exocrine differentiation (Murtaugh, Law, Dor, & Melton, 2005). Consistently, inhibition of Wnt activity by expressing a soluble form of *Fzd8* under the *Pdx1* promoter results into a reduced number of proliferating pancreatic progenitor cells and organ hypoplasia, affecting both exocrine and endocrine compartments in mice (Papadopoulou & Edlund, 2005).

Interestingly, Wnt signaling is also an essential mediator of pancreatic mesenchymal cell survival. Specific inactivation of canonical Wnt signaling in the pancreatic mesenchyme, using *β-catenin*$^{fl/fl}$;*Nkx3.2-Cre* Tg embryos, results in the loss of mesenchymal cells, which in turn affects the epithelium (Landsman et al., 2011). Consistently, a more recent study has identified Wnt7b ligand to be expressed in the epithelium and act as trophic factor for the mesenchyme *via* a paracrine mechanism from epithelium-to-mesenchyme (Afelik, Pool, Schmerr, Penton, & Jensen, 2015). Overall, because of the myriad of Wnt and Frizzled possible combinations as well as their diverse context-dependent functions, the complexity of the pathway deserves further attention.

Taken together, most of the studies discussed in this section highlight the relevance of signaling pathways, many of which act concurrently, in epithelial-mesenchymal crosstalk(s) during pancreas organogenesis. The mechanisms by which these mesenchymal signals are coordinated and integrated to stimulate epithelial progenitor proliferation and differentiation to produce functional pancreatic cells remain elusive. Also, whether is the developing pancreatic mesenchyme a homogenous tissue or composed of different mesenchymal cell populations, resembling distinct "cellular niches," like in the hematopoietic and nervous system, lungs and skin, is unclear (Crane, Jeffery, & Morrison, 2017; Kfoury & Scadden, 2015; Lattanzi, Parolisi, Barba, & Bonfanti, 2015; Roberts, Kershner, & Beachy, 2017; Zepp et al., 2017). If different niches exist, what TFs or intrinsic regulators confer spatio-temporal control of gene expression within the mesenchyme during pancreatic development is unknown.

To date, only a small number of TFs have been reported in the embryonic pancreatic mesenchyme, including *Isl1*, *Nkx3.2* and *Hoxa6*, *Hoxb6* and *Hoxc6* (Ahlgren et al., 1997; Asayesh et al., 2006; Hecksher-Sørensen et al., 2004; Landsman et al., 2011; Larsen, Hrycaj, Newman, Li, & Wellik, 2015). *Isl1* is expressed in both pancreatic epithelium and mesenchyme and is required for dorsal mesenchyme formation (Ahlgren et al., 1997). *Nkx3.2* (a.k.a. *Bapx1*) expression is in the splenopancreatic mesenchyme, but not in the epithelium (Asayesh et al., 2006; Hecksher-Sørensen et al., 2004).

Null mutant embryos for *Nkx3.2* displayed asplenia and metaplastic transformation of embryonic pancreas into intestinal-like tissue due to the loss of splenic mesenchyme condensation and separation (see above) (Asayesh et al., 2006; Hecksher-Sørensen et al., 2004). More recently, the expression of *Hox6* genes has been identified and studied in the mesenchymal compartment of the developing pancreas. Full KO embryos for all the three *Hox6* paralogs display impaired endocrine differentiation with disruption of branching morphogenesis (Larsen et al., 2015). To fully elucidate the identity and role(s) of mesenchymal genetic regulators driving pancreas development, future investigation is required using tools which enable fine-tuned gene manipulation in the mesenchyme.

2.4 Extracellular matrix molecules

The basement membrane (BM) separates the epithelium from the surrounding mesenchyme and is crucial for epithelial tissue organization and function, including in the pancreas (Hisaoka et al., 1993). The BM is a fundamental site of interaction between epithelial cells and the ECM: the ECM mainly consists of fibronectin (FN), laminins and collagen-IV components, while the epithelium expresses multiple integrins molecules, which act as molecular hubs connecting the outside environment (mesenchyme) with the inside (epithelium) (Glentis, Gurchenkov, & Vignjevic, 2014; Guo & Giancotti, 2004; Hisaoka et al., 1993).

During pancreas organogenesis, the BM controls branching morphogenesis and cell differentiation (Crisera et al., 2000; Jiang, Cram, DeAizpurua, & Harrison, 1999; Jiang, Georges-Labouesse, & Harrison, 2001; Shih et al., 2016). Disruption of cell-ECM interactions in pancreatic explant cultures using the tetrapeptide arginyl-glycyl-aspartyl-serine (RGDS), which competitively inhibits ECM-cell interactions, blocks branch formation (Shih et al., 2016). Moreover, substrates containing Laminin 1 (*a.k.a.* Laminin 111) have been shown to induce pancreatic cell differentiation in ex vivo cultures (Jiang et al., 1999, 2001). Consistent with this, several studies have suggested the requirement of integrins for initiation of branching morphogenesis, delamination of endocrine progenitors and islet formation. For instance, pancreatic-specific deletion of the β_1-integrin subunit leads to impaired branching morphogenesis, due to BM defects and aberrant organization of the actin cytoskeleton (Shih et al., 2016). Furthermore, $\alpha_6\beta_4$ and $\alpha_5\beta_1$ integrin, together with Netrin, a Laminin soluble form, are important for pancreatic epithelial cell adhesion and migration in vitro (Yebra et al., 2003).

Other integrins, such as $\alpha_V\beta_5$ and $\alpha_V\beta_3$, have been shown to interact with Fibronectin, Collagen-IV and Vitronectin to maintain ECM anchorage and allow cell migration in the human fetal pancreas (Cirulli et al., 2000).

Taken together, these studies highlight the relevance of ECM and cell adhesion molecules as interfaces of signaling transmission within different compartments of the pancreas during the entire duration of its development. Microenvironmental signals transmitted *via* the ECM might establish "niches" and mediate localized cellular behaviors in the embryonic pancreas.

2.5 Blood vessels

The physical interaction between the pancreatic epithelium and blood vessels is highly dynamic and changes throughout development (Azizoglu & Cleaver, 2016). During the primary transition (E9.5–E12), the pancreatic epithelium is avascular, with blood vessels only surrounding but not penetrating the epithelial buds (Pierreux et al., 2010). Starting with the secondary transition, blood vessels intercalate between the forming epithelial branches, either actively through angiogenesis or, passively, as the epithelium remodels (Pierreux et al., 2010). Moreover, the vessels follow a stereotypical pattern in their organization along the branches, becoming preferentially located near the central (trunk) epithelial cells and at a distance from the branch tips where acinar cells differentiate. This correlates with a predominant expression of the angiogenic factor Vascular Endothelial Growth Factor-A (VEGF-A) in trunk cells (Pierreux et al., 2010). In the mature organ, the branches of the pancreatic epithelium and the vascular systems closely intercalate, with densely vascularized islets connected to large vessels that run across the pancreas (Azizoglu & Cleaver, 2016). This constantly evolving relationship between blood vessels and pancreatic epithelium seems to influence different aspects of pancreas development (Azizoglu & Cleaver, 2016). While the early interaction with the aorta is crucial for pancreas fate specification (see above), at later stages endothelial cells restrict the growth of embryonic pancreas as wells as inhibit branching and differentiation (Jacquemin et al., 2006; Lammert et al., 2001; Magenheim, Ilovich, et al., 2011; Pierreux et al., 2010; Sand et al., 2011; Yoshitomi & Zaret, 2004). For instance, in vitro treatment of wild-type mouse pancreatic explants with VEGFA or VEGFR2-inhibitor leads to decreased exocrine proliferation or increased exocrine differentiation, respectively (Magenheim, Ilovich, et al., 2011; Pierreux et al., 2010). Moreover, pancreatic explant cultures isolated from mouse transgenic embryos overexpressing *Vegfa* display hypervascularization with reduced exocrine differentiation (Pierreux et al., 2010). Similar effects have been reported in vivo in pancreata from E12.5 embryos overexpressing *Vegfa* (Magenheim, Ilovich, et al., 2011). In summary, blood vessels- and blood-derived signals play dynamic roles during pancreatic development either by direct signaling to the endoderm or indirectly through the mesenchyme.

2.6 Neural crest-derived cells

Neural crest (NC) cells are multipotent embryonic cells that arise from the dorsal neural tube and migrate to generate diverse ectodermal cell lineages (Young & Newgreen, 2001). Concomitant with the evagination of the pancreatic buds at E9, migrating NC cells reach the rostral foregut and, in a rostral-to-caudal migratory wave, populate the entire gut with progenitors of glial and neural cells (Young & Newgreen, 2001). Approximately at E10, NC cells enter the pancreatic mesenchyme and intermingle with the pancreatic epithelium as it branches into the mesenchyme (Plank et al., 2011). Subsequently, these NC cells differentiate into glial Schwann cells that surround mature pancreatic islets and into neurons of the autonomic nervous system, both sympathetic and parasympathetic, that regulate islet function, such as insulin and glucagon secretion (Plank et al., 2011). *Glial Cell Derived Neurotrophic Factor* (*GDNF*), released from the pancreatic epithelium, has been proposed to act as a neurotrophic factor guiding the migration of neural progenitors into the pancreas (Muñoz-Bravo et al., 2013).

To date, the role of NC-derived cells during pancreas development is not fully understood. Mouse mutants for two NC-specific TFs *Forkhead Box D3* (*Foxd3*) and *Paired-Like Homeobox 2b* (*Phox2b*) display loss of NC cells, which is accompanied by increased β-cell proliferation (Nekrep et al., 2008; Plank et al., 2011). Although the number is increased, mutant β-cells display immature features, such as decreased expression of *Mafa*, *Pdx1* and *Glut2* (Plank et al., 2011). Additionally, *Phox2b(−/−)* mutant embryos display upregulation of *Nkx2.2* expression in the epithelium, suggesting a non-cell-autonomous feedback loop that links the neural crest with the pancreatic epithelium (Nekrep et al., 2008). Indeed, analysis of *Nkx2.2* KO embryos showed that signals from the pancreatic epithelium silence the expression of *Phox2b* in the NC-derived cells as they migrate into the pancreas (Nekrep et al., 2008). Perhaps, such inhibition is required for NC cells differentiation into specific cell types, like in the central nervous system. Altogether, these studies suggest that NC-derivatives negatively regulate β-cell proliferation and promote β-cell maturation.

3. Implications for directed differentiation of pluripotent stem cells into pancreatic β-cells

Diabetes is a degenerative disease that affects over 400 million persons worldwide (http://www.who.int/diabetes/) (Ellis et al., 2017; Sneddon et al., 2018). In type I diabetic (T1D) patients, the autoimmune destruction of pancreatic beta-cells leads to insufficient production of insulin and hyperglycemia ensues (Pociot & Lernmark, 2016; Weir & Bonner-Weir, 2013). T2D is the most common form of the disease; it develops during adulthood and is characterized by β-cell dysfunction, insulin resistance and metabolic stress, which eventually lead to β-cell mass reduction (Nolan, Damm, & Prentki, 2011; Weir & Bonner-Weir, 2013). Despite the availability of insulin as treatment to temporarily restore the glucostasis in diabetic patients, this remedy is unable to avoid both the acute dangers of hypoglycemia or the long-term complications of hyperglycemia. Eventually, the challenge of treating diabetes centers on the replacement of the destroyed β-cells. Indeed, islet and whole pancreas transplantation have become gold-standard procedures in achieving glucose control in diabetic patients (Shapiro et al., 2000; Shapiro, Pokrywczynska, & Ricordi, 2017). However, the extremely limited number of suitable donor tissues severely hampers the broad application of these therapies, underscoring the need for an unlimited source of insulin-producing cells that could be used in cell-based replacement therapy.

Over the last decade, remarkable progress has been made with regard to the generation of functional β-cell equivalents from pluripotent stem cells (hES and iPS cells) and other cellular sources (Ellis et al., 2017; Sneddon et al., 2018). It has also become clear that the key to success is to closely recapitulate the path that pluripotent stem cells undertake during embryogenesis to become first definitive endoderm, then pancreatic progenitors, and finally pancreatic islet cells (Ellis et al., 2017; Sneddon et al., 2018).

In a first landmark study by D'Amour et al. (2005), hESC cultures were differentiated into endoderm upon combined exposure to Wnt and Activin/TGFβ signalings and, subsequently, into endocrine progenitor cells following the activation of FGF10, RA and BMP inhibition, similar to what happens in the mouse embryo. This study has not only paved the way for making β-like cells in the lab, but also revealed that pancreatic signals, which create the micro-environment controlling β-cell development, are conserved between mouse and human (D'Amour et al., 2005). Subsequent studies have improved each step of this differentiation protocol, including the initial fate specification, maturation and proliferation (D'Amour et al., 2006; Kroon et al., 2008; Nostro et al., 2011; Pagliuca et al., 2014;

Rezania et al., 2013; Russ et al., 2015). For example, compound screens of molecules have been performed to direct ES cells to the endodermal lineage and to induce pancreatic progenitor specification (Borowiak et al., 2009; Chen et al., 2009). Also, co-culturing human and murine ESC-derived pancreatic and endocrine progenitors with primary pancreatic mesenchymal cell has been shown to promote progenitor expansion without altering differentiation potential (Sneddon, Borowiak, & Melton, 2012). The same study subsequently tested the ability of 16 growth factors to mimic the co-culture effect but none of them, alone or in combination, reproduced the co-culture effect at the same magnitude, suggesting the involvement of cell-cell contact in this effect (Sneddon et al., 2012).

To date, the ex vivo generation of β-cell equivalents that possess physiological features of their native counterparts remains a challenge: hESC-derived insulin-producing cells often display features of immature β-cells and are frequently polyhormonal (Sneddon et al., 2018). Nevertheless, once grafted in vivo in the mouse, these immature β-like cells are able to further differentiate and protect animals from chemical-induced diabetes (Kroon et al., 2008; Sneddon et al., 2018). More recent efforts have been placed on generating ex vivo single hormone-positive β-cells capable of glucose-stimulated insulin secretion and perfectly replicating all aspects of endogenous β-cells (Pagliuca et al., 2014; Rezania et al., 2013; Russ et al., 2015).

Beside further refinement of the cytokine cocktail, recent studies have underscored the benefits of three-dimensional cell culture system, which allows the generation of endocrine clusters that are perhaps closer to the endogenous islet (Pagliuca et al., 2014; Rezania et al., 2013; Russ et al., 2015). Indeed, a key determinant of β-cell function is cell-cell communication among endocrine cells as well as with cells belonging to the islet niche, which is composed of multiple non-endocrine cell types, including endothelial, pericytes, neuronal, and mesenchymal cells (Aamodt & Powers, 2017; Brissova et al., 2014; Coronel & Stabler, 2013; Sasson et al., 2016). Given the specialized relationship between β-cells and supporting cells in their natural environment, future efforts are required to define elements of the endogenous islet niche for improving survival and function of hESC-derived β-cells in vitro.

4. Open questions and future directions

The interplay between pancreatic epithelium and all cellular components of the pancreatic microenvironment has been shown to be critical for the formation of a proper functional pancreas. The understanding of the effectors and mechanisms underlying the crosstalk has tremendously helped in developing the current pancreatic differentiation protocols of pluripotent stem cells into β-cell equivalents for cell transplantation. Additionally, a comprehensive knowledge of the pancreatic microenvironment composition and its various signaling molecules will be relevant for further elucidating the pathogenesis underlying pancreatic diseases, such as cancer. PDAC is histologically characterized by an abundant stroma surrounding malignant epithelial cells that consists of a dynamic assortment of ECM components and fibroblasts, endothelial cells, immune cells, and neurons (Feig et al., 2012; Sinha & Leach, 2016). Accumulating evidence indicates that the epithelial and stromal compartments interact to enhance the aggressive nature of this disease (Feig et al., 2012; Sinha & Leach, 2016). Aberrant activation of embryonic signaling pathways is frequent in PDAC, making developmental regulators therapeutically attractive (Roy et al., 2016). Specifically, a number of pathways, including Hedgehog, Notch and TGF-β, which are at play in the pancreatic mesenchyme during development, are also involved in mediating the cross-talk between the malignant pancreatic epithelium and its associated stroma, making them unique pathways to target for the treatment of pancreatic cancer (Bailey & Leach, 2012). How similar is the composition of the pancreatic microenvironment in the embryo and in the tumor is an open question that needs to be addressed. Only very recently, transcriptome analyses of human embryonic pancreatic mesenchyme have become available (Ramond et al., 2017), making such comparison possible.

Many other questions deserve further investigation with regard to the pancreatic microenvironment in development and diseases. For instance, few examples of how mechanistically signaling pathways from the environment converge into the transcriptional cascade governing pancreatic development have been reported so far. Distinct types of interactions are likely to occur between different pancreatic cell types and surrounding cells or ECM components, and these sets of interactions are dynamic and change with time. To date, the information on the identity and spatial distribution of these different types of interaction is elusive. Also, the degree of heterogeneity within the pancreatic mesenchyme is a critical open question that starts only now to be unraveled (Byrnes et al., 2018). Single-cell transcriptome analysis suggests indeed the existence of different mesenchymal lineages creating distinct "cellular niches" (Byrnes et al., 2018). This heterogeneity and potential region-specific cross-talks are awaiting to be further characterized by lineage tracing and functional studies. Finally, such information will enable the discovery of new paradigms on how fine-tuning the activity of different signaling pathways to achieve better differentiation and greater number of β-cell equivalents.

Acknowledgments

We thank all members of the Spagnoli lab for helpful discussion. This work was supported by funds from the Helmholtz Association, Berlin Institute of Health (BIH) and German Israeli Foundation (GIF) Grant. The authors declare that they have no competing financial interests.

References

Aamodt, K. I., & Powers, A. C. (2017). Signals in the pancreatic islet microenvironment influence β-cell proliferation. *Diabetes, Obesity and Metabolism*, *19*, 124–136.

Afelik, S., Pool, B., Schmerr, M., Penton, C., & Jensen, J. (2015). Wnt7b is required for epithelial progenitor growth and operates during epithelial-to-mesenchymal signaling in pancreatic development. *Developmental Biology*, *399*, 204–217.

Ahlgren, U., Pfaff, S. L., Jessell, T. M., Edlund, T., & Edlund, H. (1997). Independent requirement for ISL1 in formation of pancreatic mesenchyme and islet cells. *Nature*, *385*, 257–260.

Ahnfelt-Rønne, J., Hald, J., Bødker, A., Yassin, H., Serup, P., & Hecksher-Sørensen, J. (2007). Preservation of proliferating pancreatic progenitor cells by Delta-Notch signaling in the embryonic chicken pancreas. *BMC Developmental Biology*, *7*, 63.

Ahnfelt-Rønne, J., Jørgensen, M. C., Klinck, R., Jensen, J. N., Füchtbauer, E.-M., Deering, T., et al. (2012). Ptf1a-mediated control of Dll1 reveals an alternative to the lateral inhibition mechanism. *Development*, *139*, 33–45.

Ahnfelt-Rønne, J., Ravassard, P., Pardanaud-Glavieux, C., Scharfmann, R., & Serup, P. (2010). Mesenchymal bone morphogenetic protein signaling is required for normal pancreas development. *Diabetes*, *59*, 1948–1956.

Anderson, R. M., Delous, M., Bosch, J. A., Ye, L., Robertson, M. A., Hesselson, D., et al. (2013). Hepatocyte growth factor signaling in intrapancreatic ductal cells drives pancreatic morphogenesis. *PLoS Genetics*, *9*, e1003650.

Angelo, J. R., Guerrero-Zayas, M. I., & Tremblay, K. D. (2012). A fate map of the murine pancreas buds reveals a multipotent ventral foregut organ progenitor. *PLoS ONE*, *7*, e40707.

Angelo, J. R., & Tremblay, K. D. (2018). Identification and fate mapping of the pancreatic mesenchyme. *Developmental Biology*, *435*, 15–25.

Apelqvist, A., Ahlgren, U., & Edlund, H. (1997). Sonic hedgehog directs specialised mesoderm differentiation in the intestine and pancreas. *Current Biology*, *7*, 801–804.

Apelqvist, A., Li, H., Sommer, L., Beatus, P., Anderson, D. J., Honjo, T., et al. (1999). Notch signalling controls pancreatic cell differentiation. *Nature*, *400*, 877–881.

Arda, H. E., Benitez, C. M., & Kim, S. K. (2013). Gene regulatory networks governing pancreas development. *Developmental Cell*, *25*, 5–13.

Arnaud-Dabernat, S., Kritzik, M., Kayali, A. G., Zhang, Y.-Q., Liu, G., Ungles, C., et al. (2007). FGFR3 is a negative regulator of the expansion of pancreatic epithelial cells. *Diabetes*, *56*, 96–106.

Arregi, I., Climent, M., Iliev, D., Strasser, J., Gouignard, N., Johansson, J. K., et al. (2016). Retinol dehydrogenase-10 regulates pancreas organogenesis and endocrine cell differentiation via paracrine retinoic acid signalling. *Endocrinology*, *157*(12), 4615–4631.

Asayesh, A., Sharpe, J., Watson, R. P., Hecksher-Sørensen, J., Hastie, N. D., Hill, R. E., et al. (2006). Spleen versus pancreas: Strict control of organ interrelationship revealed by analyses of Bapx1 −/− mice. *Genes & Development*, *20*, 2208–2213.
Attali, M., Stetsyuk, V., Basmaciogullari, A., Aiello, V., Zanta-Boussif, M. A., Duvillie, B., et al. (2007). Control of β-cell differentiation by the pancreatic mesenchyme. *Diabetes*, *56*, 1248–1258.
Azizoglu, D. B., & Cleaver, O. (2016). Blood vessel crosstalk during organogenesis—Focus on pancreas and endothelial cells. *Wiley Interdisciplinary Reviews: Developmental Biology*, *5*, 598–617.
Bailey, J. M., & Leach, S. D. (2012). Signaling pathways mediating epithelial-mesenchymal crosstalk in pancreatic cancer: Hedgehog, Notch and TGFβ. In P. J. Grippo & H. G. Munshi (Eds.), *Pancreatic cancer and tumor microenvironment*. Trivandrum (India): Transworld Research Network Chapter 7.
Bankaitis, E. D., Bechard, M. E., & Wright, C. V. E. (2015). Feedback control of growth, differentiation, and morphogenesis of pancreatic endocrine progenitors in an epithelial plexus niche. *Genes & Development*, *29*, 2203–2216.
Bhushan, A., Itoh, N., Kato, S., Thiery, J. P., Czernichow, P., Bellusci, S., et al. (2001). Fgf10 is essential for maintaining the proliferative capacity of epithelial progenitor cells during early pancreatic organogenesis. *Development*, *128*, 5109–5117.
Biankin, A. V., Waddell, N., Kassahn, K. S., Gingras, M.-C., Muthuswamy, L. B., Johns, A. L., et al. (2012). Pancreatic cancer genomes reveal aberrations in axon guidance pathway genes. *Nature*, *491*, 399–405.
Borowiak, M., Maehr, R., Chen, S., Chen, A. E., Tang, W., Fox, J. L., et al. (2009). Small molecules efficiently direct endodermal differentiation of mouse and human embryonic stem cells. *Cell Stem Cell*, *4*, 348–358.
Bort, R., Martinez-Barbera, J. P., Beddington, R. S. P., & Zaret, K. S. (2004). Hex homeobox gene-dependent tissue positioning is required for organogenesis of the ventral pancreas. *Development*, *131*, 797–806.
Brissova, M., Aamodt, K., Brahmachary, P., Prasad, N., Hong, J.-Y., Dai, C., et al. (2014). Islet microenvironment, modulated by vascular endothelial growth factor-a signaling, promotes β cell regeneration. *Cell Metabolism*, *19*, 498–511.
Byrnes, L. E., Wong, D. M., Subramaniam, M., Meyer, N. P., Gilchrist, C. L., Knox, S. M., et al. (2018). Lineage dynamics of murine pancreatic development at single-cell resolution. *Nature Communications*, *9*, 3922.
Chen, S., Borowiak, M., Fox, J. L., Maehr, R., Osafune, K., Davidow, L., et al. (2009). A small molecule that directs differentiation of human ESCs into the pancreatic lineage. *Nature Chemical Biology*, *5*, 258–265.
Chen, Y., Pan, F. C., Brandes, N., Afelik, S., Sölter, M., & Pieler, T. (2004). Retinoic acid signaling is essential for pancreas development and promotes endocrine at the expense of exocrine cell differentiation in Xenopus. *Developmental Biology*, *271*, 144–160.
Chung, W.-S., Shin, C. H., & Stainier, D. Y. R. (2008). Bmp2 signaling regulates the hepatic versus pancreatic fate decision. *Developmental Cell*, *15*, 738–748.
Cirulli, V., Beattie, G. M., Klier, G., Ellisman, M., Ricordi, C., Quaranta, V., et al. (2000). Expression and function of αvβ3 and αvβ5 integrins in the developing pancreas. *The Journal of Cell Biology*, *150*, 1445–1460.
Coronel, M. M., & Stabler, C. L. (2013). Engineering a local microenvironment for pancreatic islet replacement. *Current Opinion in Biotechnology*, *24*, 900–908.
Crane, G. M., Jeffery, E., & Morrison, S. J. (2017). Adult haematopoietic stem cell niches. *Nature Reviews Immunology*, *17*, 573–590.
Crisera, C. A., Kadison, A. S., Breslow, G. D., Maldonado, T. S., Longaker, M. T., & Gittes, G. K. (2000). Expression and role of laminin-1 in mouse pancreatic organogenesis. *Diabetes*, *49*, 936–944.

Dai, C., Huh, C.-G., Thorgeirsson, S. S., & Liu, Y. (2005). β-cell-specific ablation of the hepatocyte growth factor receptor results in reduced islet size, impaired insulin secretion, and glucose intolerance. *The American Journal of Pathology*, *167*, 429–436.

D'Amour, K. A., Agulnick, A. D., Eliazer, S., Kelly, O. G., Kroon, E., & Baetge, E. E. (2005). Efficient differentiation of human embryonic stem cells to definitive endoderm. *Nature Biotechnology*, *23*, 1534–1541.

D'Amour, K. A., Bang, A. G., Eliazer, S., Kelly, O. G., Agulnick, A. D., Smart, N. G., et al. (2006). Production of pancreatic hormone-expressing endocrine cells from human embryonic stem cells. *Nature Biotechnology*, *24*, 1392–1401.

Demeterco, C., Beattie, G. M., Dib, S. A., Lopez, A. D., & Hayek, A. (2000). A role for activin A and betacellulin in human fetal pancreatic cell differentiation and growth. *The Journal of Clinical Endocrinology & Metabolism*, *85*, 3892–3897.

Deutsch, G., Jung, J., Zheng, M., Lóra, J., & Zaret, K. S. (2001). A bipotential precursor population for pancreas and liver within the embryonic endoderm. *Development*, *128*, 871–881.

Dichmann, D. S., Miller, C. P., Jensen, J., Heller, R. S., & Serup, P. (2003). Expression and misexpression of members of the FGF and TGFbeta families of growth factors in the developing mouse pancreas. *Developmental Dynamics*, *226*, 663–674.

Dong, P. D. S., Munson, C. A., Norton, W., Crosnier, C., Pan, X., Gong, Z., et al. (2007). Fgf10 regulates hepatopancreatic ductal system patterning and differentiation. *Nature Genetics*, *39*, 397–402.

Drut, R. M., Drut, R., Gilbert-Barness, E., & Reynolds, J. F. (1993). Abnormal spleen lobulation and short pancreas. *Birth Defects Original Article Series*, *29*, 345–352.

Ellis, C., Ramzy, A., & Kieffer, T. J. (2017). Regenerative medicine and cell-based approaches to restore pancreatic function. *Nature Reviews Gastroenterology & Hepatology*, *14*, 612–628.

Esni, F., Ghosh, B., Biankin, A. V., Lin, J. W., Albert, M. A., Yu, X., et al. (2004). Notch inhibits Ptf1 function and acinar cell differentiation in developing mouse and zebrafish pancreas. *Development*, *131*, 4213–4224.

Esni, F., Johansson, B. R., Radice, G. L., & Semb, H. (2001). Dorsal pancreas agenesis in N-cadherin-deficient mice. *Developmental Biology*, *238*, 202–212.

Feig, C., Gopinathan, A., Neesse, A., Chan, D. S., Cook, N., & Tuveson, D. A. (2012). The pancreas cancer microenvironment. *Clinical Cancer Research: An Official Journal of the American Association for Cancer Research*, *18*, 4266–4276.

Gao, T., McKenna, B., Li, C., Reichert, M., Nguyen, J., Singh, T., et al. (2014). Pdx1 maintains β cell identity and function by repressing an α cell program. *Cell Metabolism*, *19*, 259–271.

Garcia-Ocaña, A., Takane, K. K., Syed, M. A., Philbrick, W. M., Vasavada, R. C., & Stewart, A. F. (2000). Hepatocyte growth factor overexpression in the islet of transgenic mice increases beta cell proliferation, enhances islet mass, and induces mild hypoglycemia. *Journal of Biological Chemistry*, *275*, 1226–1232.

Georgia, S., Soliz, R., Li, M., Zhang, P., & Bhushan, A. (2006). p57 and Hes1 coordinate cell cycle exit with self-renewal of pancreatic progenitors. *Developmental Biology*, *298*, 22–31.

Gittes, G. K. (2009). Developmental biology of the pancreas: A comprehensive review. *Developmental Biology*, *326*, 4–35.

Glentis, A., Gurchenkov, V., & Vignjevic, D. M. (2014). Assembly, heterogeneity, and breaching of the basement membranes. *Cell Adhesion & Migration*, *8*, 236–245.

Golosow, N., & Grobstein, C. (1962). Epitheliomesenchymal interaction in pancreatic morphogenesis. *Developmental Biology*, *4*, 242–255.

Goulley, J., Dahl, U., Baeza, N., Mishina, Y., & Edlund, H. (2007). BMP4-BMPR1A signaling in β cells is required for and augments glucose-stimulated insulin secretion. *Cell Metabolism*, *5*, 207–219.

Gouzi, M., Kim, Y. H., Katsumoto, K., Johansson, K., & Grapin-Botton, A. (2011). Neurogenin3 initiates stepwise delamination of differentiating endocrine cells during pancreas development. *Developmental Dynamics, 240*, 589–604.

Gualdi, R., Bossard, P., Zheng, M., Hamada, Y., Coleman, J. R., & Zaret, K. S. (1996). Hepatic specification of the gut endoderm in vitro: Cell signaling and transcriptional control. *Genes & Development, 10*, 1670–1682.

Guo, W., & Giancotti, F. G. (2004). Integrin signalling during tumour progression. *Nature Reviews Molecular Cell Biology, 5*, 816–826.

Hadar, H., Gadoth, N., Herskovitz, P., & Heifetz, M. (1991). Short pancreas in polysplenia syndrome. *Acta Radiologica (Stockholm, Sweden: 1987), 32*, 299–301.

Hart, A., Papadopoulou, S., & Edlund, H. (2003). Fgf10 maintains notch activation, stimulates proliferation, and blocks differentiation of pancreatic epithelial cells. *Developmental Dynamics, 228*, 185–193.

Haumaitre, C., Barbacci, E., Jenny, M., Ott, M. O., Gradwohl, G., & Cereghini, S. (2005). Lack of TCF2/vHNF1 in mice leads to pancreas agenesis. *Proceedings of the National Academy of Sciences of the United States of America, 102*, 1490–1495.

Hebrok, M., Kim, S. K., Jacques, B. S., McMahon, A. P., & Melton, D. A. (2000). Regulation of pancreas development by hedgehog signaling. *Development, 127*, 4905–4913.

Hebrok, M., Kim, S. K., & Melton, D. A. (1998). Notochord repression of endodermal Sonic hedgehog permits pancreas development. *Genes & Development, 12*, 1705–1713.

Hecksher-Sørensen, J., Watson, R. P., Lettice, L. A., Serup, P., Eley, L., Angelis, C. D., et al. (2004). The splanchnic mesodermal plate directs spleen and pancreatic laterality, and is regulated by Bapx1/Nkx3.2. *Development, 131*, 4665–4675.

Heiser, P. W., Lau, J., Taketo, M. M., Herrera, P. L., & Hebrok, M. (2006). Stabilization of beta-catenin impacts pancreas growth. *Development, 133*, 2023–2032.

Heller, R. S., Dichmann, D. S., Jensen, J., Miller, C., Wong, G., Madsen, O. D., et al. (2002). Expression patterns of Wnts, Frizzleds, sFRPs, and misexpression in transgenic mice suggesting a role for Wnts in pancreas and foregut pattern formation. *Developmental Dynamics, 225*, 260–270.

Henseleit, K. D., Nelson, S. B., Kuhlbrodt, K., Hennings, J. C., Ericson, J., & Sander, M. (2005). NKX6 transcription factor activity is required for alpha- and beta-cell development in the pancreas. *Development, 132*, 3139–3149.

Herman, T. E., & Siegel, M. J. (1991). Polysplenia syndrome with congenital short pancreas. *American Journal of Roentgenology, 156*, 799–800.

Hibsher, D., Epshtein, A., Oren, N., & Landsman, L. (2016). Pancreatic mesenchyme regulates islet cellular composition in a patched/hedgehog-dependent manner. *Scientific Reports, 6*, 38008.

Hick, A., van Eyll, J., Cordi, S., Forez, C., Passante, L., Kohara, H., et al. (2009). Mechanism of primitive duct formation in the pancreas and submandibular glands: A role for SDF-1. *BMC Developmental Biology, 9*, 66.

Hisaoka, M., Haratake, J., & Hashimoto, H. (1993). Pancreatic morphogenesis and extracellular matrix organization during rat development. *Differentiation; Research in Biological Diversity, 53*, 163–172.

Hogan, B. L. M. (1999). Morphogenesis. *Cell, 96*, 225–233.

Huotari, M.-A., Miettinen, P. J., Palgi, J., Koivisto, T., Ustinov, J., Harari, D., et al. (2002). ErbB signaling regulates lineage determination of developing pancreatic islet cells in embryonic organ culture. *Endocrinology, 143*, 4437–4446.

Jacquemin, P., Yoshitomi, H., Kashima, Y., Rousseau, G. G., Lemaigre, F. P., & Zaret, K. S. (2006). An endothelial–mesenchymal relay pathway regulates early phases of pancreas development. *Developmental Biology, 290*, 189–199.

Jensen, J., Pedersen, E. E., Galante, P., Hald, J., Heller, R. S., Ishibashi, M., et al. (2000). Control of endodermal endocrine development by Hes-1. *Nature Genetics, 24*, 36–44.

Jiang, F. X., Cram, D. S., DeAizpurua, H. J., & Harrison, L. C. (1999). Laminin-1 promotes differentiation of fetal mouse pancreatic beta-cells. *Diabetes*, *48*, 722–730.
Jiang, F. X., Georges-Labouesse, E., & Harrison, L. C. (2001). Regulation of laminin 1-induced pancreatic beta-cell differentiation by alpha6 integrin and alpha-dystroglycan. *Molecular Medicine*, 7, 107–114.
Jørgensen, M. C., Ahnfelt-Rønne, J., Hald, J., Madsen, O. D., Serup, P., & Hecksher-Sørensen, J. (2006). An illustrated review of early pancreas development in the mouse. *Endocrine Reviews*, *28*, 685–705.
Kawaguchi, Y., Cooper, B., Gannon, M., Ray, M., MacDonald, R. J., & Wright, C. V. E. (2002). The role of the transcriptional regulator Ptf1a in converting intestinal to pancreatic progenitors. *Nature Genetics*, *32*, 128–134.
Kawahira, H., Ma, N. H., Tzanakakis, E. S., McMahon, A. P., Chuang, P.-T., & Hebrok, M. (2003). Combined activities of hedgehog signaling inhibitors regulate pancreas development. *Development*, *130*, 4871–4879.
Kawahira, H., Scheel, D. W., Smith, S. B., German, M. S., & Hebrok, M. (2005). Hedgehog signaling regulates expansion of pancreatic epithelial cells. *Developmental Biology*, *280*, 111–121.
Kesavan, G., Sand, F. W., Greiner, T. U., Johansson, J. K., Kobberup, S., Wu, X., et al. (2009). Cdc42-mediated tubulogenesis controls cell specification. *Cell*, *139*, 791–801.
Kfoury, Y., & Scadden, D. T. (2015). Mesenchymal cell contributions to the stem cell niche. *Cell Stem Cell*, *16*, 239–253.
Kim, S. K., Hebrok, M., Li, E., Oh, S. P., Schrewe, H., Harmon, E. B., et al. (2000). Activin receptor patterning of foregut organogenesis. *Genes & Development*, *14*, 1866–1871.
Kim, S. K., Hebrok, M., & Melton, D. A. (1997). Notochord to endoderm signaling is required for pancreas development. *Development*, *124*, 4243–4252.
Kim, S. W., Shin, H. C., Kim, I. Y., Bae, S.-B., & Park, J. M. (2009). Duplication of the spleen with a short pancreas. *The British Journal of Radiology*, *82*, e42–e43.
Koss, M., Bolze, A., Brendolan, A., Saggese, M., Capellini, T. D., Bojilova, E., et al. (2012). Congenital asplenia in mice and humans with mutations in a Pbx/Nkx2-5/p15 module. *Developmental Cell*, *22*, 913–926.
Kritzik, M. R., Krahl, T., Good, A., Gu, D., Lai, C., Fox, H., et al. (2000). Expression of ErbB receptors during pancreatic islet development and regrowth. *Journal of Endocrinology*, *165*, 67–77.
Kroon, E., Martinson, L. A., Kadoya, K., Bang, A. G., Kelly, O. G., Eliazer, S., et al. (2008). Pancreatic endoderm derived from human embryonic stem cells generates glucose-responsive insulin-secreting cells in vivo. *Nature Biotechnology*, *26*, 443–452.
Kumar, M., Jordan, N., Melton, D., & Grapin-Botton, A. (2003). Signals from lateral plate mesoderm instruct endoderm toward a pancreatic fate. *Developmental Biology*, *259*, 109–122.
Lammert, E., Brown, J., & Melton, D. A. (2000). Notch gene expression during pancreatic organogenesis. *Mechanisms of Development*, *94*, 199–203.
Lammert, E., Cleaver, O., & Melton, D. (2001). Induction of pancreatic differentiation by signals from blood vessels. *Science*, *294*, 564–567.
Landsman, L., Nijagal, A., Whitchurch, T. J., VanderLaan, R. L., Zimmer, W. E., MacKenzie, T. C., et al. (2011). Pancreatic mesenchyme regulates epithelial organogenesis throughout development. *PLoS Biology*, *9*, e1001143.
Larsen, H. L., & Grapin-Botton, A. (2017). The molecular and morphogenetic basis of pancreas organogenesis. *Seminars in Cell & Developmental Biology*, *66*, 51–68.
Larsen, B. M., Hrycaj, S. M., Newman, M., Li, Y., & Wellik, D. M. (2015). Mesenchymal Hox6 function is required for mouse pancreatic endocrine cell differentiation. *Development*, *142*, 3859–3868.

Lattanzi, W., Parolisi, R., Barba, M., & Bonfanti, L. (2015). Osteogenic and neurogenic stem cells in their own place: Unraveling differences and similarities between niches. *Frontiers in Cellular Neuroscience*, *9*, 455.
Lee, J.-H., Tammela, T., Hofree, M., Choi, J., Marjanovic, N. D., Han, S., et al. (2017). Anatomically and functionally distinct lung mesenchymal populations marked by Lgr5 and Lgr6. *Cell*, *170*. 1149–1163.e1112.
Li, L.-C., Qiu, W.-L., Zhang, Y.-W., Xu, Z.-R., Xiao, Y.-N., Hou, C., et al. (2018). Single-cell transcriptomic analyses reveal distinct dorsal/ventral pancreatic programs. *EMBO Reports*, *19*, e46148.
Li, Y., Rankin, S. A., Sinner, D., Kenny, A. P., Krieg, P. A., & Zorn, A. M. (2008). Sfrp5 coordinates foregut specification and morphogenesis by antagonizing both canonical and noncanonical Wnt11 signaling. *Genes & Development*, *22*, 3050–3063.
Löf-Öhlin, Z. M., Nyeng, P., Bechard, M. E., Hess, K., Bankaitis, E., Greiner, T. U., et al. (2017). EGFR signalling controls cellular fate and pancreatic organogenesis by regulating apicobasal polarity. *Nature Cell Biology*, *19*, 1313–1325.
Low, J. P., Williams, D., & Chaganti, J. R. (2011). Polysplenia syndrome with agenesis of the dorsal pancreas and preduodenal portal vein presenting with obstructive jaundice—A case report and literature review. *The British Journal of Radiology*, *84*, e219–e222.
Magenheim, J., Ilovich, O., Lazarus, A., Klochendler, A., Ziv, O., Werman, R., et al. (2011). Blood vessels restrain pancreas branching, differentiation and growth. *Development*, *138*, 4743–4752.
Magenheim, J., Klein, A. M., Stanger, B. Z., Ashery-Padan, R., Sosa-Pineda, B., Gu, G., et al. (2011). Ngn3+ endocrine progenitor cells control the fate and morphogenesis of pancreatic ductal epithelium. *Developmental Biology*, *359*, 26–36.
Maldonado, T. S., Kadison, A. S., Crisera, C. A., Grau, J. B., Alkasab, S. L., Longaker, M. T., et al. (2000). Ontogeny of activin B and follistatin in developing embryonic mouse pancreas: Implications for lineage selection. *Journal of Gastrointestinal Surgery: Official Journal of the Society for Surgery of the Alimentary Tract*, *4*, 269–275.
Manfroid, I., Delporte, F., Baudhuin, A., Motte, P., Neumann, C. J., Voz, M. L., et al. (2007). Reciprocal endoderm-mesoderm interactions mediated by fgf24 and fgf10 govern pancreas development. *Development*, *134*, 4011–4021.
Martín, M., Gallego-Llamas, J., Ribes, V., Kedinger, M., Niederreither, K., Chambon, P., et al. (2005). Dorsal pancreas agenesis in retinoic acid-deficient Raldh2 mutant mice. *Developmental Biology*, *284*, 399–411.
Massagué, J., Seoane, J., & Wotton, D. (2005). Smad transcription factors. *Genes & Development*, *19*, 2783–2810.
Mastracci, T. L., Anderson, K. R., Papizan, J. B., & Sussel, L. (2013). Regulation of Neurod1 contributes to the lineage potential of Neurogenin3+ endocrine precursor cells in the pancreas. *PLoS Genetics*, *9*, e1003278.
McLin, V. A., Rankin, S. A., & Zorn, A. M. (2007). Repression of Wnt/β-catenin signaling in the anterior endoderm is essential for liver and pancreas development. *Development*, *134*, 2207–2217.
Mfopou, J. K., Chen, B., Mateizel, I., Sermon, K., & Bouwens, L. (2010). Noggin, retinoids, and fibroblast growth factor regulate hepatic or pancreatic fate of human embryonic stem cells. *Gastroenterology*, *138*. 2233-2245.e2214.
Miettinen, P. J., Huotari, M., Koivisto, T., Ustinov, J., Palgi, J., Rasilainen, S., et al. (2000). Impaired migration and delayed differentiation of pancreatic islet cells in mice lacking EGF-receptors. *Development*, *127*, 2617–2627.
Miettinen, P. J., Ustinov, J., Ormio, P., Gao, R., Palgi, J., Hakonen, E., et al. (2006). Downregulation of EGF receptor signaling in pancreatic islets causes diabetes due to impaired postnatal β-cell growth. *Diabetes*, *55*, 3299–3308.

Miralles, F., Battelino, T., Czernichow, P., & Scharfmann, R. (1998). TGF-beta plays a key role in morphogenesis of the pancreatic islets of Langerhans by controlling the activity of the matrix metalloproteinase MMP-2. *The Journal of Cell Biology, 143*, 827–836.

Miralles, F., Czernichow, P., Ozaki, K., Itoh, N., & Scharfmann, R. (1999). Signaling through fibroblast growth factor receptor 2b plays a key role in the development of the exocrine pancreas. *Proceedings of the National Academy of Sciences of the United States of America, 96*, 6267–6272.

Miralles, F., Czernichow, P., & Scharfmann, R. (1998). Follistatin regulates the relative proportions of endocrine versus exocrine tissue during pancreatic development. *Development, 125*, 1017–1024.

Miralles, F., Lamotte, L., Couton, D., & Joshi, R. L. (2006). Interplay between FGF10 and Notch signalling is required for the self-renewal of pancreatic progenitors. *The International Journal of Developmental Biology, 50*, 17–26.

Miyatsuka, T., Kosaka, Y., Kim, H., & German, M. S. (2011). Neurogenin3 inhibits proliferation in endocrine progenitors by inducing Cdkn1a. *Proceedings of the National Academy of Sciences of the United States of America, 108*, 185–190.

Molotkov, A., Molotkova, N., & Duester, G. (2005). Retinoic acid generated by Raldh2 in mesoderm is required for mouse dorsal endodermal pancreas development. *Developmental Dynamics, 232*, 950–957.

Muñoz-Bravo, J. L., Hidalgo-Figueroa, M., Pascual, A., López-Barneo, J., Leal-Cerro, A., & Cano, D. A. (2013). GDNF is required for neural colonization of the pancreas. *Development, 140*, 3669–3679.

Murtaugh, L. C., Law, A. C., Dor, Y., & Melton, D. A. (2005). Beta-catenin is essential for pancreatic acinar but not islet development. *Development, 132*, 4663–4674.

Murtaugh, L. C., Stanger, B. Z., Kwan, K. M., & Melton, D. A. (2003). Notch signaling controls multiple steps of pancreatic differentiation. *Proceedings of the National Academy of Sciences of the United States of America, 100*, 14920–14925.

Nekrep, N., Wang, J., Miyatsuka, T., & German, M. S. (2008). Signals from the neural crest regulate beta-cell mass in the pancreas. *Development, 135*, 2151–2160.

Nolan, C. J., Damm, P., & Prentki, M. (2011). Type 2 diabetes across generations: From pathophysiology to prevention and management. *The Lancet, 378*, 169–181.

Norgaard, G. A., Jensen, J. N., & Jensen, J. (2003). FGF10 signaling maintains the pancreatic progenitor cell state revealing a novel role of Notch in organ development. *Developmental Biology, 264*, 323–338.

Nostro, M. C., Sarangi, F., Ogawa, S., Holtzinger, A., Corneo, B., Li, X., et al. (2011). Stage-specific signaling through TGFβ family members and WNT regulates patterning and pancreatic specification of human pluripotent stem cells. *Development, 138*, 861–871.

Ober, E. A., Verkade, H., Field, H. A., & Stainier, D. Y. R. (2006). Mesodermal Wnt2b signalling positively regulates liver specification. *Nature, 442*, 688–691.

Offield, M. F., Jetton, T. L., Labosky, P. A., Ray, M., Stein, R. W., Magnuson, M. A., et al. (1996). PDX-1 is required for pancreatic outgrowth and differentiation of the rostral duodenum. *Development, 122*, 983–995.

Osipovich, A. B., Long, Q., Manduchi, E., Gangula, R., Hipkens, S. B., Schneider, J., et al. (2014). Insm1 promotes endocrine cell differentiation by modulating the expression of a network of genes that includes Neurog3 and Ripply3. *Development, 141*, 2939–2949.

Otonkoski, T., Cirulli, V., Beattie, M., Mally, M. I., Soto, G., Rubin, J. S., et al. (1996). A role for hepatocyte growth factor/scatter factor in fetal mesenchyme-induced pancreatic beta-cell growth. *Endocrinology, 137*, 3131–3139.

Pagliuca, F. W., Millman, J. R., Gürtler, M., Segel, M., Van Dervort, A., Ryu, J. H., et al. (2014). Generation of functional human pancreatic β cells in vitro. *Cell, 159*, 428–439.

Pan, F. C., & Wright, C. (2011). Pancreas organogenesis: From bud to plexus to gland. *Developmental Dynamics, 240*, 530–565.

Papadopoulou, S., & Edlund, H. (2005). Attenuated Wnt signaling perturbs pancreatic growth but not pancreatic function. *Diabetes*, *54*, 2844–2851.

Petzold, K. M., Naumann, H., & Spagnoli, F. M. (2013). Rho signalling restriction by the RhoGAP Stard13 integrates growth and morphogenesis in the pancreas. *Development*, *140*, 126–135.

Pierreux, C. E., Cordi, S., Hick, A.-C., Achouri, Y., Ruiz de Almodovar, C., Prévot, P.-P., et al. (2010). Epithelial: Endothelial cross-talk regulates exocrine differentiation in developing pancreas. *Developmental Biology*, *347*, 216–227.

Plank, J. L., Mundell, N. A., Frist, A. Y., Legrone, A. W., Kim, T., Musser, M. A., et al. (2011). Influence and timing of arrival of murine neural crest on pancreatic beta cell development and maturation. *Developmental Biology*, *349*, 321–330.

Pociot, F., & Lernmark, Å. (2016). Genetic risk factors for type 1 diabetes. *The Lancet*, *387*, 2331–2339.

Pulkkinen, M.-A., Spencer-Dene, B., Dickson, C., & Otonkoski, T. (2003). The IIIb isoform of fibroblast growth factor receptor 2 is required for proper growth and branching of pancreatic ductal epithelium but not for differentiation of exocrine or endocrine cells. *Mechanisms of Development*, *120*, 167–175.

Puri, S., & Hebrok, M. (2007). Dynamics of embryonic pancreas development using real-time imaging. *Developmental Biology*, *306*, 82–93.

Qu, X., Afelik, S., Jensen, J. N., Bukys, M. A., Kobberup, S., Schmerr, M., et al. (2013). Notch-mediated post-translational control of Ngn3 protein stability regulates pancreatic patterning and cell fate commitment. *Developmental Biology*, *376*, 1–12.

Ramond, C., Glaser, N., Berthault, C., Ameri, J., Kirkegaard, J. S., Hansson, M., et al. (2017). Reconstructing human pancreatic differentiation by mapping specific cell populations during development. *eLife*, *6*, e27564.

Rezania, A., Bruin, J. E., Xu, J., Narayan, K., Fox, J. K., O'Neil, J. J., et al. (2013). Enrichment of human embryonic stem cell-derived NKX6.1-expressing pancreatic progenitor cells accelerates the maturation of insulin-secreting cells in vivo. *Stem Cells*, *31*, 2432–2442.

Ritvos, O., Tuuri, T., Erämaa, M., Sainio, K., Hildén, K., Saxén, L., et al. (1995). Activin disrupts epithelial branching morphogenesis in developing glandular organs of the mouse. *Mechanisms of Development*, *50*, 229–245.

Roberts, K. J., Kershner, A. M., & Beachy, P. A. (2017). The stromal niche for epithelial stem cells: A template for regeneration and a brake on malignancy. *Cancer Cell*, *32*, 404–410.

Roccisana, J., Reddy, V., Vasavada, R. C., Gonzalez-Pertusa, J. A., Magnuson, M. A., & Garcia-Ocaña, A. (2005). Targeted inactivation of hepatocyte growth factor receptor c-met in β-cells leads to defective insulin secretion and GLUT-2 downregulation without alteration of β-cell mass. *Diabetes*, *54*, 2090–2102.

Rodríguez-Seguel, E., Mah, N., Naumann, H., Pongrac, I. M., Cerdá-Esteban, N., Fontaine, J.-F., et al. (2013). Mutually exclusive signaling signatures define the hepatic and pancreatic progenitor cell lineage divergence. *Genes & Development*, *27*, 1932–1946.

Romer, A. I., & Sussel, L. (2015). Pancreatic islet cell development and regeneration. *Current Opinion in Endocrinology, Diabetes and Obesity*, *22*, 255.

Rossi, J. M., Dunn, N. R., Hogan, B. L. M., & Zaret, K. S. (2001). Distinct mesodermal signals, including BMPs from the septum transversum mesenchyme, are required in combination for hepatogenesis from the endoderm. *Genes & Development*, *15*, 1998–2009.

Roy, N., Takeuchi, K. K., Ruggeri, J. M., Bailey, P., Chang, D., Li, J., et al. (2016). PDX1 dynamically regulates pancreatic ductal adenocarcinoma initiation and maintenance. *Genes & Development*, *30*, 2669–2683.

Russ, H. A., Parent, A. V., Ringler, J. J., Hennings, T. G., Nair, G. G., Shveygert, M., et al. (2015). Controlled induction of human pancreatic progenitors produces functional beta-like cells in vitro. *The EMBO Journal*, *34*, 1759–1772.

Sand, F. W., Hörnblad, A., Johansson, J. K., Lorén, C., Edsbagge, J., Ståhlberg, A., et al. (2011). Growth-limiting role of endothelial cells in endoderm development. *Developmental Biology*, *352*, 267–277.

Sander, M., Sussel, L., Conners, J., Scheel, D., Kalamaras, J., Cruz, F. D., et al. (2000). Homeobox gene Nkx6.1 lies downstream of Nkx2.2 in the major pathway of beta-cell formation in the pancreas. *Development*, *127*, 5533–5540.

Sanvito, F., Nichols, A., Herrera, P. L., Huarte, J., Wohlwend, A., Vassalli, J. D., et al. (1995). TGF-β1 overexpression in murine pancreas induces chronic pancreatitis and together with TNF-α, triggers insulin-dependent diabetes. *Biochemical and Biophysical Research Communications*, *217*, 1279–1286.

Sasson, A., Rachi, E., Sakhneny, L., Baer, D., Lisnyansky, M., Epshtein, A., et al. (2016). Islet pericytes are required for β-cell maturity. *Diabetes*, *65*, 3008–3014.

Sener, R. N., & Alper, H. (1994). Polysplenia syndrome: A case associated with transhepatic portal vein, short pancreas, and left inferior vena cava with hemiazygous continuation. *Abdominal Imaging*, *19*, 64–66.

Seymour, P. A., Freude, K. K., Tran, M. N., Mayes, E. E., Jensen, J., Kist, R., et al. (2007). SOX9 is required for maintenance of the pancreatic progenitor cell pool. *Proceedings of the National Academy of Sciences of the United States of America*, *104*, 1865–1870.

Seymour, P. A., Shih, H. P., Patel, N. A., Freude, K. K., Xie, R., Lim, C. J., et al. (2012). A Sox9/Fgf feed-forward loop maintains pancreatic organ identity. *Development*, *139*, 3363–3372.

Shapiro, A. M. J., Lakey, J. R. T., Ryan, E. A., Korbutt, G. S., Toth, E., Warnock, G. L., et al. (2000). Islet transplantation in seven patients with type 1 diabetes mellitus using a glucocorticoid-free immunosuppressive regimen. *New England Journal of Medicine*, *343*, 230–238.

Shapiro, A. M. J., Pokrywczynska, M., & Ricordi, C. (2017). Clinical pancreatic islet transplantation. *Nature Reviews Endocrinology*, *13*, 268–277.

Shen, C.-N., Marguerie, A., Chien, C.-Y., Dickson, C., Slack, J. M. W., & Tosh, D. (2007). All-trans retinoic acid suppresses exocrine differentiation and branching morphogenesis in the embryonic pancreas. *Differentiation*, *75*, 62–74.

Shih, H. P., Panlasigui, D., Cirulli, V., & Sander, M. (2016). ECM signaling regulates collective cellular dynamics to control pancreas branching morphogenesis. *Cell Reports*, *14*, 169–179.

Shih, H. P., Wang, A., & Sander, M. (2013). Pancreas organogenesis: From lineage determination to morphogenesis. *Annual Review of Cell and Developmental Biology*, *29*, 81–105.

Sinha, S., & Leach, S. D. (2016). New insights in the development of pancreatic cancer. *Current Opinion in Gastroenterology*, *32*, 394.

Slack, J. M. (1995). Developmental biology of the pancreas. *Development*, *121*, 1569–1580.

Sneddon, J. B., Borowiak, M., & Melton, D. A. (2012). Self-renewal of embryonic-stem-cell-derived progenitors by organ-matched mesenchyme. *Nature*, *491*, 765–768.

Sneddon, J. B., Tang, Q., Stock, P., Bluestone, J. A., Roy, S., Desai, T., et al. (2018). Stem cell therapies for treating diabetes: Progress and remaining challenges. *Cell Stem Cell*, *22*, 810–823.

Solar, M., Cardalda, C., Houbracken, I., Martín, M., Maestro, M. A., De Medts, N., et al. (2009). Pancreatic exocrine duct cells give rise to insulin-producing β cells during embryogenesis but not after birth. *Developmental Cell*, *17*, 849–860.

Spagnoli, F. M. (2007). From endoderm to pancreas: A multistep journey. *Cellular and Molecular Life Sciences*, *64*, 2378–2390.

Spagnoli, F. M., & Brivanlou, A. H. (2008). The Gata5 target, TGIF2, defines the pancreatic region by modulating BMP signals within the endoderm. *Development*, *135*, 451–461.

Sriplung, H. (1991). Polysplenia syndrome: A case with congenital heart block, infarction of a splenic mass, and a short pancreas. *Journal of the Medical Association of Thailand = Chotmaihet Thangphaet*, *74*, 355–358.

Stafford, D., & Prince, V. E. (2002). Retinoic acid signaling is required for a critical early step in zebrafish pancreatic development. *Current Biology*, *12*, 1215–1220.

Taylor, B. L., Liu, F.-F., & Sander, M. (2013). Nkx6.1 is essential for maintaining the functional state of pancreatic beta cells. *Cell Reports*, *4*, 1262–1275.

Thomas, M. K., Lee, J. H., Rastalsky, N., & Habener, J. F. (2001). Hedgehog signaling regulation of homeodomain protein islet duodenum homeobox-1 expression in pancreatic β-cells. *Endocrinology*, *142*, 1033–1040.

Thomas, M. K., Rastalsky, N., Lee, J. H., & Habener, J. F. (2000). Hedgehog signaling regulation of insulin production by pancreatic beta-cells. *Diabetes*, *49*, 2039–2047.

Tremblay, K. D., & Zaret, K. S. (2005). Distinct populations of endoderm cells converge to generate the embryonic liver bud and ventral foregut tissues. *Developmental Biology*, *280*, 87–99.

Tulachan, S. S., Doi, R., Kawaguchi, Y., Tsuji, S., Nakajima, S., Masui, T., et al. (2003). All-trans retinoic acid induces differentiation of ducts and endocrine cells by mesenchymal/epithelial interactions in embryonic pancreas. *Diabetes*, *52*, 76–84.

Tulachan, S. S., Tei, E., Hembree, M., Crisera, C., Prasadan, K., Koizumi, M., et al. (2007). TGF-beta isoform signaling regulates secondary transition and mesenchymal-induced endocrine development in the embryonic mouse pancreas. *Developmental Biology*, *305*, 508–521.

Verschueren, K., Dewulf, N., Goumans, M.-J., Lonnoy, O., Feijen, A., Grimsby, S., et al. (1995). Expression of type I and type IB receptors for activin in midgestation mouse embryos suggests distinct functions in organogenesis. *Mechanisms of Development*, *52*, 109–123.

Villasenor, A., Marty-Santos, L., Dravis, C., Fletcher, P., Henkemeyer, M., & Cleaver, O. (2012). EphB3 marks delaminating endocrine progenitor cells in the developing pancreas. *Developmental Dynamics*, *241*, 1008–1019.

Wainwright, H., & Nelson, M. (1993). Polysplenia syndrome and congenital short pancreas. *American Journal of Medical Genetics*, *47*, 318–320.

Wandzioch, E., & Zaret, K. S. (2009). Dynamic signaling network for the specification of embryonic pancreas and liver progenitors. *Science*, *324*, 1707–1710.

Weir, G. C., & Bonner-Weir, S. (2013). Islet β cell mass in diabetes and how it relates to function, birth, and death. *Annals of the New York Academy of Sciences*, *1281*, 92–105.

Wessells, N. K., & Cohen, J. H. (1967). Early pancreas organogenesis: Morphogenesis, tissue interactions, and mass effects. *Developmental Biology*, *15*, 237–270.

Xuan, S., & Sussel, L. (2016). GATA4 and GATA6 regulate pancreatic endoderm identity through inhibition of hedgehog signaling. *Development*, *143*, 780–786.

Ye, F., Duvillié, B., & Scharfmann, R. (2005). Fibroblast growth factors 7 and 10 are expressed in the human embryonic pancreatic mesenchyme and promote the proliferation of embryonic pancreatic epithelial cells. *Diabetologia*, *48*, 277–281.

Yebra, M., Montgomery, A. M. P., Diaferia, G. R., Kaido, T., Silletti, S., Perez, B., et al. (2003). Recognition of the neural chemoattractant Netrin-1 by integrins α6β4 and α3β1 regulates epithelial cell adhesion and migration. *Developmental Cell*, *5*, 695–707.

Yoshitomi, H., & Zaret, K. S. (2004). Endothelial cell interactions initiate dorsal pancreas development by selectively inducing the transcription factor Ptf1a. *Development*, *131*, 807–817.

Young, H. M., & Newgreen, D. (2001). Enteric neural crest-derived cells: Origin, identification, migration, and differentiation. *The Anatomical Record*, *262*, 1–15.

Zepp, J. A., Zacharias, W. J., Frank, D. B., Cavanaugh, C. A., Zhou, S., Morley, M. P., et al. (2017). Distinct mesenchymal lineages and niches promote epithelial self-renewal and myofibrogenesis in the lung. *Cell*, *170*. 1134-1148.e1110.

Zheng, Y., Ley, S. H., & Hu, F. B. (2018). Global aetiology and epidemiology of type 2 diabetes mellitus and its complications. *Nature Reviews Endocrinology*, *14*, 88–98.

Zhou, Q., Law, A. C., Rajagopal, J., Anderson, W. J., Gray, P. A., & Melton, D. A. (2007). A multipotent progenitor domain guides pancreatic organogenesis. *Developmental Cell*, *13*, 103–114.

Zorn, A. M., & Wells, J. M. (2007). In B.-I. R. O. Cytology (Ed.), *Molecular basis of vertebrate endoderm development* (pp. 49–111). Academic Press.

CHAPTER EIGHT

Gametogenesis: A journey from inception to conception

Hailey Larose[a,†], Adrienne Niederriter Shami[a,†], Haley Abbott[b,†], Gabriel Manske[a], Lei Lei[b,c,*], Saher Sue Hammoud[a,d,e,*]

[a]Department of Human Genetics, University of Michigan Medical School, Ann Arbor, MI, United States
[b]Department of Cell and Developmental Biology, University of Michigan Medical School, Ann Arbor, MI, United States
[c]Department of Biomedical Engineering, University of Michigan Medical School, Ann Arbor, MI, United States
[d]Department of Obstetrics and Gynecology, University of Michigan Medical School, Ann Arbor, MI, United States
[e]Department of Urology, University of Michigan Medical School, Ann Arbor, MI, United States
*Corresponding authors: e-mail address: leile@med.umich.edu; hammou@med.umich.edu

Contents

† Equal first authors.

Current Topics in Developmental Biology, Volume 132
ISSN 0070-2153
https://doi.org/10.1016/bs.ctdb.2018.12.006

Abstract

Gametogenesis, the process of forming mature germ cells, is an integral part of both an individual's and a species' health and well-being. This chapter focuses on critical male and female genetic and epigenetic processes underlying normal gamete formation through their differentiation to fertilization. Finally, we explore how knowledge gained from this field has contributed to progress in areas with great clinical promise, such as *in vitro* gametogenesis.

1. Germ cell specification and gonadal morphogenesis

Gametes serve as a link between the past, present, and future of a species. Unlike somatic cells, germ cells undergo a series of mitotic and meiotic divisions, followed by a differentiation program giving rise to either sperm or egg. These two mature gametes come together at the process of fertilization to produce the totipotent zygote, from which all somatic lineages/tissue and the next generation of gametes will later arise. This cycle continues indefinitely from one generation to the next, highlighting the "immortality" of the germ line. Therefore, understanding the mechanisms and pathways of cell germ cell specification, formation, and differentiation is pivotal for understanding human reproduction, infertility, and evolution.

1.1 PGC specification and migration

In metazoans, primordial germ cells (PGC) are the progenitors for both male and female gametes, giving rise to spermatozoa and oocytes, respectively. Two distinct mechanisms for PGC specification have been described previously. In invertebrates (*C. elegans*, *Drosophila*) and non-mammalian vertebrates (Zebrafish, *Xenopus*), PGCs are specified by a set of maternally inherited factors, known as the germ plasm, which is comprised of RNA, proteins, and organelles that are amassed in a specific location within the oocytes, and subsequently allocated to a few cells in the embryo founding the germline (Matova & Cooley, 2001). Whereas, in mammals the precursors of PGCs arise at about embryonic day 6 (E6) from the equipotent epiblast in response to bone morphogenetic protein (BMP) signals emanating from the extraembryonic ectoderm (BMP4 and 8b) and visceral endoderm (BMP2) (Coucouvanis & Martin, 1999; de Sousa Lopes et al., 2004; Lawson et al., 1999; Loebel, Watson, De Young, & Tam, 2003; McLaren, 2003; Ohinata et al., 2009; Saitou, Barton, & Surani, 2002; Ying, Liu, Marble, Lawson, & Zhao, 2000). BMP signaling is indispensable for PGC specification, and targeted disruption of *Bmp2, Bmp4, Bmp8b*, or BMP signaling transducers *Smad1, Smad4, Smad5* or *Alk2*, all demonstrate loss or reduced numbers of PGCs (Lawson et al., 1999; Saitou et al., 2002; Saitou & Yamaji, 2012; Tremblay, Dunn, & Robertson, 2001; Ying et al., 2000; Ying & Zhao, 2001). The earliest known marker of nascent PGCs in mice is *Fragilis* (a member of the interferon-(IFN)-inducible transmembrane protein family, Saitou et al., 2002). Interestingly, *Fragilis* knockout mice do not exhibit any discernible fertility or developmental defects (Lange et al., 2008). Of the *Fragilis* positive cells, approximately only six cells will express B-lymphocyte-induced maturation protein 1 (BLIMP1), also known as PR domain-containing protein 1 (PRDM1) and two additional key transcription factors, PRDM14 and transcription factor AP-2, gamma (TFAP2C). BLIMP1, PRDM14 and TFAP2C form a tripartite transcription factor network that facilitates mouse PGC specification by suppressing somatic gene expression such as *Hoxa1, Hoxb1, Lim1, Evx1, Fgf8* and *Snail* genes, while initiating the germ cell transcriptional program and triggering genome-wide epigenetic reprogramming (Fig. 1) (Ancelin et al., 2006; Hayashi, Chuva de Sousa Lopes, & Surani, 2007; Kurimoto, Yamaji, Seki, & Saitou, 2008; Magnusdottir et al., 2013; Richardson & Lehmann, 2010; Saitou & Yamaji, 2012; Vincent et al., 2005; Yamaji et al., 2008). Knockout of any of the three factors result in defects in PGC specification process. In contrast, overexpression of these three factors together in competent epiblast like cells *in vitro* is sufficient to induce mouse germ cell formation in the absence of cytokines (Magnusdottir et al., 2013), further underscoring the importance of these three transcription factors in germ cell formation and maintenance. From ~E7 onward, the specified PGCs express the PGC-characteristic markers tissue non-specific alkaline phosphatase (TNAP), stage-specific embryonic antigen 1 (SSEA1) and developmental pluripotency associated 3 (DPPA3 or STELLA) (Chiquoine, 1954; Fox, Damjanov, Martinez-Hernandez, Knowles, & Solter, 1981; Ginsburg, Snow, & McLaren, 1990; Saitou et al., 2002; Sato et al., 2002; Tam & Zhou, 1996). Interestingly, PGCs also maintain the expression of several pluripotent genes, such as SRY (sex-determining region-Y)-box2 (*Sox2*), *Nanog* and octamer-binding transcription factor 4 (*Oct4*) (Avilion et al., 2003; Chambers et al., 2007; Schöler, Hatzopoulos, Balling, Suzuki, & Gruss, 1989; Schöler, Ruppert, Suzuki, Chowdhury, & Gruss, 1990; Yamaguchi, Kimura, Tada, Nakatsuji, & Tada, 2005).

Although the mechanism of PGC specification in mammals is best understood in mice, additional studies in non-rodent mammals in recent

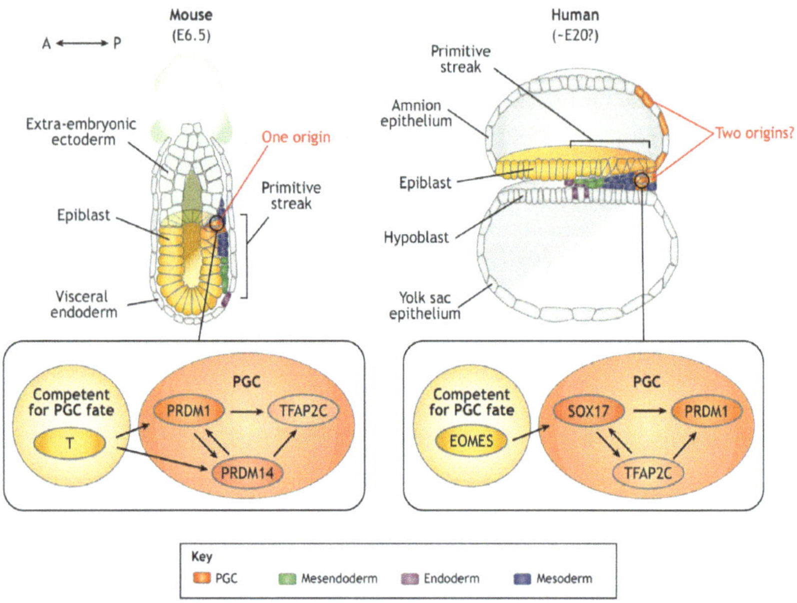

Fig. 1 PGC specification in mice and humans during gastrulation. *Adapted from Kobayashi, T., & Surani, M. A. (2018). On the origin of the human germline.* Development, 145*(16). doi:10.1242/dev.150433.*

years have identified similarities and differences between species. In humans, PGCs are first formed around the third week of gestation. *In vitro* models of human PGC specification from naïve pluripotent stem cells suggest that human PGCs originate from mesodermal precursor cells, and rely on BMP and WNT signaling pathways (Irie et al., 2015; Kojima et al., 2017; Tang, Kobayashi, Irie, Dietmann, & Surani, 2016). The PGC-competent human epiblast cells activate the expression of eomesodermin (EOMES), which subsequently induces expression of TFAP2C, SRY-box 17 (SOX17), and BLIMP1 concomitantly (Irie et al., 2015; Pastor et al., 2018; Tang et al., 2016) (Fig. 1). Unlike mouse PGCs, hPGCs lack PDRM14 and SOX2 expression (Irie et al., 2015); therefore, these subtle differences in mouse and human PGC transcriptional network circuitry may be attributed to the differences in embryonic origin or in pluripotency circuitry.

Once the PGCs are specified, they proliferate while migrating through the hindgut, and then subsequently into the future gonad (also known as the genital ridge) between ~E7.5 and E10.5 (Anderson, Copeland, Schöler, Heasman, & Wylie, 2000; Molyneaux, Stallock, Schaible, & Wylie, 2001; Richardson & Lehmann, 2010). PGC expansion and directional migration are facilitated by two germ cell–soma signaling pathways: cKIT-STEEL and SDF-CXCR4. In mice, c-KIT is expressed in germ cells, whereas STEEL is expressed in somatic cells lining the route to the gonad. The cKIT-STEEL interaction is required for PGC proliferation, survival, and migration from the primitive streak to the future hindgut and then to the genital ridge (Ewen, Baker, Wilhelm, Aitken, & Koopman, 2009; Gu, Runyan, Shoemaker, Surani, & Wylie, 2009; Ohta, Tohda, & Nishimune, 2003; Raz, 2004; Runyan et al., 2006). Homozygous *cKit* or *Steel* mutant mice are sterile because they lack spermatogonial stem cells and thus differentiated germ cells (Blume-Jensen et al., 2000; Chabot, Stephenson, Chapman, Besmer, & Bernstein, 1988; Ohta et al., 2003; Reith et al., 1990). The directionality of PGC migration is also facilitated by the chemoattractant stromal cell-derived factor 1 (SDF-1), expressed at the genital ridges in the surrounding mesenchyme, which is detected by its receptor C-X-C chemokine receptor type 4 (CXCR4) expressed on the surface of PGCs (Raz, 2004). In support of this relationship, genetic mouse models demonstrated that removal of either SDF1 or CXCR4 resulted in very few PGCs reaching the genital ridge. Similarly, ectopic expression of SDF1 causes PGCs to migrate to new locations (Ara et al., 2003; Molyneaux et al., 2003), accounting for the development of some extra-gonadal germ cell tumors in humans (Richardson & Lehmann, 2010).

During this active migration, PGCs continue to proliferate, reaching approximately 500 PGCs in each fetal gonad at E10.5 (Pepling & Spradling, 1998). Once in the gonad, PGCs are now referred to as oogonia in females or gonocytes in males. These fetal germ cells undergo approximately five additional mitotic divisions from E10.5 to E14.5 with incomplete cytokinesis to form germline cysts (Lei & Spradling, 2013). Cysts cluster together and form germ cell nests in both female and male fetal gonads (Lei & Spradling, 2013; Mork et al., 2012; Pepling & Spradling, 1998). These nests eventually resolve and give rise to primary oocytes and prospermatogonia in the respective differentiated gonads.

1.2 Gonadogenesis and sex determination

In mice, the future gonad first appears between E10 and 10.5 as regional thickenings of the epithelium overlying the ventromedial surfaces of the mesonephros. Through the use of numerous loss-of-function mouse models, many genes including Lim homeobox 9 (*Lhx9*), Empty spiracles homeobox 2 (*Emx2*), Wilms tumor 1 (*Wt1*), Chromobox 2 (*Cbx2*), Nuclear receptor superfamily 5 group A (*Nr5a1*), Six homeobox (*Six1/4*), and genes encoding the insulin receptors have all been implicated in establishing the bipotential genital ridges in both male and females (Birk et al., 2000; Fujimoto et al., 2013; Katoh-Fukui et al., 1998; Kreidberg et al., 1993; Luo, Ikeda, & Parker, 1994; Miyamoto, Yoshida, Kuratani, Matsuo, & Aizawa, 1997; Nef et al., 2003). This bipotential gonad remains morphologically and molecular indistinguishable until sex determination, when the supporting somatic cells commit to ovarian or testis lineages (Brennan & Capel, 2004; DeFalco & Capel, 2009). Consistently, many genes involved in establishing sexual dimorphism, such as dosage-sensitive sex reversal, adrenal hypoplasia critical region, on chromosome X, gene 1 (*Dax1*), SRY-box9 (*Sox9*), fibroblast growth factor 9 (*Fgf9*) and Wnt family member 4 (*Wnt4*) are initially expressed in both XX and XY gonads, but later become developmentally restricted (Hoyle, Narvaez, Alldus, Lovell-Badge, & Swain, 2002; Kim et al., 2006a, 2006b; Morais da Silva et al., 1996; Schmahl, Kim, Colvin, Ornitz, & Capel, 2004; Swain, Narvaez, Burgoyne, Camerino, & Lovell-Badge, 1998; Vainio, Heikkila, Kispert, Chin, & McMahon, 1999).

In males, testis morphogenesis is initiated at E11.5 when the testis determining factor SRY is first expressed in the center of the XY genital ridges (Gubbay et al., 1990; Hawkins et al., 1992; Koopman, Gubbay, Vivian, Goodfellow, & Lovell-Badge, 1991; Lovell-Badge & Robertson, 1990), and subsequently expands toward the gonadal poles (Albrecht & Eicher, 2001; Bullejos & Koopman, 2005). The expression of SRY in the supportive somatic cells specifies these cells to differentiate into SOX9+ Sertoli cells, which in turn proliferate in response to FGF9 (Chaboissier et al., 2004; Kim et al., 2006a, 2006b; Palmer & Burgoyne, 1991; Schmahl et al., 2004; Willerton, Smith, Russell, & Mackay, 2004). Interestingly, SRY and SOX9 appear to regulate a significant number of shared targets (Li, Zheng, & Lau, 2014), consistent with the earlier genetic finding demonstrating that ectopic expression of SOX9 is necessary and sufficient for testis differentiation (Chaboissier et al., 2004; Vidal, Chaboissier, de Rooij, & Schedl, 2001). However, in the absence of SRY, FGF9 or SOX9, pregranulosa cells are specified and ovarian development is initiated in XY males (Capel, 2006; Chaboissier et al., 2004; Colvin, Green, Schmahl, Capel, & Ornitz, 2001; Jeays-Ward, Dandonneau, & Swain, 2004; Jeays-Ward et al., 2003; Kent, Wheatley, Andrews, Sinclair, & Koopman, 1996; Koopman, 1999; Koopman et al., 1991; Morais da Silva et al., 1996; Schmahl et al., 2004; Vainio et al., 1999). During normal male testis development and shortly after *Sox9* activation, the Sertoli cells begin to aggregate around clusters of germ cells and form testis cords (Combes et al., 2009). The testis cords are a hallmark structure that separates Sertoli cells and germ cells from the interstitium (Brennan & Capel, 2004). The testis embryonic interstitium arises from two cellular precursors migrating into the gonad from either the coelomic epithelium or the mesonephros in response to anti-Müllerian hormone (AMH) (Karl & Capel, 1998; Martineau, Nordqvist, Tilmann, Lovell-Badge, & Capel, 1997; Ross, Tilman, Yao, MacLaughlin, & Capel, 2003). The interstitium is comprised of the steroidogenic fetal Leydig cells, peritubular myoid cells, macrophages, vasculature, and other more poorly characterized cell types such as fibroblasts and vascular associated cells (Brennan & Capel, 2004; DeFalco, Bhattacharya, Williams, Sams, & Capel, 2014).

Female sex differentiation was originally thought to be the default pathway, during which supportive somatic cells differentiate into pregranulosa cells. However, it is now clear that ovarian differentiation is a coordinated event driven by secreted factors including R-spondin1 (RSPO1), WNT4, and Follistatin, as well as transcriptional regulators such as β-catenin (Biason-Lauber et al., 2007; Biason-Lauber, Konrad, Navratil, & Schoenle, 2004; Chassot et al., 2008; Crisponi et al., 2001; Pailhoux et al., 2001, 2002; Parma et al., 2006; Tomizuka et al., 2008; Vainio et al., 1999). In mice, RSPO and WNT4 are expressed in XX gonads from E11.5 onward, and actively suppress testis vasculature formation, steroidogenic cell migration from the mesonephros, and germ cell loss (Jeays-Ward et al., 2003; Nicol & Yao, 2014). Global loss-of-function models of WNT4 or RSPO1 suggest both factors work through a common signaling pathway, and loss of either factor leads to testis formation in females (Chassot et al., 2008; Tomizuka et al., 2008; Vainio et al., 1999). Duplication of part of human chromosome 1p, which includes the *Wnt4* gene, was shown to lead to XY male to female sex reversal (Jordan et al., 2001). Similarly, genetic loss of the downstream targets of WNT4 and RSPO, such as β-catenin or Follistatin, results in testis development in XX gonads as well (Yao et al., 2004). However, upon normal differentiation of the fetal ovary, nested germ cells and pregranulosa cells organize into an ovigerous cord structure—a structure more pronounced in the fetal ovary of large mammals, such as sheep and humans (Konishi, Fujii, Okamura, Parmley, & Mori, 1986; Wilhelm, Palmer, & Koopman, 2007). Surrounding the ovigerous cords is an interstitium comprised of theca cells and poorly defined stromal cells that arise from either WT1+ endogenous cells of the ovary or glioma-associated oncogene 1 (GLI1) +mesenchymal cells migrating from the mesonephros (Liu, Peng, Matzuk, & Yao, 2015; Martineau et al., 1997).

Although somatic cells in the fetal gonad commit to female or male lineages during sex differentiation, the somatic sex differences require active maintenance. In males, Doublesex—also known as DMRT1—is expressed in both Sertoli cells and germ cells in adult testis. Loss of *Dmrt1* in Sertoli cells leads to the transdifferentiation of Sertoli cells to granulosa cells. Furthermore, ectopic expression of *Dmrt1* in the ovary resulted in morphological changes that are reminiscent of an ovary to testis transdifferentiation (Lindeman et al., 2015). Similarly, the forkhead box L2 (FOXL2) transcription factor maintains granulosa cell fate in adult ovaries. Loss of *Foxl2* in the adult granulosa cells results in granulosa to Sertoli cell transdifferentiation (Uhlenhaut et al., 2009). Thus, it has been proposed that specification and maintenance of the male and female gonadal somatic cells are an active and constant balance between active female (*Wnt4/Rspo/B-catenin/Foxl2*) and male (*Fgf9/SRY/Sox9*) gene regulation (Chassot et al., 2008).

2. Germ cell sex determination

As somatic cells commit to testis or ovary lineages, germ cells also undergo sex differentiation in the fetal gonad and become developmentally restricted (Adams & McLaren, 2002). Meiotic entry is generally considered a hallmark of female sex differentiation. In mice, female germ cells begin to enter meiosis at ~E13.5, and progress through leptotene, zygotene, and pachytene stages, and become arrested at the diplotene stage between E17.5 and P5 (Borum, 1961). In the male gonad ~E14.5, gonocytes commit to male fate and exit the cell cycle, arresting at G0, and remain quiescent until after birth (Western, Miles, van den Bergen, Burton, & Sinclair, 2008).

In mammals, meiotic competency and entry appear to be regulated both intrinsically and extrinsically. Deleted in azoospermia-like (DAZL) is a key intrinsic factor enabling PGCs to respond to cues from the somatic environment and commit to gametogenesis. DAZL is an RNA binding protein required for meiotic chromosome condensation and meiotic prophase protein expression (Lin & Page, 2005). *Dazl* mRNA is detected in both female and male germ cells between E10.5 and E11.5 (Seligman & Page, 1998). *Dazl* deficient mice are infertile due to germ cell differentiation defects (Lin & Page, 2005; Ruggiu et al., 1997; Saunders et al., 2003; Schrans-Stassen, Saunders, Cooke, & de Rooij, 2001).

A well-established extrinsic regulator of meiotic entry in both male and females is retinoic acid (RA). In females, RA is produced by the fetal ovary somatic cells and the mesonephros (Bowles et al., 2016; Koubova et al., 2006; Wilhelm et al., 2007). RA induces the expression of stimulated by retinoic acid gene 8 (STRA8) and meiotic recombination protein 8 (REC8), which are required for meiotic DNA replication, and the subsequent processes of meiotic prophase, and for chromosome synapsis and segregation, respectively (Koubova et al., 2006). In the fetal ovary, *Stra8* expression occurs in an anterior-to-posterior wave. This is followed by anterior-to-posterior waves of expression of meiotic markers DMC1 (a meiotic recombinase) and synaptonemal complex protein 3 (SYCP3) (Bullejos & Koopman, 2004; Menke, Koubova, & Page, 2003; Yao, DiNapoli, & Capel, 2003). Mouse mutations disrupting the genes involved in recombination and mismatch repair, including *Spo11*, *Dmc1*, *Atm*, *Msh4*, *Mlh1*, and *Msh5*, result in both male and female infertility (Barlow et al., 1998; Baudat, Manova, Yuen, Jasin, & Keeney, 2000; Edelmann et al., 1999; Kneitz et al., 2000; Pittman et al., 1998; Romanienko & Camerini-Otero, 2000; Yoshida et al., 1998). Interestingly, in the *Stra8* mutant mouse ovary, fetal germ cells differentiate into primary oocytes without entering meiosis. These primary oocytes could develop in the adult ovary, but fail to mature properly due to meiotic defects (Dokshin, Baltus, Eppig, & Page, 2013). In the embryonic male gonad, RA is degraded by cytochrome P450 26B1 (CYP26B1) produced by Sertoli cells, thereby preventing gonocytes from entering meiosis and inducing arrest at G0/G1 phase of mitosis around E14.5. However, in the postnatal testis, RA (initially provided by the surrounding niche) induces *Stra8* expression and is required for spermatogonial stem cell differentiation and meiotic progression (see Section 4.3) (Zhou et al., 2008). Taken together, despite the differences in gametogenesis timing, the processes regulating meiotic licensing and onset may be conserved between the two sexes.

3. Oogenesis

In mammalian females, primordial follicles that form during the fetal stage serve as the only source for egg production in adulthood. Oogenesis is comprised of two stages: oocyte differentiation (embryonic) and oocyte development (postpuberty). During oocyte differentiation, PGCs differentiate into primary oocytes, which are further encapsulated by a single layer of

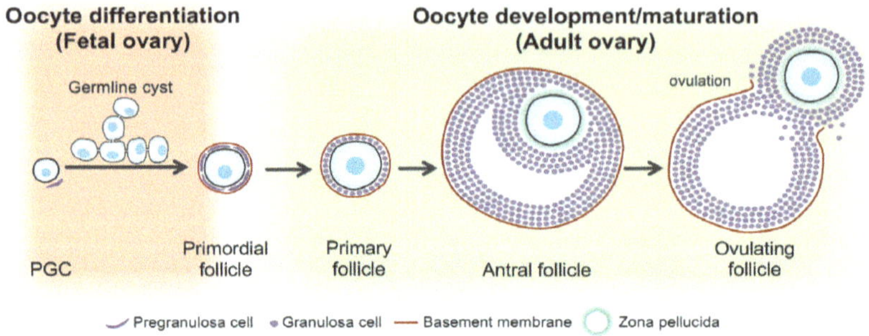

Fig. 2 The process of oogenesis in mammals. *Adapted from Ikami, K., Nuzhat, N., & Lei, L. (2017). Organelle transport during mouse oocyte differentiation in germline cysts.* Current Opinion in Cell Biology, 44, *14–19. doi:10.1016/j.ceb.2016.12.002.*

pregranulosa cells to form primordial follicles. During oocyte development, primordial follicles develop into mature follicles that ultimately produce mature and fertilizable oocytes (Fig. 2).

3.1 Oocyte differentiation and primordial follicle formation

In mice, oocyte differentiation takes place from E13.5 to P4. During this process, germline cysts/germ cell nests fragment and ~20% of the germ cells differentiate into primary oocytes and assemble into primordial follicles (Fig. 2). This process of primordial follicle formation is complex, requiring the presence of germ cells (McLaren, 1984), and is regulated by several transcription factors (factor in germline α (FIG α) and FOXL2) and signaling pathways (Notch, TGFβ, and Kit) (Pepling, 2006, 2012). Interestingly, steroid hormones have been shown to play an inhibitory role in germ cell nest breakdown and primordial follicle formation. Specifically, neonatal mice or rats treated with estrogen (including synthetic estrogen), bisphenol-A (BPA), testosterone, or progesterone delayed primordial follicle formation or led to multi-oocyte follicles (MOFs) in adult ovaries (Chen, Jefferson, Newbold, Padilla-Banks, & Pepling, 2007; Iguchi, Fukazawa, Uesugi, & Takasugi, 1990; Iguchi & Takasugi, 1986; Iguchi, Takasugi, Bern, & Mills, 1986; Iguchi, Todoroki, Takasugi, & Petrow, 1988; Iguchi-Ariga & Schaffner, 1989; Jefferson, Couse, Padilla-Banks, Korach, & Newbold, 2002; Suzuki et al., 2002). The primordial follicles formed in the embryonic/neonatal phase remain quiescent until puberty, with only a few primordial follicles recruited immediately—giving rise to the so called the first wave of folliculogenesis (Cordeiro et al., 2015; Kerr, Myers, & Anderson, 2013).

A phenomenon unique to mammalian oocyte differentiation is the extensive germ cell loss throughout this process. In humans, the number of germ cells in the ovary peaks at 5 months of gestation with ~3.5 million germ cells per ovary. Germ cell number decreases to ~1 million by the time oocyte differentiation is completed in the neonatal ovary (Baker, 1963). In mice, germ cell number starts to decline from E14.5. Studies from mutant mouse models suggest that the reduction in germ cell number may be important for clearing oocytes with chromosomal abnormalities or mitochondrial defects (Bergeron et al., 1998; Flaws, Hirshfield, Hewitt, Babus, & Furth, 2001; Gawriluk et al., 2011; Greenfeld, Pepling, Babus, Furth, & Flaws, 2007; Malki, van der Heijden, O'Donnell, Martin, & Bortvin, 2014; Perez et al., 1999; Ratts, Flaws, Kolp, Sorenson, & Tilly, 1995). However, a study by Lei and Spradling suggests that germ cell loss is a consequence of the "nursing process" within the germline cyst during oocyte differentiation. This work showed that ~20% of the E14.5 germ cells in cysts collect cytoplasmic content and become primary oocytes, whereas the remaining 80% of the germ cells undergo apoptosis preferentially after donating their cytoplasm (Lei & Spradling, 2016). Which fetal germ cells become primary oocytes and what determines the number of primary oocyte during oocyte differentiation are interesting open questions.

3.2 Primordial follicle activation

To maintain normal ovarian function, quiescent primordial follicles are continually recruited to developmental phase, *via* primordial follicle activation. About 10–30 primordial follicles initiate development every day in mice (Faddy & Gosden, 1995). As primordial follicles initiate development, oocytes become transcriptionally and translationally active and grow in size (Moore & Lintern-Moore, 1979). Pregranulosa cells become cuboidal and begin to proliferate, facilitating the growth of the follicles (Lintern-Moore & Moore, 1979). Follicles that have just initiated follicular development are called primary follicles, which are recognized structurally by a single layer of cuboidal granulosa cells.

The precise mechanism by which quiescent primordial follicles initiate development is not well understood. An inverse relationship between the proportion of growing follicles and the size of primordial follicle pool has been observed. This correlation led to the hypothesis that primordial follicles experience inhibitory factors from neighboring primordial follicles and growing follicles (Da Silva-Buttkus et al., 2008; Gougeon & Chainy, 1987). Two oocyte extrinsic factors implicated in primordial follicle quiescence include AMH (Durlinger et al., 2002) and FOXL2 (Schmidt et al., 2004; Uda et al., 2004). *In vitro*, ovaries treated with AMH contained

40% fewer growing follicles compared with control ovaries. Whereas, loss of FOXL2 in granulosa cells resulted in premature follicle activation and ovarian failure (Schmidt et al., 2004; Uda et al., 2004).

Additionally, regulation of primordial follicle activation is regulated intrinsically by signaling through the phosphoinositide 3-kinase (PI3K) pathway and various transcription factors, i.e., spermatogenesis and oogenesis basic helix-loop-helix transcription factor 1 (SOHLH1), SOHLH2, newborn ovary homeobox protein (NOBOX), LIM-homeobox transcription factor 8 (LHX8), and forkhead box O3 (FOXO3). Oocyte-specific deletion of *Pten* (a negative regulator of PI3K) resulted in follicular overactivation, leading to premature ovarian failure in young adult mice (Reddy et al., 2008). Similar phenotypes are also observed in the null mutants all of the above four oocyte-specific transcription factors (Choi, Ballow, Xin, & Rajkovic, 2008; Choi, Yuan, & Rajkovic, 2008; Pangas et al., 2006; Rajkovic, Pangas, Ballow, Suzumori, & Matzuk, 2004). Similarly, in humans, mutations in *Nobox*, *Sohlh1* and *Foxl2* are associated with human premature ovarian failure (De Baere et al., 2003; Jagarlamudi & Rajkovic, 2012; Ren et al., 2015).

3.3 Follicle development, selection, and oocyte maturation

Follicle development in mammals can be divided into two phases. The first phase is termed the preantral, or gonadotropin-independent phase, which is regulated by autocrine and paracrine signaling. The second phase of follicle development, termed the antral or gonadotropin-dependent phase, is characterized by the rapid proliferation and secretive activity of granulosa cells (Fig. 2). The follicular fluid secreted by granulosa cells in this stage leads to the formation of the antrum in the follicle. Surrounding each developing follicle are several layers of squamous theca cells immediately outside the basal lamina of the follicle, while interstitial stromal cells distribute in the spaces between follicles (Zeleznik, 2004).

Follicle development is under complex regulation by cytokines and hormones. In particular, the TGF-β family plays a major role in supporting early follicle development. Growth differentiation factor 9 (GDF-9) and BMP-15 are two key oocyte-secreted members of the TGF-β superfamily. GDF-9 mRNA is synthesized only in the oocytes that have reached the primary follicle stage and beyond. In agreement with this expression pattern, *Gdf9*-null female mice are sterile due to the blockage of follicle development at the primary follicle stage (Dong et al., 1996; Günesdogan & Surani, 2016).

A similar follicle development defect was observed in sheep with a mutation in *Bmp15* (Chang, Brown, & Matzuk, 2002), while *Bmp15* null mutant mice are sub-fertile with ovulation defects and decreased egg quality (Yan et al., 2001). Disruptions in these genes have also been implicated in human disease. Dysregulation of GDF-9 expression in oocytes was associated with polycystic ovarian syndrome (PCOS), and mutations in *Bmp15* were found in patients with premature ovarian failure (Di Pasquale et al., 2006; Teixeira Filho et al., 2002). Additional members of the TGF-β family with roles in follicle development include activins and inhibins (Namwanje & Brown, 2016). Conditional inactivation of *Inhba* (βA subunit gene) in developing ovarian granulosa cells in mice reduced female fertility by 38%, while complete loss of *Inhba* and *Inhbb* (activins A and B) in granulosa cells caused infertility (Pangas et al., 2007).

Follicular-stimulating hormone (FSH) and Luteinizing hormone (LH) are two major gonadotropins that promote follicle development and oocyte maturation (Williams & Erickson, 2000). Granulosa cells start to express FSH receptors at the primary follicle stage. FSH promotes granulosa cell proliferation, estrogen production, and is also required for follicle development from primary to preantral stage (Oktay, Briggs, & Gosden, 1997; Oktay, Newton, Mullan, & Gosden, 1998). LH receptors are expressed specifically in theca cells in early stage follicles and are essential for steroidogenesis (progesterone and testosterone) in theca cells. Because later stage follicle development is FSH-dependent, the fluctuation of the circulating FSH level during each menstrual cycle facilitates follicle selection. During this process, only a small fraction (one in humans) of antral follicles mature and become ovulatory follicles. According to the "FSH threshold" model, a quick elevation of FSH level due to the regression of the corpus luteum from the previous cycle allows one of the many antral follicles to quickly become the dominant follicle. The growth of the dominant follicle decreases the amount of circulating FSH due to the negative feedback. This lack of FSH causes the rest of the preantral follicles to regress (Zeleznik, 2004). Based on this mechanism of follicle selection, FSH and its analogs have been used to promote ovulation at IVF clinics.

During follicle development, granulosa cells and the oocyte actively communicate with each other using cellular extrusions called the trans-zona pellucida. These extrusions are connected by gap junctions (connexin 43 and connexin 37) that allow small molecules to be exchanged during follicle development. In the developing follicle, the oocyte remains arrested at prophase I, which is characterized by the large nucleus, referred to as the germinal vesicle (GV). The prophase arrest is maintained *via* the combination of the cyclic adenosine monophosphate (cAMP) produced by the oocyte and the cyclic guanosine monophosphate (cGMP) from cumulus cells (granulosa cells that immediately surround the oocyte in the antral follicle) (Jaffe & Egbert, 2017; Mehlmann, 2005). In mice, the hormone LH activates the MAPK pathway in cumulus cells, which leads to decreases in gap junction permeability and cGMP/cAMP in the oocyte (Li & Albertini, 2013). These events further promote the resumption of meiosis and maturation, which can be recognized by the disappearance of the oocyte nucleus, namely "germinal vesicle breakdown" (Jaffe & Egbert, 2017; Lintern-Moore & Moore, 1979; Norris et al., 2008; Shuhaibar et al., 2015). Oocyte enter the second round of meiosis immediately and arrest at metaphase at the time ovulation occurs. The second meiotic arrest is maintained by high cyclin-dependent protein kinase 1 (CDK1) activity in the oocyte and resumes only after fertilization. During meiosis, homologous chromosomes are separated during the first round of meiotic division (Meiosis I), and sister chromatids are separated during the second round of division (Meiosis II). Meiosis during oogenesis undergoes asymmetric cytokinesis. Thus, by the end of meiosis, the initial primary oocyte produces a mature oocyte and three polar bodies that will eventually degenerate. Meiosis I appears to be predisposed toward aneuploidy, especially with aging or under environmental perturbations. The surprisingly high incidence of chromosome segregation errors observed in human oocytes has been linked to defects in embryonic development and acquired birth defects such as Down syndrome (Li & Albertini, 2013).

Besides chromosome reduction and haploidization (so-called nuclear maturation) during oocyte maturation, the oocyte cytoplasm must also mature. In addition to collecting cytoplasmic content during oocyte differentiation at the fetal stage (nursing process), the maturing oocyte expands its cytoplasmic content tremendously through active transcription and translation (Lei & Spradling, 2016; Lintern-Moore & Moore, 1979). Interestingly, as oocytes resume meiosis and undergo the final stages of maturation, they become transcriptionally quiescent (Oh, Hwang, McLaughlin, Solter, & Knowles, 2000). The major translational, posttranslational, and organelle modification events observed during oocyte maturation are essential for the completion of meiosis, fertilization, and early embryonic development (Morgan et al., 2017; Oh et al., 2000; Svoboda, Franke, Schultz, 2015).

Mature oocytes have a unique ability to support early embryogenesis, yet are produced in a limited number throughout a female's reproductive life span. The drastic decline in oocyte quantity and quality in women in their late 30s leads to the increased chance of infertility and birth defects. Uncovering the mechanisms underlying primordial follicle activation, as well as, the formation and maturation of oocyte cytoplasm will provide new insights in solving age-related decline in ovarian function in females.

4. Spermatogenesis

The production of sperm is a continuous process throughout an individual's lifespan, which relies on the constant supply of a rare population of cells with long-term renewal potential, called spermatogonial stem cells (SSCs), located along the basement membrane of the seminiferous tubules of the testes. In rodents, these cells are largely described by their clonal arrangement and expression of a variety of heterogeneous molecular markers. Spermatogonial cell proliferation and their ultimate differentiation into mature spermatozoa are highly regulated developmental processes, and alterations in these processes can result in infertility.

4.1 First wave of spermatogenesis

SSCs are derived from gonocytes, transitionary cells that migrate to the basement membrane of the seminiferous tubules from the center of the testicular cords postnatally, between postnatal days 1 and 4 in mice, and 8 and 12 weeks in humans, in response to still unknown signals. This gonocyte population is heterogeneous. The portion of the gonocytes that express neurogenin 3 (*Ngn3+*) transition to form the founding SSC population in mice between postnatal days 3–6. Whereas gonocytes lacking neurogenin 3 (*Ngn−*) expression directly differentiate into A_2 spermatogonia, initiating the first wave of spermatogenesis at approximately postnatal day 3 (Yoshida et al., 2006, 2004).

The rate at which the first wave of spermatogenesis proceeds is much quicker than that of subsequent adult cycles (Kluin, Kramer, & de Rooij, 1982). Furthermore, a significant fraction of spermatogonia and pachytene spermatocytes from the first wave undergo apoptosis (Kluin et al., 1982; Mori et al., 1997). The apoptosis observed is mediated by caspase 3 activity and is correlated with incomplete blood-testis barrier formation and not necessarily an intrinsic germ cell problem (Jahnukainen et al., 2004; Morales, Mohamed, & Cavicchia, 2007; Moreno, Lizama, Urzúa, Vergara, & Reyes, 2006). Consistent with this observation, the very few remaining first sperm produced during this prepubertal wave still retain the capacity to produce fertile offspring (Miki et al., 2004; Yoshida et al., 2006). Following this first wave of spermatogenesis, regular cycles of asynchronous sperm production begin, each lasting ~35–36 days in mouse. This highly orchestrated developmental process relies on germ cell intrinsic and extrinsic signals from somatic support cells.

4.2 Models for spermatogenesis

The most widely accepted model for germ cell development is the "A_{single} model." In postnatal rodent testes, the self-renewing SSCs are believed to exist in the form of a single spermatogonia (A_s). A_s SSCs divide asymmetrically to maintain the A_s pool or symmetrically to generate an A paired (A_{pr}) spermatogonia that undergo a series of synchronized division to produce A aligned (A_{al}) spermatogonia comprised of 8, 16 or 32 cells. Together, the A_s, A_{pr}, and A_{al} cells make up the total population of undifferentiated spermatogonia, which is estimated to be ~0.3% of the total mouse testis. Only 10% of this population (or 0.03% of the testis) are A_s SSCs (Huckins, 1971; Oakberg, 1971; Phillips, Gassei, & Orwig, 2010; Valli et al., 2014).

Much effort has been invested in determining markers to specifically distinguish A_s, A_p, and A_{al} SSCs (Fayomi & Orwig, 2018). Interestingly, the majority of markers identified cover most or all undifferentiated spermatogonia (ZBTB16, SALL4, LIN28, CDH1 and FOXO1) (Buaas et al., 2004; Costoya et al., 2004; Eildermann et al., 2012; Gassei & Orwig, 2013; Goertz, Wu, Gallardo, Hamra, & Castrillon, 2011; Hobbs et al., 2012; Tokuda, Kadokawa, Kurahashi, & Marunouchi, 2007), while some were restricted to A_s spermatogonia (ID4, PAX7, BMI1) (Aloisio et al., 2014; Komai et al., 2014; Oatley, Kaucher, Racicot, & Oatley, 2011) or extend only to A_{al} (GFRa1 and NANOS2) (Meng et al., 2000; Suzuki, Sada, Yoshida, & Saga, 2009; van Bragt et al., 2008). Surprisingly, the loss of function experiments from many of the undifferentiated SSC markers, including those most specific to A_s, show progressive infertility rather than the complete germ cell loss, suggesting there may not be a strict stem cell hierarchy (Aloisio et al., 2014; Buaas et al., 2004; Costoya et al., 2004; Eildermann et al., 2012; Fayomi & Orwig, 2018; Gassei & Orwig, 2013; Goertz et al., 2011; Hobbs et al., 2012; Komai et al., 2014; Meng et al., 2000; Oatley et al., 2011; Suzuki et al., 2009; Tokuda et al., 2007; van Bragt et al., 2008). Rather, in the absence of the ultimate SSC, other populations can compensate. From an evolutionary standpoint, a plastic stem and progenitor pool would ensure continuation of spermatogenesis, and thus fertility, even in the presence of genetic lesions.

An alternative model for germ cell development is the "fragmentation model" which relies on a series of experiments performed by the Yoshida group, who developed live-imaging to follow the behavior of spermatogonia on the basal lamina of mouse seminiferous tubules for a period of about 3 days (Hara et al., 2014; Nakagawa, Sharma, Nabeshima, Braun, & Yoshida, 2010). *In vivo* monitoring of *Ngn3*+ or *GFRα1*+ single-cell dynamics and biophysical modeling of SSC kinetics during steady state or regeneration after injury suggests that the fragmented SSCs from $A_{aligned}$ chains have the capacity to repopulate the spermatogonial niche (Hara et al., 2014; Nakagawa et al., 2010). Consistent with this model, early single cell RNA sequencing studies from mice show that the undifferentiated SSC population is fairly heterogeneous and does not reveal distinct functional subtypes or hierarchy, thus, supporting a plastic stem/progenitor model rather than a hierarchal model within the undifferentiated spermatogonia (Green et al., 2018). However, it is also possible that the developmental hierarchy among the undifferentiated SPG cells may be maintained at the level of protein content or cell-cell interaction, or alternatively, is dependent on very subtle transcriptomic differences which would require much deeper sequencing to discern. Therefore, future single cell transcriptomes/proteomes on selected A_s, A_{pr}, and A_{al} cells will help reconcile the two models of SSC differentiation.

Regardless of preferred model of stem cell divisions, spermatogenesis proceeds to A_{al} (A aligned) spermatogonia which differentiate into A_1 spermatogonia upon gaining expression of the cKIT receptor, a highly conserved marker among both rodents and primates. At this stage, it is believed that the germ cells have fully committed to the complete differentiation process. Further sequential mitotic divisions produce A_2, A_3, A_4, Intermediate and B spermatogonia. The type B spermatogonia each divide to produce two primary spermatocytes and undergo meiotic reduction division to generate secondary spermatocytes which enter spermiogenesis—a process of biochemical and morphological maturation—to produce functional, flagellated spermatozoa. Consequently, due to numerous transit amplifying divisions, one mouse SSC has the potential to produce up to 4096 sperm cells in a single spermatogenic cycle (Russell, Ettlin, Sinha Hikim, & Clegg, 1990), although this has been shown to be a significant overestimate due to apoptosis occurring at the Type A spermatogonia stage (de Rooij, 1973; Huckins, 1978; Huckins & Oakberg, 1978; Russell, Chiarini-Garcia, Korsmeyer, & Knudson, 2002).

In humans, the SSC population does not undergo clonal expansion to the extent seen in rodents, but rather the human testis maintains a significantly larger percentage of undifferentiated spermatogonia (~22% of all cells *vs.* 0.3% in mice) (Paniagua, Codesal, Nistal, Rodriguez, & Santamaria, 1987). These SSCs have been classified histologically using hematoxylin staining into A pale (A_{pale}) or A dark (A_{dark}) spermatogonia (Clermont, 1966; Clermont & Antar, 1973). Historically, the A_{pale} spermatogonia are thought to be the actively cycling population, while the A_{dark} acts as a reserve spermatogonia population (Clermont, 1966; Clifton & Bremner, 1983; Oakberg, 1968; van Alphen, van de Kant, & de Rooij, 1988). However, subsequent labeling studies have arrived at conflicting results (Buageaw et al., 2005; Clermont & Antar, 1973; de Rooij, van Alphen, & van de Kant, 1986; Ehmcke, Simorangkir, & Schlatt, 2005; Fouquet & Dadoune, 1986; Kluin, Kramer, & de Rooij, 1983; Schlatt & Weinbauer, 1994; Simorangkir, Marshall, & Plant, 2009), and whether these A_{dark} are truly quiescent or just in a different phase of the cell cycle remains to be determined. Like mice, human spermatogonia-specific markers are diverse with some overlap between species, but to date these markers do not correlate with an A dark/A pale nuclear morphology (reviewed in Fayomi & Orwig, 2018; von Kopylow & Spiess, 2017). Unlike rodents, the human undifferentiated spermatogonial pool undergoes only one or two mitotic divisions prior to differentiation into type B spermatogonia, entering into meiosis, and completing spermiogenesis.

4.3 Germ cell—Soma communication: Genes/pathways regulating spermatogonial cells

Spermatogenesis is a continuous and asynchronous developmental process strictly controlled by the microenvironment in the seminiferous tubules. Multiple endocrine and paracrine signaling pathways have been implicated in maintaining SSC self-renewal and differentiation (Fig. 3A and B).

4.3.1 GnRH-FSH-LH signaling

Gonadotropin-releasing hormone (GnRH), a hormone produced by the hypothalamus, induces the release of FSH and LH from gonadotrophic cells in the anterior pituitary. Both LH and FSH travel through the circulatory system to the testis where they stimulate testosterone and glial cell line-derived neurotrophic factor (GDNF) production by Leydig and Sertoli cells, respectively. FSH binds to the FSH receptor in Sertoli cells, and stimulates GDNF synthesis (see below), as well as stimulates Sertoli cell proliferation in the prepubertal testis (Dierich et al., 1998; Haywood et al., 2003; Heckert &

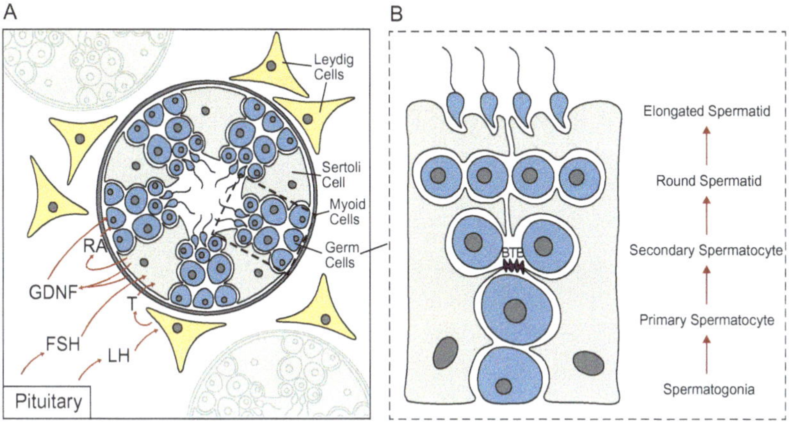

Fig. 3 (A) Schematic overview of testis cross section with major somatic and germ cell types shown, along with select paracrine and endocrine signaling molecules. (B) Progression of spermatogenesis. *BTB*, blood-testis barrier; *FSH*, follicle stimulating hormone; *GDNF*, Glial derived neutrophic factor; *LH*, luteinizing hormone; *RA*, Retinoic acid; *T*, testosterone.

Griswold, 2002; O'Shaughnessy, Monteiro, Verhoeven, De Gendt, & Abel, 2010). Postpuberty, FSH has been implicated in meiosis maintenance; reduction in FSH levels during meiosis resulted in increased pachytene spermatocyte apoptosis and it has been postulated that FSH may act as an apoptosis suppressor (Ruwanpura, McLachlan, Matthiesson, & Meachem, 2008). On the other hand, LHR knockout (LHRKO) males were born phenotypically normal, with testes and genital structures indistinguishable from their wild-type littermates. Postnatally, testicular growth, external genital and accessory sex organ maturation were blocked in LHRKO males, and spermatogenesis was arrested at the round spermatid stage (Zhang, Poutanen, Wilbertz, & Huhtaniemi, 2001). Furthermore, the number and size of Leydig cells were dramatically reduced. Transplanting mesenchymal stem cells into the adult testis of LHRKO mice, restored serum testosterone levels and spermatogenesis (Lo, Lei, Rao Ch, Beck, & Lamb, 2004). Despite the importance of hormonal regulation for the development of the testis and the maintenance of spermatogenesis, the molecular mechanisms through which these hormones act are still poorly understood.

4.3.2 GDNF-RET-GFRA1 signaling

GDNF is secreted by Sertoli and peritubular myoid cells and is a well-defined paracrine factor promoting SSC renewal and maintenance. GDNF is perceived by GFRA1, a glycosylphosphatidylinositol anchored cell surface receptor for GDNF, and RET tyrosine kinase receptor within undifferentiated spermatogonia (Jing et al., 1996; Naughton, Jain, Strickland, Gupta, & Milbrandt, 2006). Loss of GDNF signals from Sertoli cells or peritubular myoid cells *in vivo* results in a loss of the undifferentiated germ cells, whereas overexpression leads to an expansion of the undifferentiated SSCs and the development of tumors (Chen, Willis, & Eddy, 2016; Meng et al., 2000). Similarly, the loss of GDNF receptors on germ cells (*RET* or *GFRA1)* phenocopies loss of GDNF ligand (Jain et al., 2004; Jijiwa et al., 2008; Naughton et al., 2006). Microarray analysis of cultured Thy1 + SSCs *in vitro* demonstrate that GDNF withdrawal reduces the expression of a number of transcription factors known to promote SSC self-renewal (including: *BCL6B*, *LHX1*, *ERM*, *BRACHYURY*, *ETV5*, *POU3F1* and *ID4*). Whereas, the expression of these transcription factors is restored upon addition of GDNF (Oatley, Avarbock, & Brinster, 2007; Oatley, Avarbock, Telaranta, Fearon, & Brinster, 2006; Oatley et al., 2011; Wu, Zhang, & Cairns, 2011). Through a series of molecular experiments, it has been shown that the downstream effects of GDNF signaling are mediated by PI3K/AKT-dependent pathway and the SRC family kinase (SFK) (Jijiwa et al., 2008; Lee et al., 2007; Oatley et al., 2007).

4.3.3 FGF2-MAP2K1 signaling

Fibroblast growth factor 2 (FGF2) belongs to a large family of over 15 Fgf members known to mediate epithelial-mesenchymal cell interactions (Li et al., 2016). *In vitro*, FGF2 induces phosphorylation of MAPK1/3 and AKT in germ cells and stimulates the expression of *Etv5*, *Bcl6b*, and *Lhx1* genes, alongside GDNF (Ishii, Kanatsu-Shinohara, Toyokuni, & Shinohara, 2012). Mouse models have demonstrated that in FGF2-depleted testis, both GDNF levels and SSC numbers increase, suggesting that a balance between FGF2 and GDNF influences SSC self-renewal and highlights the complexity in SSC regulation (Takashima et al., 2015). More research is needed to further understand how FGF2 works to regulate SSC renewal, as well as its interaction with other known SSC maintenance pathways.

4.3.4 Chemokine signaling

Numerous chemokines are produced by Sertoli cells, some of which have unexplored functions in spermatogenesis. One chemokine/receptor pair tested *in vitro* and *in vivo* is the C-X-C motif chemokine 12 (CXCL12)/CXCR4 receptor. CXCL12 is secreted from Sertoli cells and binds to the CXCR4 cell-surface receptor on SSCs. Studies using primary cell cultures of undifferentiated SSCs demonstrated that treatment with CXCR4-specific inhibitor results in SSC loss (Yang, Kim, Kaucher, Oatley, & Oatley, 2013). Additionally, injection of a specific CXCR4 inhibitor directly into the testis of adult mice leads to SSC loss and eventual loss of the germline (Yang et al., 2013).

4.3.5 Retinoic acid (RA) signaling

Progression through spermatogenesis is regulated by intrinsic and extrinsic pulses and stage specific functions of RA signaling. Genetic and molecular studies have elegantly demonstrated that RA signaling is important for spermatogonia differentiation, meiotic initiation, spermatid elongation, and sperm release (Van Pelt & De Rooij, 1990a, 1990b, 1991). During the first wave of spermatogenesis, RA is produced by Sertoli cells and is required spermatogonial cell differentiation (Raverdeau et al., 2012). In subsequent spermatogenic cycles, RA production is germ cell autonomous (Raverdeau et al., 2012), and the downstream targets of RA appear to be cell type and stage specific. However, the molecular targets and interactors of retinoic acid receptors (RARs) at various stages of germ cell development remain unknown.

4.3.6 Androgen signaling

Androgen biosynthesis occurs specifically within the Leydig cells that populate the interstitial tissue of the testis (Luo, Chen, Stocco, & Zirkin, 1998). Testosterone released from Leydig cells binds the *nuclear hormone androgen receptor* (AR) in Sertoli cells, peritubular myoid cells, Leydig cells, vascular smooth muscle and vascular endothelial tissue (Zhou, Kudo, Kawakami, & Hirano, 1996; Zhou et al., 2002). Androgen signaling in Sertoli cells plays an essential role in the maintenance of the blood-testes barrier (Meng, Holdcraft, Shima, Griswold, & Braun, 2005; Willems et al., 2010). Loss of AR signaling in Sertoli cells, as seen in the SCARKO mice, results in pachytene spermatocytes arrest (De Gendt et al., 2004; Tsai et al., 2006). Despite the importance of androgen regulation/signaling in spermatogenesis, the molecular mechanisms through which testosterone acts remain poorly understood.

In summary, germ cell development is complex process due to the reciprocal interactions between germ cells, their somatic supporting cells, and the endocrine system; all of which are required for proper sperm development and fertility. Therefore, understanding this process is not only important for potentially diagnosing and treating male infertility conditions but also for potentially developing male contraceptive targets.

5. Epigenetic reprogramming in PGCs, gametes, and early embryos

The epigenome consists of chemical modifications on either histone proteins or DNA. The interplay between the epigenome, signal-transduction pathways, and transcription factors confers gene expression patterns in a cell, which are stably maintained in subsequent mitotic divisions. Unlike somatic cells, the sperm and egg genomes undergo two waves of epigenetic reprogramming. The first wave occurs in the embryo, but the erasure during this phase is incomplete—leaving maternally and paternally imprinted genes intact in all somatic lineages of the embryo. The second and more comprehensive wave of epigenetic reprogramming occurs upon specification of the germline to reflect the sex of the embryo (Fig. 4). Despite these two global waves of epigenetic erasure, some epigenetic information can be passed from one generation to the next. However, the molecular mechanism by which epigenetic information is conveyed from one generation to the next is a long-standing biological mystery that requires elucidation.

5.1 Epigenetic reprogramming of germ cells

PGCs arise from a small population of cells in the posterior epiblast (Magnusdottir et al., 2013; Ohinata et al., 2005; Weber et al., 2010; Yamaji et al., 2008). Upon specification, PGCs undergo DNA methylation erasure in two mechanistically distinct waves. The first wave is passive DNA demethylation and occurs between E6.5 and E10.5 as a result of repressing *de novo* DNA methyltransferases DNMT3a/b (Kagiwada, Kurimoto, Hirota, Yamaji, & Saitou, 2013; Kurimoto et al., 2008; Seisenberger et al., 2012; Seki et al., 2007; Shirane et al., 2016), and the rapid proliferation of PGCs. The contribution of an active DNA demethylation process during this developmental window has been ruled out since the known DNA demethylases, i.e., AID, APOBEC, or TET family proteins are either not transcribed in early PGCs or their enzymatic activity is dispensable for the global decrease in DNA methylation levels noted (Kagiwada et al., 2013; Popp et al., 2010; Vincent et al., 2013). However, during this passive

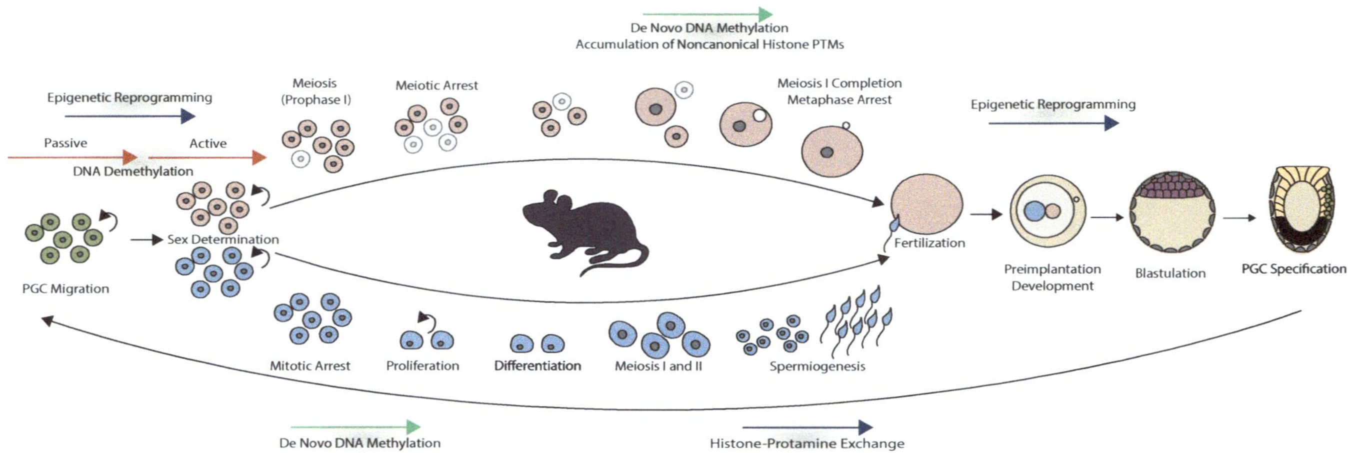

Fig. 4 Epigenetic dynamics during germ cell specification, gametogenesis, and early development.

demethylation developmental time period the maintenance DNA methyltransferase 1 (DNMT1) prevents the dilution of DNA methylation modifications at imprinted loci and meiotic gene promoters (Hargan-Calvopina et al., 2016). Supporting this model, loss of *Dnmt1* in PGCs results in infertility due to premature meiotic entry in females and precocious differentiation of prospermatogonia in males (Hargan-Calvopina et al., 2016).

The second wave of demethylation takes place between E10.5 and E12.5, as PGCs migrate from the hindgut to the genital ridge and begin sex determination. DNA methylation reaches its lowest level during this period, as the enzymes TET1 and TET2 begin to actively remove DNA methylation (Hackett et al., 2013; Vincent et al., 2013; Yamaguchi et al., 2013). This DNA demethylation window ensures erasure of maternally/paternally imprinted loci and resistant promoter regions, leaving E13.5 germ cells with 96% of the genome hypomethylated (Kobayashi et al., 2013; Yamaguchi, Shen, Liu, Sendler, & Zhang, 2013). Surprisingly, loss of TET1 and 2 in germ cells has no effects on fertility, and only a handful of loci exhibit altered epigenetic states (Dawlaty et al., 2013; Popp et al., 2010; Yamaguchi et al., 2012; Yamaguchi, Shen, et al., 2013). These experiments indicate that the multiple families of active DNA demethylase enzymes may have redundant functions during this window.

Loss of DNA methylation in somatic lineages usually results in ectopic expression of retrotransposons, uncontrolled proliferation, and eventual cell death (Smith & Meissner, 2013). Strikingly, PGC development is not affected by the absence of DNA methylation. It has been hypothesized that the reorganization of PGC chromatin may enhance genome stability and maintain global transcriptional quiescence during the period of DNA hypomethylation. Indeed, between E8.5 and E11.5, the repressive histone marks H3 lysine 27 trimethylation (H3K27me3) and dimethylation of arginine 3 on H2A and H4 (H2A/H4R3me2) were found to increase dramatically in PGCs (Ancelin et al., 2006; Hajkova et al., 2008; Seki et al., 2005). In support of their role as modifications that prevent ectopic transcription, an early germline-specific deletion of the H3K9 methyltransferase *SETDB1* results in a loss of H3K9me3 and H3K27me3 and reactivation of the expression of retroviral elements in both sexes (Liu et al., 2014). Similarly, loss of *PRMT5* (the arginine methyltransferase for H2A/H4R3me2) in PGCs also causes reexpression of retroviral elements, and results in complete sterility of both males and females (Ng et al., 2013).

This mechanism of germ cell reprogramming is a conserved process in multiple species. *In vivo* experiments from human PGCs have shown that the dynamics of DNA demethylation mirrors that found in rodents (Gkountela et al., 2015; Guo et al., 2015; Tang et al., 2015). Migratory human PGCs exhibit lower levels of DNA methylation, and several weeks later, the genome of PGCs residing in the hindgut is completely hypomethylated (Guo et al., 2015; Tang et al., 2015). Imprinting regions in humans are also demethylated prior to PGCs taking up residence in the gonad, though it is not known whether this is due to lack of DNA methylation maintenance at these regions or an active DNA demethylation process seen in mice (Tang et al., 2015).

5.2 Chromatin dynamics from sex determination through gametogenesis

Most epigenetic marks are erased in PGCs by E12.5, and are reset to reflect the sex of the embryo (Fig. 4). *De novo* DNA methylation in males resumes in prospermatogonia at E14.5, and is fully established at birth (Kato et al., 2007; Li, Lees-Murdock, Xu, & Walsh, 2004). In females, DNA remethylation does not begin until the postnatal oocyte development, where *de novo* methylation is enriched at sites of active transcription (Bourc'his, Xu, Lin, Bollman, & Bestor, 2001; Hiura, Obata, Komiyama, Shirai, & Kono, 2006; Lucifero, Mann, Bartolomei, & Trasler, 2004; Stewart, Veselovska, & Kelsey, 2016). DNMT3A or 3B and DNMT3L have been identified as the essential factors that mediate DNA methylation at imprinted loci in both the male and female germ lines (Bestor & Bourc'his, 2004; Bourc'his & Bestor, 2004; Bourc'his et al., 2001; Kaneda et al., 2004). Although the sequence identity and characteristics of imprinted regions have been well characterized in gametes, the mechanism which targets the *de novo* methyltransferases to imprinting regions is still largely unexplored (Smallwood et al., 2011).

After germline specification, the next epigenetic hurdle during gametogenesis comes during meiosis. During the first meiotic division, germ cell development stalls in an extended prophase, allowing parental genomes to exchange genetic information through meiotic recombination (Marston & Amon, 2004). In this process, homologous chromosomes become paired (Kota & Feil, 2010; Marston & Amon, 2004), and DNA double-strand breaks (DSBs) are initiated by Spo11 which are ultimately repaired by non-homologous end joining (NHEJ) or homologous recombination (HR) to generate either non-crossover or crossover events, respectively (Hunter, 2006). Maturation of crossover events and repair by HR is necessary to ensure euploidy in gametes, and its loss is implicated in infertility and aneuploidy in the offspring (Baudat et al., 2000; Romanienko & Camerini-Otero, 2000; Zelazowski et al., 2017). These recombination events occur at genomic hotspots, and are enriched at regions outside promoters that bear histone H3K4me3 peaks—established by PRDM9 (Baudat et al., 2010; Hayashi, Yoshida, & Matsui, 2005; Jeffreys, Kauppi, & Neumann, 2001; Mcvean, Myers, & Hunt, 2004; Myers, 2005; Myers, Freeman, Auton, Donnelly, & McVean, 2008; Parvanov, Petkov, & Paigen, 2010). Additional histone modifications such as methylation, acetylation, ubiquitination, and phosphorylation on H3, H4, H2B, and H2A.X, respectively, have all been detected at DSB sites (Borde et al., 2009; Buard, Barthès, Grey, & De Massy, 2009; Lee, Lee, Legube, & Haber, 2014; Yamada et al., 2004). Furthermore, mutations in enzymes involved in establishing meiotic histone posttranslational modifications (PTMs) alter DSB activity, suggesting an active role for histone modifications in initiation/repair process (Gong & Miller, 2013; Hirota, Mizuno, Shibata, & Ohta, 2008; Sollier et al., 2004; Yamashita, Shinohara, & Shinohara, 2004). The progression through meiosis is characterized by a series of transient histone variants including H1t, macroH2A, H2A.X, H2A.Z, and H3.3, all of which have been shown to have roles in the regulation of meiosis (Celeste et al., 2002; Greaves, Rangasamy, Devoy, Marshall Graves, & Tremethick, 2006; Van Der Heijden et al., 2007; Wu et al., 2014).

The chromatin reorganization that accompanies meiosis is largely transient, and the most extensive changes in chromatin state, structure or composition occurs after meiosis in both males and females (Kota & Feil, 2010). In females, histone modifications remain similar between sex determination and primordial follicle formation (between embryonic day 18.5 and postnatal day 10) (Stewart et al., 2015). Toward the later stages of oocyte development, histone 3 lysine 4 and lysine 27 methylation (H3K4me/H3K27me) marks are reorganized into broad domains (Dahl et al., 2016; Liu et al., 2016; Zhang et al., 2016; Zheng et al., 2016). The H3K4 methylation spreads out from its usual enrichment at promoters to cover more distal promoter regions and other intergenic loci (Dahl et al., 2016; Liu et al., 2016; Zhang et al., 2016). Interestingly, this non-canonical pattern of H3K4methylation seems to function as a repressive mark in oocytes. A conditional oocyte knockout of the major H3K4 methyltransferase MLL2 resulted in defects in genome silencing, and overexpression of the H3K4me3 demethylase KDM5b reactivated transcription in quiescent, fully grown oocytes (Andreu-Vieyra et al., 2010; Zhang et al., 2016).

Unlike somatic cells, H3K27me3 in oocytes was present in low levels at canonical polycomb (PRC2) targets, whereas the majority of H3K27me3 is enriched at intergenic regions and gene deserts with low CG density (Liu et al., 2016; Zheng et al., 2016). The distal H3K27me3 marks present in the oocytes escape the epigenetic reprogramming window, and are inherited/maintained for a few cell divisions in the embryo. Unlike the oocyte, the sperm H3K27me3 modifications present at promoters and intergenic regions are erased from the paternal genome in the 1-cell zygote. These allele-specific (maternal/paternal) differences in H3K27me3 patterns in intergenic regions persist to the blastocyst stage, where both H3K4me3/H3K27me3 bivalent promoter marks are restored at developmental genes. The factors that control the reestablishment timing and location of these chromatin marks remain unknown.

In males, postmeiotic germ cells undergo global changes in chromatin structure/packaging and the vast majority of histones in sperm are replaced with sperm-specific nuclear proteins known as protamines (Rando, 2016). The histone-to-protamine exchange is facilitated by series of intermediate steps: the hyperacetylation of histones in round spermatids, which is believed to weaken histone-DNA interactions and enabling the eviction and replacement of histones first by testes-specific histone variants, then by transition proteins, and finally by protamines (Bao & Bedford, 2016; Barral et al., 2017; Hoghoughi, Barral, Vargas, Rousseaux, & Khochbin, 2018). Protamines package the sperm DNA into an endonuclease-inaccessible toroid structure that is 10-fold more compact than the heterochromatin found in somatic nuclei (Furrer, 1976; Hud, Allen, Downing, Lee, & Balhorn, 1993).

The small number of histones retained in mature sperm were presumed to be remnants of incomplete histone-to-protamine exchange; however, earlier work by us and others has demonstrated that histones are enriched at key developmental gene promoters/enhancers in mature sperm. These sperm-retained histones bear both active and repressive histone modification—a unique "poised" chromatin state termed bivalency (Brykczynska et al., 2010; Hammoud et al., 2009). Bivalency was first observed in embryonic stem cells (ESCs) (Bernstein et al., 2006), an attribute presumed to be limited to pluripotent or totipotent cells of the zygote, and important for the maintenance of the pluripotent/totipotent state. However, the presence of bivalent domains at key developmental loci suggested that competency for totipotency is already embedded in sperm chromatin (Hammoud et al., 2014; Lesch, Dokshin, Young, McCarrey, & Page, 2013; Ng et al., 2013; Seisenberger et al., 2012).

This poised chromatin state observed at developmental genes was subsequently shown to be evolutionarily conserved between the *D. rerio* and *M. musculus* male germline, spanning more than 450 million years of evolution (Arpanahi et al., 2009; Brykczynska et al., 2010; Carone et al., 2014; Erkek et al., 2013; Hammoud et al., 2014; Wu et al., 2011). Therefore, the programmatic retention and evolutionary conservation of histone localization has changed our notion of the paternal contribution, and has implied that epigenetic information can be passed through the paternal lineage. Furthermore, altering histone levels, or chromatin regulators during spermatogenesis leads to developmental defects that can be passed on from one generation to the next (Ihara et al., 2014; Siklenka et al., 2015). Taken together, these studies suggest that histones retained in sperm may serve as molecular carriers of epigenetic memory, but whether histones are inherited or instructive for development is not yet known.

5.3 Early embryos

Fertilization of an egg by sperm creates a totipotent zygote. During most of the first cell cycle, both the maternal and paternal haploid genomes are sequestered in separate nuclear compartments called pronuclei (Clift & Schuh, 2013). The separate genomes are replicated and epigenetically reprogrammed as the pronuclei move to meet in the middle of the single cell zygote, break down their nuclear envelopes, and unite for the embryo's first mitosis (Adenot, Mercier, Renard, & Thompson, 1997; Clift & Schuh, 2013). Epigenetic remodeling and DNA replication occur asynchronously between the two compartments, and major epigenetic differences are observed between both pronuclei throughout the first cell cycle (Fig. 4) (Puschendorf et al., 2008).

As both genomes prepare for replication, their chromatin continues to be remodeled. The maternal chromatin retains its non-canonical patterns of H3K4 and H3K27 methylation, but the H3K27me3 is specifically removed from gene promoters (Zhang et al., 2016; Zheng et al., 2016). Protamine proteins are immediately removed from the male pronucleus through an unknown mechanism and replaced by maternal histones. Several hours after fertilization, the paternal pronucleus is formed and enriched with acetylated maternal H3.3 histone variants and H3K27me3 at pericentromeric heterochromatin (Adenot et al., 1997; Burton & Torres-Padilla, 2014; Lin, Koh, Wong, Conti, & Ramalho-Santos, 2014; Loppin et al., 2005; Santenard et al., 2010; Tardat et al., 2015; Torres-Padilla, Bannister, Hurd, Kouzarides, & Zernicka-Goetz, 2006). The differences in chromatin landscape remain through the 8-cell-stage embryo (Ke et al., 2017). Consistent with the differences in epigenetic landscape, allele-specific chromatin capture assays report that the maternal chromatin is highly unstructured, while the paternal chromatin displays a slightly more defined architecture (Du et al., 2017; Flyamer et al., 2017; Ke et al., 2017). It has been shown that the high-order chromatin structure in the embryo is gradually reestablished during embryogenesis, and that it requires DNA replication but not zygotic genome activation (Du et al., 2017; Flyamer et al., 2017; Ke et al., 2017).

Once the male pronucleus forms, the paternal genome experiences rapid and active demethylation by the dioxygenase enzyme TET3 (Gu et al., 2011; Kobayashi et al., 2012; Mayer, Niveleau, Walter, Fundele, & Haaf, 2000; Oswald et al., 2000). Active demethylation by TET3 is necessary for development, as loss-of-function mutants exhibit early embryonic lethality (Gu et al., 2011). Paternally imprinted loci, and the entirety of the maternal genome, are protected from TET3 demethylation by the DNA-binding protein STELLA (Nakamura et al., 2007). This protein binds dimethylated histone H3 at lysine 9 (H3K9me2), a mark that is enriched on the maternal genome and at genomic imprinting regions (Nakamura et al., 2012). After active TET3-mediated DNA methylation, both the paternal and maternal epigenomes undergo passive demethylation throughout the remainder of the preimplantation period (Smith et al., 2012).

During the first cell cycle, maternally inherited factors are preparing the embryo for the transition from an egg-driven to a zygotically controlled developmental fate (Jukam, Shariati, & Skotheim, 2017; Yartseva & Giraldez, 2015). In mice, the first major wave of transcription from the zygotic genome occurs in the late 2-cell embryo (Jukam et al., 2017). This transcription reforms canonical patterns of H3K4me3 at the promoters of actively transcribed genes in both parental alleles; however, the noncanonical patterns of H3K27me3 remain until the blastocyst stage (Dahl et al., 2016; Liu et al., 2016; Zhang et al., 2016). In general, activation of zygotic transcription begins the process of development anew. The epigenetic signatures that are acquired during the very first somatic cell fate decisions overwrite the zygote's totipotent state and restart the cycle of life.

6. *In vitro* gametogenesis

Producing mature eggs or sperm using *in vitro* culture has been a research focus of reproductive biology since the early 1900s (Nagamatsu & Hayashi, 2017). Early studies of *in vitro* gametogenesis mostly focused on establishing organ culture conditions that produce mature oocytes from immature follicles or mature sperm from testicular tissue (Champy, 1920; Cortvrindt, Smitz, & Van Steirteghem, 1996; Eppig, 1977; Eppig & O'Brien, 1996; Eppig & Schroeder, 1989; Martinovitch, 1937). By using fetal ovary organ culture, mature oocytes were acquired from cultured E12.5 mouse ovaries (Morohaku et al., 2016; Obata, Kono, & Hatada, 2002).

As research techniques in embryonic stem cells (ESCs) and induced pluripotent stem cells (iPSCs) continue to improve, recent studies have been focusing on deriving mature eggs from stem cells. The "two-step strategy" established by Hayashi et al. successfully generated mature oocytes from both ESCs and iPSCs (Hayashi et al., 2012; Hayashi, Ohta, Kurimoto, Aramaki, & Saitou, 2011; Hikabe et al., 2016). This culture strategy first induces mouse ESCs/iPSCs to epiblast-like cells by adding basic fibroblast growth factor (bFGF) and ActivinA. Epiblast-like cells can be induced into PGC-like cells by subsequently culturing the cells with media containing BMP4, stem cell factor (SCF), leukemia inhibitory factor (LIF) and epidermal growth factor (EGF) (also known as 4i media) (Hayashi et al., 2012). PGC-like cells differentiate into oocytes by using an organ culture system, in which these cells are aggregated with somatic cells that are isolated from E12.5 fetal ovaries. ESCs/iPSC-derived MII oocytes were capable of fertilization and full-term development. However, pups were produced with 20-fold lower efficiency from ESC-derived eggs than from eggs *in vivo*, suggesting that the quality of the *in vitro*-derived oocytes is partially compromised (Hikabe et al., 2016). A similar 4i culture strategy led to the formation of human PGC-like cells (Irie et al., 2015; Sasaki et al., 2015). In addition to producing fertilizable eggs, the *in vitro* PGC differentiation system allowed scientists to identify key genes that are essential for human PGC specification, such as *Sox17*. Enforced expression of *Sox17* in human ESCs/iPSCs is sufficient for PGC-like cell specification. SOX17 acts upstream of BLIMP1 and other genes to inhibit somatic lineage and promote germline lineage differentiation (Irie et al., 2015). Recently, human oogonia were derived successfully from iPSC during a long-term (~4 months) *in vitro* culture after being reconstituted with mouse fetal ovarian somatic cells. These oogonia display hallmarks of epigenetic reprogramming, such as, genome-wide DNA demethylation and imprint erasure observed during germ cell differentiation *in vivo* (Yamashiro et al., 2018).

The *in vitro* gametogenesis of male germ cells has focused on differentiating ESCs/iPSCs into SSCs and haploid cells that can fertilize eggs. The production of mature sperm from ESCs/iPSCs in culture has not been successfully achieved. Using a similar "two-step strategy," Ishikura et al. derived mouse ESCs into SSCs. In their culture system, PGC-like cells were aggregated with somatic cells dissociated from male E12.5 gonads, which reformed seminiferous tubular structures. In the reconstituted tubules, PGC-like cells differentiated into spermatogonia that express promyelocytic leukemia zinc finger protein (PLZF), a marker of SSCs. These spermatogonia were able to proliferate infinitely *in vitro* and produced functional sperms once transplanted into testes (Ishikura et al., 2016). Geijsen et al. reported a culture system that produced haploid cells from ESCs through embryoid bodies (Geijsen et al., 2004). Later Nayernia et al. developed a strategy for establishing SSC lines from embryonic stem cells (Nayernia et al., 2006). These cells are able to undergo meiosis and generate haploid male gametes *in vitro*. The haploid male gametes were able to fertilize oocytes after being injected into mouse oocytes. However, the resultant offspring were not completely healthy. The offspring were either smaller or larger than controls and died between 5 days and 5 months after birth (Nayernia et al., 2006). Zhou et al. reported a more robust system, in which haploid round spermatids could be derived from mouse ESCs (Zhou et al., 2016). One concern has been raised is that the differentiation process was accelerated in all the culture systems. It usually takes about 6 weeks to differentiate mature sperm from PGCs *via* prospermatogonia *in vivo*, whereas it requires only 2–3 weeks to progress from PGC-like cells to haploid spermatid in culture (Zhou et al., 2016).

7. Conclusion

The production of genetically (and epigenetically) competent gametes is necessary for normal fertilization and early embryonic development. Thus, understanding the underlying processes *in vivo* will allow us to better recapitulate this process *in vitro,* and potentially aid in better defining and understanding idiopathic infertility. The research highlighted over the past few decades has not only broadened our knowledge of gamete and embryo development, and related disorders like infertility and birth defects, but also hold great promise for potentially developing novel reproductive therapies.

References

Adams, I. R., & McLaren, A. (2002). Sexually dimorphic development of mouse primordial germ cells: Switching from oogenesis to spermatogenesis. *Development*, *129*(5), 1155–1164.

Adenot, P. G., Mercier, Y., Renard, J. P., & Thompson, E. M. (1997). Differential H4 acetylation of paternal and maternal chromatin precedes DNA replication and differential transcriptional activity in pronuclei of 1-cell mouse embryos. *Development (Cambridge, England)*, *124*(22), 4615–4625.

Albrecht, K. H., & Eicher, E. M. (2001). Evidence that Sry is expressed in pre-Sertoli cells and Sertoli and granulosa cells have a common precursor. *Developmental Biology*, *240*(1), 92–107. https://doi.org/10.1006/dbio.2001.0438.

Aloisio, G. M., Nakada, Y., Saatcioglu, H. D., Pena, C. G., Baker, M. D., Tarnawa, E. D., et al. (2014). PAX7 expression defines germline stem cells in the adult testis. *The Journal of Clinical Investigation*, *124*(9), 3929–3944. https://doi.org/10.1172/JCI75943.

Ancelin, K., Lange, U. C., Hajkova, P., Schneider, R., Bannister, A. J., Kouzarides, T., et al. (2006). Blimp1 associates with Prmt5 and directs histone arginine methylation in mouse germ cells. *Nature Cell Biology*, *8*, 623–630. https://doi.org/10.1038/ncb1413.

Anderson, R., Copeland, T. K., Schöler, H., Heasman, J., & Wylie, C. (2000). The onset of germ cell migration in the mouse embryo. *Mechanisms of Development*, *91*(1), 61–68. https://doi.org/10.1016/S0925-4773(99)00271-3.

Andreu-Vieyra, C. V., Chen, R., Agno, J. E., Glaser, S., Anastassiadis, K., Stewart Francis, A., et al. (2010). MLL2 is required in oocytes for bulk histone 3 lysine 4 trimethylation and transcriptional silencing. *PLoS Biology*, *8*, 53–54. https://doi.org/10.1371/journal.pbio.1000453.

Ara, T., Nakamura, Y., Egawa, T., Sugiyama, T., Abe, K., Kishimoto, T., et al. (2003). Impaired colonization of the gonads by primordial germ cells in mice lacking a chemokine, stromal cell-derived factor-1 (SDF-1). *Proceedings of the National Academy of Sciences of the United States of America*, *100*(9), 5319–5323.

Arpanahi, A., Brinkworth, M., Iles, D., Krawetz, S. A., Paradowska, A., Platts, A. E., et al. (2009). Endonuclease-sensitive regions of human spermatozoal chromatin are highly enriched in promoter and CTCF binding sequences. *Genome Research*, *19*(8), 1338–1349. https://doi.org/10.1101/gr.094953.109.

Avilion, A. A., Nicolis, S. K., Pevny, L. H., Perez, L., Vivian, N., & Lovell-Badge, R. (2003). Multipotent cell lineages in early mouse development depend on SOX2 function. *Genes & Development*, *17*(1), 126–140. https://doi.org/10.1101/gad.224503.

Baker, T. G. (1963). A quantitative and cytological study of germ cells in human ovaries. *Proceedings of the Royal Society of London, Series B: Biological Sciences*, *158*, 417–433.

Bao, J., & Bedford, M. T. (2016). Epigenetic regulation of the histone-to-protamine transition during spermiogenesis. *Reproduction (Cambridge, England)*, *151*, R55–R70. https://doi.org/10.1530/REP-15-0562.

Barlow, C., Liyanage, M., Moens, P. B., Tarsounas, M., Nagashima, K., Brown, K., et al. (1998). Atm deficiency results in severe meiotic disruption as early as leptonema of prophase I. *Development (Cambridge, England)*, *125*(20), 4007–4017.

Barral, S., Morozumi, Y., Tanaka, H., Montellier, E., Govin, J., de Dieuleveult, M., et al. (2017). Histone variant H2A.L.2 guides transition protein-dependent protamine assembly in male germ cells. *Molecular Cell*, *66*(1). https://doi.org/10.1016/j.molcel.2017.02.025. 89-101.e108.

Baudat, F., Buard, J., Grey, C., Fledel-Alon, A., Ober, C., Przeworski, M., et al. (2010). PRDM9 is a major determinant of meiotic recombination hotspots in humans and mice. *Science*, *327*, 836–840. https://doi.org/10.1126/science.1183439.

Baudat, F., Manova, K., Yuen, J. P., Jasin, M., & Keeney, S. (2000). Chromosome synapsis defects and sexually dimorphic meiotic progression in mice lacking Spo11. *Molecular Cell*, *6*(5), 989–998. https://doi.org/10.1016/S1097-2765(00)00098-8.

Bergeron, L., Perez, G. I., Macdonald, G., Shi, L., Sun, Y., Jurisicova, A., et al. (1998). Defects in regulation of apoptosis in caspase-2-deficient mice. *Genes & Development*, *12*(9), 1304–1314.

Bernstein, B. E., Mikkelsen, T. S., Xie, X., Kamal, M., Huebert, D. J., Cuff, J., et al. (2006). A bivalent chromatin structure marks key developmental genes in embryonic stem cells. *Cell*, *125*(2), 315–326. https://doi.org/10.1016/j.cell.2006.02.041.

Bestor, T. H., & Bourc'his, D. (2004). Transposon silencing and imprint establishment in mammalian germ cells. *Cold Spring Harbor Symposia on Quantitative Biology*, *69*, 381–387. https://doi.org/10.1101/sqb.2004.69.381.

Biason-Lauber, A., De Filippo, G., Konrad, D., Scarano, G., Nazzaro, A., & Schoenle, E. J. (2007). WNT4 deficiency—A clinical phenotype distinct from the classic Mayer-Rokitansky-Kuster-Hauser syndrome: A case report. *Human Reproduction*, *22*(1), 224–229. https://doi.org/10.1093/humrep/del360.

Biason-Lauber, A., Konrad, D., Navratil, F., & Schoenle, E. J. (2004). A WNT4 mutation associated with Mullerian-duct regression and virilization in a 46,XX woman. *The New England Journal of Medicine*, *351*(8), 792–798. https://doi.org/10.1056/NEJMoa040533.

Birk, O. S., Casiano, D. E., Wassif, C. A., Cogliati, T., Zhao, L., Zhao, Y., et al. (2000). The LIM homeobox gene Lhx9 is essential for mouse gonad formation. *Nature*, *403*(6772), 909–913. https://doi.org/10.1038/35002622.

Blume-Jensen, P., Jiang, G., Hyman, R., Lee, K. F., O'Gorman, S., & Hunter, T. (2000). Kit/stem cell factor receptor-induced activation of phosphatidylinositol 3′-kinase is essential for male fertility. *Nature Genetics*, *24*(2), 157–162. https://doi.org/10.1038/72814.

Borde, V., Robine, N., Lin, W., Bonfils, S., Géli, V., & Nicolas, A. (2009). Histone H3 lysine 4 trimethylation marks meiotic recombination initiation sites. *EMBO Journal*, *28*, 99–111. https://doi.org/10.1038/emboj.2008.257.

Borum, K. (1961). Oogenesis in the mouse. A study of the meiotic prophase. *Experimental Cell Research*, *24*, 495–507.

Bourc'his, D., & Bestor, T. H. (2004). Meiotic catastrophe and retrotransposon reactivation in male germ cells lacking Dnmt3L. *Nature*, *431*(7004), 96–99. https://doi.org/10.1038/nature02886.

Bourc'his, D., Xu, G. L., Lin, C. S., Bollman, B., & Bestor, T. H. (2001). Dnmt3L and the establishment of maternal genomic imprints. *Science*, *294*(5551), 2536–2539. https://doi.org/10.1126/science.1065848.

Bowles, J., Feng, C. W., Miles, K., Ineson, J., Spiller, C., & Koopman, P. (2016). ALDH1A1 provides a source of meiosis-inducing retinoic acid in mouse fetal ovaries. *Nature Communications*, 7, 10845. https://doi.org/10.1038/ncomms10845.

Brennan, J., & Capel, B. (2004). One tissue, two fates: Molecular genetic events that underlie testis versus ovary development. *Nature Reviews. Genetics*, *5*(7), 509–521. https://doi.org/10.1038/nrg1381.

Brykczynska, U., Hisano, M., Erkek, S., Ramos, L., Oakeley, E. J., Roloff, T. C., et al. (2010). Repressive and active histone methylation mark distinct promoters in human and mouse spermatozoa. *Nature Structural & Molecular Biology*, *17*(6), 679–687. https://doi.org/10.1038/nsmb.1821.

Buaas, F. W., Kirsh, A. L., Sharma, M., McLean, D. J., Morris, J. L., Griswold, M. D., et al. (2004). Plzf is required in adult male germ cells for stem cell self-renewal. *Nature Genetics*, *36*(6), 647–652. https://doi.org/10.1038/ng1366.

Buageaw, A., Sukhwani, M., Ben-Yehudah, A., Ehmcke, J., Rawe, V. Y., Pholpramool, C., et al. (2005). GDNF family receptor alpha1 phenotype of spermatogonial stem cells in immature mouse testes. *Biology of Reproduction*, *73*(5), 1011–1016. https://doi.org/10.1095/biolreprod.105.043810.

Buard, J., Barthès, P., Grey, C., & De Massy, B. (2009). Distinct histone modifications define initiation and repair of meiotic recombination in the mouse. *EMBO Journal*, *28*, 2616–2624. https://doi.org/10.1038/emboj.2009.207.

Bullejos, M., & Koopman, P. (2004). Germ cells enter meiosis in a rostro-caudal wave during development of the mouse ovary. *Molecular Reproduction and Development, 68*(4), 422–428. https://doi.org/10.1002/mrd.20105.

Bullejos, M., & Koopman, P. (2005). Delayed Sry and Sox9 expression in developing mouse gonads underlies B6-Y(DOM) sex reversal. *Developmental Biology, 278*(2), 473–481. https://doi.org/10.1016/j.ydbio.2004.11.030.

Burton, A., & Torres-Padilla, M.-E. (2014). Chromatin dynamics in the regulation of cell fate allocation during early embryogenesis. *Nature Reviews. Molecular Cell Biology, 15*, 723–735. https://doi.org/10.1038/nrm3885.

Capel, B. (2006). R-spondin1 tips the balance in sex determination. *Nature Genetics, 38*(11), 1233–1234. https://doi.org/10.1038/ng1106-1233.

Carone, B. R., Hung, J. H., Hainer, S. J., Chou, M. T., Carone, D. M., Weng, Z., et al. (2014). High-resolution mapping of chromatin packaging in mouse embryonic stem cells and sperm. *Developmental Cell, 30*(1), 11–22. https://doi.org/10.1016/j.devcel.2014.05.024.

Celeste, A., Petersen, S., Romanienko, P. J., Fernandez-Capetillo, O., Chen, H. T., Sedelnikova, O. a., et al. (2002). Genomic instability in mice lacking histone H2AX. *Science (New York, N.Y.), 296*, 922–927. https://doi.org/10.1126/science.1069398.

Chaboissier, M. C., Kobayashi, A., Vidal, V. I., Lutzkendorf, S., van de Kant, H. J., Wegner, M., et al. (2004). Functional analysis of Sox8 and Sox9 during sex determination in the mouse. *Development, 131*(9), 1891–1901. https://doi.org/10.1242/dev.01087.

Chabot, B., Stephenson, D. A., Chapman, V. M., Besmer, P., & Bernstein, A. (1988). The proto-oncogene c-kit encoding a transmembrane tyrosine kinase receptor maps to the mouse W locus. *Nature, 335*(6185), 88–89. https://doi.org/10.1038/335088a0.

Chambers, I., Silva, J., Colby, D., Nichols, J., Nijmeijer, B., Robertson, M., et al. (2007). Nanog safeguards pluripotency and mediates germline development. *Nature, 450*, 1230.

Champy, C. (1920). Quelques résultats de la méthode de culture des tissues. *Archives de Zoologie Expérimentale et Générale, 60*, 461–500.

Chang, H., Brown, C. W., & Matzuk, M. M. (2002). Genetic analysis of the mammalian transforming growth factor-beta superfamily. *Endocrine Reviews, 23*(6), 787–823. https://doi.org/10.1210/er.2002-0003.

Chassot, A. A., Ranc, F., Gregoire, E. P., Roepers-Gajadien, H. L., Taketo, M. M., Camerino, G., et al. (2008). Activation of beta-catenin signaling by Rspo1 controls differentiation of the mammalian ovary. *Human Molecular Genetics, 17*(9), 1264–1277. https://doi.org/10.1093/hmg/ddn016.

Chen, Y., Jefferson, W. N., Newbold, R. R., Padilla-Banks, E., & Pepling, M. E. (2007). Estradiol, progesterone, and genistein inhibit oocyte nest breakdown and primordial follicle assembly in the neonatal mouse ovary in vitro and in vivo. *Endocrinology, 148*(8), 3580–3590. https://doi.org/10.1210/en.2007-0088.

Chen, L. Y., Willis, W. D., & Eddy, E. M. (2016). Targeting the GDNF gene in peritubular myoid cells disrupts undifferentiated spermatogonial cell development. *Proceedings of the National Academy of Sciences of the United States of America, 113*(7), 1829–1834. https://doi.org/10.1073/pnas.1517994113.

Chiquoine, A. D. (1954). The identification, origin, and migration of the primordial germ cells in the mouse embryo. *The Anatomical Record, 118*(2), 135–146.

Choi, Y., Ballow, D. J., Xin, Y., & Rajkovic, A. (2008). Lim homeobox gene, lhx8, is essential for mouse oocyte differentiation and survival. *Biology of Reproduction, 79*(3), 442–449. https://doi.org/10.1095/biolreprod.108.069393.

Choi, Y., Yuan, D., & Rajkovic, A. (2008). Germ cell-specific transcriptional regulator sohlh2 is essential for early mouse folliculogenesis and oocyte-specific gene expression. *Biology of Reproduction, 79*(6), 1176–1182. https://doi.org/10.1095/biolreprod.108.071217.

Clermont, Y. (1966). Renewal of spermatogonia in man. *American Journal of Anatomy, 118*, 509–524. https://doi.org/10.1002/aja.1001180211.
Clermont, Y., & Antar, M. (1973). Duration of the cycle of the seminiferous epithelium and the spermatogonial renewal in the monkey *Macaca arctoides*. *American Journal of Anatomy, 136*(2), 153–165. https://doi.org/10.1002/aja.1001360204.
Clift, D., & Schuh, M. (2013). Restarting life: Fertilization and the transition from meiosis to mitosis. *Nature Reviews. Molecular Cell Biology, 14*, 549–562. https://doi.org/10.1038/nrm3643.
Clifton, D. K., & Bremner, W. J. (1983). The effect of testicular x-irradiation on spermatogenesis in man. A comparison with the mouse. *Journal of Andrology, 4*(6), 387–392.
Colvin, J. S., Green, R. P., Schmahl, J., Capel, B., & Ornitz, D. M. (2001). Male-to-female sex reversal in mice lacking fibroblast growth factor 9. *Cell, 104*(6), 875–889.
Combes, A. N., Wilhelm, D., Davidson, T., Dejana, E., Harley, V., Sinclair, A., et al. (2009). Endothelial cell migration directs testis cord formation. *Developmental Biology, 326*(1), 112–120. https://doi.org/10.1016/j.ydbio.2008.10.040.
Cordeiro, M. H., Kim, S. Y., Ebbert, K., Duncan, F. E., Ramalho-Santos, J., & Woodruff, T. K. (2015). Geography of follicle formation in the embryonic mouse ovary impacts activation pattern during the first wave of folliculogenesis. *Biology of Reproduction, 93*(4), 88. https://doi.org/10.1095/biolreprod.115.131227.
Cortvrindt, R., Smitz, J., & Van Steirteghem, A. C. (1996). In-vitro maturation, fertilization and embryo development of immature oocytes from early preantral follicles from prepuberal mice in a simplified culture system. *Human Reproduction, 11*(12), 2656–2666.
Costoya, J. A., Hobbs, R. M., Barna, M., Cattoretti, G., Manova, K., Sukhwani, M., et al. (2004). Essential role of Plzf in maintenance of spermatogonial stem cells. *Nature Genetics, 36*(6), 653–659. https://doi.org/10.1038/ng1367.
Coucouvanis, E., & Martin, G. R. (1999). BMP signaling plays a role in visceral endoderm differentiation and cavitation in the early mouse embryo. *Development, 126*(3), 535–546.
Crisponi, L., Deiana, M., Loi, A., Chiappe, F., Uda, M., Amati, P., et al. (2001). The putative forkhead transcription factor FOXL2 is mutated in blepharophimosis/ptosis/epicanthus inversus syndrome. *Nature Genetics, 27*(2), 159–166. https://doi.org/10.1038/84781.
Da Silva-Buttkus, P., Jayasooriya, G. S., Mora, J. M., Mobberley, M., Ryder, T. A., Baithun, M., et al. (2008). Effect of cell shape and packing density on granulosa cell proliferation and formation of multiple layers during early follicle development in the ovary. *Journal of Cell Science, 121*(Pt. 23), 3890–3900. https://doi.org/10.1242/jcs.036400.
Dahl, J. A., Jung, I., Aanes, H., Greggains, G. D., Manaf, A., Lerdrup, M., et al. (2016). Broad histone H3K4me3 domains in mouse oocytes modulate maternal-to-zygotic transition. *Nature, 537*(7621), 548–552. https://doi.org/10.1038/nature19360.
Dawlaty, M. M., Breiling, A., Le, T., Raddatz, G., Barrasa, M. I., Cheng, A. W., et al. (2013). Combined deficiency of Tet1 and Tet2 causes epigenetic abnormalities but is compatible with postnatal development. *Developmental Cell, 24*, 310–323. https://doi.org/10.1016/j.devcel.2012.12.015.
De Baere, E., Beysen, D., Oley, C., Lorenz, B., Cocquet, J., De Sutter, P., et al. (2003). FOXL2 and BPES: Mutational hotspots, phenotypic variability, and revision of the genotype-phenotype correlation. *American Journal of Human Genetics, 72*(2), 478–487. https://doi.org/10.1086/346118.
De Gendt, K., Swinnen, J. V., Saunders, P. T. K., Schoonjans, L., Dewerchin, M., Devos, A., et al. (2004). A Sertoli cell-selective knockout of the androgen receptor causes spermatogenic arrest in meiosis. *Proceedings of the National Academy of Sciences of the United States of America, 101*, 1327–1332. https://doi.org/10.1073/pnas.0308114100.
de Rooij, D. G. (1973). Spermatogonial stem cell renewal in the mouse. I. Normal situation. *Cell and Tissue Kinetics, 6*(3), 281–287.

de Rooij, D. G., van Alphen, M. M., & van de Kant, H. J. (1986). Duration of the cycle of the seminiferous epithelium and its stages in the rhesus monkey (*Macaca mulatta*). *Biology of Reproduction*, *35*(3), 587–591.
de Sousa Lopes, S. M., Roelen, B. A., Monteiro, R. M., Emmens, R., Lin, H. Y., Li, E., et al. (2004). BMP signaling mediated by ALK2 in the visceral endoderm is necessary for the generation of primordial germ cells in the mouse embryo. *Genes & Development*, *18*(15), 1838–1849. https://doi.org/10.1101/gad.294004.
DeFalco, T., Bhattacharya, I., Williams, A. V., Sams, D. M., & Capel, B. (2014). Yolk-sac-derived macrophages regulate fetal testis vascularization and morphogenesis. *Proceedings of the National Academy of Sciences of the United States of America*, *111*(23), E2384–E2393. https://doi.org/10.1073/pnas.1400057111.
DeFalco, T., & Capel, B. (2009). Gonad morphogenesis in vertebrates: Divergent means to a convergent end. *Annual Review of Cell and Developmental Biology*, *25*, 457–482. https://doi.org/10.1146/annurev.cellbio.042308.13350.
Di Pasquale, E., Rossetti, R., Marozzi, A., Bodega, B., Borgato, S., Cavallo, L., et al. (2006). Identification of new variants of human BMP15 gene in a large cohort of women with premature ovarian failure. *The Journal of Clinical Endocrinology and Metabolism*, *91*(5), 1976–1979. https://doi.org/10.1210/jc.2005-2650.
Dierich, A., Sairam, M. R., Monaco, L., Fimia, G. M., Gansmuller, A., LeMeur, M., et al. (1998). Impairing follicle-stimulating hormone (FSH) signaling in vivo: Targeted disruption of the FSH receptor leads to aberrant gametogenesis and hormonal imbalance. *Proceedings of the National Academy of Sciences of the United States of America*, *95*, 13612–13617. https://doi.org/10.1073/pnas.95.23.13612.
Dokshin, G. A., Baltus, A. E., Eppig, J. J., & Page, D. C. (2013). Oocyte differentiation is genetically dissociable from meiosis in mice. *Nature Genetics*, *45*(8), 877–883. https://doi.org/10.1038/ng.2672.
Dong, J., Albertini, D. F., Nishimori, K., Kumar, T. R., Lu, N., & Matzuk, M. M. (1996). Growth differentiation factor-9 is required during early ovarian folliculogenesis. *Nature*, *383*(6600), 531–535.
Du, Z., Zheng, H., Huang, B., Ma, R., Wu, J., Zhang, X., et al. (2017). Allelic reprogramming of 3D chromatin architecture during early mammalian development. *Nature*, *547*, 232–235. https://doi.org/10.1038/nature23263.
Durlinger, A. L., Gruijters, M. J., Kramer, P., Karels, B., Ingraham, H. A., Nachtigal, M. W., et al. (2002). Anti-Mullerian hormone inhibits initiation of primordial follicle growth in the mouse ovary. *Endocrinology*, *143*(3), 1076–1084. https://doi.org/10.1210/endo.143.3.8691.
Edelmann, W., Cohen, P. E., Kneitz, B., Winand, N., Lia, M., Heyer, J., et al. (1999). Mammalian MutS homologue 5 is required for chromosome pairing in meiosis. *Nature Genetics*, *21*(1), 123–127. https://doi.org/10.1038/5075.
Ehmcke, J., Simorangkir, D. R., & Schlatt, S. (2005). Identification of the starting point for spermatogenesis and characterization of the testicular stem cell in adult male rhesus monkeys. *Human Reproduction*, *20*(5), 1185–1193. https://doi.org/10.1093/humrep/deh766.
Eildermann, K., Aeckerle, N., Debowski, K., Godmann, M., Christiansen, H., Heistermann, M., et al. (2012). Developmental expression of the pluripotency factor sal-like protein 4 in the monkey, human and mouse testis: Restriction to premeiotic germ cells. *Cells, Tissues, Organs*, *196*(3), 206–220. https://doi.org/10.1159/000335031.
Eppig, J. J. (1977). Mouse oocyte development in vitro with various culture systems. *Developmental Biology*, *60*(2), 371–388.
Eppig, J. J., & O'Brien, M. J. (1996). Development in vitro of mouse oocytes from primordial follicles. *Biology of Reproduction*, *54*(1), 197–207.
Eppig, J. J., & Schroeder, A. C. (1989). Capacity of mouse oocytes from preantral follicles to undergo embryogenesis and development to live young after growth, maturation, and fertilization in vitro. *Biology of Reproduction*, *41*(2), 268–276.

Erkek, S., Hisano, M., Liang, C. Y., Gill, M., Murr, R., Dieker, J., et al. (2013). Molecular determinants of nucleosome retention at CpG-rich sequences in mouse spermatozoa. *Nature Structural & Molecular Biology, 20*(10), 1236. https://doi.org/10.1038/nsmb1013-1236b.

Ewen, K., Baker, M., Wilhelm, D., Aitken, R. J., & Koopman, P. (2009). Global survey of protein expression during gonadal sex determination in mice. *Molecular & Cellular Proteomics, 8*(12), 2624–2641. https://doi.org/10.1074/mcp.M900108-MCP200.

Faddy, M. J., & Gosden, R. G. (1995). A mathematical model of follicle dynamics in the human ovary. *Human Reproduction, 10*(4), 770–775.

Fayomi, A. P., & Orwig, K. E. (2018). Spermatogonial stem cells and spermatogenesis in mice, monkeys and men. *Stem Cell Research, 29*, 207–214. https://doi.org/10.1016/j.scr.2018.04.009.

Flaws, J. A., Hirshfield, A. N., Hewitt, J. A., Babus, J. K., & Furth, P. A. (2001). Effect of bcl-2 on the primordial follicle endowment in the mouse ovary. *Biology of Reproduction, 64*(4), 1153–1159.

Flyamer, I. M., Gassler, J., Imakaev, M., Brandão, H. B., Ulianov, S. V., Abdennur, N., et al. (2017). Single-nucleus Hi-C reveals unique chromatin reorganization at oocyte-to-zygote transition. *Nature, 544*, 110–114. https://doi.org/10.1038/nature21711.

Fouquet, J. P., & Dadoune, J. P. (1986). Renewal of spermatogonia in the monkey (*Macaca fascicularis*). *Biology of Reproduction, 35*(1), 199–207.

Fox, N., Damjanov, I., Martinez-Hernandez, A., Knowles, B. B., & Solter, D. (1981). Immunohistochemical localization of the early embryonic antigen (SSEA-1) in postimplantation mouse embryos and fetal and adult tissues. *Developmental Biology, 83*(2), 391–398. https://doi.org/10.1016/0012-1606(81)90487-5.

Fujimoto, Y., Tanaka, S. S., Yamaguchi, Y. L., Kobayashi, H., Kuroki, S., Tachibana, M., et al. (2013). Homeoproteins Six1 and Six4 regulate male sex determination and mouse gonadal development. *Developmental Cell, 26*(4), 416–430. https://doi.org/10.1016/j.devcel.2013.06.018.

Furrer, R. (1976). Physical characteristics of mouse sperm nuclei. *Biophysical Journal, 16*, 811–825. https://doi.org/10.1007/978-3-662-47337-5.

Gassei, K., & Orwig, K. E. (2013). SALL4 expression in gonocytes and spermatogonial clones of postnatal mouse testes. *PLoS One, 8*(1), e53976. https://doi.org/10.1371/journal.pone.0053976.

Gawriluk, T. R., Hale, A. N., Flaws, J. A., Dillon, C. P., Green, D. R., & Rucker, E. B., 3rd. (2011). Autophagy is a cell survival program for female germ cells in the murine ovary. *Reproduction, 141*(6), 759–765. https://doi.org/10.1530/REP-10-0489.

Geijsen, N., Horoschak, M., Kim, K., Gribnau, J., Eggan, K., & Daley, G. Q. (2004). Derivation of embryonic germ cells and male gametes from embryonic stem cells. *Nature, 427*(6970), 148–154. https://doi.org/10.1038/nature02247.

Ginsburg, M., Snow, M., & McLaren, A. (1990). Primordial germ cells in the mouse embryo during gastrulation. *Development, 528*(2), 521–528.

Gkountela, S., Zhang, K. X., Shafiq, T. A., Liao, W. W., Hargan-Calvopiña, J., Chen, P. Y., et al. (2015). DNA demethylation dynamics in the human prenatal germline. *Cell, 161*, 1425–1436. https://doi.org/10.1016/j.cell.2015.05.012.

Goertz, M. J., Wu, Z., Gallardo, T. D., Hamra, F. K., & Castrillon, D. H. (2011). Foxo1 is required in mouse spermatogonial stem cells for their maintenance and the initiation of spermatogenesis. *The Journal of Clinical Investigation, 121*(9), 3456–3466. https://doi.org/10.1172/JCI57984.

Gong, F., & Miller, K. M. (2013). Mammalian DNA repair: HATs and HDACs make their mark through histone acetylation. *Mutation Research, 750*(1–2), 23–30. https://doi.org/10.1016/j.mrfmmm.2013.07.002.

Gougeon, A., & Chainy, G. B. (1987). Morphometric studies of small follicles in ovaries of women at different ages. *Journal of Reproduction and Fertility, 81*(2), 433–442.

Greaves, I. K., Rangasamy, D., Devoy, M., Marshall Graves, J. A., & Tremethick, D. J. (2006). The X and Y chromosomes assemble into H2A.Z, containing facultative heterochromatin, following meiosis. *Molecular and Cellular Biology*, *26*, 5394–5405. https://doi.org/10.1128/MCB.00519-06.

Green, C. D., Ma, Q., Manske, G. L., Shami, A. N., Zheng, X., Marini, S., et al. (2018). A comprehensive roadmap of murine spermatogenesis defined by single-cell RNA-Seq. *Developmental Cell*, *46*(5). https://doi.org/10.1016/j.devcel.2018.07.025. 651-667.e10.

Greenfeld, C. R., Pepling, M. E., Babus, J. K., Furth, P. A., & Flaws, J. A. (2007). BAX regulates follicular endowment in mice. *Reproduction*, *133*(5), 865–876. https://doi.org/10.1530/REP-06-0270.

Gu, T.-P., Guo, F., Yang, H., Wu, H.-P., Xu, G.-F., Liu, W., et al. (2011). The role of Tet3 DNA dioxygenase in epigenetic reprogramming by oocytes. *Nature*, *477*, 606–612. https://doi.org/10.1038/nature10443.

Gu, Y., Runyan, C., Shoemaker, A., Surani, A., & Wylie, C. (2009). Steel factor controls primordial germ cell survival and motility from the time of their specification in the allantois, and provides a continuous niche throughout their migration. *Development*, *136*(8), 1295–1303. https://doi.org/10.1242/dev.030619.

Gubbay, J., Collignon, J., Koopman, P., Capel, B., Economou, A., Munsterberg, A., et al. (1990). A gene mapping to the sex-determining region of the mouse Y chromosome is a member of a novel family of embryonically expressed genes. *Nature*, *346*(6281), 245–250. https://doi.org/10.1038/346245a0.

Günesdogan, U., & Surani, M. A. (2016). Developmental competence for primordial germ cell fate. *Current Topics in Developmental Biology*, *117*, 471–496. https://doi.org/10.1016/BS.CTDB.2015.11.007.

Guo, F., Yan, L., Guo, H., Li, L., Hu, B., Zhao, Y., et al. (2015). The transcriptome and DNA methylome landscapes of human primordial germ cells. *Cell*, *161*, 1437–1452. https://doi.org/10.1016/j.cell.2015.05.015.

Hackett, J. A., Sengupta, R., Zylicz, J. J., Murakami, K., Lee, C., Down, T. A., et al. (2013). Germline DNA demethylation dynamics and imprint erasure through 5-hydroxymethylcytosine. *Science*, *339*, 448–452. https://doi.org/10.1126/science.1229277.

Hajkova, P., Ancelin, K., Waldmann, T., Lacoste, N., Lange, U. C., Cesari, F., et al. (2008). Chromatin dynamics during epigenetic reprogramming in the mouse germ line. *Nature*, *452*, 877–881. https://doi.org/10.1038/nature06714.

Hammoud, S. S., Low, D. H., Yi, C., Carrell, D. T., Guccione, E., & Cairns, B. R. (2014). Chromatin and transcriptional transitions in adult germline stem cells and spermatogenesis. *Cell Stem Cell*, *15*(2), 239–253.

Hammoud, S. S., Nix, D. A., Zhang, H., Purwar, J., Carrell, D. T., & Cairns, B. R. (2009). Distinctive chromatin in human sperm packages genes for embryo development. *Nature*, *460*(7254), 473–478. https://doi.org/10.1038/nature08162.

Hara, K., Nakagawa, T., Enomoto, H., Suzuki, M., Yamamoto, M., Simons, B. D., et al. (2014). Mouse spermatogenic stem cells continually interconvert between equipotent singly isolated and syncytial states. *Cell Stem Cell*, *14*(5), 658–672. https://doi.org/10.1016/j.stem.2014.01.019.

Hargan-Calvopina, J., Taylor, S., Cook, H., Hu, Z., Lee, S. A., Yen, M.-R., et al. (2016). Stage-specific demethylation in primordial germ cells safeguards against precocious differentiation. *Developmental Cell*, *39*, 75–86. https://doi.org/10.1016/j.devcel.2016.07.019.

Hawkins, J. R., Taylor, A., Berta, P., Levilliers, J., Van der Auwera, B., & Goodfellow, P. N. (1992). Mutational analysis of SRY: Nonsense and missense mutations in XY sex reversal. *Human Genetics*, *88*(4), 471–474.

Hayashi, K., Chuva de Sousa Lopes, S. M., & Surani, M. A. (2007). Germ cell specification in mice. *Science*, *316*, 394–396.

Hayashi, K., Ogushi, S., Kurimoto, K., Shimamoto, S., Ohta, H., & Saitou, M. (2012). Offspring from oocytes derived from in vitro primordial germ cell-like cells in mice. *Science*, *338*(6109), 971–975. https://doi.org/10.1126/science.1226889.

Hayashi, K., Ohta, H., Kurimoto, K., Aramaki, S., & Saitou, M. (2011). Reconstitution of the mouse germ cell specification pathway in culture by pluripotent stem cells. *Cell*, *146*(4), 519–532. https://doi.org/10.1016/j.cell.2011.06.052.

Hayashi, K., Yoshida, K., & Matsui, Y. (2005). A histone H3 methyltransferase controls epigenetic events required for meiotic prophase. *Nature*, *438*, 374–378. https://doi.org/10.1038/nature04112.

Haywood, M., Spaliviero, J., Jimemez, M., King, N. J., Handelsman, D. J., & Allan, C. M. (2003). Sertoli and germ cell development in hypogonadal (hpg) mice expressing transgenic follicle-stimulating hormone alone or in combination with testosterone. *Endocrinology*, *144*(2), 509–517. https://doi.org/10.1210/en.2002-220710.

Heckert, L. L., & Griswold, M. D. (2002). The expression of the follicle-stimulating hormone receptor in spermatogenesis. *Recent Progress in Hormone Research*, *57*, 129–148.

Hikabe, O., Hamazaki, N., Nagamatsu, G., Obata, Y., Hirao, Y., Hamada, N., et al. (2016). Reconstitution in vitro of the entire cycle of the mouse female germ line. *Nature*, *539*(7628), 299–303. https://doi.org/10.1038/nature20104.

Hirota, K., Mizuno, K., Shibata, T., & Ohta, K. (2008). Distinct chromatin modulators regulate the formation of accessible and repressive chromatin at the fission yeast recombination hotspot ade6-M26. *Molecular Biology of the Cell*, *19*(3), 1162–1173. https://doi.org/10.1091/mbc.E07-04-0377.

Hiura, H., Obata, Y., Komiyama, J., Shirai, M., & Kono, T. (2006). Oocyte growth-dependent progression of maternal imprinting in mice. *Genes to Cells: Devoted to Molecular & Cellular Mechanisms*, *11*, 353–361. https://doi.org/10.1111/j.1365-2443.2006.00943.x.

Hobbs, R. M., Fagoonee, S., Papa, A., Webster, K., Altruda, F., Nishinakamura, R., et al. (2012). Functional antagonism between Sall4 and Plzf defines germline progenitors. *Cell Stem Cell*, *10*(3), 284–298. https://doi.org/10.1016/j.stem.2012.02.004.

Hoghoughi, N., Barral, S., Vargas, A., Rousseaux, S., & Khochbin, S. (2018). Histone variants: Essential actors in male genome programming. *Journal of Biochemistry*, *163*(2), 97–103. https://doi.org/10.1093/jb/mvx079.

Hoyle, C., Narvaez, V., Alldus, G., Lovell-Badge, R., & Swain, A. (2002). Dax1 expression is dependent on steroidogenic factor 1 in the developing gonad. *Molecular Endocrinology*, *16*(4), 747–756. https://doi.org/10.1210/mend.16.4.0802.

Huckins, C. (1971). The spermatogonial stem cell population in adult rats. I. Their morphology, proliferation and maturation. *The Anatomical Record*, *169*(3), 533–557. https://doi.org/10.1002/ar.1091690306.

Huckins, C. (1978). The morphology and kinetics of spermatogonial degeneration in normal adult rats: An analysis using a simplified classification of the germinal epithelium. *The Anatomical Record*, *190*(4), 905–926. https://doi.org/10.1002/ar.1091900410.

Huckins, C., & Oakberg, E. F. (1978). Morphological and quantitative analysis of spermatogonia in mouse testes using whole mounted seminiferous tubules, I. The normal testes. *The Anatomical Record*, *192*(4), 519–528. https://doi.org/10.1002/ar.1091920406.

Hud, N. V., Allen, M. J., Downing, K. H., Lee, J., & Balhorn, R. (1993). Identification of the elemental packing unit of DNA in mammalian sperm cells by atomic force microscopy. *Biochemical and Biophysical Research Communications*, *193*, 1347–1354.

Hunter, N. (2006). Meiotic recombination. In R. Rothstein & A. Aguilera (Eds.), *Molecular Genetics of Recombination*: *Vol. 17*. Berlin, Heidelberg: Springer.

Iguchi, T., Fukazawa, Y., Uesugi, Y., & Takasugi, N. (1990). Polyovular follicles in mouse ovaries exposed neonatally to diethylstilbestrol in vivo and in vitro. *Biology of Reproduction, 43*(3), 478–484.

Iguchi, T., & Takasugi, N. (1986). Polyovular follicles in the ovary of immature mice exposed prenatally to diethylstilbestrol. *Anatomy and Embryology (Berlin), 175*(1), 53–55.

Iguchi, T., Takasugi, N., Bern, H. A., & Mills, K. T. (1986). Frequent occurrence of polyovular follicles in ovaries of mice exposed neonatally to diethylstilbestrol. *Teratology, 34*(1), 29–35. https://doi.org/10.1002/tera.1420340105.

Iguchi, T., Todoroki, R., Takasugi, N., & Petrow, V. (1988). The effects of an aromatase inhibitor and a 5 alpha-reductase inhibitor upon the occurrence of polyovular follicles, persistent anovulation, and permanent vaginal stratification in mice treated neonatally with testosterone. *Biology of Reproduction, 39*(3), 689–697.

Iguchi-Ariga, S. M., & Schaffner, W. (1989). CpG methylation of the cAMP-responsive enhancer/promoter sequence TGACGTCA abolishes specific factor binding as well as transcriptional activation. *Genes & Development, 3*(5), 612–619.

Ihara, M., Meyer-Ficca, M. L., Leu, N. A., Rao, S., Li, F., Gregory, B. D., et al. (2014). Paternal poly (ADP-ribose) metabolism modulates retention of inheritable sperm histones and early embryonic gene expression. *PLoS Genetics, 10*, https://doi.org/10.1371/journal.pgen.1004317.

Irie, N., Weinberger, L., Tang, W. W., Kobayashi, T., Viukov, S., Manor, Y. S., et al. (2015). SOX17 is a critical specifier of human primordial germ cell fate. *Cell, 160*(1–2), 253–268. https://doi.org/10.1016/j.cell.2014.12.013.

Ishii, K., Kanatsu-Shinohara, M., Toyokuni, S., & Shinohara, T. (2012). FGF2 mediates mouse spermatogonial stem cell self-renewal via upregulation of Etv5 and Bcl6b through MAP2K1 activation. *Development, 139*(10), 1734–1743.

Ishikura, Y., Yabuta, Y., Ohta, H., Hayashi, K., Nakamura, T., Okamoto, I., et al. (2016). In vitro derivation and propagation of spermatogonial stem cell activity from mouse pluripotent stem cells. *Cell Reports, 17*(10), 2789–2804. https://doi.org/10.1016/j.celrep.2016.11.026.

Jaffe, L. A., & Egbert, J. R. (2017). Regulation of mammalian oocyte meiosis by intercellular communication within the ovarian follicle. *Annual Review of Physiology, 79*, 237–260. https://doi.org/10.1146/annurev-physiol-022516-034102.

Jagarlamudi, K., & Rajkovic, A. (2012). Oogenesis: Transcriptional regulators and mouse models. *Molecular and Cellular Endocrinology, 356*(1–2), 31–39. https://doi.org/10.1016/j.mce.2011.07.049.

Jahnukainen, K., Chrysis, D., Hou, M., Parvinen, M., Eksborg, S., & Söder, O. (2004). Increased apoptosis occurring during the first wave of spermatogenesis is stage-specific and primarily affects midpachytene spermatocytes in the rat testis. *Biology of Reproduction, 70*, 290–296. https://doi.org/10.1095/biolreprod.103.018390.

Jain, S., Naughton, C. K., Yang, M., Strickland, A., Vij, K., Encinas, M., et al. (2004). Mice expressing a dominant-negative Ret mutation phenocopy human Hirschsprung disease and delineate a direct role of Ret in spermatogenesis. *Development, 131*(21), 5503–5513.

Jeays-Ward, K., Dandonneau, M., & Swain, A. (2004). Wnt4 is required for proper male as well as female sexual development. *Developmental Biology, 276*(2), 431–440. https://doi.org/10.1016/j.ydbio.2004.08.049.

Jeays-Ward, K., Hoyle, C., Brennan, J., Dandonneau, M., Alldus, G., Capel, B., et al. (2003). Endothelial and steroidogenic cell migration are regulated by WNT4 in the developing mammalian gonad. *Development, 130*(16), 3663–3670.

Jefferson, W. N., Couse, J. F., Padilla-Banks, E., Korach, K. S., & Newbold, R. R. (2002). Neonatal exposure to genistein induces estrogen receptor (ER)alpha expression and multioocyte follicles in the maturing mouse ovary: Evidence for ERbeta-mediated and nonestrogenic actions. *Biology of Reproduction, 67*(4), 1285–1296.

Jeffreys, A., Kauppi, L., & Neumann, R. (2001). Intensely punctate meiotic recombination in the class II region of the major histocompatibility complex. *Nature Genetics, 29*, 217–222. https://doi.org/10.1038/ng1001-217.

Jijiwa, M., Kawai, K., Fukihara, J., Nakamura, A., Hasegawa, M., Suzuki, C., et al. (2008). GDNF-mediated signaling via RET tyrosine 1062 is essential for maintenance of spermatogonial stem cells. *Genes to Cells: Devoted to Molecular & Cellular Mechanisms, 13*(4), 365–374. https://doi.org/10.1111/j.1365-2443.2008.01171.x.

Jing, S., Wen, D., Yu, Y., Holst, P. L., Luo, Y., Fang, M., et al. (1996). GDNF-induced activation of the ret protein tyrosine kinase is mediated by GDNFR-alpha, a novel receptor for GDNF. *Cell, 85*(7), 1113–1124.

Jordan, B. K., Mohammed, M., Ching, S. T., Delot, E., Chen, X. N., Dewing, P., et al. (2001). Up-regulation of WNT-4 signaling and dosage-sensitive sex reversal in humans. *American Journal of Human Genetics, 68*(5), 1102–1109. https://doi.org/10.1086/320125.

Jukam, D., Shariati, S. A. M., & Skotheim, J. M. (2017). Zygotic genome activation in vertebrates. *Developmental Cell, 42*, 316–332. https://doi.org/10.1016/j.devcel.2017.07.026.

Kagiwada, S., Kurimoto, K., Hirota, T., Yamaji, M., & Saitou, M. (2013). Replication-coupled passive DNA demethylation for the erasure of genome imprints in mice. *EMBO Journal, 32*, 340–353. https://doi.org/10.1038/emboj.2012.331.

Kaneda, M., Okano, M., Hata, K., Sado, T., Tsujimoto, H., Li, E., et al. (2004). Essential role for de novo DNA methyltransferase Dnmt3a in paternal and maternal imprinting. *Nature, 429*, 900–903. https://doi.org/10.1038/nature02633.

Karl, J., & Capel, B. (1998). Sertoli cells of the mouse testis originate from the coelomic epithelium. *Developmental Biology, 203*(2), 323–333. https://doi.org/10.1006/dbio.1998.9068.

Kato, Y., Kaneda, M., Hata, K., Kumaki, K., Hisano, M., Kohara, Y., et al. (2007). Role of the Dnmt3 family in de novo methylation of imprinted and repetitive sequences during male germ cell development in the mouse. *Human Molecular Genetics, 16*, 2272–2280. https://doi.org/10.1093/hmg/ddm179.

Katoh-Fukui, Y., Tsuchiya, R., Shiroishi, T., Nakahara, Y., Hashimoto, N., Noguchi, K., et al. (1998). Male-to-female sex reversal in M33 mutant mice. *Nature, 393*(6686), 688–692. https://doi.org/10.1038/31482.

Ke, Y., Xu, Y., Chen, X., Feng, S., Liu, Z., Sun, Y., et al. (2017). 3D chromatin structures of mature gametes and structural reprogramming during mammalian embryogenesis. *Cell, 170*. https://doi.org/10.1016/j.cell.2017.06.029. 367-381.e320.

Kent, J., Wheatley, S. C., Andrews, J. E., Sinclair, A. H., & Koopman, P. (1996). A male-specific role for SOX9 in vertebrate sex determination. *Development, 122*(9), 2813–2822.

Kerr, J. B., Myers, M., & Anderson, R. A. (2013). The dynamics of the primordial follicle reserve. *Reproduction, 146*(6), R205–R215. https://doi.org/10.1530/REP-13-0181.

Kim, Y., Kobayashi, A., Sekido, R., DiNapoli, L., Brennan, J., Chaboissier, M. C., et al. (2006a). Fgf9 and Wnt4 act as antagonistic signals to regulate mammalian sex determination. *PLoS Biology, 4*(6), 1000–1009. https://doi.org/10.1007/11558958_121.

Kim, Y., Kobayashi, A., Sekido, R., DiNapoli, L., Brennan, J., Chaboissier, M. C., et al. (2006b). Fgf9 and Wnt4 act as antagonistic signals to regulate mammalian sex determination. *PLoS Biology, 4*(6), e187. https://doi.org/10.1371/journal.pbio.0040187.

Kluin, P. M., Kramer, M. F., & de Rooij, D. G. (1982). Spermatogenesis in the immature mouse proceeds faster than in the adult. *International Journal of Andrology, 5*, 282–294. https://doi.org/10.1111/j.1365-2605.1982.tb00257.x.

Kluin, P. M., Kramer, M. F., & de Rooij, D. G. (1983). Testicular development in Macaca irus after birth. *International Journal of Andrology, 6*(1), 25–43.

Kneitz, B., Cohen, P. E., Avdievich, E., Zhu, L., Kane, M. F., Hou, H., Jr., et al. (2000). MutS homolog 4 localization to meiotic chromosomes is required for chromosome pairing during meiosis in male and female mice. *Genes & Development, 14*(9), 1085–1097.

Kobayashi, H., Sakurai, T., Imai, M., Takahashi, N., Fukuda, A., Yayoi, O., et al. (2012). Contribution of intragenic DNA methylation in mouse gametic DNA methylomes to establish oocyte-specific heritable marks. *PLoS Genetics 8*, e1002440. https://doi.org/10.1371/journal.pgen.1002440.

Kobayashi, H., Sakurai, T., Miura, F., Imai, M., Mochiduki, K., Yanagisawa, E., et al. (2013). High-resolution DNA methylome analysis of primordial germ cells identifies gender-specific reprogramming in mice. *Genome Research*, *23*, 616–627. https://doi.org/10.1101/gr.148023.112.

Kojima, Y., Sasaki, K., Yokobayashi, S., Sakai, Y., Nakamura, T., Yabuta, Y., et al. (2017). Evolutionarily distinctive transcriptional and signaling programs drive human germ cell lineage specification from pluripotent stem cells. *Cell Stem Cell*, *21*(4). https://doi.org/10.1016/j.stem.2017.09.005. 517-532.e515.

Komai, Y., Tanaka, T., Tokuyama, Y., Yanai, H., Ohe, S., Omachi, T., et al. (2014). Bmi1 expression in long-term germ stem cells. *Scientific Reports*, *4*, 6175. https://doi.org/10.1038/srep06175.

Konishi, I., Fujii, S., Okamura, H., Parmley, T., & Mori, T. (1986). Development of interstitial cells and ovigerous cords in the human fetal ovary: An ultrastructural study. *Journal of Anatomy*, *148*, 121–135.

Koopman, P. (1999). Sry and Sox9: Mammalian testis-determining genes. *Cellular and Molecular Life Sciences*, *55*(6–7), 839–856.

Koopman, P., Gubbay, J., Vivian, N., Goodfellow, P., & Lovell-Badge, R. (1991). Male development of chromosomally female mice transgenic for Sry. *Nature*, *351*(6322), 117–121. https://doi.org/10.1038/351117a0.

Kota, S. K., & Feil, R. (2010). Epigenetic transitions in germ cell development and meiosis. *Developmental Cell*, *19*, 675–686. https://doi.org/10.1016/j.devcel.2010.10.009.

Koubova, J., Menke, D. B., Zhou, Q., Capel, B., Griswold, M. D., & Page, D. C. (2006). Retinoic acid regulates sex-specific timing of meiotic initiation in mice. *Proceedings of the National Academy of Sciences of the United States of America*, *103*(8), 2474–2479. https://doi.org/10.1073/pnas.0510813103.

Kreidberg, J. A., Sariola, H., Loring, J. M., Maeda, M., Pelletier, J., Housman, D., et al. (1993). WT-1 is required for early kidney development. *Cell*, *74*(4), 679–691. https://doi.org/10.1016/0092-8674(93)90515-R.

Kurimoto, K., Yamaji, M., Seki, Y., & Saitou, M. (2008). Specification of the germ cell lineage in mice: A process orchestrated by the PR-domain proteins, Blimp1 and Prdm14. *Cell Cycle*, *7*, 3514–3518. https://doi.org/10.4161/cc.7.22.6979.

Lange, U. C., Adams, D. J., Lee, C., Barton, S., Schneider, R., Bradley, A., et al. (2008). Normal germ line establishment in mice carrying a deletion of the Ifitm/Fragilis gene family cluster. *Molecular Cell Biology*, *28*(15), 4688–4696. https://doi.org/10.1128/MCB.00272-08.

Lawson, K. A., Dunn, N. R., Roelen, B. A., Zeinstra, L. M., Davis, A. M., Wright, C. V., et al. (1999). Bmp4 is required for the generation of primordial germ cells in the mouse embryo. *Genes & Development*, *13*(4), 424–436.

Lee, J., Kanatsu-Shinohara, M., Inoue, K., Ogonuki, N., Miki, H., Toyokuni, S., et al. (2007). Akt mediates self-renewal division of mouse spermatogonial stem cells. *Development*, *134*(10), 1853–1859. https://doi.org/10.1242/dev.003004.

Lee, C. S., Lee, K., Legube, G., & Haber, J. E. (2014). Dynamics of yeast histone H2A and H2B phosphorylation in response to a double-strand break. *Nature Structural and Molecular Biology*, *21*, 103–109. https://doi.org/10.1038/nsmb.2737.

Lei, L., & Spradling, A. C. (2013). Mouse primordial germ cells produce cysts that partially fragment prior to meiosis. *Development*, *140*(10), 2075–2081.

Lei, L., & Spradling, A. C. (2016). Mouse oocytes differentiate through organelle enrichment from sister cyst germ cells. *Science*, *352*(6281), 95–99. https://doi.org/10.1126/science.aad2156.

Lesch, B. J., Dokshin, G. A., Young, R. A., McCarrey, J. R., & Page, D. C. (2013). A set of genes critical to development is epigenetically poised in mouse germ cells from fetal stages through completion of meiosis. *Proceedings of the National Academy of Sciences of the United States of America*, *110*(40), 16061–16066. https://doi.org/10.1073/pnas.1315204110.

Li, R., & Albertini, D. F. (2013). The road to maturation: Somatic cell interaction and self-organization of the mammalian oocyte. *Nature Reviews. Molecular Cell Biology*, *14*(3), 141–152. https://doi.org/10.1038/nrm3531.

Li, J. Y., Lees-Murdock, D. J., Xu, G. L., & Walsh, C. P. (2004). Timing of establishment of paternal methylation imprints in the mouse. *Genomics*, *84*, 952–960. https://doi.org/10.1016/j.ygeno.2004.08.012.

Li, Y., Sun, C., Yates, E. A., Jiang, C., Wilkinson, M. C., & Fernig, D. G. (2016). Heparin binding preference and structures in the fibroblast growth factor family parallel their evolutionary diversification. *Open Biology*, *6*(3), 150275. https://doi.org/10.1098/rsob.150275.

Li, Y., Zheng, M., & Lau, Y. F. (2014). The sex-determining factors SRY and SOX9 regulate similar target genes and promote testis cord formation during testicular differentiation. *Cell Reports*, *8*(3), 723–733. https://doi.org/10.1016/j.celrep.2014.06.055.

Lin, C. J., Koh, F. M., Wong, P., Conti, M., & Ramalho-Santos, M. (2014). Hira-mediated H3.3 incorporation is required for DNA replication and ribosomal RNA transcription in the mouse zygote. *Developmental Cell*, *30*(3), 268–279. https://doi.org/10.1016/j.devcel.2014.06.022.

Lin, Y., & Page, D. C. (2005). Dazl deficiency leads to embryonic arrest of germ cell development in XY C57BL/6 mice. *Developmental Biology*, *288*(2), 309–316. https://doi.org/10.1016/j.ydbio.2005.06.032.

Lindeman, R. E., Gearhart, M. D., Minkina, A., Krentz, A. D., Bardwell, V. J., & Zarkower, D. (2015). Sexual cell-fate reprogramming in the ovary by DMRT1. *Current Biology*, *25*(6), 764–771. https://doi.org/10.1016/j.cub.2015.01.034.

Lintern-Moore, S., & Moore, G. P. (1979). The initiation of follicle and oocyte growth in the mouse ovary. *Biology of Reproduction*, *20*(4), 773–778.

Liu, S., Brind'Amour, J., Karimi, M. M., Shirane, K., Bogutz, A., Lefebvre, L., et al. (2014). Setdb1 is required for germline development and silencing of H3K9me3-marked endogenous retroviruses in primordial germ cells. *Genes and Development*, *28*, 2041–2055. https://doi.org/10.1101/gad.244848.114.

Liu, C., Peng, J., Matzuk, M. M., & Yao, H. H. (2015). Lineage specification of ovarian theca cells requires multicellular interactions via oocyte and granulosa cells. *Nature Communications*, *6*, 6934. https://doi.org/10.1038/ncomms7934.

Liu, X., Wang, C., Liu, W., Li, J., Li, C., Kou, X., et al. (2016). Distinct features of H3K4me3 and H3K27me3 chromatin domains in pre-implantation embryos. *Nature*, *537*, 558–562. https://doi.org/10.1038/nature19362.

Lo, K. C., Lei, Z., Rao, C. V., Beck, J., & Lamb, D. J. (2004). De novo testosterone production in luteinizing hormone receptor knockout mice after transplantation of Leydig stem cells. *Endocrinology*, *145*(9), 4011–4015. https://doi.org/10.1210/en.2003-1729.

Loebel, D. A., Watson, C. M., De Young, R. A., & Tam, P. P. (2003). Lineage choice and differentiation in mouse embryos and embryonic stem cells. *Developmental Biology*, *264*(1), 1–14.

Loppin, B., Bonnefoy, E., Anselme, C., Laurencon, A., Karr, T. L., & Couble, P. (2005). The histone H3.3 chaperone HIRA is essential for chromatin assembly in the male pronucleus. *Nature*, *437*(7063), 1386–1390. https://doi.org/10.1038/nature04059.

Lovell-Badge, R., & Robertson, E. (1990). XY female mice resulting from a heritable mutation in the primary testis-determining gene, Tdy. *Development*, *109*(3), 635–646.

Lucifero, D., Mann, M. R. W., Bartolomei, M. S., & Trasler, J. M. (2004). Gene-specific timing and epigenetic memory in oocyte imprinting. *Human Molecular Genetics*, *13*, 839–849. https://doi.org/10.1093/hmg/ddh104.

Luo, L., Chen, H., Stocco, D. M., & Zirkin, B. R. (1998). Leydig cell protein synthesis and steroidogenesis in response to acute stimulation by luteinizing hormone in rats 1. *Biology of Reproduction, 59*, 263–270. https://doi.org/10.1095/biolreprod59.2.263.

Luo, X., Ikeda, Y., & Parker, K. L. (1994). A cell-specific nuclear receptor is essential for adrenal and gonadal development and sexual differentiation. *Cell*, 77(4), 481–490.

Magnusdottir, E., Dietmann, S., Murakami, K., Gunesdogan, U., Tang, F., Bao, S., et al. (2013). A tripartite transcription factor network regulates primordial germ cell specification in mice. *Nature Cell Biology, 15*(8), 905–915. https://doi.org/10.1038/ncb2798.

Malki, S., van der Heijden, G. W., O'Donnell, K. A., Martin, S. L., & Bortvin, A. (2014). A role for retrotransposon LINE-1 in fetal oocyte attrition in mice. *Developmental Cell, 29*(5), 521–533. https://doi.org/10.1016/j.devcel.2014.04.027.

Marston, A. L., & Amon, A. (2004). Meiosis: Cell-cycle controls shuffle and deal. *Nature Reviews. Molecular Cell Biology, 5*, 983–997. https://doi.org/10.1038/nrm1526.

Martineau, J., Nordqvist, K., Tilmann, C., Lovell-Badge, R., & Capel, B. (1997). Male-specific cell migration into the developing gonad. *Current Biology*, 7(12), 958–968.

Martinovitch, P. N. (1937). Development in vitro of the mammalian gonad. *Nature, 139*, 413. https://doi.org/10.1038/139413a0.

Matova, N., & Cooley, L. (2001). Comparative aspects of animal oogenesis. *Developmental Biology, 231*(2), 291–320. https://doi.org/10.1006/dbio.2000.0120.

Mayer, W., Niveleau, A., Walter, J., Fundele, R., & Haaf, T. (2000). Embryogenesis: Demethylation of the zygotic paternal genome. *Nature, 403*, 501–502. https://doi.org/10.1038/35000656.

McLaren, A. (1984). Meiosis and differentiation of mouse germ cells. *Symposia of the Society for Experimental Biology, 38*, 7–23.

McLaren, A. (2003). Primordial germ cells in the mouse. *Developmental Biology, 262*(1), 1–15.

Mcvean, G. A. T., Myers, S. R., & Hunt, S. (2004). The fine-scale structure of recombination rate variation in the human genome. *Science, 304*, 581–584. https://doi.org/10.1126/science.1092500.

Mehlmann, L. M. (2005). Stops and starts in mammalian oocytes: Recent advances in understanding the regulation of meiotic arrest and oocyte maturation. *Reproduction, 130*(6), 791–799. https://doi.org/10.1530/rep.1.00793.

Meng, J., Holdcraft, R. W., Shima, J. E., Griswold, M. D., & Braun, R. E. (2005). Androgens regulate the permeability of the blood-testis barrier. *Proceedings of the National Academy of Sciences of the United States of America, 102*, 16696–16700. https://doi.org/10.1073/pnas.0506084102.

Meng, X., Lindahl, M., Hyvonen, M. E., Parvinen, M., de Rooij, D. G., Hess, M. W., et al. (2000). Regulation of cell fate decision of undifferentiated spermatogonia by GDNF. *Science, 287*(5457), 1489–1493.

Menke, D. B., Koubova, J., & Page, D. C. (2003). Sexual differentiation of germ cells in XX mouse gonads occurs in an anterior-to-posterior wave. *Developmental Biology, 262*(2), 303–312.

Miki, H., Lee, J., Inoue, K., Ogonuki, N., Noguchi, Y., Mochida, K., et al. (2004). Microinsemination with first-wave round spermatids from immature male mice. *The Journal of Reproduction and Development, 50*, 131–137. https://doi.org/10.1262/jrd.50.131.

Miyamoto, N., Yoshida, M., Kuratani, S., Matsuo, I., & Aizawa, S. (1997). Defects of urogenital development in mice lacking Emx2. *Development, 124*(9), 1653–1664.

Molyneaux, K. A., Stallock, J., Schaible, K., & Wylie, C. (2001). Time-lapse analysis of living mouse germ cell migration. *Developmental Biology, 240*(2), 488–498. https://doi.org/10.1006/dbio.2001.0436.

Molyneaux, K. A., Zinszner, H., Kunwar, P. S., Schaible, K., Stebler, J., Sunshine, M. J., et al. (2003). The chemokine SDF1/CXCL12 and its receptor CXCR4 regulate mouse germ cell migration and survival. *Development, 130*(18), 4279–4286.

Moore, G. P., & Lintern-Moore, S. (1979). Stimulation of endogenous RNA polymerase I activity in the mouse oocyte after PMSG treatment. *Biology of Reproduction*, *21*(2), 373–377.

Morais da Silva, S., Hacker, A., Harley, V., Goodfellow, P., Swain, A., & Lovell-Badge, R. (1996). Sox9 expression during gonadal development implies a conserved role for the gene in testis differentiation in mammals and birds. *Nature Genetics*, *14*(1), 62–68. https://doi.org/10.1038/ng0996-62.

Morales, A., Mohamed, F., & Cavicchia, J. C. (2007). Apoptosis and blood-testis barrier during the first spermatogenic wave in the pubertal rat. *Anatomical Record*, *290*, 206–214. https://doi.org/10.1002/ar.20417.

Moreno, R. D., Lizama, C., Urzúa, N., Vergara, S. P., & Reyes, J. G. (2006). Caspase activation throughout the first wave of spermatogenesis in the rat. *Cell and Tissue Research*, *325*, 533–540. https://doi.org/10.1007/s00441-006-0186-4.

Morgan, M., Much, C., DiGiacomo, M., Azzi, C., Ivanova, I., Vitsios, D. M., et al. (2017). mRNA 3′ uridylation and poly(A) tail length sculpt the mammalian maternal transcriptome. *Nature*, *548*(7667), 347–351.

Mori, C., Nakamura, N., Dix, D. J., Fujioka, M., Nakagawa, S., Shiota, K., et al. (1997). Morphological analysis of germ cell apoptosis during postnatal testis development in normal and Hsp 70-2 knockout mice. *Developmental Dynamics: An Official Publication of the American Association of the Anatomists*, *208*, 125–136. https://doi.org/10.1002/(SICI)1097-0177(199701)208:1<125::AID-AJA12>3.0.CO;2-5.

Mork, L., Maatouk, D. M., McMahon, J. A., Guo, J. J., Zhang, P., McMahon, A. P., et al. (2012). Temporal differences in granulosa cell specification in the ovary reflect distinct follicle fates in mice. *Biology of Reproduction*, *86*(2), 37.

Morohaku, K., Tanimoto, R., Sasaki, K., Kawahara-Miki, R., Kono, T., Hayashi, K., et al. (2016). Complete in vitro generation of fertile oocytes from mouse primordial germ cells. *Proceedings of the National Academy of Sciences of the United States of America*, *113*(32), 9021–9026. https://doi.org/10.1073/pnas.1603817113.

Myers, S. (2005). A fine-scale map of recombination rates and hotspots across the human genome. *Science*, *310*, 321–324. https://doi.org/10.1126/science.1117196.

Myers, S., Freeman, C., Auton, A., Donnelly, P., & McVean, G. (2008). A common sequence motif associated with recombination hot spots and genome instability in humans. *Nature Genetics*, *40*, 1124–1129. https://doi.org/10.1038/ng.213.

Nagamatsu, G., & Hayashi, K. (2017). Stem cells, in vitro gametogenesis and male fertility. *Reproduction*, *154*(6), F79–F91. https://doi.org/10.1530/REP-17-0510.

Nakagawa, T., Sharma, M., Nabeshima, Y., Braun, R. E., & Yoshida, S. (2010). Functional hierarchy and reversibility within the murine spermatogenic stem cell compartment. *Science*, *328*(5974), 62–67. https://doi.org/10.1126/science.1182868.

Nakamura, T., Arai, Y., Umehara, H., Masuhara, M., Kimura, T., Taniguchi, H., et al. (2007). PGC7/Stella protects against DNA demethylation in early embryogenesis. *Nature Cell Biology*, *9*, 64–71. https://doi.org/10.1038/ncb1519.

Nakamura, T., Liu, Y. J., Nakashima, H., Umehara, H., Inoue, K., Matoba, S., et al. (2012). PGC7 binds histone H3K9me2 to protect against conversion of 5mC to 5hmC in early embryos. *Nature*, *486*, 415–419. https://doi.org/10.1038/nature11093.

Namwanje, M., & Brown, C. W. (2016). Activins and inhibins: Roles in development, physiology, and disease. *Cold Spring Harbor Perspectives in Biology*, *8*(7). https://doi.org/10.1101/cshperspect.a021881.

Naughton, C. K., Jain, S., Strickland, A. M., Gupta, A., & Milbrandt, J. (2006). Glial cell-line derived neurotrophic factor-mediated RET signaling regulates spermatogonial stem cell fate. *Biology of Reproduction*, *74*(2), 314–321. https://doi.org/10.1095/biolreprod.105.047365.

Nayernia, K., Nolte, J., Michelmann, H. W., Lee, J. H., Rathsack, K., Drusenheimer, N., et al. (2006). In vitro-differentiated embryonic stem cells give rise to male gametes that can generate offspring mice. *Developmental Cell*, *11*(1), 125–132. https://doi.org/10.1016/j.devcel.2006.05.010.

Nef, S., Verma-Kurvari, S., Merenmies, J., Vassalli, J. D., Efstratiadis, A., Accili, D., et al. (2003). Testis determination requires insulin receptor family function in mice. *Nature*, *426*(6964), 291–295. https://doi.org/10.1038/nature02059.

Ng, J. H., Kumar, V., Muratani, M., Kraus, P., Yeo, J. C., Yaw, L. P., et al. (2013). In vivo epigenomic profiling of germ cells reveals germ cell molecular signatures. *Developmental Cell*, *24*(3), 324–333. https://doi.org/10.1016/j.devcel.2012.12.011.

Nicol, B., & Yao, H. H. (2014). Building an ovary: Insights into establishment of somatic cell lineages in the mouse. *Sexual Development*, *8*(5), 243–251. https://doi.org/10.1159/000358072.

Norris, R. P., Freudzon, M., Mehlmann, L. M., Cowan, A. E., Simon, A. M., Paul, D. L., et al. (2008). Luteinizing hormone causes MAP kinase-dependent phosphorylation and closure of connexin 43 gap junctions in mouse ovarian follicles: One of two paths to meiotic resumption. *Development*, *135*(19), 3229–3238. https://doi.org/10.1242/dev.025494.

Oakberg, E. F. (1968). Mammalian gametogenesis and species comparisons in radiation response of the gonads. In *Effects of radiation on meiotic systems. Report of a study group* (pp. 3–15). Vienna: International Atomic Energy Agency.

Oakberg, E. F. (1971). Spermatogonial stem-cell renewal in the mouse. *The Anatomical Record*, *169*(3), 515–531. https://doi.org/10.1002/ar.1091690305.

Oatley, J. M., Avarbock, M. R., & Brinster, R. L. (2007). Glial cell line-derived neurotrophic factor regulation of genes essential for self-renewal of mouse spermatogonial stem cells is dependent on Src family kinase signaling. *Journal of Biological Chemistry*, *282*, 25842–25851. https://doi.org/10.1074/jbc.M703474200.

Oatley, J. M., Avarbock, M. R., Telaranta, A. I., Fearon, D. T., & Brinster, R. L. (2006). Identifying genes important for spermatogonial stem cell self-renewal and survival. *Proceedings of the National Academy of Sciences of the United States of America*, *103*(25), 9524–9529.

Oatley, M. J., Kaucher, A. V., Racicot, K. E., & Oatley, J. M. (2011). Inhibitor of DNA binding 4 is expressed selectively by single spermatogonia in the male germline and regulates the self-renewal of spermatogonial stem cells in mice. *Biology of Reproduction*, *85*(2), 347–356. https://doi.org/10.1095/biolreprod.111.091330.

Obata, Y., Kono, T., & Hatada, I. (2002). Gene silencing: Maturation of mouse fetal germ cells in vitro. *Nature*, *418*(6897), 497. https://doi.org/10.1038/418497a.

Oh, B., Hwang, S., McLaughlin, J., Solter, D., & Knowles, B. B. (2000). Timely translation during the mouse oocyte-to-embryo transition. *Development*, *127*(17), 3795–3803.

Ohinata, Y., Ohta, H., Shigeta, M., Yamanaka, K., Wakayama, T., & Saitou, M. (2009). A signaling principle for the specification of the germ cell lineage in mice. *Cell*, *137*(3), 571–584. https://doi.org/10.1016/j.cell.2009.03.014.

Ohinata, Y., Payer, B., O'Carroll, D., Ancelin, K., Ono, Y., Sano, M., et al. (2005). Blimp1 is a critical determinant of the germ cell lineage in mice. *Nature*, *436*, 207–213. https://doi.org/10.1038/nature03813.

Ohta, H., Tohda, A., & Nishimune, Y. (2003). Proliferation and differentiation of spermatogonial stem cells in the w/wv mutant mouse testis. *Biology of Reproduction*, *69*(6), 1815–1821. https://doi.org/10.1095/biolreprod.103.019323.

Oktay, K., Briggs, D., & Gosden, R. G. (1997). Ontogeny of follicle-stimulating hormone receptor gene expression in isolated human ovarian follicles. *The Journal of Clinical Endocrinology and Metabolism*, *82*(11), 3748–3751. https://doi.org/10.1210/jcem.82.11.4346.

Oktay, K., Newton, H., Mullan, J., & Gosden, R. G. (1998). Development of human primordial follicles to antral stages in SCID/hpg mice stimulated with follicle stimulating hormone. *Human Reproduction*, *13*(5), 1133–1138.

O'Shaughnessy, P. J., Monteiro, A., Verhoeven, G., De Gendt, K., & Abel, M. H. (2010). Effect of FSH on testicular morphology and spermatogenesis in gonadotrophin-deficient hypogonadal mice lacking androgen receptors. *Reproduction, 139*, 177–184. https://doi.org/10.1530/REP-09-0377.

Oswald, J., Engemann, S., Lane, N., Mayer, W., Olek, A., Fundele, R., et al. (2000). Active demethylation of the paternal genome in the mouse zygote. *Current Biology, 10*, 475–478. https://doi.org/10.1016/S0960-9822(00)00448-6.

Pailhoux, E., Vigier, B., Chaffaux, S., Servel, N., Taourit, S., Furet, J. P., et al. (2001). A 11.7-kb deletion triggers intersexuality and polledness in goats. *Nature Genetics, 29*(4), 453–458. https://doi.org/10.1038/ng769.

Pailhoux, E., Vigier, B., Vaiman, D., Servel, N., Chaffaux, S., Cribiu, E. P., et al. (2002). Ontogenesis of female-to-male sex-reversal in XX polled goats. *Developmental Dynamics, 224*(1), 39–50. https://doi.org/10.1002/dvdy.10083.

Palmer, S. J., & Burgoyne, P. S. (1991). In situ analysis of fetal, prepuberal and adult XX——XY chimaeric mouse testes: Sertoli cells are predominantly, but not exclusively, XY. *Development, 112*(1), 265–268.

Pangas, S. A., Choi, Y., Ballow, D. J., Zhao, Y., Westphal, H., Matzuk, M. M., et al. (2006). Oogenesis requires germ cell-specific transcriptional regulators Sohlh1 and Lhx8. *Proceedings of the National Academy of Sciences of the United States of America, 103*(21), 8090–8095. https://doi.org/10.1073/pnas.0601083103.

Pangas, S. A., Jorgez, C. J., Tran, M., Agno, J., Li, X., Brown, C. W., et al. (2007). Intraovarian activins are required for female fertility. *Molecular Endocrinology, 21*(10), 2458–2471. https://doi.org/10.1210/me.2007-0146.

Paniagua, R., Codesal, J., Nistal, M., Rodriguez, M. C., & Santamaria, L. (1987). Quantification of cell types throughout the cycle of the human seminiferous epithelium and their DNA content. A new approach to the spermatogonial stem cell in man. *Anatomy and Embryology (Berlin), 176*(2), 225–230.

Parma, P., Radi, O., Vidal, V., Chaboissier, M. C., Dellambra, E., Valentini, S., et al. (2006). R-spondin1 is essential in sex determination, skin differentiation and malignancy. *Nature Genetics, 38*(11), 1304–1309. https://doi.org/10.1038/ng1907.

Parvanov, E. D., Petkov, P. M., & Paigen, K. (2010). Prdm9 controls activation of mammalian recombination hotspots. *Science, 327*, 835. https://doi.org/10.1126/science.1181495.

Pastor, W. A., Liu, W., Chen, D., Ho, J., Kim, R., Hunt, T. J., et al. (2018). TFAP2C regulates transcription in human naive pluripotency by opening enhancers. *Nature Cell Biology, 20*(5), 553–564. https://doi.org/10.1038/s41556-018-0089-0.

Pepling, M. E. (2006). From primordial germ cell to primordial follicle: Mammalian female germ cell development. *Genesis, 44*(12), 622–632. https://doi.org/10.1002/dvg.20258.

Pepling, M. E. (2012). Follicular assembly: Mechanisms of action. *Reproduction, 143*(2), 139–149. https://doi.org/10.1530/REP-11-0299.

Pepling, M. E., & Spradling, A. C. (1998). Female mouse germ cells form synchronously dividing cysts. *Development, 125*(17), 3323–3328.

Perez, G. I., Robles, R., Knudson, C. M., Flaws, J. A., Korsmeyer, S. J., & Tilly, J. L. (1999). Prolongation of ovarian lifespan into advanced chronological age by Bax-deficiency. *Nature Genetics, 21*(2), 200–203. https://doi.org/10.1038/5985.

Phillips, B. T., Gassei, K., & Orwig, K. E. (2010). Spermatogonial stem cell regulation and spermatogenesis. *Philosophical Transactions of the Royal Society of London. Series B, Biological Sciences, 365*(1546), 1663–1678. https://doi.org/10.1098/rstb.2010.0026.

Pittman, D. L., Cobb, J., Schimenti, K. J., Wilson, L. A., Cooper, D. M., Brignull, E., et al. (1998). Meiotic prophase arrest with failure of chromosome synapsis in mice deficient for Dmc1, a germline-specific RecA homolog. *Molecular Cell, 1*(5), 697–705.

Popp, C., Dean, W., Feng, S., Cokus, S. J., Andrews, S., Pellegrini, M., et al. (2010). Genome-wide erasure of DNA methylation in mouse primordial germ cells is affected by AID deficiency. *Nature, 463*, 1101–1105. https://doi.org/10.1038/nature08829.

Puschendorf, M., Terranova, R., Boutsma, E., Mao, X., Isono, K.-I., Brykczynska, U., et al. (2008). PRC1 and Suv39h specify parental asymmetry at constitutive heterochromatin in early mouse embryos. *Nature Genetics, 40*, 411–420. https://doi.org/10.1038/ng.99.

Rajkovic, A., Pangas, S. A., Ballow, D., Suzumori, N., & Matzuk, M. M. (2004). NOBOX deficiency disrupts early folliculogenesis and oocyte-specific gene expression. *Science, 305*(5687), 1157–1159. https://doi.org/10.1126/science.1099755.

Rando, O. J. (2016). Intergenerational transfer of epigenetic information in sperm. *Cold Spring Harbor Perspectives in Medicine, 6*, 1–14. https://doi.org/10.1101/cshperspect.a022988.

Ratts, V. S., Flaws, J. A., Kolp, R., Sorenson, C. M., & Tilly, J. L. (1995). Ablation of bcl-2 gene expression decreases the numbers of oocytes and primordial follicles established in the post-natal female mouse gonad. *Endocrinology, 136*(8), 3665–3668. https://doi.org/10.1210/endo.136.8.7628407.

Raverdeau, M., Gely-Pernot, A., Feret, B., Dennefeld, C., Benoit, G., Davidson, I., et al. (2012). Retinoic acid induces Sertoli cell paracrine signals for spermatogonia differentiation but cell autonomously drives spermatocyte meiosis. *Proceedings of the National Academy of Sciences of the United States of America, 109*, 16582–16587. https://doi.org/10.1073/pnas.1214936109.

Raz, E. (2004). Guidance of primordial germ cell migration. *Current Opinion in Cell Biology, 16*(2), 169–173. https://doi.org/10.1016/j.ceb.2004.01.004.

Reddy, P., Liu, L., Adhikari, D., Jagarlamudi, K., Rajareddy, S., Shen, Y., et al. (2008). Oocyte-specific deletion of Pten causes premature activation of the primordial follicle pool. *Science, 319*(5863), 611–613. https://doi.org/10.1126/science.1152257.

Reith, A. D., Rottapel, R., Giddens, E., Brady, C., Forrester, L., & Bernstein, A. (1990). W mutant mice with mild or severe developmental defects contain distinct point mutations in the kinase domain of the c-kit receptor. *Genes & Development, 4*(3), 390–400.

Ren, Y., Suzuki, H., Jagarlamudi, K., Golnoski, K., McGuire, M., Lopes, R., et al. (2015). Lhx8 regulates primordial follicle activation and postnatal folliculogenesis. *BMC Biology, 13*, 39. https://doi.org/10.1186/s12915-015-0151-3.

Richardson, B. E., & Lehmann, R. (2010). Mechanisms guiding primordial germ cell migration: Strategies from different organisms. *Nature Reviews. Molecular Cell Biology, 11*(1), 37–49. https://doi.org/10.1038/nrm2815.

Romanienko, P. J., & Camerini-Otero, R. D. (2000). The mouse Spo11 gene is required for meiotic chromosome synapsis. *Molecular Cell, 6*(5), 975–987. https://doi.org/10.1016/S1097-2765(00)00097-6.

Ross, A. J., Tilman, C., Yao, H., MacLaughlin, D., & Capel, B. (2003). AMH induces mesonephric cell migration in XX gonads. *Molecular and Cellular Endocrinology, 211*(1–2), 1–7.

Ruggiu, M., Speed, R., Taggart, M., McKay, S. J., Kilanowski, F., Saunders, P., et al. (1997). The mouse Dazla gene encodes a cytoplasmic protein essential for gametogenesis. *Nature, 389*(6646), 73–77. https://doi.org/10.1038/37987.

Runyan, C., Schaible, K., Molyneaux, K., Wang, Z., Levin, L., & Wylie, C. (2006). Steel factor controls midline cell death of primordial germ cells and is essential for their normal proliferation and migration. *Development, 133*(24), 4861–4869. https://doi.org/10.1242/dev.02688.

Russell, L. D., Chiarini-Garcia, H., Korsmeyer, S. J., & Knudson, C. M. (2002). Bax-dependent spermatogonia apoptosis is required for testicular development and spermatogenesis. *Biology of Reproduction, 66*(4), 950–958.

Russell, L. D., Ettlin, R. A., Sinha Hikim, A. P., & Clegg, E. D. (1990). *Histological and histopathological evaluation of the testis.* Clearwater, FL: Cache River Press.

Ruwanpura, S. M., McLachlan, R. I., Matthiesson, K. L., & Meachem, S. J. (2008). Gonadotrophins regulate germ cell survival, not proliferation, in normal adult men. *Human Reproduction*, *23*(2), 403–411. https://doi.org/10.1093/humrep/dem376.

Saitou, M., Barton, S. C., & Surani, M. A. (2002). A molecular programme for the specification of germ cell fate in mice. *Nature*, *418*(6895), 293–300. https://doi.org/10.1038/nature00927.

Saitou, M., & Yamaji, M. (2012). Primordial germ cells in mice. *Cold Spring Harbor Perspectives in Biology*, *4*(11). https://doi.org/10.1101/cshperspect.a008375.

Santenard, A., Ziegler-Birling, C., Koch, M., Tora, L., Bannister, A. J., & Torres-Padilla, M. E. (2010). Heterochromatin formation in the mouse embryo requires critical residues of the histone variant H3.3. *Nature Cell Biology*, *12*(9), 853–862. https://doi.org/10.1038/ncb2089.

Sasaki, K., Yokobayashi, S., Nakamura, T., Okamoto, I., Yabuta, Y., Kurimoto, K., et al. (2015). Robust in vitro induction of human germ cell fate from pluripotent stem cells. *Cell Stem Cell*, *17*(2), 178–194. https://doi.org/10.1016/j.stem.2015.06.014.

Sato, M., Kimura, T., Kurokawa, K., Fujita, Y., Abe, K., Masuhara, M., et al. (2002). Identification of PGC7, a new gene expressed specifically in preimplantation embryos and germ cells. *Mechanisms of Development*, *113*(1), 91–94. https://doi.org/10.1016/S0925-4773(02)00002-3.

Saunders, P. T., Turner, J. M., Ruggiu, M., Taggart, M., Burgoyne, P. S., Elliott, D., et al. (2003). Absence of mDazl produces a final block on germ cell development at meiosis. *Reproduction*, *126*(5), 589–597.

Schlatt, S., & Weinbauer, G. F. (1994). Immunohistochemical localization of proliferating cell nuclear antigen as a tool to study cell proliferation in rodent and primate testes. *International Journal of Andrology*, *17*(4), 214–222.

Schmahl, J., Kim, Y., Colvin, J. S., Ornitz, D. M., & Capel, B. (2004). Fgf9 induces proliferation and nuclear localization of FGFR2 in Sertoli precursors during male sex determination. *Development*, *131*(15), 3627–3636. https://doi.org/10.1242/dev.01239.

Schmidt, D., Ovitt, C. E., Anlag, K., Fehsenfeld, S., Gredsted, L., Treier, A. C., et al. (2004). The murine winged-helix transcription factor Foxl2 is required for granulosa cell differentiation and ovary maintenance. *Development*, *131*(4), 933–942. https://doi.org/10.1242/dev.00969.

Schöler, H. R., Hatzopoulos, A. K., Balling, R., Suzuki, N., & Gruss, P. (1989). A family of octamer-specific proteins present during mouse embryogenesis: Evidence for germline-specific expression of an Oct factor. *The EMBO Journal*, *8*(9), 2543–2550.

Schöler, H. R., Ruppert, S., Suzuki, N., Chowdhury, K., & Gruss, P. (1990). New type of POU domain in germ line-specific protein Oct-4. *Nature*, *344*, 435.

Schrans-Stassen, B. H., Saunders, P. T., Cooke, H. J., & de Rooij, D. G. (2001). Nature of the spermatogenic arrest in Dazl −/− mice. *Biology of Reproduction*, *65*(3), 771–776.

Seisenberger, S., Andrews, S., Krueger, F., Arand, J., Walter, J., Santos, F., et al. (2012). The dynamics of genome-wide DNA methylation reprogramming in mouse primordial germ cells. *Molecular Cell*, *48*(6), 849–862. https://doi.org/10.1016/j.molcel.2012.11.001.

Seki, Y., Hayashi, K., Itoh, K., Mizugaki, M., Saitou, M., & Matsui, Y. (2005). Extensive and orderly reprogramming of genome-wide chromatin modifications associated with specification and early development of germ cells in mice. *Developmental Biology*, *278*, 440–458. https://doi.org/10.1016/j.ydbio.2004.11.025.

Seki, Y., Yamaji, M., Yabuta, Y., Sano, M., Shigeta, M., Matsui, Y., et al. (2007). Cellular dynamics associated with the genome-wide epigenetic reprogramming in migrating primordial germ cells in mice. *Development*, *134*, 2627–2638. https://doi.org/10.1242/dev.005611.

Seligman, J., & Page, D. C. (1998). The Dazh gene is expressed in male and female embryonic gonads before germ cell sex differentiation. *Biochemical and Biophysical Research Communications*, *245*(3), 878–882. https://doi.org/10.1006/bbrc.1998.8530.

Shirane, K., Kurimoto, K., Yabuta, Y., Yamaji, M., Satoh, J., Ito, S., et al. (2016). Global landscape and regulatory principles of DNA methylation reprogramming for germ cell specification by mouse pluripotent stem cells. *Developmental Cell*, *39*, 87–103. https://doi.org/10.1016/j.devcel.2016.08.008.

Shuhaibar, L. C., Egbert, J. R., Norris, R. P., Lampe, P. D., Nikolaev, V. O., Thunemann, M., et al. (2015). Intercellular signaling via cyclic GMP diffusion through gap junctions restarts meiosis in mouse ovarian follicles. *Proceedings of the National Academy of Sciences of the United States of America*, *112*(17), 5527–5532. https://doi.org/10.1073/pnas.1423598112.

Siklenka, K., Erkek, S., Godmann, M., Lambrot, R., McGraw, S., Lafleur, C., et al. (2015). Disruption of histone methylation in developing sperm impairs offspring health transgenerationally. *Science*, *350*(6261), aab2006. https://doi.org/10.1126/science.aab2006.

Simorangkir, D. R., Marshall, G. R., & Plant, T. M. (2009). A re-examination of proliferation and differentiation of type A spermatogonia in the adult rhesus monkey (*Macaca mulatta*). *Human Reproduction*, *24*(7), 1596–1604. https://doi.org/10.1093/humrep/dep051.

Smallwood, S. A., Tomizawa, S. I., Krueger, F., Ruf, N., Carli, N., Segonds-Pichon, A., et al. (2011). Dynamic CpG island methylation landscape in oocytes and preimplantation embryos. *Nature Genetics*, *43*, 811–814. https://doi.org/10.1038/ng.864.

Smith, Z. D., Chan, M. M., Mikkelsen, T. S., Gu, H., Gnirke, A., Regev, A., et al. (2012). A unique regulatory phase of DNA methylation in the early mammalian embryo. *Nature*, *484*, 339–344. https://doi.org/10.1038/nature10960.

Smith, Z. D., & Meissner, A. (2013). DNA methylation: Roles in mammalian development. *Nature Reviews. Genetics*, *14*, 204–220. https://doi.org/10.1038/nrg3354.

Sollier, J., Lin, W., Soustelle, C., Suhre, K., Nicolas, A., Geli, V., et al. (2004). Set1 is required for meiotic S-phase onset, double-strand break formation and middle gene expression. *The EMBO Journal*, *23*(9), 1957–1967. https://doi.org/10.1038/sj.emboj.7600204.

Stewart, K. R., Veselovska, L., & Kelsey, G. (2016). Establishment and functions of DNA methylation in the germline. *Epigenomics*, *8*, 1399–1413. https://doi.org/10.2217/epi-2016-0056.

Stewart, K. R., Veselovska, L., Kim, J., Huang, J., Saadeh, H., Tomizawa, S. I., et al. (2015). Dynamic changes in histone modifications precede de novo DNA methylation in oocytes. *Genes and Development*, *29*, 2449–2462. https://doi.org/10.1101/gad.271353.115.

Suzuki, H., Sada, A., Yoshida, S., & Saga, Y. (2009). The heterogeneity of spermatogonia is revealed by their topology and expression of marker proteins including the germ cell-specific proteins Nanos2 and Nanos3. *Developmental Biology*, *336*(2), 222–231. https://doi.org/10.1016/j.ydbio.2009.10.002.

Suzuki, A., Sugihara, A., Uchida, K., Sato, T., Ohta, Y., Katsu, Y., et al. (2002). Developmental effects of perinatal exposure to bisphenol-A and diethylstilbestrol on reproductive organs in female mice. *Reproductive Toxicology*, *16*(2), 107–116.

Svoboda, P., Franke, V., & Schultz, R. M. (2015). Sculpting the transcriptome during the oocyte-to-embryo transition in mouse. *Current Topics in Developmental Biology*, *113*, 305–349.

Swain, A., Narvaez, V., Burgoyne, P., Camerino, G., & Lovell-Badge, R. (1998). Dax1 antagonizes Sry action in mammalian sex determination. *Nature*, *391*(6669), 761–767. https://doi.org/10.1038/35799.

Takashima, S., Kanatsu-Shinohara, M., Tanaka, T., Morimoto, H., Inoue, K., Ogonuki, N., et al. (2015). Functional differences between GDNF-dependent and FGF2-dependent mouse spermatogonial stem cell self-renewal. *Stem Cell Reports*, *4*(3), 489–502. https://doi.org/10.1016/j.stemcr.2015.01.010.

Tam, P. P. L., & Zhou, S. X. (1996). The allocation of epiblast cells to ectodermal and germ-line lineages is influenced by the position of the cells in the gastrulating mouse embryo. *Developmental Biology*, *178*(1), 124–132. https://doi.org/10.1006/dbio.1996.0203.

Tang, W. W. C., Dietmann, S., Irie, N., Leitch, H. G., Floros, V. I., Bradshaw, C. R., et al. (2015). A unique gene regulatory network resets the human germline epigenome for development. *Cell*, *161*, 1453–1467. https://doi.org/10.1016/j.cell.2015.04.053.

Tang, W. W., Kobayashi, T., Irie, N., Dietmann, S., & Surani, M. A. (2016). Specification and epigenetic programming of the human germ line. *Nature Reviews. Genetics*, *17*, 585–600. https://doi.org/10.1038/nrg.2016.88.

Tardat, M., Albert, M., Kunzmann, R., Liu, Z., Kaustov, L., Thierry, R., et al. (2015). Cbx2 targets PRC1 to constitutive heterochromatin in mouse zygotes in a parent-of-origin-dependent manner. *Molecular Cell*, *58*(1), 157–171. https://doi.org/10.1016/j.molcel.2015.02.013.

Teixeira Filho, F. L., Baracat, E. C., Lee, T. H., Suh, C. S., Matsui, M., Chang, R. J., et al. (2002). Aberrant expression of growth differentiation factor-9 in oocytes of women with polycystic ovary syndrome. *The Journal of Clinical Endocrinology and Metabolism*, *87*(3), 1337–1344. https://doi.org/10.1210/jcem.87.3.8316.

Tokuda, M., Kadokawa, Y., Kurahashi, H., & Marunouchi, T. (2007). CDH1 is a specific marker for undifferentiated spermatogonia in mouse testes. *Biology of Reproduction*, *76*(1), 130–141. https://doi.org/10.1095/biolreprod.106.053181.

Tomizuka, K., Horikoshi, K., Kitada, R., Sugawara, Y., Iba, Y., Kojima, A., et al. (2008). R-spondin1 plays an essential role in ovarian development through positively regulating Wnt-4 signaling. *Human Molecular Genetics*, *17*(9), 1278–1291. https://doi.org/10.1093/hmg/ddn036.

Torres-Padilla, M. E., Bannister, A. J., Hurd, P. J., Kouzarides, T., & Zernicka-Goetz, M. (2006). Dynamic distribution of the replacement histone variant H3.3 in the mouse oocyte and preimplantation embryos. *The International Journal of Developmental Biology*, *50*(5), 455–461. https://doi.org/10.1387/ijdb.052073mt.

Tremblay, K. D., Dunn, N. R., & Robertson, E. J. (2001). Mouse embryos lacking Smad1 signals display defects in extra-embryonic tissues and germ cell formation. *Development*, *128*(18), 3609–3621.

Tsai, M.-Y., Yeh, S.-D., Wang, R.-S., Yeh, S., Zhang, C., Lin, H.-Y., et al. (2006). Differential effects of spermatogenesis and fertility in mice lacking androgen receptor in individual testis cells. *Proceedings of the National Academy of Sciences of the United States of America*, *103*, 18975–18980. https://doi.org/10.1073/pnas.0608565103.

Uda, M., Ottolenghi, C., Crisponi, L., Garcia, J. E., Deiana, M., Kimber, W., et al. (2004). Foxl2 disruption causes mouse ovarian failure by pervasive blockage of follicle development. *Human Molecular Genetics*, *13*(11), 1171–1181. https://doi.org/10.1093/hmg/ddh124.

Uhlenhaut, N. H., Jakob, S., Anlag, K., Eisenberger, T., Sekido, R., Kress, J., et al. (2009). Somatic sex reprogramming of adult ovaries to testes by FOXL2 ablation. *Cell*, *139*(6), 1130–1142. https://doi.org/10.1016/j.cell.2009.11.021.

Vainio, S., Heikkila, M., Kispert, A., Chin, N., & McMahon, A. P. (1999). Female development in mammals is regulated by Wnt-4 signalling. *Nature*, *397*(6718), 405–409. https://doi.org/10.1038/17068.

Valli, H., Sukhwani, M., Dovey, S. L., Peters, K. A., Donohue, J., Castro, C. A., et al. (2014). Fluorescence- and magnetic-activated cell sorting strategies to isolate and enrich human spermatogonial stem cells. *Fertility and Sterility*, *102*(2). https://doi.org/10.1016/j.fertnstert.2014.04.036. 566-580.e567.

van Alphen, M. M., van de Kant, H. J., & de Rooij, D. G. (1988). Depletion of the spermatogonia from the seminiferous epithelium of the rhesus monkey after X irradiation. *Radiation Research*, *113*(3), 473–486.

van Bragt, M. P., Roepers-Gajadien, H. L., Korver, C. M., Bogerd, J., Okuda, A., Eggen, B. J., et al. (2008). Expression of the pluripotency marker UTF1 is restricted to a subpopulation of early A spermatogonia in rat testis. *Reproduction, 136*(1), 33–40. https://doi.org/10.1530/REP-07-0536.

Van Der Heijden, G. W., Derijck, A. A., Pósfai, E., Giele, M., Pelczar, P., Ramos, L., et al. (2007). Chromosome-wide nucleosome replacement and H3.3 incorporation during mammalian meiotic sex chromosome inactivation. *Nature Genetics, 39*, 251–258. https://doi.org/10.1038/ng1949.

Van Pelt, A. M., & De Rooij, D. G. (1990a). The origin of the synchronization of the seminiferous epithelium in vitamin A-deficient rats after vitamin A replacement. *Biology of Reproduction, 42*, 677–682.

Van Pelt, A. M. M., & de Rooij, D. G. (1990b). Synchronization of the seminiferous epithelium after vitamin a replacement in vitamin A-deficient mice. *Biology of Reproduction, 43*, 363–367.

Van Pelt, A. M. M., & De Rooij, D. G. (1991). Retinoic acid is able to reinitiate spermatogenesis in vitamin A-deficient rats and high replicate doses support the full development of spermatogenic cells. *Endocrinology, 128*, 697–704.

Vidal, V. P., Chaboissier, M. C., de Rooij, D. G., & Schedl, A. (2001). Sox9 induces testis development in XX transgenic mice. *Nature Genetics, 28*(3), 216–217. https://doi.org/10.1038/90046.

Vincent, S. D., Dunn, N. R., Sciammas, R., Shapiro-Shalef, M., Davis, M. M., Calame, K., et al. (2005). The zinc finger transcriptional repressor Blimp1/Prdm1 is dispensable for early axis formation but is required for specification of primordial germ cells in the mouse. *Development, 132*(6), 1315–1325. https://doi.org/10.1242/dev.01711.

Vincent, J. J., Huang, Y., Chen, P. Y., Feng, S., Calvopiña, J. H., Nee, K., et al. (2013). Stage-specific roles for Tet1 and Tet2 in DNA demethylation in primordial germ cells. *Cell Stem Cell, 12*, 470–478. https://doi.org/10.1016/j.stem.2013.01.016.

von Kopylow, K., & Spiess, A. N. (2017). Human spermatogonial markers. *Stem Cell Research, 25*, 300–309. https://doi.org/10.1016/j.scr.2017.11.011.

Weber, S., Eckert, D., Nettersheim, D., Gillis, A. J. M., Schäfer, S., Kuckenberg, P., et al. (2010). Critical function of AP-2gamma/TCFAP2C in mouse embryonic germ cell maintenance. *Biology of Reproduction, 82*, 214–223. https://doi.org/10.1095/biolreprod.109.078717.

Western, P. S., Miles, D. C., van den Bergen, J. A., Burton, M., & Sinclair, A. H. (2008). Dynamic regulation of mitotic arrest in fetal male germ cells. *Stem Cells, 26*(2), 339–347. https://doi.org/10.1634/stemcells.2007-0622.

Wilhelm, D., Palmer, S., & Koopman, P. (2007). Sex determination and gonadal development in mammals. *Physiological Reviews, 87*(1), 1–28. https://doi.org/10.1152/physrev.00009.2006.

Willems, A., Batlouni, S. R., Esnal, A., Swinnen, J. V., Saunders, P. T. K., Sharpe, R. M., et al. (2010). Selective ablation of the androgen receptor in mouse sertoli cells affects sertoli cell maturation, barrier formation and cytoskeletal development. *PLoS One, 5*. https://doi.org/10.1371/journal.pone.0014168.

Willerton, L., Smith, R. A., Russell, D., & Mackay, S. (2004). Effects of FGF9 on embryonic Sertoli cell proliferation and testicular cord formation in the mouse. *The International Journal of Developmental Biology, 48*(7), 637–643. https://doi.org/10.1387/ijdb.031778lw.

Williams, C. J., & Erickson, G. F. (2000). Morphology and physiology of the ovary. In L. J. De De Groot, G. Chrousos, K. Dungan, K. R. Feingold, A. Grossman, J. M. Hershman, C. Koch, M. Korbonits, R. McLachlan, M. New, J. Purnell, R. Rebar, F. Singer, & A. Vinik (Eds.), *Endotext*. South Dartmouth (MA): MDText.com, Inc.

Wu, B. J., Dong, F. L., Ma, X. S., Wang, X. G., Lin, F., & Liu, H. L. (2014). Localization and expression of histone H2A variants during mouse oogenesis and preimplantation embryo development. *Genetics and Molecular Research, 13*, 5929–5939. https://doi.org/10.4238/2014.August.7.8.

Wu, S. F., Zhang, H., & Cairns, B. R. (2011). Genes for embryo development are packaged in blocks of multivalent chromatin in zebrafish sperm. *Genome Research, 21*(4), 578–589. https://doi.org/10.1101/gr.113167.110.

Yamada, T., Mizuno, K., Hirota, K., Kon, N., Wahls, W. P., Hartsuiker, E., et al. (2004). Roles of histone acetylation and chromatin remodeling factor in a meiotic recombination hotspot. *The EMBO Journal, 23*(8), 1792–1803. https://doi.org/10.1038/sj.emboj.7600138.

Yamaguchi, S., Hong, K., Liu, R., Inoue, A., Shen, L., Zhang, K., et al. (2013). Dynamics of 5-methylcytosine and 5-hydroxymethylcytosine during germ cell reprogramming. *Cell Research, 23*, 329–339. https://doi.org/10.1038/cr.2013.22.

Yamaguchi, S., Hong, K., Liu, R., Shen, L., Inoue, A., Diep, D., et al. (2012). Tet1 controls meiosis by regulating meiotic gene expression. *Nature, 492*, 443–447. https://doi.org/10.1038/nature11709.

Yamaguchi, S., Kimura, H., Tada, M., Nakatsuji, N., & Tada, T. (2005). Nanog expression in mouse germ cell development. *Gene Expression Patterns, 5*(5), 639–646. https://doi.org/10.1016/j.modgep.2005.03.001.

Yamaguchi, S., Shen, L., Liu, Y., Sendler, D., & Zhang, Y. (2013). Role of Tet1 in erasure of genomic imprinting. *Nature, 504*, 460–464. https://doi.org/10.1038/nature12805.

Yamaji, M., Seki, Y., Kurimoto, K., Yabuta, Y., Yuasa, M., Shigeta, M., et al. (2008). Critical function of Prdm14 for the establishment of the germ cell lineage in mice. *Nature Genetics, 40*(8), 1016–1022. https://doi.org/10.1038/ng.186.

Yamashiro, C., Sasaki, K., Yabuta, Y., Kojima, Y., Nakamura, T., Okamoto, I., et al. (2018). Generation of human oogonia from induced pluripotent stem cells in vitro. *Science, 362*, 356–360. https://doi.org/10.1126/science.aat1674.

Yamashita, K., Shinohara, M., & Shinohara, A. (2004). Rad6-Bre1-mediated histone H2B ubiquitylation modulates the formation of double-strand breaks during meiosis. *Proceedings of the National Academy of Sciences of the United States of America, 101*(31), 11380–11385. https://doi.org/10.1073/pnas.0400078101.

Yan, C., Wang, P., DeMayo, J., DeMayo, F. J., Elvin, J. A., Carino, C., et al. (2001). Synergistic roles of bone morphogenetic protein 15 and growth differentiation factor 9 in ovarian function. *Molecular Endocrinology, 15*(6), 854–866. https://doi.org/10.1210/mend.15.6.0662.

Yang, Q. E., Kim, D., Kaucher, A., Oatley, M. J., & Oatley, J. M. (2013). CXCL12-CXCR4 signaling is required for the maintenance of mouse spermatogonial stem cells. *Journal of Cell Science, 126*(4), 1009–1020. https://doi.org/10.1242/jcs.119826.

Yao, H. H., DiNapoli, L., & Capel, B. (2003). Meiotic germ cells antagonize mesonephric cell migration and testis cord formation in mouse gonads. *Development, 130*(24), 5895–5902. https://doi.org/10.1242/dev.00836.

Yao, H. H., Matzuk, M. M., Jorgez, C. J., Menke, D. B., Page, D. C., Swain, A., et al. (2004). Follistatin operates downstream of Wnt4 in mammalian ovary organogenesis. *Developmental Dynamics, 230*(2), 210–215. https://doi.org/10.1002/dvdy.20042.

Yartseva, V., & Giraldez, A. J. (2015). The maternal-to-zygotic transition during vertebrate development: A model for reprogramming. *Current Topics in Developmental Biology, 113*, 191–232. https://doi.org/10.1016/bs.ctdb.2015.07.020.

Ying, Y., Liu, X. M., Marble, A., Lawson, K. A., & Zhao, G. Q. (2000). Requirement of Bmp8b for the generation of primordial germ cells in the mouse. *Molecular Endocrinology, 14*(7), 1053–1063. https://doi.org/10.1210/mend.14.7.0479.

Ying, Y., & Zhao, G. Q. (2001). Cooperation of endoderm-derived BMP2 and extraembryonic ectoderm-derived BMP4 in primordial germ cell generation in the mouse. *Developmental Biology, 232*(2), 484–492. https://doi.org/10.1006/dbio.2001.0173.

Yoshida, K., Kondoh, G., Matsuda, Y., Habu, T., Nishimune, Y., & Morita, T. (1998). The mouse RecA-like gene Dmc1 is required for homologous chromosome synapsis during meiosis. *Molecular Cell*, *1*(5), 707–718.

Yoshida, S., Sukeno, M., Nakagawa, T., Ohbo, K., Nagamatsu, G., Suda, T., et al. (2006). The first round of mouse spermatogenesis is a distinctive program that lacks the self-renewing spermatogonia stage. *Development*, *133*, 1495–1505. https://doi.org/10.1242/dev.02316.

Yoshida, S., Takakura, A., Ohbo, K., Abe, K., Wakabayashi, J., Yamamoto, M., et al. (2004). Neurogenin3 delineates the earliest stages of spermatogenesis in the mouse testis. *Developmental Biology*, *269*, 447–458. https://doi.org/10.1016/j.ydbio.2004.01.036.

Zelazowski, M. J., Sandoval, M., Paniker, L., Hamilton, H. M., Han, J., Gribbell, M. A., et al. (2017). Age-dependent alterations in meiotic recombination cause chromosome segregation errors in spermatocytes. *Cell*, *171*(3). https://doi.org/10.1016/j.cell.2017.08.042. 601-614.e613.

Zeleznik, A. J. (2004). The physiology of follicle selection. *Reproductive Biology and Endocrinology*, *2*, 31. https://doi.org/10.1186/1477-7827-2-31.

Zhang, F. P., Poutanen, M., Wilbertz, J., & Huhtaniemi, I. (2001). Normal prenatal but arrested postnatal sexual development of luteinizing hormone receptor knockout (LuRKO) mice. *Molecular Endocrinology*, *15*(1), 172–183. https://doi.org/10.1210/mend.15.1.0582.

Zhang, B., Zheng, H., Huang, B., Li, W., Xiang, Y., Peng, X., et al. (2016). Allelic reprogramming of the histone modification H3K4me3 in early mammalian development. *Nature*, *537*, 553–557. https://doi.org/10.1038/nature19361.

Zheng, H., Huang, B., Zhang, B., Xiang, Y., Du, Z., Xu, Q., et al. (2016). Resetting epigenetic memory by reprogramming of histone modifications in mammals. *Molecular Cell*, *63*, 1066–1079. https://doi.org/10.1016/j.molcel.2016.08.032.

Zhou, X., Kudo, A., Kawakami, H., & Hirano, H. (1996). Immunohistochemical localization of androgen receptor in mouse testicular germ cells during fetal and postnatal development. *The Anatomical Record*, *245*, 509–518. https://doi.org/10.1002/(SICI)1097-0185(199607)245:3<509::AID-AR7>3.0.CO;2-M.

Zhou, Q., Nie, R., Li, Y., Friel, P., Mitchell, D., Hess, R. A., et al. (2008). Expression of stimulated by retinoic acid gene 8 (Stra8) in spermatogenic cells induced by retinoic acid: An in vivo study in vitamin A-sufficient postnatal murine testes. *Biology of Reproduction*, *79*(1), 35–42. https://doi.org/10.1095/biolreprod.107.066795.

Zhou, Q., Nie, R., Prins, G. S., Saunders, P. T. K., Katzenellenbogen, B. S., & Hess, R. A. (2002). Localization of androgen and estrogen receptors in adult male mouse reproductive tract. *Journal of Andrology*, *23*, 870–881. https://doi.org/10.1002/j.1939-4640.2002.tb02345.x.

Zhou, Q., Wang, M., Yuan, Y., Wang, X., Fu, R., Wan, H., et al. (2016). Complete meiosis from embryonic stem cell-derived germ cells in vitro. *Cell Stem Cell*, *18*(3), 330–340. https://doi.org/10.1016/j.stem.2016.01.017.

CHAPTER NINE

Genetics of scapula and pelvis development: An evolutionary perspective

Mariel Young[a], Licia Selleri[b,c,1], Terence D. Capellini[a,d,1]

[a]Department of Human Evolutionary Biology, Harvard University, Cambridge, MA, United States
[b]Program in Craniofacial Biology, Department of Orofacial Sciences, Eli and Edythe Broad Center of Regeneration Medicine & Stem Cell Research, University of California, Institute of Human Genetics, San Francisco, CA, United States
[c]Program in Craniofacial Biology, Department of Anatomy, Eli and Edythe Broad Center of Regeneration Medicine & Stem Cell Research, University of California, Institute of Human Genetics, San Francisco, CA, United States
[d]Broad Institute of Harvard and MIT, Cambridge, MA, United States
[1]Corresponding authors: e-mail address: licia.selleri@ucsf.edu; tcapellini@fas.harvard.edu

Contents

Abstract

In tetrapods, the scapular and pelvic girdles perform the important function of anchoring the limbs to the trunk of the body and facilitating the movement of each appendage. This shared function, however, is one of relatively few similarities between the

Current Topics in Developmental Biology, Volume 132
ISSN 0070-2153
https://doi.org/10.1016/bs.ctdb.2018.12.007

scapula and pelvis, which have significantly different morphologies, evolutionary histories, embryonic origins, and underlying genetic pathways. The scapula evolved in jawless fish prior to the pelvis, and its embryonic development is unique among bones in that it is derived from multiple progenitor cell populations, including the dermomyotome, somatopleure, and neural crest. Conversely, the pelvis evolved several million years later in jawed fish, and it develops from an embryonic somatopleuric cell population. The genetic networks controlling the formation of the pelvis and scapula also share similarities and differences, with a number of genes shaping only one or the other, while other gene products such as PBX transcription factors act as hierarchical developmental regulators of both girdle structures. Here, we provide a detailed review of the cellular processes and genetic networks underlying pelvis and scapula formation in tetrapods, while also highlighting unanswered questions about girdle evolution and development.

1. Introduction

As key components of the girdles, the scapula and pelvis are unique bones with multiple functions, the most important of which involves anchoring the appendicular to axial skeleton in tetrapods and facilitating movement of each limb. Though they share these general functional similarities, their morphology, evolutionary history, and development markedly differ. Despite extensive inquiry into each girdle's functional anatomy and evolution across vertebrates, studies on their prenatal development and genetic architecture are in their infancy, especially compared to all that is known on the development of their adjoining limbs. Here, we present a detailed review of the current knowledge of the embryonic, fetal, and postnatal development of both girdle skeletal elements, focusing on the tissues and genetic networks that play a role in their patterning and morphogenesis. We depart from the traditional developmental biology review as we view these bones and their development through an evolutionary context. We believe this is highly relevant because selective pressures in different vertebrates have shaped girdle phenotypes by directly altering developmental processes and their underlying genetic networks. We begin with a description of each bone's form and function, followed by a review of their evolutionary history in tetrapods and mammals. We then discuss the embryonic and early postnatal developmental processes shaping the scapula and pelvis, as well as review the genetic networks underlying the development of each bone.

1.1 Anatomy and function

At the level of the anterior brachiocephalic region, the scapula connects the forelimb to the trunk, whereas at its posterior lumbar-sacral region, the pelvis connects the hindlimb. Because of the different functions of each limb type in tetrapods, the scapula and pelvis exhibit drastically different morphologies. In general, the scapula has a triangular form and is positioned over the dorsolateral region of the thorax (Fig. 1) (Romer & Parsons, 1986). The scapula consists medially of a large, flat, thin plane known as the blade, which is divided dorsally into the supraspinatus and infraspinatus fossae by a bony spine. This spine extends laterally, and depending on whether the animal is orthograde or pronograde, forms a laterally or inferiorly oriented acromion, respectively (Young, 2004). The acromion forms a ligamentous articulation with the clavicle, which is anchored to the sternum ventrally via additional ligaments. Thus, the clavicle and scapula constitute the pectoral girdle in mammals, albeit some species have lost the clavicle (Romer & Parsons, 1986; see Sears, Capellini, & Diogo, 2015 for review on girdle evolution). The blade also laterally extends into the neck and head, the latter of which possesses a laterally facing cavity called the glenoid fossa for articulation with the humeral head at the shoulder joint. Superior to the glenoid is the coracoid process, which, like the blade, serves as an attachment site for muscles that envelop the scapula and the shoulder (Romer & Parsons, 1986). Thus, the scapula proper has no true bony articulation with the axial skeleton and is attached to the rib cage and vertebral column only via skeletal muscles (Valasek et al., 2010). This has led to the suggestion that the scapula arose by acting as a large sesamoid bone, increasing the lever advantages of muscles that traverse it and insert into the forelimb (Sears et al., 2015; Williams, 2003). While debatable, the scapula blade's envelopment by muscles is aligned with its developmental origins from the somites (see below).

The pelvis, consisting of right and left os coxae, does not share the scapula's flat, triangular shape. Rather, each coxae is comprised of three fused elements, which extend in distinct planes: a superiorly positioned ilium, an inferiorly/ventrally positioned pubis, and an inferiorly/dorsally positioned ischium (Fig. 2) (Romer & Parsons, 1986; Young & Capellini, 2015). These three bones centrally fuse at the acetabulum, which articulates with the femoral head of the hindlimb, thus anchoring the limb to the trunk (Romer & Parsons, 1986). Both os coxae articulate with each other ventrally via the pubic symphysis, and dorsally at the sacrum via the sacroiliac joints (Verbruggen & Nowlan, 2017). The various rami, fossa, and spines

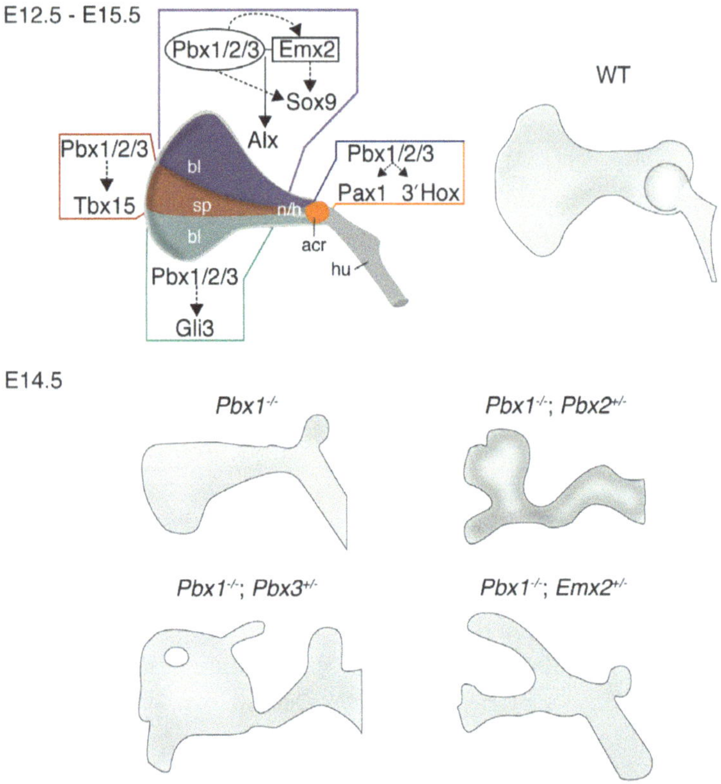

Fig. 1 Genetic regulation of scapula development, including Pbx hierarchical control on effectors of scapular girdle morphogenesis. (Top left) Schematic illustration depicting genetic networks that govern murine pectoral girdle development. Colored areas correspond to specific anatomical structures (displayed at E12.5–E15.5) of the developing girdle and to critical domains of gene expression during earlier patterning stages (i.e., E10.5 and E11.5). Superior scapular blade (bl) domain, where expression of *Aristaless* genes such as *Alx1*, *Alx3*, and *Alx4* is pronounced, is represented in purple; the spine (sp) and central scapula blade (bl), corresponding to a *Tbx15* expression territory is represented in red/brown; the inferior scapular blade (bl) domain, expressing *Gli3* is represented in green; the acromion (acr), that exhibits *Pax1* and *Hox* gene expression, is represented in orange. The proximal limb element, humerus (hu) is represented in gray. Pbx factors work as upstream regulators of domain-specific gene networks that pattern all of these structures. In the superior scapula blade, genetic and molecular interaction of Pbx with Emx2 directly regulate the transcription of *Alx1* (solid arrow), responsible for superior scapular blade patterning. However, Pbx also regulates Emx2 and Sox9, either directly or indirectly, and this control influences mesenchymal condensation and chondrogenesis across the entire blade and head/neck region. In the central scapular blade, including the root of the spine, Pbx-dependent control of *Tbx15* expression is highlighted (within the brown box). In the inferior scapular blade, *Gli3* (within the teal box) is also regulated hierarchically by Pbx family members. *Pax1* and 3′ *Hox* genes in the acromion and neck/head structures (within the box outlined in purple and orange) are additionally regulated upstream by Pbx. Solid arrows represent direct protein binding and molecular interactions; dashed arrows indicate genetic control. (Top right and Bottom) Pectoral girdle phenotypes in allelic series of E14.5 WT (Top right) and *Pbx* compound mutant embryos and *Pbx1;Emx2* mutants (Bottom), depicting the respective reductions and malformations in scapular neck/head, spine, and blade of each mutant, as compared to wild type (see text for details).

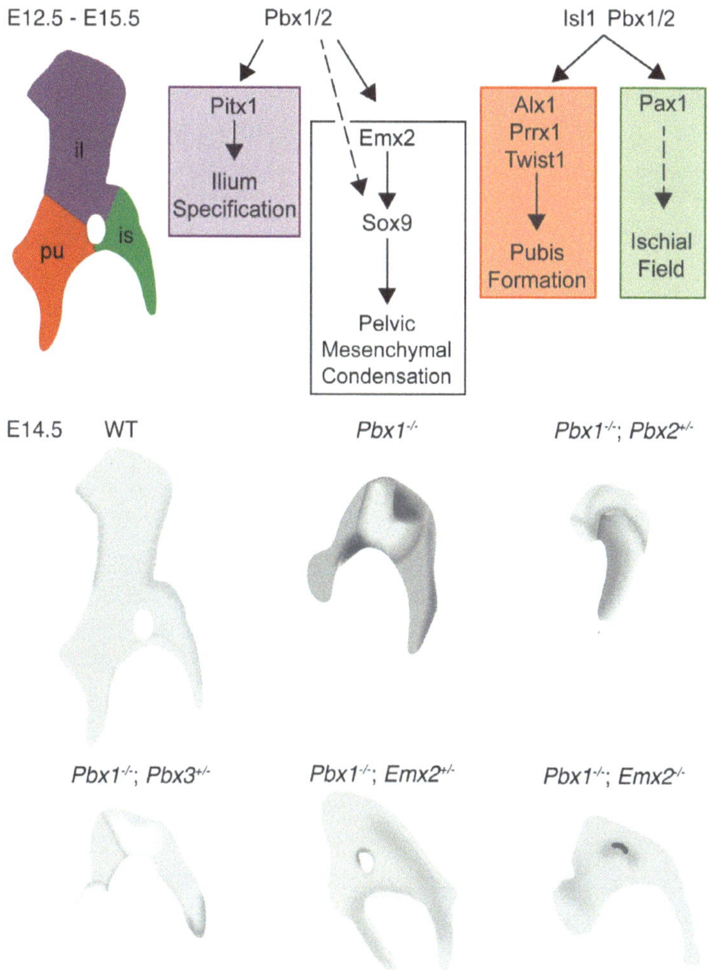

Fig. 2 Genetic regulation of pelvic girdle development, including Pbx hierarchical control on effectors of girdle morphogenesis. (Top left) Schematic illustration depicting the murine pelvic girdle at E12.5–E15.5. Colored areas correspond to specific anatomical structures and to critical domains of gene expression during earlier patterning stages. Ilium (il) domain is represented in purple; ischium (is) in green; and pubis (pu) in red. The panels that comprise genes involved in the formation of each one of these elements are depicted on the right side of the pelvic anatomical illustration with the same colors of the anatomical elements. Pbx1/Pbx2 hierarchically regulate *Pitx1* during the specification of the ilium (purple panel); *Alx1*, *Prrx1*, and *Twist1* in pubis development (red panel); and *Pax1* during ischial field establishment (green panel). *Islet1* (*Isl1*) additionally regulates in parallel pubis and ischium specification, thus dividing the pelvis into superior (ilium) and inferior (pubis and ischium) portions. In addition, *Pbx1/Pbx2* hierarchically regulates *Emx2* during the mesenchymal condensation of the entire pelvic girdle, as well as *Sox9* (white panel). Solid arrows represent direct molecular interactions; dashed arrows indicate genetic control. (Bottom) Illustrations depicting the WT pelvic girdle at E14.5 and the phenotypes of compound Pbx family mutants and *Pbx1;Emx2* mutants (see text for details).

on the ilium, ischium, and pubis provide attachment sites for muscles of the pelvic floor and hindlimb and are thus necessary for positional movement and locomotion.

The primary function of the scapula and pelvis is to anchor the limbs to the trunk and provide integral attachment sites for many limb and axial muscles. In quadrupedal animals (e.g., mice and monkeys), the pectoral girdle aids in forelimb movement and in transmitting force to the body axis during locomotion (Romer & Parsons, 1986). In bipedal animals (e.g., humans), the scapula instead prioritizes mobility and a broader range of motion for purposes such as carrying, manipulating objects, using tools, and throwing. Conversely, in nearly all terrestrial mammals, the pelvis is chiefly involved in weight bearing and locomotion, as it serves to transmit force generated in both hindlimb propulsion and transmission to the axis. In bipeds, it plays additional roles in supporting the weight of the upper body during locomotion, and its bowl-shaped ilia, which form the superior pelvic girdle, provide a protective structure for the organs and tissues of the lower abdomen and pelvic cavity. Lastly, in humans the pelvis must remain wide enough to allow for a birth canal large enough to enable passage of a large-brained neonate (DeSilva & Rosenberg, 2017).

1.2 Evolutionary history

There is an extraordinary amount of locomotor diversity within vertebrates, which is mirrored by diversity of scapular and pelvic morphology as the bones have been shaped by natural selection to be ideally tailored to each species' form of locomotion. The different morphology of the scapula and pelvis is reflected in the separate evolutionary origins of each girdle. The pectoral girdle fin structure arose first and is found in jawless fish in the fossil record dating to ~430 million years ago (MYA) (Coates, 2003; Johanson & Trinajstic, 2014; Trinajstic, Boisvert, Long, Maksimenko, & Johanson, 2015). It existed in isolation for several million years until the pelvic girdle and fin structures surfaced in jawed fish ~413 MYA (Zhu, Yu, Choo, Wang, & Jia, 2012). During this time, the pectoral girdles anchored directly to the dermal bones of the head or cephalic region, while the pelvic fin bones had few attachments to other skeletal elements (Diogo & Ziermann, 2015; Zhu et al., 2012, 2013). Approximately 40 million years later, during the fin-to-limb transition, changes to the pectoral girdle once again preceded those of the pelvis, this time in response to the demands of terrestriality as many animals became more tetrapod-like (Boisvert, 2005;

Clack, 2009; Coates, Ruta, & Friedman, 2008). As tetrapods diversified into many clades with different forms of locomotion, the scapula and pelvis continued to evolve to suit these locomotor demands (see Romer & Parsons, 1986; Sears et al., 2015).

Within modern mammals, humans and chimpanzees are closely related species with very different scapula and pelvis morphologies. The chimpanzee scapula and pelvis are morphologically similar to those of other great apes and, in some attributes, terrestrial quadrupeds. The chimp scapula has a more cranially oriented configuration of the acromion and glenoid fossa and a narrower blade than that of modern humans (Roach, Venkadesan, Rainbow, & Lieberman, 2013), morphologies that permit greater ranges of movement important for arboreal locomotion. With regard to the pelvis, the chimp pelvis is narrow medial-laterally and tall superior-inferiorly, with the ilia extending cranially, providing stability to the thoraco-pelvic unit (Sockol, Raichlen, & Pontzer, 2007). This configuration is similar to the pattern in quadrupedal animals such as monkeys and mice, although stability in many quadrupeds is improved by the elongation of the lumbar vertebral region, a pattern not seen in apes. The human pelvis has shorter, more broadly flared ilia, and is wider medial-laterally than that of chimps (Lovejoy, 1988; Swindler & Wood, 1973). These changes are adaptations to bipedality, which evolved after the human and chimpanzee lineages split approximately 6–8 MYA (Crompton, Sellers, & Thorpe, 2010). Bipedality also freed the forelimb for other functions in hominins, such as making and using tools, which affected shoulder morphology.

2. Scapula development

Recent studies, conducted mainly in mouse and chick, have traced the developmental trajectory of the scapula in the embryonic, fetal, and early postnatal period, focusing on early formation and patterning. Here, we summarize those findings as well as provide information about the later chondrogenesis and osteogenesis stages of scapula growth during the fetal and postnatal periods, much of which was documented in 20th century studies in humans.

2.1 Embryonic and fetal scapula development

The scapula is forms from multiple progenitor cell populations from different tissue layers—a rare characteristic among appendicular bones, most of which derive from only one tissue layer or progenitor cell population.

The scapula has contributions from the dermomyotome (a dorsolateral derivative of the somite), somatopleure (somatopleuric derivative of the lateral plate mesoderm plus inner ectoderm), and neural crest (migratory pluripotent cells delaminated from the dorsal neural tube) (Durland, Sferlazzo, Logan, & Burke, 2008; Huang, Christ, & Patel, 2006; Huang, Zhi, Patel, Wilting, & Christ, 2000; Matsuoka et al., 2005). The dermomyotome gives rise to the scapula blade and proximal scapula spine, whereas the somatopleure gives rise to the glenoid fossa, coracoid, acromion, structures of the scapula neck, head, and spine, although the extent to which each of these progenitor tissues contribute to the scapula varies across species (Durland et al., 2008; Ehehalt, Wang, Christ, Patel, & Huang, 2004; Huang et al., 2000; Malashichev, Christ, & Pröls, 2008; Valasek et al., 2010). Matsuoka et al. (2005) additionally demonstrated in mice that post-otic neural crest cells contribute to the scapular blade's superior border, spine/acromion, and coracoid process.

The complexity of the dermomyotomal origin of the scapula blade is underscored by the complex organization of the dermomyotome itself, and how it is pre-patterned and differentiates during somitogenesis. Early in somitogenesis, a partially overlapping 3′ *Hox* expression domain is suggested to pre-pattern or specify (scapula) pre-dermomyotomal cells in the undifferentiated somite to become blade progenitors (Huang et al., 2006). Indeed, it appears that only specific hypaxial dermomyotomal cells of the brachio-thoracic region (somites 17–24 in chick) give rise to superior/inferior domains of the scapula blade and spine (Huang et al., 2006, 2000) and that such cells apparently do not mix from one axial level to the next (Ehehalt et al., 2004; Huang et al., 2000). When these somites are ablated the scapula blade is also absent (Malashichev et al., 2008). Importantly, at this stage, dermomyotomal cells from the cervical region also appear incapable of forming scapular elements, indicating that the brachio-thoracic dermomyotome is pre-patterned with the intrinsic ability to form the scapula blade (Ehehalt et al., 2004). These findings support a pre-patterning mechanism, although evidence for a direct role of a *Hox* code is weak (see below) and needs further testing.

Shortly after this pre-patterning stage, the dermomyotome proper forms as a dorsal subdivision of the somite (Gilbert, 2000) and further subdivides into an epaxial domain situated dorsal-medially and a hypaxial domain located ventral-laterally (Burke & Nowicki, 2003; Cheng, Alvares, Ahmed, El-Hanfy, & Dietrich, 2004). Here, dermomyotomal differentiation occurs under distinct regulatory inputs: in the epaxial domain, regulatory signals arise from the notochord and neural tube, whereas in the hypaxial region signals arise from the adjacent and more lateral plate mesoderm (Chiang et al., 1996; Ehehalt et al., 2004; Huang et al., 2006, 2000; Rong, Teillet, Ziller, & Le Douarin, 1992; Teillet, Lapointe, & Le Douarin, 1998). The hypaxial region appears important for blade differentiation since ablation of epaxial tissues results in normally patterned scapula blades (Teillet et al., 1998), whereas inhibition of signaling from the lateral plate mesoderm blocks scapula formation (Wang et al., 2005).

Once hypaxial dermomyotomal cells have been specified as scapular progenitors, additional signaling from the ectoderm and somatopleure is critical for dermomyotomal maturation, migration, and differentiation into pre-chondrogenic scapular mesenchyme. Based on the work of Moeller (2003), Rodríguez-Niedenführ, Dathe, Jacob, Pröls, and Christ (2003), and Ehehalt et al. (2004), Wang et al. (2005) argued that ectodermal WNT signaling into the hypaxial dermomyotome is critical for keeping cells in an undifferentiated epithelial state and that the cessation of such signals, which occurs normally in a cranial-to-caudal sequence during somitogenesis, leads to their differentiation into mesenchyme, an initial step into forming scapula cartilages. However, while Huang et al. (2000) demonstrated that only brachio-thoracic dermomyotome cells are capable of forming scapula blade condensations, Ehehalt et al. (2004) found that the ectoderm overlying the cervical region is able to support ectopic brachio-thoracic dermomyotome to form scapular mesenchymal and cartilage condensations. This finding suggests that signaling by the ectoderm (e.g., possibly by WNTs) might be responsible for inducing later, scapular mesenchymal- and/or cartilage-like fates, or that it plays an earlier, unappreciated role in pre-patterning cells to become scapular blade tissues. Additionally, during the transition to dermomyotomal mesenchyme, bone morphogenetic protein (BMP) signaling from the somatopleure facilitates the expression of early scapula progenitor marking genes, such as *Pax1* (Wang et al., 2005), while blocking or inhibiting BMP signaling in this domain in chick results in the downregulation of *Pax1* and the absence of the scapula blade.

The extent to which the dermomyotome contributes to the scapula blade across tetrapods remains quite unclear. In mice, fate-mapping experiments using *Prx1-Cre/Z/AP* and *Pax3-Cre/R26RYFP* reveal that dermomyotome cells give rise to only the vertebral edge of the scapula blade and minor parts of the spine and acromion (Durland et al., 2008; Valasek et al., 2010). Most blade regions are formed from the somatopleure, as is most of the head and neck, while the spine and acromion have been shown to have a neural crest cell contribution (see below) (Durland et al., 2008;

Huang et al., 2006; Matsuoka et al., 2005; Valasek et al., 2010). This difference between chick and mouse appears to reflect the differential growth of the somatopleuric portion of the blade during mouse scapulogenesis (Shearman, Tulenko, & Burke, 2011; Valasek et al., 2010).

Regardless of the extent to which various embryonic tissues contribute to the scapula, the resulting cell populations become integrated into one mesenchymal condensation by embryonic day (E) 11.5 in mice (human E43), and into a clearly recognizable pre-cartilaginous scapula by E12.5 (human E44) located adjacent to the forelimb bud at the brachio-thoracic-axial level (Durland et al., 2008; Hita-Contreras et al., 2018; Huang et al., 2006, 2000; Young & Capellini, 2015). Recent analysis by Hita-Contreras et al. (2018) has shown that this mesenchymal condensation has multiple outgrowths: one each for the scapular body, coracoid process, and the acromion and spine. These findings agree with Capellini et al. (2010) assessment using 3D Optimal Projection Tomography of the boundaries of *Sox9* expression, a marker of mesenchymal condensation formation. Subsequently, this mass shifts caudally to its final position posterior at the thorax, such that by human embryonic day 44 (E12.5–13 in mouse) it is positioned lateral to or behind the ribs, depending on the species (Bardeen & Lewis, 1901; O'Rahilly & Gardner, 1972).

Following this patterning stage, the chondrogenesis stage of bone formation begins at E12.5 in mice or the sixth gestational week in humans (Monkhouse, 1996). Like nearly all appendicular elements, the scapula forms via endochondral ossification (Karsenty & Wagner, 2002), in which the mesenchymal condensation that prefigures scapular morphology begins to differentiate into chondrocytes by secreting a collagen matrix consisting of types II, IX, and XI collagen, as well as proteoglycans (Long & Ornitz, 2013). The chondrocytes divide and differentiate, forming growth plates that ultimately elongate to form a nearly complete chondrocyte replica of the osseous scapula. During later embryonic gestation in mice (E15.5) or the second gestational month in humans (Andersen, 1963; Mall, 1906; Monkhouse, 1996; unpublished results from the Capellini laboratory), this cartilaginous scapula is invaded by blood vessels and osteoblasts allowing for the appearance of a single primary ossification center in the scapular neck region, which then facilitates bone formation (Long & Ornitz, 2013). Ossification typically extends bidirectionally across the blade medially and toward the glenohumeral joint laterally (Scheuer, Black, & Christie, 2000), reaching the base of the spine by gestational week 9 in humans (equivalent to E18.5 in mice) and the glenoid by gestational week 12 (P02 in mice) (Andersen, 1963; Ogden & Phillips, 1983).

During this period, post-otic neural crest cells also contribute to the scapula, specifically to endochondral and perichondrial tissues of the superior border, spine/acromion, and coracoid process, as well as to the connective tissues of the muscles inserting into these structures (Matsuoka et al., 2005). *Sox10-Cre ROSA GFP+* cells contribute specifically the endochondral and perichondrial tissues of these structures and by doing so they provide, and help anchor, major attachment regions for the branchial muscles (e.g., the trapezius). This process occurs after the initial patterning of scapula progenitors in early fetal development.

2.2 Postnatal scapula development

By birth much of the acromion, coracoid process, medial border, inferior angle, and glenoid articular surface remain cartilaginous in humans (Monkhouse, 1996). A primary ossification center in the coracoid appears within the first year of life (Camp & Cilley, 1931), though it can also be present before birth (Fawcett, 1910; Scheuer et al., 2000). The full ossification of the coracoid continues for several years, and its fusion to the scapula occurs around age 14–15 years (Andersen, 1963; Scheuer et al., 2000). Postnatally, the scapula also has at least seven secondary centers of ossification: one at the acromion, three at the coracoid process, one at the inferior angle, one at the inferior glenoid region, and one at the vertebral border that consists of several small ossification centers (Scheuer et al., 2000). These commence ossification later in life, with the first, the subcoracoid center, appearing around ages 8–10 (Basmajian & Slonecker, 1989; Birkner, 1978). Subsequently, other secondary ossification centers arise in the teen years (Flecker, 1942; Francis, 1940; Hodges, 1933; Johnston, 1961). The last secondary ossification centers, those at the vertebral border and inferior angle, arise at 15–17 years and fuse by 23 years, when the scapula has reached its adult morphology (Basmajian & Slonecker, 1989; Birkner, 1978; Girdany & Golden, 1952; Hodges, 1933; Stevenson, 1924).

3. Pelvic development

Although the pelvic girdle is the quintessential locomotor, weight bearing, and birthing structure in tetrapods, few studies have elucidated the structure's development during embryonic, fetal, and postnatal life. Much of our knowledge about pelvic development derives from research in chick and mouse, although studies conducted on human development in the early to mid-20th century have shed light on its chondrogenic and osteogenic progression during fetal life. Below, we summarize findings from both types of research.

3.1 Embryonic and fetal pelvic development

Developmental studies in mouse and chick demonstrate that each pelvic element (i.e., ilium, ischium, and pubis) arises from somatopleure, whereas the sacrum forms from somite-derived sclerotome. In mouse, one of the earliest fate-mapping studies using *Prx1-Cre/Z/AP* lines revealed that os coxae arise from mesenchymal cells solely from the somatopleure (Logan et al., 2002). Indeed, when posterior hindlimb somatopleure is ablated in chick no pelvis forms, whereas when lumbar-sacral somites are ablated the os coxae are unaffected but the sacrum is absent (Malashichev et al., 2008). Furthermore, when quail hindlimb field somatopleure is transplanted into the corresponding location in chick, the grafts give rise to the os coxae; whereas when lumbosacral somites are transplanted, regardless of developmental stage, no quail cellular contributions can be identified in os coxae (Malashichev et al., 2008). More detailed studies utilizing computer-animated 4D visualization also reveal that the os coxae forms from a single large condensation of somatopleuric mesenchyme that is located medially and adjacent to the lateral plate mesoderm (Pomikal & Streicher, 2010). However, in all of these studies, it remains unclear whether different subregions within the somatopleure individually contribute mesoderm to each subelement or if each forms from differential growth of a common somatopleuric precursor population (Malashichev et al., 2008; Pomikal & Streicher, 2010; Valasek et al., 2010).

Additional research in chick demonstrates that the subelements form in a distinctive sequence, beginning with the ilium, followed by the pubis, and lastly the ischium (Malashichev, Borkhvardt, Christ, & Scaal, 2005). It was also shown that pelvic formation requires signaling from laterally positioned (limb) mesenchyme and the overlying ectoderm. For example, ablation of the limb mesenchyme prior to Carnegie Stage 18 results in pelvic malformations, whereas ectodermal removal from the pre-limb field and the presumptive pelvic forming domain during Stage 12–18 results in either near complete agenesis of all girdle elements at Stage 12 or severe girdle malformations by Stage 18. In these studies, the pubis was most often affected, the ischium was affected only when ectoderm was removed by Stage 12, and the ilium was never completely absent, indicating a differential role in the ectoderm in the potential specification and/or maturation of the individual subelements. In a similar vein, studies by Malashichev et al. (2005, 2008) also indicate that the specification of pelvic skeletal subelements occurs before limb bud outgrowth in chick, which suggests that a pre-patterning stage may exist for the girdle, as has been proposed for the limb (Dudley, Ros, & Tabin, 2002), although this remains to be tested.

Based on gene expression studies and histology, this somatopleuric anlagen appears adjacent to the hindlimb bud at E28 in humans (mouse E9.5/10) (Bardeen & Lewis, 1901; Capellini et al., 2011; O'Rahilly & Gardner, 1975) and between E34–36 (mouse E9.5/10–E11.5/12) mesodermal cells condense in a specific region demarcated by the obturator, femoral, and sciatic nerves (Laurenson, 1964a). Shortly after, the mass then begins to expand to form three processes, each of which marks the precursor cell populations of the ilium, ischium, and pubis (Fazekas & Kosa, 1978; Pomikal & Streicher, 2010). The presumptive ischial and pubic masses fuse inferiorly at the level of the obturator nerve, forming the obturator foramen (Scheuer et al., 2000). Subsequently, the iliac mass grows toward the mesenchymal progenitor cells of the pre-sacral vertebrae, eventually fusing with the sacrum at the sacroiliac joint (Bardeen, 1905; Fazekas & Kosa, 1978; Scheuer et al., 2000). Inferiorly, right and left pubic progenitor cell populations extend medially and fuse to form the pubic symphysis (Scheuer et al., 2000). This process is dependent on molecular signaling from nearby tissues of the axial mesoderm, body wall ectoderm, and limb bud (Malashichev et al., 2005, 2008).

This early patterning stage of pelvic development lasts several weeks in humans (1–2 days in mice), until the endochondral ossification commences at about embryonic weeks 6–7 (mouse E12) (Bardeen, 1905; O'Rahilly & Gardner, 1975; Okumura et al., 2017). Accordingly, chondrogenesis begins in a region near the acetabulum, followed in the ischium and then pubis (Laurenson, 1964a), and these three centers expand such that by embryonic week 8 (mouse E14) they have fused at the acetabulum (Adair, 1918).

Although the morphology of the cartilaginous precursor of each os coxae is not as complex as the mature form, many adult features of the pelvis are already present early during chondrogenesis. Previous research suggested that features of the human ilium, such as the greater sciatic notch, could be observed by 4–5 gestational months in humans (Laurenson, 1964a). However, recent experimental work on human developmental tissues by Okumura et al. (2017), as well as research in the Capellini laboratory indicates that the distinct shape of the ilium and its features, such as the greater sciatic notch and early iliac spines, are visible as early as 59 days gestation. Additionally, it has been reported that the ischium is identifiable by the third trimester of pregnancy and that the pubis also acquires its adult form around that time (Scheuer et al., 2000; Verbruggen & Nowlan, 2017). It is possible that these reports referred to the shape of bones in their ossified state, rather than their cartilaginous state, since our recent observations have revealed that at 59 days gestation, well before the third trimester or even the second, the cartilages of both the ischium and pubis are already reminiscent of their adult morphology. Indeed, the entire os coxae and much of the skeleton at this stage appear as miniaturized versions of adult forms. Okumura et al. (2017) have also demonstrated that many features of adult pelvis morphology, such as the ischial tuberosity, pubic symphysis, and sacroiliac joint are present by early fetal stages of human development. These observations highlight how important chondrogenesis is in specifying the future functional morphology of bones.

Ossification begins after chondrogenesis, but the two processes can occur in different regions of the developing pelvis simultaneously. Indeed, when primary ossification centers arise in central regions of the ilium, pubis, and ischium the outer areas of each bone are still undergoing chondrogenesis. The primary ossification centers mirror the chondrification centers in both location and order, with the first arising in the ilium at the end of the second gestational month in humans in the perichondrium of the acetabular roof (Adair, 1918; Bardeen, 1905; Laurenson, 1964a, 1964b). This ossification center extends cranially, first covering the exterior perichondral surface of the ilium, and ultimately invading the underlying cartilage. The ischium's primary ossification center is next to form at 4–5 gestational months (Francis, 1951) and is located inferior and posterior to the acetabulum. Lastly, the pubis ossification center develops by 5–6 gestational months, forming inferiorly and anteriorly to the acetabulum in the region of the superior pubic ramus, and expanding to envelop the entire pubis (Scheuer et al., 2000).

3.2 Postnatal pelvic development: Secondary ossification and biomechanical responses

At birth, all three primary ossification centers are described as "well developed," with each extending enough to form its section of the acetabular wall (Scheuer et al., 2000). In the first three postnatal months the ilium, ischium, and pubis undergo rapid growth in size, but little morphological change in shape (Reynolds, 1945). The growth continues at a slower rate until age three, when it slows even further until puberty, when the adolescent growth spurt occurs and secondary sexually related growth changes take place (Miles & Bulman, 1994). During this slow developmental course, the three primary ossification centers fuse by age six (Scheuer et al., 2000; Verbruggen & Nowlan, 2017).

During teen years, several secondary centers of ossification appear in the iliac crest, acetabulum, pubic ramus, and ischial tuberosity, among other variable centers (Scheuer et al., 2000). Three epiphyses, one each from the ilium, pubis, and ischium, form within the acetabulum and ultimately extend to form its outer rim and articular surface (Verbruggen & Nowlan, 2017). These centers appear earliest, around age 10, and the complete fusion of the acetabulum takes place by mid-puberty (Freedman, 1934; Stevenson, 1924). Additional ossification centers for each subelement continue to develop, often with different timing between males and females, yet these epiphyses will not fully fuse until the early 20s (Scheuer et al., 2000; Verbruggen & Nowlan, 2017). The earlier maturation of the acetabulum compared to the rest of the os coxae is likely due to its need to withstand significant body mass forces, which increases during puberty (Scheuer et al., 2000).

Because the pelvis is a weight bearing structure in humans and other bipeds, mechanical loading plays a major role in the development of pelvic morphology. The characteristic shape of the human ilium, with its bowl-like structure and unique concavities and convexities, does not appear until age two, when adult-like bipedal locomotion begins (Scheuer et al., 2000; Verbruggen & Nowlan, 2017). In this vein, it is possible that mechanical stimulation as early as in utero initiates pelvic bone remodeling, as the internal trabecular bone organization patterns in the fetal ilium resemble patterns associated with bipedal locomotion in adults (Cunningham & Black, 2009). Experiments in model organisms have also demonstrated the need for structural and mechanical integrity of the cartilage and ligaments surrounding the sacroiliac joint to enable proper fusion (Harrison, 1958a, 1958b). Development of the acetabulum as a cup-shaped structure—named for ancient Roman and Greek vinegar vessels—is also dependent on the presence of a rounded femoral head; if the femoral head is missing or abnormally shaped, the acetabulum will bear a corresponding malformation presumably due to altered signaling and/or biomechanical influences (Harrison, 1961; Ponseti, 1978). This has been supported using rat and chick models, in which in utero limb movements were restrained or immobilized, resulting in alterations to acetabular articular surface morphology (Hashimoto, Kihara, & Otani, 2002), or acetabular cup morphology and pelvic orientation (Nowlan, Chandaria, & Sharpe, 2014).

3.3 Postnatal pelvic development: Sexual dimorphism

Mammals display fundamental sex differences in pelvic shapes, in part relating to obstetrical factors and differences in body size. In mice, these differences are evident in the size and shape of the pubis, which is thinner in adult females, together with other subtle differences in the os coxae between sexes (Iguchi, Irisawa, Fukazawa, Uesugi, & Takasugi, 1989). The mature human pelvis shows marked sexual dimorphism due to its fundamental role in the birth of neonates, which are comparatively much larger brained than other great apes and mice. These sex differences include enlarged acetabular cavities in males and broader sciatic notches and larger pelvic inlets and outlets in females to create a wider birth canal (Leong, 2006).

These differences are chiefly believed to arise during puberty in humans. While some have claimed subtle pelvic sexual dimorphism as early as the fetal period (Boucher, 1957; Holcomb & Konigsberg, 1995), others, including a more recent quantitative assessment based on reconstructed fetal ilia CT images, found no evidence supporting this assertion (Mokrane et al., 2013; Weaver, 1980). Indeed, during human puberty, males show significantly greater acetabular growth to support the increased mass of the male upper body (LaVelle, 1995), whereas females display more dramatic morphological changes across the ilium, ischium, and pubis (Bilfeld et al., 2013; Wilson, Ives, Cardoso, & Humphrey, 2015). These changes begin when females produce higher levels of circulating estrogen, which then bind to estrogen receptors (Purves et al., 2004). When these receptors are activated, downstream transcription factors expressed in the pelvis bind to currently unidentified genes to influence gene transcription and ultimately bone growth in specific domains to produce the female adult pelvic phenotype. When the full suite of genes expressed in the pelvis during prenatal chondrogenesis is identified, it is likely that a subset of these same genes will be activated by estrogen receptor binding during puberty.

4. Genetic networks

Much of our current knowledge of the genes involved in girdle development derives from genetic studies in mouse and tissue transplantation experiments in chick. The vast majority of genes known to be active in either developing girdle have been identified as a result of research primarily focused on the limb proper; thus, we lack a full understanding of the genetic networks that govern the development and morphogenesis of the girdles sensu stricto. Despite this limitation, most research indicates substantial divergence in the genetic networks that underlie scapula versus pelvic development, and this likely reflects each girdle's distinct evolutionary and developmental origin (Sears et al., 2015). It is also important to note that many of the genes discussed below are active in the patterning stage of girdle development, and while many are likely to remain active during chondrogenesis, fewer genes specific to the latter stages of girdle formation have been identified.

4.1 Genetic control of scapula development

To date, most genes involved in scapulogenesis have been shown to act modularly (i.e., influencing the blade versus the head), which likely reflects the scapula's unique development from multiple progenitor cell populations, although several appear to hierarchically regulate scapulogenesis. These genes, along with the majority of identified scapula effectors, have been found to act within the somatopleure, with very few identified in the dermomyotome or encoding signaling molecules responsible for progenitor cell migration from the dermomyotome.

Regarding the pre-patterning and specification of early blade progenitors in the dermomyotome, a 3′ *Hox* code has been implicated. While evidence for this is scarce, mouse embryos with homozygous loss of function (LOF) of *Hoxb5* (*Hoxb5*$^{-/-}$) display a rostral shift in the axial location of scapular formation and this shift is more severe in compound *Hox5* mutants (Rancourt, Tsuzuki, & Capecchi, 1995). Therefore, it is only after multiple *Hox* paralogs are removed that scapula blade defects become evident, indicating that HOX transcription factors may pre-pattern the dermomyotome during somitogenesis or even operate within the lateral plate mesoderm collaboratively. However, it is important to note that only 3′ *Hox* are likely to have roles in scapulogenesis since experimental compound loss of posterior *Hox9*, *Hox10*, and *Hox11* paralogs does not result in scapula defects (McIntyre et al., 2007; Wellik & Capecchi, 2003; Xu & Wellik, 2011).

Another *Hox* gene, *Hoxc6*, is strongly expressed in the somatopleure/lateral plate mesoderm and anterior limb field in chicks and mice. On the one hand, alterations in *Hoxc6* expression cause shoulder malformations in chicks (Oliver, De Robertis, Wolpert, & Tickle, 1990) and retinoic acid (RA) application to chick proximal limbs results in duplications of the distal scapula blade, neck, and head region with an expansion of the *Hoxc6* domain (Oliver et al., 1990). This finding suggests that, as in the limb, RA is a critical upstream regulator of HOXC6-driven patterning of the scapula at least in chicks and possibly mouse. Supporting evidence also comes from compound *RARα; RARχ* mouse mutants which have scapular agenesis (Lohnes et al., 1994), suggesting that interactions of HOX proteins and RA may be essential. Given the known effects of RA on axial *Hox* expression (see Gilbert, 2000), a scenario can be envisaged whereby cranial-to-caudal RA signaling and its effects on 3′ HOX-directed patterning of somites, somatopleuric, and lateral plate mesoderm lie upstream of girdle formation. On the other hand, *Hoxc6* LOF, as well as loss of function of all paralogous *Hox6* genes (McIntyre et al., 2007) does not result in scapula defects, indicating that if these genes are important in patterning the scapula or its parts they may only become functionally noticeable when additional 3′ *Hox* family members (e.g., *Hox5* paralogs) are experimentally deleted on the backdrop of *Hox6* paralog LOF.

As a HOX cofactor, albeit one that has also HOX-independent regulatory functions, PBX1 has been shown to dually regulate the development of the scapula and proximal limb (Selleri et al., 2001). *Pbx1* is expressed throughout the somatopleure and lateral plate mesoderm, and *Pbx1*$^{-/-}$ mouse mutants display fusions of the humeral head to the scapula and alterations to the deltoid tuberosity and humeral shaft, indicating roles in scapular head and limb formation. Yet, *Pbx1*$^{-/-}$ embryos also exhibit scapular blades that are superiorly-to-inferiorly constricted and thinner with a hypoplastic dorsal spine and acromion process (Selleri et al., 2001). These data suggest that *Pbx1* expression in the somites and/or in their derivatives, or in the somatopleure, drive blade formation possibly by cooperating with other family members (see below).

With respect to patterning of the scapular blade, neck, and head, *Tbx5*$^{-/-}$ mouse embryos lack forelimbs and scapulae. Similar to *Pbx1*, *Tbx5* is expressed throughout the early limb and somatopleuric mesenchyme in the presumptive scapula-forming domain (E9.5/10), but PBX acts upstream of *Tbx5*, at least in the limb (Capellini and Selleri, unpublished data). The role of TBX5 may be to trigger scapular progenitor differentiation and maturation in the somatopleure, and/or it may influence partial differentiation of the dermomyotomal cells once they have migrated to more lateral domains adjacent to, or within, the somatopleure. Part of the mechanism of action for TBX5 may be through its direct activation of *Fgf10* expression in the lateral plate mesoderm (Agarwal et al., 2003; Ng et al., 2002). However, *Fgf10*$^{-/-}$ embryos do exhibit a milder scapula phenotype consisting of the absence of the posterior one-third of the scapula (Min et al., 1998; Sekine et al., 1999), indicating that the role of TBX5 in *Fgf10* activation bears farther-reaching effects than the sole regulation of limb bud outgrowth (Agarwal et al., 2003; Ng et al., 2002; Rallis, 2003), and that TBX5 transcriptional regulation goes beyond *Fgf10* to likely target many somatopleuric genes. It is also unclear whether the *Fgf10* LOF from the lateral plate mesoderm is responsible directly (i.e., in a cell autonomous manner in the mesenchyme) or indirectly (i.e., in a non-cell autonomous manner through its signaling to the ectoderm) for the overlapping scapula blade defects in both *Fgf10* and *Tbx5* homozygous mutants. With respect to its direct role, if FGF10 function is disrupted during early somatopleuric signaling, blade progenitors may not be induced to properly differentiate; if the effect is indirect, FGF10 might have a more global effect on ectodermal signaling over both the lateral plate mesoderm and somatopleure with the latter having more apparent bearings on scapula blade progenitor cell differentiation.

Research on *Emx2* in the chick and mouse has aided our understanding of scapula formation. In the chick, *Emx2* expression precedes expression of *Sox9*, a critical transcription factor required for mesenchymal cell condensation in both somite and limb derivatives (Bi et al., 2001). *Emx2* and *Sox9* are co-expressed in the somatopleure as well as in the prechondrogenic condensed mesenchyme of the scapula (Pellegrini, Pantano, Fumi, Lucchini, & Forabosco, 2001). Interestingly, *Emx2*$^{-/-}$ mice exhibit skeletal elements of the scapular head but lack blade cartilages, with reduced *Sox9* expression specifically in pre-blade domains (Capellini et al., 2010). This finding suggests that blade loss in *Emx2*$^{-/-}$ mutants does not result from the failure to form scapular blade progenitors (Pröls et al., 2004) but from defects in cell migration and/or signaling, mesenchymal cell condensation, or chondrogenesis. It has also been reported that lateral plate mesoderm-specific inactivation of *Beta-catenin*, a critical factor involved in the WNT signaling pathway, leads to *Emx2* downregulation with concomitant loss and/or reduction of the murine scapular blade. Furthermore, these *Beta-catenin* mutants exhibit a distal shift in *Hoxc6* expression (see above) and marked downregulation of *Pax1* (see below) (Hill, Taketo, Birchmeier, & Hartmann, 2006). Since *Beta-catenin* is also expressed in limb field ectoderm and that ectoderm removal leads to scapula blade agenesis (Ehehalt et al., 2004), our understanding of the effects of this tissue-specific gene loss on *Emx2*, *Hoxc6*, and *Pax1* expression is vague, in light of the required mesenchymal/ectodermal interactions in scapula and limb formation. For example, WNT signaling in the mesenchyme may act directly on *Emx2* (Hill et al., 2006) and/or act indirectly through its effects on ectodermal signaling that, in turn, alter the expression of *Emx2* and other scapula effectors.

A number of other transcription factors expressed in the proximal anterior limb field is known to pattern just the scapular blade with minimal impacts on other scapula structures. These genes include *Tbx15*, *Gli3*, *Alx1* (formerly *Cart1*) *Alx3*, and *Alx4* (Kuijper, Beverdam, et al., 2005; Kuijper, Feitsma, et al., 2005). In most cases, the deletion of any of these genes results in minor blade malformation, whereas their compound LOF in the mouse demonstrates that various genetic interactions are required for normal scapula morphogenesis. A summary of the key findings by Kuijper, Beverdam, et al. (2005), Kuijper, Feitsma, et al. (2005), and others is presented here:

First, *Tbx15*$^{-/-}$ mutants exhibit the presence of a central or infraspinatous blade foramen (Kuijper, Beverdam, et al., 2005). This phenotype is similar to that found in mutants for the polycomb homolog *M33* (Coré et al., 1997) and in mice homozygous for a *Gli3* mutant allele, *Extratoes* (*Xt*) (Hui & Joyner, 1993). The similarity between the *Xt* and *Tbx15* null mutants led Kuijper, Beverdam, et al. (2005) to examine scapula formation in compound *Gli3;Tbx15* mutant mice, where they found that the foramen is more severe in compound mutants (i.e., in *Gli3*$^{+/-}$;*Tbx15*$^{-/-}$ the foramen is enlarged, while in *Gli3*$^{-/-}$;*Tbx15*$^{+/-}$ it is shifted posteriorly). Furthermore, in *Gli3*$^{-/-}$;*Tbx15*$^{-/-}$ mutant mice, only the superior scapular blade remains and appears as a single longitudinal skeletal element. These findings demonstrate that *Gli3* and *Tbx15* genetically interact to form the central blade and that *Gli3* specifically affects inferior (posterior) blade formation (Fig. 1). Interestingly, embryos concurrently lacking *RAR*α and *RAR*χ display a range of phenotypic abnormalities including the loss of the inferior scapula border, a superior reduction in the blade, and also a slight foramen (Lohnes et al., 1994).

Second, *Alx3* and *Alx4* were also shown to genetically interact during scapular blade patterning. While *Alx3*$^{-/-}$ and *Alx4*$^{-/-}$ embryos exhibit relatively normal scapulae, compound mutant *Alx3*$^{-/-}$;*Alx4*$^{-/-}$ embryos display shortening of the scapular blade. This genetic interaction is complicated by the presence of the additional co-expressed Aristaless family member *Alx1*. For example, while *Alx1*$^{-/-}$ mutant scapulae have a shortened superior border, this defect is exacerbated in *Alx1*$^{-/-}$;*Alx3*$^{-/-}$ and *Alx1*$^{-/-}$;*Alx4*$^{-/-}$ double mutants (Fig. 1) (Kuijper, Feitsma, et al., 2005).

Third, given the extensive spatiotemporal overlap in their expression, genetic interactions between *Alx1*, *Alx3*, *Alx4*, and *Tbx15* were also examined. Indeed, analysis of double and triple mutants (i.e., *Alx1*$^{-/-}$;*Tbx15*$^{-/-}$, *Alx4*$^{-/-}$;*Tbx15*$^{-/-}$, *Alx1*$^{-/-}$;*Alx4*$^{-/-}$;*Tbx15*$^{+/-}$, and *Alx1*$^{-/-}$;*Alx4*$^{-/-}$;*Tbx15*$^{-/-}$) demonstrated that *Alx1* and *Alx4* (or *Alx3*) are essential for superior (anterior) blade formation. For example, when all three genes (*Alx1;Alx4;Tbx15*) are absent a single inferiorly positioned longitudinal skeletal element remains in the scapula, revealing their importance to superior blade formation (Fig. 1).

Fourth, Kuijper, Beverdam, et al. (2005) and Kuijper, Feitsma, et al. (2005) demonstrated that *Emx2*'s effects on scapular blade formation occur independently from, or in parallel with *Tbx15*, *Gli3*, *Alx1*, *Alx3*, and *Alx4*, since the expression of *Emx2* in mice with single and compound LOF for these genes remained relatively unaltered. These findings suggest that several different genetic pathways impinge on blade development, although their spatiotemporal order is unclear (Fig. 1).

The modular control of the scapula also extends to the patterning of the scapular spine and acromion. For example, PAX1 has been shown to aid the specification of progenitor cells of the anterior limb bud which have been fated to shoulder regions in the developing avian skeleton (Bowen, Hinchliffe, Horder, & Reeve, 1989; Saunders, 1948; Vargesson et al., 1997), and expression studies in mice have corroborated these results (Timmons, Wallin, Rigby, & Balling, 1994). Moreover, several mouse mutants that bear alterations in, or loss of, the *Pax1* locus exhibit scapular spine defects (Adham et al., 2005; Balling, 1994; Dietrich & Gruss, 1995; Timmons et al., 1994; Wilm, Dahl, Peters, Balling, & Imai, 1998); natural *Pax1* allelic variants (e.g., *undulated* (*un*) mice, Timmons et al., 1994); and targeted *Pax1* null mutants (Wilm et al., 1998) exhibit specific spine defects including spine absence. Interestingly, in approximately 60% of *Hoxa5*$^{-/-}$ mutant embryos, there is a residual hypoplastic or missing acromion process without defects in the scapular blade or limb (Aubin, Lemieux, Tremblay, Behringer, & Jeannotte, 1998; Aubin, Lemieux, Tremblay, Bérard, & Jeannotte, 1997). Given additional axial phenotypes that these mutants share with *Pax1* homozygous LOF, Aubin, Lemieux, Moreau, Lapointe, and Jeannotte (2002) investigated the interaction of both genes during formation of the scapular head and spine and found that compound *Hoxa5;Pax1 (un/un)* mutants display more severe spine and acromion defects than those seen in either single mutant. They additionally demonstrated that while PAX1 is a key regulator of scapular prechondrogenic mesenchymal formation, HOXA5 is necessary for the specification of cell lineages during chondrogenesis in a manner that also relies on the presence of SOX9. It is argued that *Hoxa5* may provide essential regional cues for acromion formation by ensuring proper initiation of *Pax1* expression.

In addition to blade loss, *Emx2*$^{-/-}$ embryos also lack the spine and the proximal acromion process, suggesting that it lies upstream or genetically interacts with other critical regulators of scapular spine formation. Indeed, both *Pax1* and *Emx2* are co-expressed in the anterior and proximal limb field, and while *Pax1* expression is not reduced in *Emx2*$^{-/-}$ scapula domains it is extended anterior-dorsally, suggesting that both genes may genetically interact in the lateral plate mesoderm to pattern the spine (Pellegrini et al., 2001).

Therefore, EMX2 may function in patterning the spine and acromion similarly to how it functions in the formation of the blade, wherein it affects progenitor cells maturation into skeletal tissues. In a similar vein, *Emx2* may also genetically interact with *Hoxc6* to pattern the scapular head, although the underlying mechanisms remain unclear (Pellegrini et al., 2001).

4.2 Genetic control of pelvis development

Relatively less is known about the developmental genetic control of pelvicogenesis, although the modularity observed for the scapula is also observed for the different subelements of the os coxae. Accordingly, most genes appear to divide the os coxae into superior (ilium) and inferior (pubis and ischium) domains, albeit some have more specific inferior effects, and others hierarchically regulate both domains. Additionally, much less is known about the molecular mechanisms that pre-pattern or specify cells to become pelvic tissues. In regard to the latter, similar to the scapula, *Hox* genes may play significant roles, but the underlying mechanisms are unclear. While *Hox*-encoded protein products have been shown to be major effectors of limb development and positioning, the latter of which influences pelvic/sacral location, misexpression or loss of *Hox* genes only subtly affects pelvis formation, typically in individual element alterations. Misexpression of *Hoxd12* in mice alters lateral plate derivatives, including the pelvis (Knezevic et al., 1997), whereas *Hoxc10* LOF mutants show minor abnormalities of the pelvic bones and sacrum (Hostikka, Gong, & Carpenter, 2009). Interestingly, when multiple 5′ *Hox* paralogs are experimentally deleted, element phenotypes become more evident. For example, Wellik and Capecchi (2003) demonstrated that the loss of *Hox10* family members cause reductions in lumbar and sacral vertebrae, with ribbed, thoracic vertebrae extending more posteriorly along the axial column. Additionally, the morphology of the pelvis itself is impacted with the pubis and ischium appearing more curved and dysmorphic. Similarly, loss of *Hox11* results in sacral vertebrae assuming a lumbar identity, and the perturbations of pelvic morphology are more severe, with the loss of the inferior portion of the pubis. These studies suggest that multiple 5′ HOX act redundantly to pattern the pelvis.

Regarding the superior os coxae, several genes expressed within somatopleure execute critical roles in ilium formation. *Emx2*$^{-/-}$ and *Pbx1*$^{-/-}$ embryos display specific reductions in the ilium and/or losses of the articulation between this element and the inferior os coxae subelements (Pellegrini et al., 2001; Selleri et al., 2001). These factors likely help specify progenitor cell commitment, as the expression of *Sox9*, a marker of mesenchymal condensation, is markedly reduced in single mutants; *Sox9*$^{-/-}$ embryos additionally display iliac reductions, indicating that it is downstream of PBX and EMX2 in during ilium formation (Bi et al., 2001; Capellini et al., 2010; Malashichev et al., 2008). However, there is also evidence of more complex genetic and biochemical interactions between PBX family members and EMX2, which affects scapula and pelvic girdle development in general. As *Pbx1*, *Emx2*, and *Sox9* are expressed in the lateral plate mesoderm and somatopleure, they may also have a major impact on the expression of *Fgf10*, which encodes a critical signaling molecule for pelvic morphogenesis. Indeed, *Fgf10*$^{-/-}$ mice exhibit a near-total lack of pelvic structures (Sekine et al., 1999), and *Fgf10* expression is reduced in *Pbx/Emx2* mutants (Capellini et al., 2011, 2010; Capellini and Selleri, unpublished results).

Ilium loss is also observed in *Pitx1*$^{-/-}$ embryos (Lanctôt, Moreau, Chamberland, Tremblay, & Drouin, 1999) indicating that this gene is essential not only in the limb but also in pelvic field formation. This finding is further corroborated by the discovery that loss of a distinct *Pitx1* pelvic enhancer in freshwater stickleback fish removes aspects of the pelvic fin, including the superior anchoring girdle spines (Chan et al., 2010). The enhancer additionally contains a number of HOX, PBX, and MEIS binding sites, factors important for limb and pelvic formation. Likewise, *Tbx4*, a downstream target of PITX1 in the limb field and somatopleure, regulates aspects of ilium formation—i.e., conditional *Tbx4*$^{-/-}$ embryos also exhibit reduced and dysmorphic iliac blades (Naiche & Papaioannou, 2007). These findings indicate a regulatory network hierarchically controlled by PBX, but cascading down through PITX1, TBX4, and likely other effectors (Fig. 2). Two potential targets are *Tbx15* and *Lmx1b*. *Tbx15* is expressed in the hindlimb field and somatopleure and LOF of *Tbx15* in mice results in ilium defects (Singh et al., 2005). Interestingly, patients with TBX15 coding mutations have Cousin syndrome, and exhibit a number of skeletal defects including alterations to iliac shape (Lausch et al., 2008). *Lmx1b* is expressed in the dorsal limb/somatopleure, and LOF mutations in *Lmx1b* result in iliac blade defects (Chen et al., 1998).

A number of genes have specific modular roles in inferior os coxae development. For example, Aristaless family genes *Alx1*, *Alx3*, and *Alx4* are expressed in the proximal hindlimb field and interactions between these family members result in pubic defects; compound *Alx1;Alx4* mutants display loss of the anterior pubic ramus, which is also evident in *Alx3;Alx4* double null mice (Kuijper, Feitsma, et al., 2005). The effects of their interaction are similar to their effects during anterior scapula blade formation, although they were not exacerbated by the concomitant loss of *Tbx15* (i.e., *Alx1;Alx3;Tbx15*, *Alx1;Alx4;Tbx15*, and *Alx3;Alx4;Tbx15* compound mutant mice lack additional inferior ox coxae defects). Similarly, despite their roles in inferior scapula blade patterning, neither *Gli3*$^{-/-}$ nor compound *Gli3;Tbx15* mutant mice (i.e., *Gli3*$^{+/-}$;*Tbx15*$^{-/-}$, *Gli3*$^{-/-}$; *Tbx15*$^{+/-}$, *Gli3*$^{-/-}$;*Tbx15*$^{-/-}$) display os coxae defects, indicating a departure in the genetic control of the scapula versus the pelvis (Kuijper, Beverdam, et al., 2005). Interestingly, null mutations in *Prx1* and *Prx2* (ten Berge, Brouwer, Korving, Martin, & Meijlink, 1998), other genes of the Aristaless family, result in the loss of the pubis. These findings suggest that complex interactions have similarly evolved among Aristaless genes during both pelvic and pectoral girdle development. In addition to these key factors, *Twist1* (Krawchuk et al., 2010), *Cv2* (Ikeya et al., 2006), *Msx1–2* (Lallemand, 2005), *Islet1* (Itou et al., 2012), *Fgfr1-IIIb* (De Moerlooze et al., 2000), *Pbx1–3* (Capellini et al., 2006, 2011), and members of the WNT pathway (Lee & Behringer, 2007) influence pubis or pubic symphysis formation.

To date, comparatively fewer genes have been described to influence ischium development, albeit none have specific modularized effects on the ischium, but rather more generally influence inferior os coxae development. These include *Islet1* (Itou et al., 2012), *Pax1* (Timmons et al., 1994), and *Pbx* family members (Capellini et al., 2011). The chief hierarchical regulators of inferior (i.e., pubis and ischium) pelvic development appears to be members of the *Islet* family specifically during early limb field development (Itou et al., 2012). *Islet1*$^{-/-}$ mice display substantially reduced and dysmorphic ischial and pubic elements. Future research addressing gene function specifically in specific girdle tissues during the patterning and chondrogenesis stages will shed light on their complex genetic control.

4.3 Hierarchical roles of Pbx genes in girdle patterning and morphogenesis

It has been established that one family of homeodomain transcription factor-encoding genes, the *Pbx* family, act as hierarchical regulators of development programs in the morphogenesis of many organs, including the face (Ferretti et al., 2011; Losa et al., 2018), heart (Stankunas et al., 2008), lung (McCulley et al., 2018), pancreas (Kim et al., 2002), spleen (Brendolan et al., 2005;

Koss et al., 2012), cerebral cortex (Golonzhka et al., 2015), spinal motor neurons (Hanley et al., 2016), limbs (Capellini et al., 2006), and both girdles (Capellini et al., 2011, 2010). In general, these genes are expressed in the somatopleure of the forelimb (*Pbx1–3*) in specific regions of the scapula and hindlimb (*Pbx1–2*) in the progenitor populations of the pelvis (Bi et al., 2001; Capellini et al., 2010; Pellegrini et al., 2001). Originally, it was noted that *Pbx1* homozygous LOF resulted in lethality of mutant mouse embryos at gestational day 15/16 due to widespread defects in multiple organs, as well as abnormal patterning of the axial and appendicular skeleton (Selleri et al., 2001). Conversely, mice with *Pbx2* and *Pbx3* single LOF displayed no overt skeletal or limb abnormalities (Rhee et al., 2004; Selleri et al., 2004). Further studies explored the complex relationships between multiple *Pbx* family members. Embryos with compound mutations of *Pbx1* and *Pbx2* exhibited proximal limb defects, as well as severe distal limb abnormalities (Capellini et al., 2006). Subsequently, *Pbx* genes were also shown to hierarchically regulate scapula and pelvis formation in mice (Capellini et al., 2011, 2010).

In the murine scapula, PBX1–3 transcription factors are necessary for normal patterning of the blade, with LOF of *Pbx1* either alone or in concert with *Pbx2* or *Pbx3* resulting in marked blade reductions, bifurcations, and foramina (Capellini et al., 2010) (Fig. 1). *Pbx* genes also affect progenitor cell patterning and mesenchymal condensation via interaction with *Emx2*. *Pbx; Emx2* compound mutant mice display marked defects in blade patterning, yet these are distinct from those present in compound *Pbx* mutants. These findings, together with studies in compound *Pbx1;Pbx2* and *Pbx1;Pbx3* mice, demonstrate hierarchical regulation of most known scapular patterning genes by *Pbx/Emx2* interactions (Fig. 1). This hierarchical control occurs at the transcriptional level, as we revealed (Capellini et al., 2010) that PBX1 and EMX2 also biochemically interact in proximal anterior forelimb bud tissues to physically bind to a regulatory enhancer upstream of *Alx1*, a critical gene in blade patterning.

These genetic interactions are not only confined to the blade but also lead to mis-patterning of the scapular spine, acromion, and head/neck. *Pbx1;Pbx2* and *Pbx1;Pbx3* double null mice exhibit head/neck duplications (Fig. 1), a likely consequence of the misregulation of a number of *Hox* genes in the early forelimb field, including a marked dorsal-ventral expansion of *Hoxc6*, whose misexpression in chick appears to result in scapula duplications (see above). In compound mutant mice, most scapular patterning genes analyzed that drive spine, acromion, and head/neck development are either downregulated and/or exhibit spatially altered expression patterns. Overall, these findings reveal the important hierarchical control of *Pbx* family members on scapular formation.

Pbx genes also act as hierarchical regulators in the pelvis, but unlike in the scapula only *Pbx1* and *Pbx2* are expressed in the somatopleure of the pelvic field (Capellini et al., 2011). In contrast to the observations in the scapula, loss of *Pbx1* in mouse results in the complete absence of the ilium (Capellini et al., 2011). When *Pbx2* and *Pbx3* are simultaneously lost, the pubis and ischium are additionally reduced or absent, suggesting that *Pbx* genes play key roles in the formation of pelvic mesenchymal condensations and/or the specification of pelvic progenitors (Fig. 2). These studies also uncovered that a number of genes involved in the development of individual pelvic elements are downregulated or spatially altered in compound *Pbx1;Pbx2* mice.

As in scapulogenesis, *Pbx* genes also hierarchically regulate *Emx2* and *Sox9* expression, particularly in the formation of superior pelvic structures. When *Emx2* or *Sox9* are lost, much of the scapula is also reduced or absent; however, in the pelvis, the resulting defects are restricted to the ilium rather than across the entire os coxae (Bi et al., 2001; Capellini et al., 2010; Malashichev et al., 2008). This suggests that the genetic networks comprising *Pbx*, *Emx2*, and *Sox9* differ between pelvic and pectoral girdles. Furthermore, *Pbx* and *Emx2* also cooperatively regulate developmental programs of the superior pelvis, albeit in compound double mutant mice for either *Pbx1;Pbx2* or *Pbx1;Emx2* select inferior pelvic structures are compromised as is the femoral-acetabular (hip) joint (Fig. 2). Indeed, there are a number of putative transcriptional targets across the genome of a PBX1-EMX2 complex, whose locations are near genes with functions enriched in hindlimb and pelvic morphogenesis (Capellini et al., 2011). The emerging understanding is that *Pbx* family members overall control superior pelvic development, with some influence on inferior pelvic formation, whereas *Islet* genes orchestrate inferior pelvic development (Fig. 2).

Recently, it has also been noted that PBX1 may act as a pioneer transcription factor, which is a type of transcription factor that can recognize and bind to their target sites even when chromatin is tightly wound rather than loose and easily accessible (Donaghey et al., 2018; Grebbin & Schulte, 2017). Few pioneer factors have been identified and thoroughly studied, but evidence from in vivo and in vitro studies suggests that PBX1 might belong to this category. The three hallmarks of pioneer factor function reside in: (a) binding to target sites in closed chromatin, (b) increasing DNA access for other proteins, and (c) being actively involved in cell fate specification or cellular (re)programming (Grebbin & Schulte, 2017; Iwafuchi-Doi & Zaret, 2016). With respect to (a), it has been shown that PBX1 binds to target sites in promoter/enhancer regions of genes such as *Dcx* and *Th* before they become transcriptionally active, suggesting that PBX1 target sites for these genes could be located in closed chromatin at the time of binding (Brill et al., 2008; Grebbin & Schulte, 2017). However, it remains to be unequivocally established whether PBX proteins have the intrinsic ability to act as pioneer factors.

4.4 Cis-regulatory evolution of the pelvis and scapula

The genes described above represent a small subset of those that drive girdle development in mice, humans, and other tetrapods. Currently, transcriptomic approaches are underway to identify the suite of genes that are involved in the patterning and chondrogenesis of both girdles. However, it will be equally important to investigate how the non-coding portion of the genome affects these developmental processes. It is now well understood that developmentally encoded traits such as skeletal shape are controlled at the level of gene regulation, rather than by variants in coding exons (King & Wilson, 1975; Spitz & Furlong, 2012). Several recent studies have demonstrated that regulatory elements such as enhancers, promoters, and repressors allow for fine-tuned, modular control of gene expression in time and space to mediate specific anatomical outcomes (Serfling, Jasin, & Schaffner, 1985; Spitz & Furlong, 2012). This research has identified distinct enhancer sequences that act as key musculoskeletal regulators, including specific long bone and joint regulatory sequences for genes such as *Gdf5* (Capellini et al., 2017), *Gdf6* (Mortlock, Guenther, & Kingsley, 2003), *Bmp5* (Guenther, Pantalena-Filho, & Kingsley, 2008), *Fgf8* (Marinić, Aktas, Ruf, & Spitz, 2013), and *Myf5* and *Myf6* (Summerbell et al., 2000; Vinagre et al., 2010). However, aside from the developmental work on the biochemical and genetic interaction of *Pbx1* and *Emx2* on an *Alx1* regulatory element during scapula patterning (Capellini et al., 2010) and studies on the cis-regulatory structure of the *Pitx1* locus (Chan et al., 2010; Sarro et al., 2018; Spielmann et al., 2012), few tissue-specific regulatory elements have been identified so far during scapula and pelvic development.

It has been hypothesized that evolutionary changes in regulatory sequences that control gene expression likely drive key phenotypic differences between organisms (King & Wilson, 1975). This is because most genes are highly conserved and largely pleiotropic, meaning that an alteration to their protein-coding sequences could result in negative effects in multiple locations in the body, or to a phenotype that significantly lowers the organism's fitness (Carroll, 2008). On the other hand, regulatory sequences such as enhancers and promoters often drive a gene's expression in only one subdomain of that gene's entire expression pattern. Thus, enhancers allow for modular control of targeted genes, which is spatially and temporally specific, and genetic variants in these sequences have more localized effects and are therefore more likely to serve as material for natural selection to operate upon. Recent research has therefore focused on uncovering the regulatory profile of chondrocytes that form the component parts of the scapula and pelvis in order to identify functional sequences that orchestrate the development of each girdle element during chondrogenesis. Ascertainment of the species-specific epigenomic profiles of each girdle structure will enhance not only our understanding of how the pelvic and scapula develop prenatally but also will provide a catalog of potentially significant evolutionary targets that selection may have acted upon to form the unique girdle morphologies of humans, chimps, and mice.

5. Conclusions and future directions

The research discussed here comprises various approaches spanning multiple disciplines—including functional anatomy, paleoanthropology, embryology, developmental biology, and genetics—in an attempt to understand the formation of two unique bones and the underlying mechanisms that shape and direct morphological variation during evolution and in disease. Comparative biomechanical, anatomical, and paleontological studies of vertebrate taxa including hominin species can provide us with an evolutionary context of why the scapula and pelvis have their distinctive morphologies across the animal kingdom and, in particular, in humans. They also shed light on the adaptive nature of the observed changes in morphology. Developmental studies have also shown that in many vertebrates, but especially modern humans, the core morphologies of each girdle are present well before birth, meaning that natural selection acted on developmental processes to generate adult functional phenotypes. Therefore, studying the underlying developmental pathways, as well as the genetic architectures that control them, will allow us to elucidate both how these structures form, and how natural selection has shaped the genomic control of key skeletal traits. In time, it will shed light on how pathological mutations generate abnormal girdle morphologies and birth defects.

Here we have summarized over a century's worth of studies that have built a foundation on which future research will answer these questions. Early and mid-20th century description of anatomical specimens of prenatal humans have proved invaluable to our understanding of embryonic and fetal girdle development. More recent experimental work in chicks and mice has allowed for a clearer understanding of the early patterning and tissue origins of both structures. Genetic studies in mice have also identified genes that are involved in pelvis and scapula patterning and morphogenesis and are active only in specific subdomains of each girdle, thus leading to a basic understanding of core networks. However, the genes known so far to control girdle formation are only a small subset of the hundreds that are involved in the prenatal development of each girdle, and very little is known about their regulatory interactions. Therefore, in order to fully understand patterning and morphogenesis of the pelvis and scapula in animals as disparate as chicks, mice, and humans further work is necessary to identify both the genes and regulatory elements active in the specific components of each girdle. This approach will ultimately determine which of these genetic elements have been under natural selection to produce the tremendous anatomical variation in the girdles across tetrapods.

Acknowledgments

Part of this research was funded by NIH grant R01HD043997 to L.S. and NSF grant BCS-1518596 to T.D.C. L.S. is personally grateful to Drs. Rolf Zeller, Javier Lopez Rios, Marco Osterwalder, Xin Sun, Deneen Wellik, and Susan Mackem for inspiring discussions. M.Y. and T.D.C. are additionally grateful to Neil Roach, Nathan Young, Heather Dingwall, and Evelyn Jagoda for support and scientific discussion. The authors would also like to thank Robert Aho for generating all artwork.

References

Adair, F. (1918). The ossification centers of the fetal pelvis. *Transactions of the American Gynecological Society*, *43*(89), 516–517.

Adham, I. M., Gille, M., Gamel, A. J., Reis, A., Dressel, R., Steding, G., et al. (2005). The scoliosis (sco) mouse: A new allele of Pax1. *Cytogenetic and Genome Research*, *111*, 16–26. https://doi.org/10.1159/000085665.

Agarwal, P., Wylie, J. N., Galceran, J., Arkhitko, O., Li, C., Deng, C., et al. (2003). Tbx5 is essential for forelimb bud initiation following patterning of the limb field in the mouse embryo. *Development*, *130*, 623–633. https://doi.org/10.1242/dev.00191.

Andersen, H. (1963). Histochemistry and development of the human shoulder and acromioclavicular joints with particular reference to the early development of the clavicle. *Acta Anatomica*, *55*, 124–165. https://doi.org/10.1159/000142467.

Aubin, J., Lemieux, M., Moreau, J., Lapointe, J., & Jeannotte, L. (2002). Cooperation of Hoxa5 and Pax1 genes during formation of the pectoral girdle. *Developmental Biology*, *244*, 96–113. https://doi.org/10.1006/dbio.2002.0596.

Aubin, J., Lemieux, M., Tremblay, M., Behringer, R. R., & Jeannotte, L. (1998). Transcriptional interferences at the Hoxa4/Hoxa5 locus: Importance of correct Hoxa5 expression for the proper specification of the axial skeleton. *Developmental Dynamics*, *212*, 141–156.

Aubin, J., Lemieux, M., Tremblay, M., Bérard, J., & Jeannotte, L. (1997). Early postnatal lethality in Hoxa-5 mutant mice is attributable to respiratory tract defects. *Developmental Biology*, *192*, 432–445. https://doi.org/10.1006/dbio.1997.8746.

Balling, R. (1994). The undulated mouse and the development of the vertebral column. Is there a human PAX-1 homologue? *Clinical Dysmorphology*, *3*(3), 185–191.

Bardeen, C. R. (1905). Studies of the development of the human skeleton. (A). The development of the lumbar, sacral and coccygeal vertebrae. (B). The curves and the proportionate regional lengths of the spinal column during the first three months of embryonic developnent. (C). T. *American Journal of Anatomy*, *4*, 265–302. https://doi.org/10.1002/aja.1000040302.

Bardeen, C. R., & Lewis, W. H. (1901). Development of the limbs, body-wall and back in man. *American Journal of Anatomy*, *1*, 1–35. https://doi.org/10.1002/aja.1000010102.

Basmajian, J. V., & Slonecker, C. E. (1989). *Grant's method of anatomy. A clinical problem-solving approach.* Baltimore, MD: Williams and Wilkins, pp. 1–615.

Bi, W., Huang, W., Whitworth, D. J., Deng, J. M., Zhang, Z., Behringer, R. R., et al. (2001). Haploinsufficiency of Sox9 results in defective cartilage primordia and premature skeletal mineralization. *Proceedings of the National Academy of Sciences of the United States of America*, *98*(12), 6698–6703. https://doi.org/10.1073/pnas.111092198.

Bilfeld, M. F., Dedouit, F., Sans, N., Rousseau, H., Rougé, D., & Telmon, N. (2013). Ontogeny of size and shape sexual dimorphism in the ilium: A multislice computed tomography study by geometric morphometry. *Journal of Forensic Sciences*, *58*, 303–310. https://doi.org/10.1111/1556-4029.12037.

Birkner, R. (1978). *Normal radiographic patterns and variances of the human skeleton—An X-ray atlas of adults and children.* Baltimore: Urban and Schwarzenberg, pp. 1–536.

Boisvert, C. A. (2005). The pelvic fin and girdle of Panderichthys and the origin of tetrapod locomotion. *Nature*, *438*, 1145–1147. https://doi.org/10.1038/nature04119.

Boucher, B. J. (1957). Sex differences in the foetal pelvis. *American Journal of Physical Anthropology*, *15*, 581–600. https://doi.org/10.1002/ajpa.1330150409.

Bowen, J., Hinchliffe, J. R., Horder, T. J., & Reeve, A. M. F. (1989). The fate map of the chick forelimb-bud and its bearing on hypothesized developmental control mechanisms. *Anatomy and Embryology*, *179*, 269–283. https://doi.org/10.1007/BF00326592.

Brendolan, A., Ferretti, E., Salsi, V., Moses, K., Quaggin, S., Blasi, F., et al. (2005). A Pbx1-dependent genetic and transcriptional network regulates spleen ontogeny. *Development (Cambridge, England)*, *132*, 3113–3126. https://doi.org/10.1242/dev.01884.

Brill, M. S., Snapyan, M., Wohlfrom, H., Ninkovic, J., Jawerka, M., Mastick, G. S., et al. (2008). A Dlx2- and Pax6-dependent transcriptional code for periglomerular neuron specification in the adult olfactory bulb. *Journal of Neuroscience*, *28*, 6439–6452. https://doi.org/10.1523/JNEUROSCI.0700-08.2008.

Burke, A. C., & Nowicki, J. L. (2003). A new view of patterning domains in the vertebrate mesoderm. *Developmental Cell*, *4*, 159–165. https://doi.org/10.1016/S1534-5807(03)00033-9.

Camp, J. D., & Cilley, E. I. L. (1931). Diagrammatic chart showing time of appearance of the various centres of ossification and period of union. *American Journal of Roentgenology*, *26*, 905.

Capellini, T. D., Chen, H., Cao, J., Doxey, A. C., Kiapour, A. M., Schoor, M., et al. (2017). Ancient selection for derived alleles at a GDF5 enhancer influencing human growth and osteoarthritis risk. *Nature Genetics*, *49*, 1202–1210. https://doi.org/10.1038/ng.3911.

Capellini, T. D., Di Giacomo, G., Salsi, V., Brendolan, A., Ferretti, E., Srivastava, D., et al. (2006). Pbx1/Pbx2 requirement for distal limb patterning is mediated by the hierarchical control of Hox gene spatial distribution and Shh expression. *Development (Cambridge, England)*, *133*(11), 2263–2273. https://doi.org/10.1242/dev.02395.

Capellini, T. D., Handschuh, K., Quintana, L., Ferretti, E., Di Giacomo, G., Fantini, S., et al. (2011). Control of pelvic girdle development by genes of the Pbx family and Emx2. *Developmental Dynamics: An Official Publication of the American Association of the Anatomists*, *240*(5), 1173–1189.

Capellini, T. D., Vaccari, G., Ferretti, E., Fantini, S., He, M., Pellegrini, M., et al. (2010). Scapula development is governed by genetic interactions of Pbx1 with its family members and with Emx2 via their cooperative control of Alx1. *Development (Cambridge, England)*, *137*(15), 2559–2569.

Carroll, S. B. (2008). Evo-devo and an expanding evolutionary synthesis: A genetic theory of morphological evolution. *Cell*, *134*(1), 25–36. https://doi.org/10.1016/j.cell.2008.06.030.

Chan, Y. F., Marks, M. E., Jones, F. C., Villarreal, G., Shapiro, M. D., Brady, S. D., et al. (2010). Adaptive evolution of pelvic reduction in sticklebacks by recurrent deletion of a pitxl enhancer. *Science*, *327*, 302–305. https://doi.org/10.1126/science.1182213.

Chen, H., Lun, Y., Ovchinnikov, D., Kokubo, H., Oberg, K. C., Pepicelli, C. V., et al. (1998). Limb and kidney defects in Lmx1b mutant mice suggest an involvement of LMX1B in human nail patella syndrome. *Nature Genetics*, *19*, 51–55. https://doi.org/10.1038/ng0598-51.

Cheng, L., Alvares, L. E., Ahmed, M. U., El-Hanfy, A. S., & Dietrich, S. (2004). The epaxial-hypaxial subdivision of the avian somite. *Developmental Biology*, *274*, 348–369. https://doi.org/10.1016/j.ydbio.2004.07.020.

Chiang, C., Litingtung, Y., Lee, E., Young, K. E., Corden, J. L., Westphal, H., et al. (1996). Cyclopia and defective axial patterning in mice lacking sonic hedgehog gene function. *Nature*, *383*, 407–413. https://doi.org/10.1038/383407a0.

Clack, J. A. (2009). The fin to limb transition: New data, interpretations, and hypotheses from paleontology and developmental biology. *Annual Review of Earth and Planetary Sciences*, *37*, 163–179. https://doi.org/10.1146/annurev.earth.36.031207.124146.

Coates, M. I. (2003). The evolution of paired fins. *Theory in Biosciences*, *122*, 266–287. https://doi.org/10.1078/1431-7613-00087.

Coates, M. I., Ruta, M., & Friedman, M. (2008). Ever since Owen: Changing perspectives on the early evolution of tetrapods. *Annual Review of Ecology, Evolution, and Systematics*, *39*, 571–592. https://doi.org/10.1146/annurev.ecolsys.38.091206.095546.

Coré, N., Bel, S., Gaunt, S. J., Aurrand-Lions, M., Pearce, J., Fisher, A., et al. (1997). Altered cellular proliferation and mesoderm patterning in polycomb-M33-deficient mice. *Development (Cambridge, England)*, *124*, 721–729. https://doi.org/10.1038/374724a0.

Crompton, R. H., Sellers, W. I., & Thorpe, S. K. S. (2010). Arboreality, terrestriality and bipedalism. *Philosophical Transactions of the Royal Society of London. Series B, Biological Sciences*, *365*(1556), 3301–3314. https://doi.org/10.1098/rstb.2010.0035.

Cunningham, C. A., & Black, S. M. (2009). Iliac cortical thickness in the neonate—the gradient effect. *Journal of Anatomy*, *215*(3), 364–370. https://doi.org/10.1111/j.1469-7580.2009.01112.x.

DeSilva, J. M., & Rosenberg, K. R. (2017). Anatomy, development, and function of the human pelvis. *Anatomical Record*, *300*, 628–632. https://doi.org/10.1002/ar.23561.

Dietrich, S., & Gruss, P. (1995). Undulated phenotypes suggest a role of Pax-1 for the development of vertebral and extravertebral structures. *Developmental Biology*, *167*, 529–548. https://doi.org/10.1006/dbio.1995.1047.

Diogo, R., & Ziermann, J. M. (2015). Muscles of chondrichthyan paired appendages: Comparison with osteichthyans, deconstruction of the fore-hindlimb serial homology dogma, and new insights on the evolution of the vertebrate neck. *Anatomical Record, 298*, 513–530. https://doi.org/10.1002/ar.23047.

Donaghey, J., Thakurela, S., Charlton, J., Chen, J. S., Smith, Z. D., Gu, H., et al. (2018). Genetic determinants and epigenetic effects of pioneer-factor occupancy. *Nature Genetics, 50*, 250–258. https://doi.org/10.1038/s41588-017-0034-3.

Dudley, A. T., Ros, M. A., & Tabin, C. J. (2002). A re-examination of proximodistal patterning during vertebrate limb development. *Nature, 418*, 539–544. https://doi.org/10.1038/nature00945.

Durland, J. L., Sferlazzo, M., Logan, M., & Burke, A. C. (2008). Visualizing the lateral somitic frontier in the Prx1Cre transgenic mouse. *Journal of Anatomy, 212*(5), 590–602. https://doi.org/10.1111/j.1469-7580.2008.00879.x.

Ehehalt, F., Wang, B., Christ, B., Patel, K., & Huang, R. (2004). Intrinsic cartilage-forming potential of dermomyotomal cells requires ectodermal signals for the development of the scapula blade. *Anatomy and Embryology, 208*, 431–437. https://doi.org/10.1007/s00429-004-0415-0.

Fawcett, E. (1910). Anatomical notes (on a fetal clavicle and scapula). *Journal of Anatomy and Physiology, 44*, 204–205.

Fazekas, I., & Kosa, F. (1978). *Forensic fetal osteology*. Akademiai Kiado, pp. 1–413.

Ferretti, E., Li, B., Zewdu, R., Wells, V., Hebert, J. M., Karner, C., et al. (2011). A conserved Pbx-Wnt-p63-Irf6 regulatory module controls face morphogenesis by promoting epithelial apoptosis. *Developmental Cell, 21*, 627–641. https://doi.org/10.1016/j.devcel.2011.08.005.

Flecker, H. (1942). Time of appearance and fusion of ossification centres as observed by roentgenographic methods. *American Journal of Roentgenology, 47*, 97–159.

Francis, C. C. (1940). The appearance of centers of ossification from 6 to 15 years. *American Journal of Physical Anthropology, 27*, 127–138. https://doi.org/10.1002/ajpa.1330270132.

Francis, C. C. (1951). Appearance of centres os ossification in human pelvis before birth. *American Journal of Roentgenology, 65*, 778–783.

Freedman, E. (1934). Os acetabuli. *American Journal of Roentgenology, 1934*, 492–495.

Gilbert, S. F. (2000). *Developmental biology*. Sinauer Associates. Retrieved from: *https://www.ncbi.nlm.nih.gov/books/NBK9983/*.

Girdany, B. R., & Golden, R. (1952). Centres of ossification of the skeleton. *American Journal of Roentgenology, 68*, 922–924.

Golonzhka, O., Nord, A., Tang, P. L. F., Lindtner, S., Ypsilanti, A. R., Ferretti, E., et al. (2015). Pbx regulates patterning of the cerebral cortex in progenitors and postmitotic neurons. *Neuron, 88*, 1192–1207. https://doi.org/10.1016/j.neuron.2015.10.045.

Grebbin, B. M., & Schulte, D. (2017). PBX1 as pioneer factor: A case still open. *Frontiers in Cell and Development Biology, 5*, 9. https://doi.org/10.3389/fcell.2017.00009.

Guenther, C., Pantalena-Filho, L., & Kingsley, D. M. (2008). Shaping skeletal growth by modular regulatory elements in the Bmp5 gene. *PLoS Genetics, 4*(12), 1–15. https://doi.org/10.1371/journal.pgen.1000308.

Hanley, O., Zewdu, R., Cohen, L. J., Jung, H., Lacombe, J., Philippidou, P., et al. (2016). Parallel Pbx-dependent pathways govern the coalescence and fate of motor columns. *Neuron, 91*, 1005–1020. https://doi.org/10.1016/j.neuron.2016.07.043.

Harrison, T. (1958a). An experimental study of pelvic growth in the rat. *Journal of Anatomy, 92*, 483.

Harrison, T. (1958b). The growth of the pelvis in the rat—A mensural and morphology study. *Journal of Anatomy, 92*, 236.

Harrison, T. J. (1961). The influence of the femoral head on pelvic growth and acetabular form in the rat. *Journal of Anatomy*, *95*, 12–24.

Hashimoto, R., Kihara, I., & Otani, H. (2002). Perinatal development of the rat hip joint with restrained fetal movement. *Congenital Anomalies*, *42*, 135–142. https://doi.org/10.1111/j.1741-4520.2002.tb00863.x.

Hill, T. P., Taketo, M. M., Birchmeier, W., & Hartmann, C. (2006). Multiple roles of mesenchymal -catenin during murine limb patterning. *Development (Cambridge, England)*, *133*(7), 1219–1229. https://doi.org/10.1242/dev.02298.

Hita-Contreras, F., Sánchez-Montesinos, I., Martínez-Amat, A., Cruz-Díaz, D., Barranco, R. J., & Roda, O. (2018). Development of the human shoulder joint during the embryonic and early fetal stages: Anatomical considerations for clinical practice. *Journal of Anatomy*, *232*(3), 422–430. https://doi.org/10.1111/joa.12753.

Hodges, P. C. (1933). An epiphyseal chart. *American Journal of Roentgenology*, *30*, 809–810.

Holcomb, S. M., & Konigsberg, L. W. (1995). Statistical study of sexual dimorphism in the human fetal sciatic notch. *American Journal of Physical Anthropology*, *97*, 113–125. https://doi.org/10.1002/ajpa.1330970204.

Hostikka, S. L., Gong, J., & Carpenter, E. M. (2009). Axial and appendicular skeletal transformations, ligament alterations, and motor neuron loss in Hoxc10 mutants. *International Journal of Biological Sciences*, *5*, 397–410.

Huang, R., Christ, B., & Patel, K. (2006). Regulation of scapula development. *Anatomy and Embryology*, *211*, 65–71. https://doi.org/10.1007/s00429-006-0126-9.

Huang, R., Zhi, Q., Patel, K., Wilting, J., & Christ, B. (2000). Dual origin and segmental organisation of the avian scapula. *Development (Cambridge, England)*, *127*(17), 3789–3794.

Hui, C. C., & Joyner, A. L. (1993). A mouse model of greig cephalo–polysyndactyly syndrome: The extra–toesJ mutation contains an intragenic deletion of the Gli3 gene. *Nature Genetics*, *3*, 241–246. https://doi.org/10.1038/ng0393-241.

Iguchi, T., Irisawa, S., Fukazawa, Y., Uesugi, Y., & Takasugi, N. (1989). Morphometric analysis of the development of sexual dimorphism of the mouse pelvis. *The Anatomical Record*, *224*(4), 490–494.

Ikeya, M., Kawada, M., Kiyonari, H., Sasai, N., Nakao, K., Furuta, Y., et al. (2006). Essential pro-Bmp roles of crossveinless 2 in mouse organogenesis. *Development (Cambridge, England)*, *133*, 4463–4473. https://doi.org/10.1242/dev.02647.

Itou, J., Kawakami, H., Quach, T., Osterwalder, M., Evans, S. M., Zeller, R., et al. (2012). Islet1 regulates establishment of the posterior hindlimb field upstream of the Hand2-Shh morphoregulatory gene network in mouse embryos. *Development (Cambridge, England)*, *139*(9), 1620–1629. https://doi.org/10.1242/dev.073056.

Iwafuchi-Doi, M., & Zaret, K. S. (2016). Cell fate control by pioneer transcription factors. *Development (Cambridge, England)*, *143*, 1833–1837. https://doi.org/10.1242/dev.133900.

Johanson, Z., & Trinajstic, K. (2014). Fossilized ontogenies: The contribution of placoderm ontogeny to our understanding of the evolution of early gnathostomes. *Palaeontology*, *57*, 505–516. https://doi.org/10.1111/pala.12093.

Johnston, F. E. (1961). Sequence of epiphyseal union in a prehistoric Kentucky population from Indian Knoll. *Human Biology*, *33*, 66–81. [An International Record of Research].

Karsenty, G., & Wagner, E. F. (2002). Reaching a genetic and molecular understanding of skeletal development. *Developmental Cell*, *2*, 389–406. https://doi.org/10.1016/S1534-5807(02)00157-0.

Kim, S. K., Selleri, L., Lee, J. S., Zhang, A. Y., Gu, X., Jacobs, Y., et al. (2002). Pbx1 inactivation disrupts pancreas development and in Ipf1-deficient mice promotes diabetes mellitus. *Nature Genetics*, *30*, 430–435. https://doi.org/10.1038/ng860.

King, M. C., & Wilson, A. C. (1975). Evolution at two levels in humans and chimpanzees. *Science (New York, N.Y.)*, *188*(4184), 107–116. https://doi.org/10.1126/science.1090005.

Knezevic, V., De Santo, R., Schughart, K., Huffstadt, U., Chiang, C., Mahon, K. A., et al. (1997). Hoxd-12 differentially affects preaxial and postaxial chondrogenic branches in the limb and regulates sonic hedgehog in a positive feedback loop. *Development (Cambridge, England)*, *124*, 4523–4536.

Koss, M., Bolze, A., Brendolan, A., Saggese, M., Capellini, T. D., Bojilova, E., et al. (2012). Congenital asplenia in mice and humans with mutations in a Pbx/Nkx2-5/p15 module. *Developmental Cell*, *22*, 913–926. https://doi.org/10.1016/j.devcel.2012.02.009.

Krawchuk, D., Weiner, S. J., Chen, Y. T., Lu, B. C., Costantini, F., Behringer, R. R., et al. (2010). Twist1 activity thresholds define multiple functions in limb development. *Developmental Biology*, *347*(1), 133–146. https://doi.org/10.1016/j.ydbio.2010.08.015.

Kuijper, S., Beverdam, A., Kroon, C., Brouwer, A., Candille, S., Barsh, G., et al. (2005). Genetics of shoulder girdle formation: Roles of Tbx15 and aristaless-like genes. *Development (Cambridge, England)*, *132*(7), 1601–1610. https://doi.org/10.1242/dev.01735.

Kuijper, S., Feitsma, H., Sheth, R., Korving, J., Reijnen, M., & Meijlink, F. (2005). Function and regulation of Alx4 in limb development: Complex genetic interactions with Gli3 and Shh. *Developmental Biology*, *285*, 533–544. https://doi.org/10.1016/j.ydbio.2005.06.017.

Lallemand, Y. (2005). Analysis of Msx1; Msx2 double mutants reveals multiple roles for Msx genes in limb development. *Development (Cambridge, England)*, *132*, 3003–3014. https://doi.org/10.1242/dev.01877.

Lanctôt, C., Moreau, A., Chamberland, M., Tremblay, M. L., & Drouin, J. (1999). Hindlimb patterning and mandible development require the Ptx1 gene. *Development (Cambridge, England)*, *126*(9), 1805–1810. http://www.ncbi.nlm.nih.gov/pubmed/10101115.

Laurenson, R. D. (1964a). The primary ossification of the human ilium. *The Anatomical Record*, *148*, 209–217. https://doi.org/10.1002/ar.1091480211.

Laurenson, R. D. (1964b). The chondrification of the human ilium. *The Anatomical Record*, *148*, 197–202. http://www.ncbi.nlm.nih.gov/pubmed/14123503.

Lausch, E., Hermanns, P., Farin, H. F., Alanay, Y., Unger, S., Nikkel, S., et al. (2008). TBX15 mutations cause craniofacial dysmorphism, hypoplasia of scapula and pelvis, and short stature in Cousin syndrome. *American Journal of Human Genetics*, *83*(5), 649–655. https://doi.org/10.1016/j.ajhg.2008.10.011.

LaVelle, M. (1995). Natural selection and developmental sexual variation in the human pelvis. *American Journal of Physical Anthropology*, *98*, 59–72. https://doi.org/10.1002/ajpa.1330980106.

Lee, H. H., & Behringer, R. R. (2007). Conditional expression of Wnt4 during chondrogenesis leads to dwarfism in mice. *PLoS One*, *2*, e450. https://doi.org/10.1371/journal.pone.0000450.

Leong, A. (2006). Sexual dimorphism of the pelvic architecture: A struggling response to destructive and parsimonious forces by natural & mate selection. *McGill Journal of Medicine*, *9*, 61–66.

Logan, M., Martin, J. F., Nagy, A., Lobe, C., Olson, E. N., & Tabin, C. J. (2002). Expression of Cre recombinase in the developing mouse limb bud driven by a Prxl enhancer. *Genesis*, *33*, 77–80. https://doi.org/10.1002/gene.10092.

Lohnes, D., Mark, M., Mendelsohn, C., Dolle, P., Dierich, A., Gorry, P., et al. (1994). Function of the retinoic acid receptors (RARs) during development (I). Craniofacial and skeletal abnormalities in RAR double mutants. *Development (Cambridge, England)*, *120*, 2723–2748. https://doi.org/10.1242/dev.00463.

Long, F., & Ornitz, D. M. (2013). Development of the endochondral skeleton. *Cold Spring Harbor Perspectives in Biology*, *5*, a008334 https://doi.org/10.1101/cshperspect.a008334.

Losa, M., Risolino, M., Li, B., Hart, J., Quintana, L., Grishina, I., et al. (2018). Face morphogenesis is promoted by Pbx-dependent EMT via regulation of *Snail1* during frontonasal prominence fusion. *Development (Cambridge, England), 145* https://doi.org/10.1242/dev.157628.

Lovejoy, C. O. (1988). Evolution of human walking. *Scientific American, 259*, 118–125. https://doi.org/10.1038/scientificamerican1188-118.

Malashichev, Y., Borkhvardt, V., Christ, B., & Scaal, M. (2005). Differential regulation of avian pelvic girdle development by the limb field ectoderm. *Anatomy and Embryology, 210*(3), 187–197. https://doi.org/10.1007/s00429-005-0014-8.

Malashichev, Y., Christ, B., & Pröls, F. (2008). Avian pelvis originates from lateral plate mesoderm and its development requires signals from both ectoderm and paraxial mesoderm. *Cell and Tissue Research, 331*(3), 595–604. https://doi.org/10.1007/s00441-007-0556-6.

Mall, F. P. (1906). On ossification centers in human embryos less than one hundred days old. *American Journal of Anatomy, 5*, 433–458. https://doi.org/10.1002/aja.1000050403.

Marinić, M., Aktas, T., Ruf, S., & Spitz, F. (2013). An integrated holo-enhancer unit defines tissue and gene specificity of the Fgf8 regulatory landscape. *Developmental Cell, 24*(5), 530–542. https://doi.org/10.1016/j.devcel.2013.01.025.

Matsuoka, T., Ahlberg, P. E., Kessaris, N., Iannarelli, R., Dennehy, U., Richardson, W. D., et al. (2005). Neural crest origins of the neck and shoulder. *Nature, 436*, 347–355. https://doi.org/10.1038/nature03837.

McCulley, D. J., Wienhold, M. D., Hines, E. A., Hacker, T. A., Rogers, A., Pewowaruk, R. J., et al. (2018). PBX transcription factors drive pulmonary vascular adaptation to birth. *The Journal of Clinical Investigation, 128*, 655–667. https://doi.org/10.1172/JCI93395.

McIntyre, D. C., Rakshit, S., Yallowitz, A. R., Loken, L., Jeannotte, L., Capecchi, M. R., et al. (2007). Hox patterning of the vertebrate rib cage. *Development (Cambridge, England), 134*, 2981–2989. https://doi.org/10.1242/dev.007567.

Miles, A. E. W., & Bulman, J. S. (1994). Growth curves of immature bones from a Scottish island population of sixteenth to mid-nineteenth century: Limb-bone diaphyses and some bones of the hand and foot. *International Journal of Osteoarchaeology, 4*, 121–136. https://doi.org/10.1002/oa.1390040205.

Min, H., Danilenko, D. M., Scully, S. A., Bolon, B., Ring, B. D., Tarpley, J. E., et al. (1998). Fgf-10 is required for both limb and lung development and exhibits striking functional similarity to Drosophila branchless. *Genes & Development, 12*, 3156–3161. https://doi.org/10.1101/gad.12.20.3156.

Moeller, C. (2003). Carboxypeptidase Z (CPZ) modulates Wnt signaling and regulates the development of skeletal elements in the chicken. *Development (Cambridge, England), 130*, 5103–5111. https://doi.org/10.1242/dev.00686.

De Moerlooze, L., Spencer-Dene, B., Revest, J. M., Hajihosseini, M., Rosewell, I., & Dickson, C. (2000). An important role for the IIIb isoform of fibroblast growth factor receptor 2 (FGFR2) in mesenchymal-epithelial signalling during mouse organogenesis. *Development (Cambridge, England), 127*, 483–492. https://doi.org/10.1016/0092-8674(91)90616-7.

Mokrane, F.-Z., Dedouit, F., Gellée, S., Sans, N., Rousseau, H., Rougé, D., et al. (2013). Sexual dimorphism of the fetal ilium: A 3D geometric morphometric approach with multislice computed tomography. *Journal of Forensic Sciences, 58*, 851–858. https://doi.org/10.1111/1556-4029.12118.

Monkhouse, W. S. (1996). Last's anatomy, regional and applied. *Journal of Anatomy, 1*(1), 1–286. https://doi.org/10.1002/ejoc.201200111.

Mortlock, D. P., Guenther, C., & Kingsley, D. M. (2003). A general approach for identifying distant regulatory elements applied to the Gdf6 gene. *Genome Research, 13*(9), 2069–2081. https://doi.org/10.1101/gr.1306003.

Naiche, L. A., & Papaioannou, V. E. (2007). Tbx4 is not required for hindlimb identity or post-bud hindlimb outgrowth. *Development (Cambridge, England)*, *134*, 93–103. https://doi.org/10.1242/dev.02712.

Ng, J. K., Kawakami, Y., Büscher, D., Raya, A., Itoh, T., Koth, C. M., et al. (2002). The limb identity gene Tbx5 promotes limb initiation by interacting with Wnt2b and Fgf10. *Development (Cambridge, England)*, *129*, 5161–5170.

Nowlan, N. C., Chandaria, V., & Sharpe, J. (2014). Immobilized chicks as a model system for early-onset developmental dysplasia of the hip. *Journal of Orthopaedic Research*, *32*, 777–785. https://doi.org/10.1002/jor.22606.

O'Rahilly, R., & Gardner, E. (1972). The initial appearance of ossification in staged human embryos. *American Journal of Anatomy*, *134*, 291–307. https://doi.org/10.1002/aja.1001340303.

O'Rahilly, R., & Gardner, E. (1975). The timing and sequence of events in the development of the limbs in the human embryo. *Anatomy and Embryology*, *148*, 1–23. https://doi.org/10.1007/BF00315559.

Ogden, J. a., & Phillips, S. B. (1983). Radiology of postnatal skeletal development. VII. The scapula. *Skeletal Radiology*, *10*, 209–220. https://doi.org/10.1007/BF00357893.

Okumura, M., Ishikawa, A., Aoyama, T., Yamada, S., Uwabe, C., Imai, H., et al. (2017). Cartilage formation in the pelvic skeleton during the embryonic and early-fetal period. *PLoS One*, *12*(4), e0173852. https://doi.org/10.1371/journal.pone.0173852.

Oliver, G., De Robertis, E. M., Wolpert, L., & Tickle, C. (1990). Expression of a homeobox gene in the chick wing bud following application of retinoic acid and grafts of polarizing region tissue. *The EMBO Journal*, *9*, 3093–3099.

Pellegrini, M., Pantano, S., Fumi, M. P., Lucchini, F., & Forabosco, A. (2001). Agenesis of the scapula in Emx2 homozygous mutants. *Developmental Biology*, *232*(1), 149–156. https://doi.org/10.1006/dbio.2001.0159.

Pomikal, C., & Streicher, J. (2010). 4D-analysis of early pelvic girdle development in the mouse (Mus musculus). *Journal of Morphology*, *271*(1), 116–126. https://doi.org/10.1002/jmor.10785.

Ponseti, I. V. (1978). Growth and development of the acetabulum in the normal child. Anatomical, histological, and roentgenographic studies. *The Journal of Bone & Joint Surgery*, *60*, 575–585.

Pröls, F., Ehehalt, F., Rodriguez-Niedenführ, M., He, L., Huang, R., & Christ, B. (2004). The role of Emx2 during scapula formation. *Developmental Biology*, *275*, 315–324. https://doi.org/10.1016/j.ydbio.2004.08.003.

Purves, D., Augustine, G. J., Fitzpatrick, D., Hall, W. C., Lamantia, A.-S., Mcnamara, J. O., et al. (2004). *Neuroscience*. Sunderland. *https://doi.org/978-0878937257*.

Rallis, C. (2003). Tbx5 is required for forelimb bud formation and continued outgrowth. *Development (Cambridge, England)*, *130*, 2741–2751. https://doi.org/10.1242/dev.00473.

Rancourt, D. E., Tsuzuki, T., & Capecchi, M. R. (1995). Genetic interaction between hoxb-5 and hoxb-6 is revealed by nonallelic noncomplementation. *Genes & Development*, *9*, 108–122. https://doi.org/10.1101/gad.9.1.108.

Reynolds, E. L. (1945). The bony pelvic girdle in early infancy. A roentgenometric study. *American Journal of Physical Anthropology*, *3*, 321–354. https://doi.org/10.1002/ajpa.1330030410.

Rhee, J. W., Arata, A., Selleri, L., Jacobs, Y., Arata, S., Onimaru, H., et al. (2004). Pbx3 deficiency results in central hypoventilation. *American Journal of Pathology*, *165*, 1343–1350. https://doi.org/10.1016/S0002-9440(10)63392-5.

Roach, N. T., Venkadesan, M., Rainbow, M. J., & Lieberman, D. E. (2013). Elastic energy storage in the shoulder and the evolution of high-speed throwing in Homo. *Nature*, *498*, 483–486. https://doi.org/10.1038/nature12267.

Rodríguez-Niedenführ, M., Dathe, V., Jacob, H. J., Pröls, F., & Christ, B. (2003). Spatial and temporal pattern of Wnt-6 expression during chick development. *Anatomy and Embryology*, *206*(6), 447–451. https://doi.org/10.1007/s00429-003-0322-9.

Romer, A. S., & Parsons, T. S. (1986). *The vertebrate body*. New York: Saunders.

Rong, P. M., Teillet, M. A., Ziller, C., & Le Douarin, N. M. (1992). The neural tube/notochord complex is necessary for vertebral but not limb and body wall striated muscle differentiation. *Development (Cambridge, England)*, *115*(3), 657–672. Retrieved from: http://www.ncbi.nlm.nih.gov/pubmed/1425345.

Sarro, R., Kocher, A. A., Emera, D., Uebbing, S., Dutrow, E. V., Weatherbee, S. D., et al. (2018). Disrupting the three-dimensional regulatory topology of the *Pitx1* locus results in overtly normal development. *Development (Cambridge, England)*, *145*, 1–9. https://doi.org/10.1242/dev.158550.

Saunders, J. W. (1948). The proximo-distal sequence of origin of the parts of the chick wing and the role of the ectoderm. *Journal of Experimental Zoology*, *108*(3), 363–403. https://doi.org/10.1002/jez.1401080304.

Scheuer, L., Black, S., & Christie, A. (2000). *Developmental juvenile osteology*. New York, NY: Academic Press, pp. 1–587.

Sears, K. E., Capellini, T. D., & Diogo, R. (2015). On the serial homology of the pectoral and pelvic girdles of tetrapods. *Evolution*, *69*, 2543–2555. https://doi.org/10.1111/evo.12773.

Sekine, K., Ohuchi, H., Fujiwara, M., Yamasaki, M., Yoshizawa, T., Sato, T., et al. (1999). Fgf10 is essential for limb and lung formation. *Nature Genetics*, *21*, 138–141. https://doi.org/10.1038/5096.

Selleri, L., Depew, M. J., Jacobs, Y., Chanda, S. K., Tsang, K. Y., Cheah, K. S., et al. (2001). Requirement for Pbx1 in skeletal patterning and programming chondrocyte proliferation and differentiation. *Development (Cambridge, England)*, *128*(18), 3543–3557.

Selleri, L., DiMartino, J., van Deursen, J., Brendolan, A., Sanyal, M., Boon, E., et al. (2004). The TALE homeodomain protein Pbx2 is not essential for development and long-term survival. *Molecular and Cellular Biology*, *24*, 5324–5331. https://doi.org/10.1128/MCB.24.12.5324-5331.2004.

Serfling, E., Jasin, M., & Schaffner, W. (1985). Enhancers and eukaryotic gene transcription. *Trends in Genetics*, *1*, 224–230. https://doi.org/10.1016/0168-9525(85)90088-5.

Shearman, R. M., Tulenko, F. J., & Burke, A. C. (2011). 3D reconstructions of quail-chick chimeras provide a new fate map of the avian scapula. *Developmental Biology*, *355*, 1–11. https://doi.org/10.1016/j.ydbio.2011.03.032.

Singh, M. K., Petry, M., Haenig, B., Lescher, B., Leitges, M., & Kispert, A. (2005). The T-box transcription factor Tbx15 is required for skeletal development. *Mechanisms of Development*, *122*, 133–144. https://doi.org/10.1016/j.mod.2004.10.011.

Sockol, M. D., Raichlen, D. a., & Pontzer, H. (2007). Chimpanzee locomotor energetics and the origin of human bipedalism. *Proceedings of the National Academy of Sciences of the United States of America*, *104*(30), 12265–12269. https://doi.org/10.1073/pnas.0703267104.

Spielmann, M., Brancati, F., Krawitz, P. M., Robinson, P. N., Ibrahim, D. M., Franke, M., et al. (2012). Homeotic arm-to-leg transformation associated with genomic rearrangements at the PITX1 locus. *American Journal of Human Genetics*, *91*, 629–635. https://doi.org/10.1016/j.ajhg.2012.08.014.

Spitz, F., & Furlong, E. E. M. (2012). Transcription factors: From enhancer binding to developmental control. *Nature Reviews. Genetics*, *13*(9), 613–626. https://doi.org/10.1038/nrg3207.

Stankunas, K., Shang, C., Twu, K. Y., Kao, S.-C., Jenkins, N. A., Copeland, N. G., et al. (2008). Pbx/Meis deficiencies demonstrate multigenetic origins of congenital heart disease. *Circulation Research*, *103*(7), 702–709. https://doi.org/10.1161/CIRCRESAHA.108.175489.

Stevenson, P. H. (1924). Age order of epiphyseal union in man. *American Journal of Physical Anthropology*, 7, 53–93. https://doi.org/10.1002/ajpa.1330070115.
Summerbell, D., Ashby, P. R., Coutelle, O., Cox, D., Yee, S., & Rigby, P. W. (2000). The expression of Myf5 in the developing mouse embryo is controlled by discrete and dispersed enhancers specific for particular populations of skeletal muscle precursors. *Development (Cambridge, England)*, *127*(17), 3745–3757. http://www.ncbi.nlm.nih.gov/pubmed/10934019.
Swindler, D. R., & Wood, C. D. (1973). *An atlas of primate gross anatomy: Baboon, chimpanzee, and man*. R.E. Krieger Publishing Company.
Teillet, M.-A., Lapointe, F., & Le Douarin, N. M. (1998). The relationships between notochord and floor plate in vertebrate development revisited. *Proceedings of the National Academy of Sciences*, *95*, 11733–11738. https://doi.org/10.1073/pnas.95.20.11733.
ten Berge, D., Brouwer, A., Korving, J., Martin, J. F., & Meijlink, F. (1998). Prx1 and Prx2 in skeletogenesis: Roles in the craniofacial region, inner ear and limbs. *Development (Cambridge, England)*, *125*(19), 3831–3842. http://www.ncbi.nlm.nih.gov/pubmed/9729491.
Timmons, P. M., Wallin, J., Rigby, P. W., & Balling, R. (1994). Expression and function of Pax 1 during development of the pectoral girdle. *Development (Cambridge, England)*, *120*(10), 2773–2785. http://www.ncbi.nlm.nih.gov/pubmed/7607069.
Trinajstic, K., Boisvert, C., Long, J., Maksimenko, A., & Johanson, Z. (2015). Pelvic and reproductive structures in placoderms (stem gnathostomes). *Biological Reviews*, *90*, 467–501. https://doi.org/10.1111/brv.12118.
Valasek, P., Theis, S., Krejci, E., Grim, M., Maina, F., Shwartz, Y., et al. (2010). Somitic origin of the medial border of the mammalian scapula and its homology to the avian scapula blade. *Journal of Anatomy*, *216*(4), 482–488. https://doi.org/10.1111/j.1469-7580.2009.01200.x.
Vargesson, N., Clarke, J. D., Vincent, K., Coles, C., Wolpert, L., & Tickle, C. (1997). Cell fate in the chick limb bud and relationship to gene expression. *Development (Cambridge, England)*, *124*(10), 1909–1918.
Verbruggen, S. W., & Nowlan, N. C. (2017). Ontogeny of the human pelvis. *Anatomical Record*, *300*, 643–652. https://doi.org/10.1002/ar.23541.
Vinagre, T., Moncaut, N., Carapuço, M., Nóvoa, A., Bom, J., & Mallo, M. (2010). Evidence for a myotomal Hox/Myf cascade governing nonautonomous control of rib specification within global vertebral domains. *Developmental Cell*, *18*, 655–661. https://doi.org/10.1016/j.devcel.2010.02.011.
Wang, B., He, L., Ehehalt, F., Geetha-Loganathan, P., Nimmagadda, S., Christ, B., et al. (2005). The formation of the avian scapula blade takes place in the hypaxial domain of the somites and requires somatopleure-derived BMP signals. *Developmental Biology*, *287*, 11–18. https://doi.org/10.1016/j.ydbio.2005.08.016.
Weaver, D. S. (1980). Sex differences in the Ilia of a known sex and age sample of fetal and infant skeletons. *American Journal of Physical Anthropology*, *52*, 191–195. https://doi.org/10.1002/ajpa.1330520205.
Wellik, D., & Capecchi, M. (2003). *Hox10* and *Hox11* genes are required to globally pattern the mammalian skeleton. *Science*, *301*, 363–367. https://doi.org/10.1126/science.1085672.
Williams, M. S. (2003). Developmental anomalies of the scapula—The "omo"st forgotten bone. *American Journal of Medical Genetics*, *120A*, 583–587.
Wilm, B., Dahl, E., Peters, H., Balling, R., & Imai, K. (1998). Targeted disruption of Pax1 defines its null phenotype and proves haploinsufficiency. *Proceedings of the National Academy of Sciences of the United States of America*, *95*, 8692–8697. https://doi.org/10.1073/pnas.95.15.8692.

Wilson, L. A. B., Ives, R., Cardoso, H. F. V., & Humphrey, L. T. (2015). Shape, size, and maturity trajectories of the human ilium. *American Journal of Physical Anthropology, 156,* 19–34. https://doi.org/10.1002/ajpa.22625.

Xu, B., & Wellik, D. M. (2011). Axial Hox9 activity establishes the posterior field in the developing forelimb. *Proceedings of the National Academy of Sciences of the United States of America, 108,* 4888–4891. https://doi.org/10.1073/pnas.1018161108.

Young, N. (2004). Modularity and integration in the hominoid scapula. *Journal of Experimental Zoology, 302,* 226–240. https://doi.org/10.1002/jez.b.21003.

Young, N. M., & Capellini, T. D. (2015). Out on a limb: Development and the evolution of the human appendicular skeleton. In *Developmental approaches to human evolution* (pp. 101–137). New York, NY: John Wiley & Sons, Inc. https://doi.org/10.1002/9781118524756.ch6.

Zhu, M., Yu, X., Ahlberg, P. E., Choo, B., Lu, J., Qiao, T., et al. (2013). A Silurian placoderm with osteichthyan-like marginal jaw bones. *Nature, 502,* 188–193. https://doi.org/10.1038/nature12617.

Zhu, M., Yu, X., Choo, B., Wang, J., & Jia, L. (2012). An antiarch placoderm shows that pelvic girdles arose at the root of jawed vertebrates. *Biology Letters, 8,* 453–456. https://doi.org/10.1098/rsbl.2011.1033.

CHAPTER TEN

Eye organogenesis: A hierarchical view of ocular development

Joel B. Miesfeld, Nadean L. Brown*
Department of Cell Biology & Human Anatomy, University of California Davis School of Medicine, Davis, CA, United States
*Corresponding author: e-mail address: nlbrown@ucdavis.edu

Contents

Abstract

This chapter provides an overview of the early developmental origins of six ocular tissues: the cornea, lens, ciliary body, iris, neural retina, and retina pigment epithelium. Many of these tissue types are concurrently specified and undergo a complex set of

Current Topics in Developmental Biology, Volume 132
ISSN 0070-2153
https://doi.org/10.1016/bs.ctdb.2018.12.008

morphogenetic movements that facilitate their structural interconnection. Within the context of vertebrate eye organogenesis, we also discuss the genetic hierarchies of transcription factors and signaling pathways that regulate growth, patterning, cell type specification and differentiation.

1. Introduction

Vertebrate ocular development begins during gastrulation with the specification of a singular eye field. The eye field subsequently splits bilaterally. Each eye then proceeds synchronously through a complex set of morphogenetic movements and tissue interactions that produce both the lens and optic vesicles. The anterior eye tissues derive from the surface ectoderm and underlying mesenchyme, while posterior tissues come from the underlying neural tube. Individual tissues do not develop in isolation, but undergo coordinated morphogenetic movements, and eventually perform integrated functions in the adult eye. The anterior tissues mainly focus and refract light onto the posterior neural retina, which sends this visual information down the optic nerve to various parts of the brain. First, we will discuss cornea and lens development, followed by eye field specification, optic vesicle/cup formation and patterning, and finally cell type differentiation of predominantly posterior tissues (iris, ciliary body, neural retina, and retina pigmented epithelium). A common theme throughout this chapter is that many of the same transcription factors and signaling pathways are used reiteratively among different tissues of the eye, as well as during the early and late development stages within each tissue (Tables 1 and 2).

2. Cornea

The cornea is the anterior-most tissue of the vertebrate eye. It offers an outer layer of protection and provides the first refractive surface for focusing light onto the retina. The cornea is comprised of three cellular layers, the corneal epithelium, stroma, and endothelium, and three membranous layers, epithelial basement, Bowman's, and Descemet's. The three layers of cells are highly interactive during development and act in a concerted manner to maintain corneal structural integrity, health, and function through adulthood. The cornea has multiple embryonic origins that include the surface ectoderm and neural crest derived periocular mesenchyme.

Table 1 List of transcription factors reiteratively used during eye organogenesis.

Transcription factor	Gene name	Expression domains	Function during eye organogenesis
Pax6	Paired box 6	Eye field, surface ectoderm, optic vesicle, lens placode, lens vesicle, cornea epithelium, optic cup, neural retina, RPE, ciliary body, iris epithelium	Eye field specification, optic vesicle evagination, optic vesicle patterning, optic cup patterning, retina neurogenesis, lens development, cornea development, ciliary body development, iris development
Rax/Rx	Retina and anterior neural fold homeobox	Eye field, optic vesicle, optic cup, neural retina	Eye field specification, optic vesicle evagination, optic vesicle patterning, retina neurogenesis
Lhx2	LIM homeobox 2	Eye field, optic vesicle, optic cup, neural retina	Eye field specification, optic vesicle patterning, optic cup formation, lens development
Sox2	Sex determining region Y(SRY)-box 2	Anterior neuroectoderm, lens placode, lens vesicle, lens epithelium, optic cup, neural retina	Eye field specification, retina neurogenesis
Otx2	Orthodenticle homeobox 2	Anterior neuroectoderm, optic vesicle, neural retina, RPE	Eye field specification, RPE development, retina neurogenesis
Six3	Sine oculis-related homeobox 3	Eye field, lens placode, lens vesicle, optic vesicle, optic cup	Eye field specification, lens development
Pitx2	Paired-like homeodomain transcription factor 2	Periocular mesenchyme, corneal epithelium, corneal stroma, corneal endothelium	Cornea development, iris development
Foxc1	Forkhead box C1	Periocular mesenchyme, corneal epithelium, corneal stroma, corneal endothelium	Cornea development, iris development

Table 2 List of signaling pathways reiteratively used during eye organogenesis.

Signaling pathways	Expression domains	Function during eye organogenesis
BMP (Bone morphogenetic protein)	Anterior neuroectoderm, surface ectoderm, optic vesicle, lens pit, lens vesicle, optic cup	Eye field specification, lens induction, RPE specification, ciliary body development, iris epithelium development
TGF (Transforming growth factor)	Periocular mesenchyme	Optic vesicle evagination, RPE specification, cornea endothelium development, cornea stroma development
RA (Retinoic acid)	Periocular mesenchyme, optic vesicle	Cornea endothelium development, cornea stroma development, iris development
FGF (Fibroblast growth factor)	Surface ectoderm, lens placode, optic vesicle	Lens primary fiber cell development, optic cup patterning
WNT (*wingless*-type *MMTV* integration site family)	Anterior neuroectoderm, surface ectoderm, optic vesicle, periocular mesenchyme, ciliary body, iris epithelium	RPE specification, ciliary body development, iris epithelium development
HH (*Hedgehog Family*)	Ventral midline, RPE, neural retina	Eye field specification, optic vesicle patterning, optic cup patterning, optic stalk induction, RPE differentiation
Hippo	Optic vesicle, optic cup, retina pigmented epithelium, lens vesicle, lens epithelium	RPE specification, RPE development, ciliary body development, lens development
Notch	Lens epithelium, optic cup, neural retina	Lens development, retina neurogenesis

2.1 Epithelium

The cornea epithelium originates from cells in the surface ectoderm, adjacent to both sides of the lens placode, termed presumptive corneal epithelium (pCE; Fig. 1). At this stage both the presumptive cornea and lens placode (pLP) express *Pax6* (Fig. 1A and Table 1) (Collomb et al., 2013).

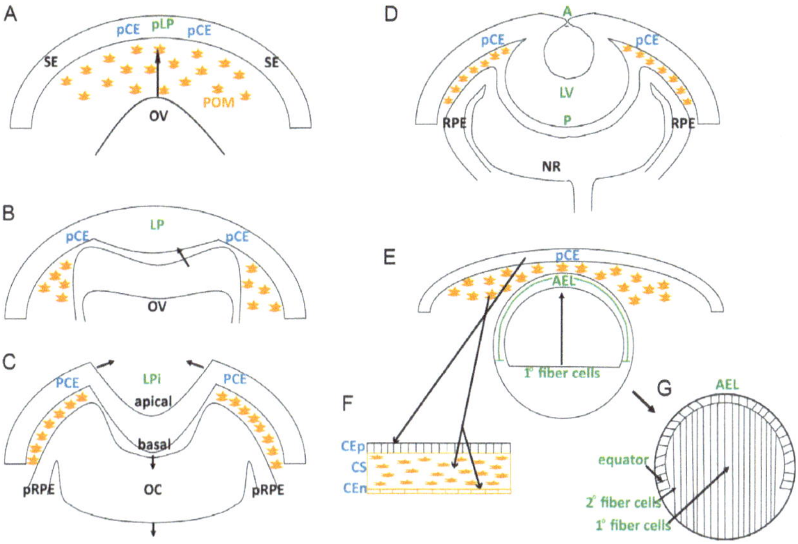

Fig. 1 Cornea and lens development. (A) The presumptive cornea epithelium (pCE) and lens placode (pLP) are located at the surface ectoderm (SE). At this stage induction of each cell type is inhibited by the presence of periocular mesenchymal cells (POM). (B) During optic vesicle (OV) morphogenesis, POM cells are relocated laterally, facilitating contact between the OV and LP, and initiation of lens induction. (C) Lens pit (LPi) invagination requires cell proliferation and apical constriction to form the spherical pit. As the pit is forming, the pCE cells are brought together. (D) The anterior side (A) of the lens vesicle (LV) fuses to form a stalk that is continuous with the pCE, but eventually pinches off to form an autonomous lens vesicle. The anterior-most cells of the lens vesicle coalesce into the epithelial cell layer, while primary fibergenesis begins on the posterior side (P) of the lens vesicle. (E) The pCE is now a contiguous cell layer at the surface, while underneath periocular mesenchyme cells once again populate the space between the pCE and lens vesicle. Meanwhile, differentiating primary lens fiber cells begin to extend toward the anterior. (F) The cornea epithelium (CEi), which starts out as 1–2 cell layers, but continues to thicken as development proceeds. The cornea stroma (CS) and endothelium (CEn) are derived from the periocular mesenchyme. (G) The mature lens is comprised of the anterior epithelial layer (AEL), equator (transition zone), secondary fiber and primary fiber cells. NR, neural retina; RPE, retina pigmented epithelium.

Corneal epithelium formation immediately follows lens placode thickening, invagination, and separation from the overlying surface ectoderm (Fig. 1F). Once the lens placode has transited into a lens vesicle and detached from the surface ectoderm, the fusion of the adjacent Pax6 positive cells fuse to create a contiguous pCE (Fig. 1E). If proper levels of Pax6 are not present at this stage several corneal defects can occur, including cornea vascularization, epithelial thinning, and improper separation of the lens and cornea

(Davis, Duncan, Robison, & Piatigorsky, 2003; Ramaesh et al., 2003). At this stage, corneal epithelial cells are multipotent, retaining an ability to develop as lens or epidermal cells, if the proper signals are not restricted. A possible mechanism for restricting lens fate is the presence of neural crest (known as periocular mesenchyme (POM)) and epiblast cells, which both express BMP inhibitors, a signal required for lens but not cornea formation (Collomb et al., 2013; Furuta & Hogan, 1998; Gerhart et al., 2009; Tzahor et al., 2003). To restrict epidermal fate the WNT inhibitor *Dkk2* represses WNT signaling emanating from the forming corneal limbus (Mukhopadhyay et al., 2006). Another molecular switch that occurs during corneal development involves cytokeratin proteins. Epithelial precursors, in the surface ectoderm, generally express K5/K14, but the pCE switches to K8/K18, and then again, to cornea-specific keratins K3/K12, with K12 expression dependent on *Pax6* (Chaloindufau, Sun, & Dhouailly, 1990; Liu, Kao, & Wilson, 1999; Shiraishi et al., 1998; Wolosin, Budak, & Akinci, 2004). Although the cornea starts out as 1–2 cell layers thick, containing flat, oval shaped cells, it grows to 6–8 cell layers, with the shape of each cell layer dependent on its final location. *Pax6* expression is maintained throughout differentiation by the entire corneal epithelium, as well as by the stem cells of the corneal limbus, which replenish the epithelium as dead cells are removed throughout life (Hanna, Bicknell, & Obrien, 1961; Koroma, Tseng, & Sundin, 1997; Li et al., 2015).

2.2 Endothelium

The endothelium is the second corneal cell layer to form during eye development. A wave of neural crest cells migrates from the dorsal neural tube, originating between the diencephalon and mesencephalon, to populate the periocular space between the lens vesicle and corneal epithelium (Creuzet, Vincent, & Couly, 2005; Johnston, Noden, Hazelton, Coulombre, & Coulombre, 1979; Lwigale, Cressy, & Bronner-Fraser, 2005). In humans, birds, and reptiles there are three waves of POM cell migration, with each wave producing different anterior segment structures. The first set of POM cells differentiate into the corneal endothelium, the second wave into corneal stroma, and third contribute to iris development. Even though the waves are in a temporally distinct temporal order, the endothelium is not established until after the second stromal lineage arrives (Hay, 1980; Rones, 1932; Wulle & Lerche, 1969). This part of anterior segment development differs in rodents, rabbits, and cats, where there is only one mesenchymal migration, with the posterior cells (lens adjacent) differentiating into endothelium and the medial cells into stroma (Cintron, Covington, & Kublin, 1983; Hay, 1980; Pei & Rhodin, 1970). In all species, *Pitx2* and *Foxc1* are expressed in the periocular mesenchyme and essential for proper anterior segment genesis, including the cornea endothelium (Table 1). These transcription factors proteins physically interact, to influence the transition from migrating mesenchyme to adherent epithelium (Berry et al., 2006). Additional evidence suggests *Pitx2* expression is downstream of, and reliant on, *FoxC1* (Silla, Naidoo, Kidson, & Sommer, 2014). Several signaling pathways influence the development of the cornea endothelium. A disruption in either *TGF-β2* or retinoic acid (RA) signaling results in anterior segment dysgenesis (Table 2) (Ittner et al., 2005; Matt et al., 2005; Saika et al., 2001). Loss of RA processing in the POM causes reduced expression of *FoxC1* and *Pitx2* (Matt et al., 2005). But, *TGF-β2* is also implicated in regulating *FoxC1*, although there is some conflicting data about this idea. In *TGF-β2* germline mutant mice, the cornea endothelium is missing, but *FoxC1* expression is unaffected in the POM (Saika et al., 2001). In contrast, a POM-specific knockout of the *TGF-β2* receptor decreases *FoxC1* expression (Ittner et al., 2005). Further, the presence of TGF-β2 can either increase (Ittner et al., 2005) or decrease (Silla et al., 2014) *FoxC1* levels in vitro, suggesting a complex role for *TGF-β2* during eye development. Overall, *Foxc1* and *Pitx2* are essential for proper cornea endothelium development but not all of the signaling molecules that regulate *Foxc1* expression are known.

2.3 Stroma

The stroma is the middle layer of the cornea that provides structural support to the entire tissue, while maintaining clarity for light passage. The main stromal cell type is keratocytes. These cells secrete matrix metalloproteinases, collagen, glycosaminoglycans, and water soluble Crystallins, which work together to maintain integrity and reduce light scatter (DelMonte & Kim, 2011). The stroma and endothelium are similarly regulated during their development, owing to their shared POM origin. Loss of *FoxC1* causes an abnormal stromal thickening, due to reduced cell packing and a morphologic change of keratinocyte cell shape, from flattened to stellate (Kidson, Kume, Deng, Winfrey, & Hogan, 1999; Seo et al., 2012). Both *Pitx2* germline and POM-specific mutant mice completely lack the corneal stroma as a consequence of mesenchymal cells accumulating between the lens and cornea epithelium without differentiating (Evans & Gage, 2005; Gage, Suh, & Camper, 1999). When

POM cells lose *TGF-β2R* expression, it induces the stromal precursors to erroneously express the pigment gene *Dct,* instead of *Pitx2,* fail to produce collagen, and assume an abnormal cell shape (Ittner et al., 2005). Loss of RA signal processing from the POM also causes complete agenesis of the stroma (Matt et al., 2005).

3. Lens

The vertebrate lens resides posterior to the cornea and functions as the second refractive surface in the eye. In order to visualize objects clearly at different depths, the lens, via a process called accommodation, changes its shape to focus light onto the retina. Structurally, the lens is comprised of the capsule, epithelium, fiber cells, and transition zone at the lens equator. The capsule is the outermost layer surrounding the entire lens. This transparent basement membrane participates in lens accommodation, protects the lens from outside pathogens, aids in lens cell differentiation, epithelial attachment, and allows for the exchange of metabolites and waste (Bassnett, Missey, & Vucemilo, 1999). The epithelial layer is comprised of lens progenitor cells that divide continuously (albeit slowly) throughout life. It abuts the transition zone where newly postmitotic progenitor cells differentiate into fiber cells. The epithelium also maintains osmotic pressure and lens volume through Na/K pumps. The terminally differentiated fiber cells produce large amounts of Crystallin structural proteins, and eventually lose their nucleus and organelles to facilitate for tissue transparency.

3.1 Lens morphogenesis

Lens formation originates with the specification of a subset of cells situated between the anterior neural plate and surface ectoderm. Initially there is a common pool of placodal progenitor cells that become subdivided into discrete cranial placodes (adenohypophyseal, olfactory, lens) (Litsiou, Hanson, & Streit, 2005; Steventon, Mayor, & Streit, 2016). In chick, preplacodal cells from the anterior neural plate border are bipotential, ultimately producing both the lens and olfactory placodes (Sjodal, Edlund, & Gunhaga, 2007). Although the preplacodal cells coexpress the transcription factors *Dlx5* and *Pax6,* each placode can be distinguished by expression of only one factor, namely, *Pax6,* in the lens placode and *Dlx5* in the olfactory placode (Bhattacharyya, Bailey, Bronner-Fraser, & Streit, 2004). Thus, the lens placode originates from a cluster of *Pax6*-positive, surface ectodermal cells (Table 1).

Lens formation is controlled by signals emanating from both the optic vesicle (positive) and periocular mesenchyme (negative) to the lens placode. This process was first described by Spemann and Mangold, in the early 1900s (Spemann & Mangold, 1924). Prior to optic vesicle evagination the periocular mesenchyme inhibits lens placode formation. This occurs along the head surface ectoderm through *Pax6* inhibition by TGFβ/Smad3 and Wnt/βcatenin signaling from the POM to the ectoderm (Grocott, Johnson, Bailey, & Streit, 2011; Smith, Miller, Song, Taketo, & Lang, 2005). Once the optic vesicle evaginates from the neural plate the vesicle physically excludes the periocular mesenchyme from the region of ectoderm where the lens placode forms (Fig. 1A). This process accomplishes two tasks. It relieves *Pax6* repression at the surface ectoderm by redistributing the POM and brings the optic vesicle close enough to the lens placode to induce lens morphogenesis through BMP signaling. The requirement for *BMP4* to be expressed at high levels in the optic vesicle has been most clearly demonstrated in the developing mouse eye. In *BMP4* mutants, lens induction fails, but this process can be rescued by the placement of BMP4-soaked beads adjacent to the optic vesicle (Furuta & Hogan, 1998). Interestingly, removing the optic vesicle and replacing it with BMP4 beads or other BMP4-expressing tissue (dorsal forebrain neural ectoderm or heart) did not rescue lens induction, suggesting that additional optic vesicle-derived factors are also important for lens induction (Furuta & Hogan, 1998). Additionally, two BMP receptors, *Bmpr1a* and *Acvr1*, act redundantly in the lens placode (Huang, Liu, Filas, Gunhaga, & Beebe, 2015; Rajagopal et al., 2009). The loss of either receptor results in an abnormally small lens, whereas double mutants ablate lens induction (Rajagopal et al., 2009). Lastly, a loss of *Lhx2*, which is specifically expressed in the optic vesicle, also blocks lens induction. Here the mechanism is indirect, since *Lhx2* mutant optic vesicle cells have reduced BMP4/7 signaling, which results in decreased *Pax6* expression in the lens placode (Tables 1 and 2) (Porter et al., 1997; Thein et al., 2016; Yun et al., 2009).

Two key regulators of lens development are the transcription factors *Ptix3* and *Foxe3*. Rodent *Foxe3* is broadly expressed in the lens placode, pit, vesicle, and anterior epithelial layer throughout lens development and required for proper lens specification, differentiation, and maintenance (Blixt, Landgren, Johansson, & Carlsson, 2007; Blixt et al., 2000; Medina-Martinez, Shah, & Jamrich, 2009). *Pitx3* expression in the lens placode is more refined than *Foxe3* but becomes expressed throughout the lens vesicle and is another key regulator of lens development (Ahmad et al., 2013; Medina-Martinez et al., 2009; Semina et al., 1998; Semina, Murray, Reiter, Hrstka, & Graw, 2000; Semina, Reiter, & Murray, 1997). Loss of both *Foxe3* and *Pitx3* causes lens aphakia (Blixt et al., 2007, 2000; Semina et al., 1998, 2000, 1997). Pitx3 transcriptionally activates *Foxe3* through a 5′ upstream regulatory element, but *Foxe3* is still expressed in *Pitx3* mutants, demonstrating that *Foxe3* expression also depends on other factors (Ahmad et al., 2013). *Pitx3* is genetically linked *Pax6*. In *Pax6* heterozygous mutants *Pitx3* expression is decreased but it is unknown if Pax6 is directly or indirectly responsible for *Pitx3* expression (Chauhan, Reed, Yang, et al., 2002; Chauhan, Reed, Zhang, et al., 2002; Chauhan, Zhang, Cveklova, Kantorow, & Cvekl, 2002). *Pax6* also forms a complex, cross-transcriptional network with *Six3* and *Sox2* within lens placode cells, which eventually induce genes crucial for lens development through cooperative regulatory sequence binding (δ-Crystallin) or individual developmental stage dependent activation (N-Cadherin) (Cvekl & Ashery-Padan, 2014).

Upon induction, the lens placode undergoes morphogenetic movements to change it into a lens pit (Fig. 1B and C). This process is initiated by a thickening and invagination of lens placode cells, which occurs coordinately with optic cup invagination (Fig. 1B and C). The factor *Nf1* induces lens placodal cell proliferation and elongation (Carbe & Zhang, 2011). The lens placode cells constrict apically at the adherens junction, through a *Shroom3*, *RhoA*, *Rac1*, and *p120-catenin* mediated process that is partially regulated by cell autonomous BMP signaling (Chauhan, Lou, Zheng, & Lang, 2011; Jidigam, Srinivasan, Patthey, & Gunhaga, 2015; Lang, Herman, Reynolds, Hildebrand, & Plageman, 2014; Maddala et al., 2011; Plageman et al., 2010; Sawyer et al., 2010). However, other molecular and cellular changes help drive lens vesicle formation, including changes to the extracellular matrix and physical interactions between the placode and optic vesicle cells. In the latter situation, placode-derived *Pax6* regulates the expression levels of multiple *collagens*, *versican*, *α5β1-integrin*, and *fibronectin* (Duncan, Kozmik, Cveklova, Piatigorsky, & Cvekl, 2000; Huang et al., 2011; Wolf et al., 2009). This influences the composition of fibronectin-containing filopodia that connect the lens pit to the optic vesicle basal lamina (Chauhan et al., 2009; Mcavoy, 1980; Porazinski et al., 2015). These processes are formed through fibronectin-β1-integrin signaling and actin-myosin contractile forces, which also require *Yap*-mediated activity (Chauhan et al., 2009; Porazinski et al., 2015). Although cell shape changes are an inherent part of lens vesicle formation, the final step is a separation of the nascent lens vesicle from the surface ectoderm through the degradation of the lens stalk (Fig. 1D–E).

Numerous factors are required for this detachment, including the transcription factors *Pax6, Sip1, Foxe3, Sox11, AP-2α*, the FGF-Spry, and Notch signaling pathways, and the cell adhesion proteins N-Cadherin/Cdh2 and *E*-Cadherin/Cdh1 (Baulmann et al., 2002; Blixt et al., 2007, 2000; Kuracha et al., 2011; Le et al., 2012; Pontoriero et al., 2008, 2009; van Raamsdonk & Tilghman, 2000; Wurm, Sock, Fuchshofer, Wegner, & Tamm, 2008; Yoshimoto, Saigou, Higashi, & Kondoh, 2005). These genes constitute a network that regulates lens stalk cell apoptosis (Ozeki, Ogura, Hirabayashi, & Shimada, 2001; Pontoriero et al., 2009; van Raamsdonk & Tilghman, 2000). Failure to complete lens stalk apoptosis leads to the clinical condition Peter's Anomaly (Pontoriero et al., 2009).

3.2 Lens differentiation

After detachment from the surface ectoderm, lens vesicle cells initiate the formation of three tissues: the lens capsule, epithelium, and fiber cells (Fig. 1F and G). The capsule is formed by the extracellular deposition of fibronectin and laminin by cells of the lens pit, starting along the posterior, as an extension of the basement membrane (Parmigiani & McAvoy, 1984). Once the lens vesicle detaches from the surface ectoderm, these extracellular matrix proteins expand their expression circumferentially to surround the entire vesicle (Parmigiani & McAvoy, 1984). An integral part of capsule closure along the anterior side of the vesicle is through integrin-linked kinase (ILK). Without ILK, the anterior capsule ruptures, accompanied by a disorganization of laminin and collagen IV (Cammas, Wolfe, Choi, Dedhar, & Beggs, 2012). As the lens vesicle continues to grow, the capsule thickens as the epithelial and fiber cells both secrete more extracellular matrix proteins (Haddad & Bennett, 1988; Young & Ocumpaugh, 1996).

During lens pit and vesicle stages, lens progenitor cells proliferate uniformly (Fig. 1C). But signals emanating from the adjacent optic cup polarize the lens vesicle, and the first (primary) fiber cells exit the cell cycle and differentiate at the posterior side, closest to the cup (Fig. 1D). As these cells elongate, the remaining progenitor cells become sequestered to the anterior side (Kallifatidis, Boros, Shin, McAvoy, & Lovicu, 2011; Lovicu & McAvoy, 1999; Modak, Morris, & Yamada, 1968). Confirmation that optic cup cells are the source of the polarizing signal first came from experiments in the chick eye. When the lens was surgically inverted along its equator, the former posterior cells (now relocated anterior), dedifferentiated to recreate the proliferating epithelium, whereas the former anterior epithelial cells (now repositioned posteriorly) rapidly differentiated into fiber cells (Coulombre & Coulombre, 1963). Two signaling pathways important for establishing lens polarity, primary fiber cell proliferation and differentiation are FGF and BMP (Faber, Robinson, Makarenkova, & Lang, 2002; Jarrin, Pandit, & Gunhaga, 2012). Through an FGF-dependent mechanism, BMP regulates cell cycle exit through maintenance of cell cycle length or progression (Jarrin et al., 2012). Additionally, FGF can promote fiber cell elongation and expression of β and γ-Crystallin, two markers of differentiated fiber cells (Mcavoy, 1978a, 1978b; Xie et al., 2016). Two transcription factors implicated in proper primary cell differentiation are *Gata3* and *Prox1*. *Gata3* is specifically expressed in the posterior lens vesicle as primary fibergenesis begins and, when absent, indirectly causes reduced progenitor cell proliferation while autonomously blocking differentiation as indicated by reduced cell cycle kinase inhibitors (CKIs) and γ-Crystallin expression (Maeda et al., 2009). By contrast, *Prox1* is expressed initially by all lens progenitor cells, but more robustly by nascent fiber cell nuclei. Its loss impacts both proliferation and primary fiber cell elongation (Wigle, Chowdhury, Gruss, & Oliver, 1999).

As the primary fiber cells accumulate in the middle of the lens vesicle, they produce Crystallin proteins and eventually attach to the coalescing anterior epithelial layer (AEL; Fig. 1E), thereby providing overall tissue structure and support. The arrangement of these two cell types with proliferating cells in a monolayer overlying the elongating fiber cells heralds the onset of secondary fibergenesis. To maintain the vertebrate lens over its lifespan, it is important for progenitor cells in the AEL to remain proliferative and produce new waves of fiber cells (termed secondary fiber cells; Fig. 1F). Multiple signals regulate both lens proliferation and differentiation during secondary fibergenesis, including the Notch and Hippo-Yap pathways (Table 2). Conditional mutations for multiple Notch pathway genes consistently reduce AEL cell growth, but owing to the localization of the ligand Jag1 expression at the transition zone, secondary fiber cell differentiation is also affected (Jia, Lin, Zhang, York, & Zhang, 2007; Le, Conley, & Brown, 2009; Rowan et al., 2008; Saravanamuthu et al., 2012). The transcriptional coactivator *Yap* is expressed throughout the AEL and transitional zone and also controls both AEL proliferation and secondary fiber cell differentiation (Dawes, Shelley, McAvoy, & Lovicu, 2018; Song et al., 2014).

The Crystallin proteins are key components of differentiated lens fiber cells, but this protein family is not uniformly expressed throughout lens development (Robinson & Overbeek, 1996). For example, both αB-Crystallin and β/γ-Crystallins are present in the lens placode, whereas αA-Crystallin is not detected until the lens pit stage (Robinson & Overbeek, 1996; Wolf et al., 2013). Crystallin gene transcription is regulated by multiple factors, already mentioned earlier in this section. They can be direct targets of *Pax6* (Kamachi, Uchikawa, Tanouchi, Sekido, & Kondoh, 2001) or the FGF pathway, as well as by *c-Maf* (also regulated by *Pax6* and FGF signaling) (Wolf, Yang, Wawrousek, & Cvekl, 2008; Xie & Cvekl, 2011; Xie et al., 2016; Yang et al., 2006). Additional regulators of the Crystallins include *Prox1*, *Hsf4*, *Sox1*, *Six3*, and *CREB* (Cui, Tomarev, Piatigorsky, Chepelinsky, & Duncan, 2004; Lengler et al., 2001; Nishiguchi, Wood, Kondoh, Lovell-Badge, & Episkopou, 1998; Somasundaram & Bhat, 2004; Yang et al., 2006). In later phases of eye development, lens fiber cells generate Aquaporin channels for passive diffusion of nutrients and waste products, rearrange their cytoskeleton and eventually eliminate their organelles to facilitate lens transparency. More details about these topics can be found in a recent review article (Cvekl & Ashery-Padan, 2014; Cvekl & Zhang, 2017).

4. Eye field specification

During gastrulation the anterior neuroectoderm adjacent to the telencephalic/diencephalic boundary forms what is termed the eye field. The iris, ciliary body, neural retina (NR) and retina pigmented epithelium (RPE) either partially (iris, ciliary body), or completely (NR, RPE) derive from the eye field. The neuroectodermal cells of the eye field express a cocktail of transcription factors that include *Rax/Rx*, *Pax6*, *Lhx2*, *Six3*, *Tll*, *Tbx3* and *Six6*, which act independently during subsequent eye field specification and optic vesicle formation, but also share a set of downstream genes (Bailey et al., 2004; Takabatake, Takabatake, Sasagawa, & Takeshima, 2002; Zuber, Gestri, Viczian, Barsacchi, & Harris, 2003). Eye field specification begins when signaling pathways and transcription factors in the presumptive telencephalon initiate expression of the eye field transcription factors. *Otx2* and *Sox2* are broadly expressed in this region and required for *Rax* activation (Danno et al., 2008), whose activity then feeds back to inhibit further *Otx2* expression in the eye field. At the same time, Sox2/Otx2 positive neuroepithelial cells secrete BMPs to shape the adjacent Rax-expression domain, thereby establishing a presumptive telencephalon/eye field boundary (Bielen & Houart, 2012; Danno et al., 2008). Another eye field transcription factor, *Six3*, performs multiple functions in the eye field, including antagonizing WNT ligand transcription (WNT1, WNT8) (Lagutin et al., 2001; Liu, Lagutin, Swindell, Jamrich, & Oliver, 2010; Oliver et al., 1995).

Precise control of WNT signaling prevents posterization of the eye field and maintains SHH expression (emanating from the ventral midline) to ensure proper eye field separation (Lagutin et al., 2003; Liu et al., 2010).

5. Optic vesicle evagination

Once the eye field has been specified from the anterior neural plate, these cells initiate the aforementioned transcription factor expression, which provides developmental potential to form eye tissues. However, this field must also split bilaterally before eye vesicle morphogenesis initiates. *Shh* and *Six3* both are critical regulators of eye field bifurcation, and their removal results in failure to split the eye field, which results in cyclopia (Chiang et al., 1996; Geng et al., 2008; Jeong et al., 2008). Six3 sustains *Shh* expression at the midline of the rostral diencephalon (Geng et al., 2008; Jeong et al., 2008). As the eye field is divided, these cells are pushed laterally by posterior diencephalon cells that migrate anteriorly along the midline. The end result is the optic vesicles (England, Blanchard, Mahadevan, & Adams, 2006; Varga, Wegner, & Westerfield, 1999). This cellular movement is driven by Tgfβ secreted by the underlying axial mesoderm (England et al., 2006; Varga et al., 1999). After bilateral eye primordia is established, the optic vesicle precursor cells continue migrating, which initiates subsequent vesicle evagination. The presence of Shh also suppresses *Pax6* midline expression, as well as induces the onset of *Pax2* (Macdonald et al., 1995). Live imaging of zebrafish embryos demonstrated that cell shape changes and migration, rather than proliferation, are the driving forces of optic vesicle evagination (Kwan et al., 2012; Rembold, Loosli, Adams, & Wittbrodt, 2006). During this process eye field cells exhibit slow, directed migration toward the midline but then turn laterally, move through the mesenchyme to their final destination where the vesicle forms (Ivanovitch, Cavodeassi, & Wilson, 2013). Similar cell shape changes occur in mammalian 3D retinal stem cell cultures or "organoids," which produce optic vesicles, although here the influence of cell migration is less considerable (Eiraku et al., 2011). For both in vitro and in vivo ocular morphogenesis, evaginating optic vesicle cells maintain expression of multiple transcription factors, interact with extracellular matrix, and respond to cell adhesion molecules.

A subset of eye field transcription factors, *Rax*, *Pax6*, and *Tll*, control optic vesicle evagination and formation (Bailey et al., 2004; Hollemann, Bellefroid, & Pieler, 1998; Rembold et al., 2006). Loss of both *Pax6* and *Tll* results in stunted and abnormal vesicles while the loss of *Rax* completely blocks eye field cells from proceeding onward with optic vesicle development (Bailey et al., 2004; Hollemann et al., 1998; Rembold et al., 2006). One proposed mechanism for *Rax*-dependent control of evagination is through repression of the cell adhesion molecule *Nlcam*, and activation of the chemokine *CXCR4*, which influences cell shape and cell movement (Bielen & Houart, 2012; Brown et al., 2010). The extracellular matrix protein Laminin also impacts cell shape changes during evagination (Ivanovitch et al., 2013). Along with cell shape changes the optic vesicle must maintain its cellular arrangement of neuroepithelial cells termed pseudo-stratified. Recent work in zebrafish suggests that *Sema6a* and *Plxna2* help maintain optic vesicle integrity. If either factor is missing, a subset of optic vesicle cells fail to migrate from the neural plate and/or delaminate from the vesicle (Ebert, Childs, Hehr, Cechmanek, & McFarlane, 2014). Once an intact optic vesicle is fully formed, vesicle cells become patterned along different axis while transitioning into another set of complex movements vital to eye development.

6. Optic vesicle/cup patterning and morphogenesis

Intrinsic and extrinsic factors give positional, patterning information to the dividing optic vesicle progenitor cells to establish dorsal-ventral, nasal-temporal, and distal-proximal identity. The establishment of these molecular boundaries is vital for subsequent steps of eye formation, as different regions of the optic vesicle give rise to RPE (dorsal), optic stalk (OS) (ventral), and NR (Fig. 2A). *Lhx2*, *Pax6*, *Pax2*, *Otx2*, *Mitf*, and *Vsx2* are initially expressed uniformly in the optic vesicle, but eventually undergo restricted expression. For example, *Pax6, Pax2, Rax, Vsx2, Lhx2, Mitf* and *Otx2* are present in NR, OS, and RPE cells of the optic cup, but are later confined to the NR (*Pax6, Lhx2, Rax, Vsx2*), OS (*Pax2*) or RPE (*Otx2, Lhx2, Mitf*) (Baumer et al., 2003; de Melo, Zibetti, et al., 2016). The activities of these early factors include cross-regulation, thereby pointing to an early ocular transcriptional regulatory hierarchy. *Lhx2* acts upstream of *Mitf*, *Vsx2*, and *Pax2* as *Lhx2* mutants are missing *Mitf* and *Vsx2* expression, with *Pax2* downregulated but not completely abolished (Yun et al., 2009). Intrinsically, *Mitf* and *Vsx2* repress each other's expression and activity through protein-protein and protein-DNA interactions in a RPE/NR domain specific manner (Bharti, Liu, Csermely, Bertuzzi, & Arnheiter, 2008; Zou & Levine, 2012). The cells of the distal/ventral portion of the optic vesicle express higher levels of *Vsx2* and form neural retina. Cells of the dorsal/posterior optic vesicle

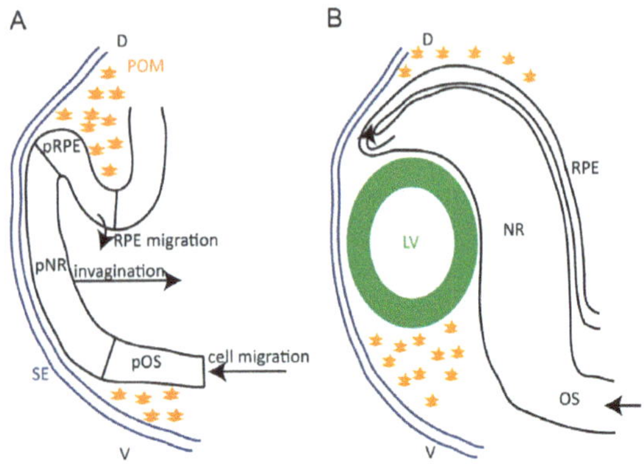

Fig. 2 Retina, RPE and optic nerve development. (A) At the end of optic vesicle morphogenesis, it contacts the surface ectoderm (SE). OV cells activate distinct subsets of genes, indicating the cells are adopting regional fates, based on their spatial location. The dorsal (D) side of the OV gives rise to the presumptive retina (pNR) and pigmented epithelium (pRPE); the central vesicle only develops into presumptive neural retina (pNR), while the ventral vesicle forms the presumptive optic stalk (pOS). Based on zebrafish live imaging studies (Heermann et al., 2015; Kwan et al., 2012; Rembold et al., 2006), cells migrate into the vesicle during OV invagination, and the RPE must migrate around the posterior (Heermann et al., 2015; Kwan et al., 2012; Rembold et al., 2006). (B) Growth of the optic cup continues with the same cell movements that began once the optic vesicle started to invaginate. (C) The optic cup is fully formed once the RPE completely surrounds it posteriorly. The optic stalk connects the optic cup to the wall of the brain. The projection neurons of the retina, the retinal ganglion cells, project their axons through the OS and into the brain. The presumptive ciliary body (pCB) and iris epithelium (pIE) are derived from peripheral optic cup progenitor cells.

maintain higher levels of *Mitf* and give rise to RPE. Additionally, *Pax6* and *Pax2* maintain RPE cell fate via *Mitf* expression in a genetically redundant mechanism (Baumer et al., 2003). After the onset of *Mitf*, and before *Pax6* downregulation in the RPE, Pax6 and Mitf directly suppress the pro-retinogenic signaling molecules *Fgf15* and *Dkk3* through *cis*-regulatory element binding (Bharti et al., 2012). In the peripheral optic cup, *Pax6* expression influences dorsal-ventral and naso-temporal patterning, restricts *Sox2* expression to the optic cup, and is essential for generating the nonneuronal ciliary and iris epithelium (Leconte, Lecoin, Martin, & Saule, 2004).

Extracellular signaling pathways also define optic cup regional boundaries. The surrounding periocular mesenchyme, surface ectoderm, and optic neuroepithelium send signals to each other that influence cell fate decisions. FGF signaling from both the surface ectoderm and NR help define the boundary between RPE and NR by restricting *Vsx2* expression

(Nguyen & Arnheiter, 2000). In contrast, TGFβ-like signals from the extraocular mesenchyme induce RPE cell fate through expression of *Mitf* (Fuhrmann, Levine, & Reh, 2000). SHH signaling from the ventral midline splits the eye field early in development (see Section 5) then acts again to promote OC proximal/distal, dorsal/ventral patterning through the induction of *Pax2* expression (Table 2). *Pax2*, an essential component of optic stalk formation, represses *Pax6*, which in turn represses *Pax2* thereby establishing a boundary between the developing optic cup and stalk (Patel & Sowden, 2017). Other signaling pathways (WNT, BMP, RA) also regulate aspects of optic cup patterning and morphogenesis, which will be highlighted in Sections 7 and 9.

Integral with optic vesicle/cup patterning are complex morphogenetic movements which impact the final position of each cell. Optic cup morphogenesis begins with the central most optic vesicle cells, closet to the surface ectoderm, migrating toward the periphery and contributing to the emerging cup shape for this tissue (Fig. 2A) (Heermann, Schutz, Lemke, Krieglstein, & Wittbrodt, 2015; Kwan et al., 2012; Rembold et al., 2006). As with the optic vesicle, cell proliferation is not the main driver of optic cup formation (Kwan et al., 2012; Miesfeld et al., 2015). Although the optic vesicle clearly induces the lens and there is coordinate morphogenesis of the lens and optic cup, it is important to note that optic vesicle/cup development occurs essentially normally in the absence of a lens (Eiraku et al., 2011; Hyer, Kuhlman, Afif, & Mikawa, 2003). The intrinsic ability of these cells to produce a retina is conserved among vertebrates, since surgical removal of the chick lens placode does not affect OC/NR formation (Hyer et al., 2003) and mammalian retinal organoids develop into a fully laminated neural network in the absence of other embryonic tissues (Eiraku et al., 2011). Concomitant with optic cup invagination, the pseudo-stratified cells of the future RPE migrate around the outside of the cup, thinning into a monolayer (Kwan et al., 2012). Two proposed mechanisms for RPE cell migration and monolayer formation involve epithelial flow of optic cup progenitors and cell shape changes (Cechmanek & McFarlane, 2017; Heermann et al., 2015).

7. Ciliary body and iris

The ciliary body is located in the peripheral tip of the optic cup and contains a ciliary epithelium (inner nonpigmented and outer pigmented), stroma, and muscles. The function of the ciliary body is to secrete aqueous humor, glycoproteins, antioxidant enzymes, neuropeptides (into the anterior chamber), and control lens accommodation (distance-based focusing) through ciliary muscle contractions (Bishop, Takanosu, Le Goff, & Mayne, 2002; Coca-Prados, Escribano, & Ortego, 1999; Hanssen, Franc, & Garrone, 2001; Raviola, 1971). The iris is connected to the ciliary body by the iris root. The iris is composed of an inner and outer pigmented epithelium, stroma, and iridial muscles (sphincter and dilator pupillae). The iris regulates the amount of light passing through the lens to the neural retina, helps focus the lens on nearby objects, and aids in regulating intraocular pressure through the circulation of aqueous humor (Davis-Silberman & Ashery-Padan, 2008). Despite these functional differences, the ciliary body and iris have the same embryonic origins. The ciliary body and iris pigmented epithelium (IPE) both derive from the non-neural progenitors at the periphery of the optic cup (ciliary margin) (Fig. 2C). The ciliary and iris stroma derive from the migrating periocular mesenchyme, similar to the cornea stroma and endothelium (Beebe, 1986; Cvekl & Tamm, 2004; Davis-Silberman & Ashery-Padan, 2008; Napier & Kidson, 2007; Thumann, 2001). In the following sections we describe transcription factors and signaling pathways that regulate ciliary and iris formation.

7.1 Epithelium

As mentioned previously, *Pax6* is expressed throughout the optic cup with high levels in the peripheral OC cells that give rise to the ciliary body and iris (Davis et al., 2009; Davis-Silberman et al., 2005; Prosser & van Heyningen, 1998; Schedl et al., 1996). The formation of both structures is *Pax6*-dependent (Davis et al., 2009; Davis-Silberman et al., 2005; Prosser & van Heyningen, 1998; Schedl et al., 1996). Pax6 regulation of ciliary/iris epithelium development is cell autonomous and some of the downstream targets include *Foxc1*, *BMP4*, and *TGFβ2* (Davis et al., 2009; Davis-Silberman et al., 2005; Wang, Shan, & Gregory-Evans, 2017). Other transcription factors present in ciliary and iris epithelial progenitor cells include *Otx1* and *Msx1*, whose expression delineates the nonneuronal ciliary marginal cells that do not contribute to the neural retina (Belanger, Robert, & Cayouette, 2017; Martinez-Morales, Signore, Acampora, Simeone, & Bovolenta, 2001; Zhao, Chen, Hung, & Overbeek, 2002). Of these two transcription factors only *Otx1* appears essential for ciliary and iris epithelium development. *Otx1* null mice lack a ciliary epithelium and have an underdeveloped iris epithelium (Acampora et al., 1996). Despite an identical spatiotemporal expression pattern to *Otx1*,

Msx1 mutants do not exhibit any abnormalities in peripherally derived optic cup tissues (Belanger et al., 2017; Houzelstein, Cohen, Buckingham, & Robert, 1997; Satokata & Maas, 1994).

In conjunction with transcription factor regulation, several signaling pathways orchestrate ciliary body and iris epithelium development. *BMP4* and *BMP7* are both dynamically expressed during ciliary/iris epithelium development (Chang et al., 2001; Zhao et al., 2002). Inhibition of BMP signaling with noggin or germline loss of *BMP4* causes iridocorneal attachment, iris hypoplasia, malformed (missing ciliary processes) or absent ciliary body (Chang et al., 2001; Zhao et al., 2002). Canonical WNT signaling (β-catenin mediated) is also active and important to the developing ciliary/iris epithelium. WNT gain of function (WNT/β-catenin overexpression) leads to ectopic expression of ciliary/iris-specific genes in the neural retina, while loss of function (dominant negative Lef-1) causes iris hypoplasia (Cho & Cepko, 2006; Liu et al., 2007; Trimarchi, Cho, & Cepko, 2009). In addition to BMP and WNT signaling, the Hippo signaling pathway is also important for proper ciliary/iris epithelium development. *NF2*, an upstream Hippo signaling component, regulates proper proliferation of progenitors through the transcriptional coactivators Yap/Taz in the ciliary margin (Moon et al., 2018).

7.2 Stroma

The ciliary and iris stroma originate from migrating POM cells which use the existing ciliary/iris epithelium as a migratory track (Cvekl & Tamm, 2004). These POM cells also give rise to the cornea endothelium and stroma (see above). Therefore, the corneal endothelium, corneal stroma, ciliary stroma, and iris stroma are under control of the same transcription factors and signaling pathways required for proper specification and differentiation. *Ptix2*, *FoxC1*, *FoxC2* and *Lmx1b* are all expressed in the POM, and mutations in each gene can cause ciliary body and iris hypoplasia (Berry et al., 2006; Hjalt, Semina, Amendt, & Murray, 2000; Liu & Johnson, 2010; McMahon, Gestri, Wilson, & Link, 2009; Pressman, Chen, & Johnson, 2000; Pressman, Martin, & Johnson, 2000; Semina et al., 1996; Seo et al., 2017; Smith et al., 2000). Even though *Pitx2* and *FoxC1* are coexpressed in the POM and can form protein dimers, the activities of these genes again suggest a genetic hierarchy with *Foxc1/2* upstream of *Pitx2* (Berry et al., 2006; Seo et al., 2017) (Table 1). The combined loss of *Foxc1/2* downregulates *Pitx2* expression, and a subset of Pitx2 target genes, *Dkk2* and *Tfap2b* (Seo et al., 2017). The RA signaling also regulates ciliary body and iris development (Table 2), by activating *Pitx2* (see Section 2), *FoxC1*, and *Lmx1b* expression in the POM (Lupo et al., 2011; Matt et al., 2005; Matt, Ghyselinck, Pellerin, & Dupe, 2008). Interestingly, three retinoic acid receptors, RARα, RARβ, and RARγ, directly bind to a retinoic acid response element in the *Pitx2* promoter (Kumar & Duester, 2010). The loss of RA signaling at this developmental stage includes iris hypoplasia, which can be rescued with human *PITX2* in a *Pitx2* knockdown zebrafish model (Bohnsack, Kasprick, Kish, Goldman, & Kahana, 2012).

8. Neural retina

The neural retina is composed of seven major cell classes, six neuronal (retinal ganglion cells (RGCs), amacrine, bipolar, horizontal, cones, rods) and one glial (Müller). All cell classes derive from multipotent retinal progenitor cells (RPC) within the optic cup. Although there is a stereotypical order to the birth of each cell class, there is also considerable temporal overlap in the appearance of multiple cell types. Among vertebrates, RGCs arise first and Muller glia last (Carter-Dawson & LaVail, 1979; Rapaport, Wong, Wood, Yasumura, & LaVail, 2004; Turner & Cepko, 1987; Turner, Snyder, & Cepko, 1990; Young, 1985). As RPCs transition from proliferative to terminal cell divisions, their ability to adopt particular fates becomes established. The competence state of these cells to produce the appropriate type of neuron or glia changes across developmental time, in part due to a lengthening of the cell cycle. Some of the factors controlling the timing of retinal neurogenesis include *Ikaros*, *Castor*, and various microRNAs (Elliott, Jolicoeur, Ramamurthy, & Cayouette, 2008; Georgi & Reh, 2011; La Torre, Georgi, & Reh, 2013; Mattar, Ericson, Blackshaw, & Cayouette, 2015). Other transcription factors such as *Rax*, *Vsx2*, *Pax6*, *Sufu*, *Six6*, and *Sox2* regulate RPC proliferation and/or multipotency (Burmeister et al., 1996; Cwinn et al., 2011; Li, Perissi, Liu, Rose, & Rosenfeld, 2002; Marquardt et al., 2001; Mathers, Grinberg, Mahon, & Jamrich, 1997; Taranova et al., 2006). As an example, conditional loss of *Pax6* in the early developing optic cup results in a hypocellular retina where RPCs maintain their identify but fail to fully differentiated into any particular cell types (Klimova & Kozmik, 2014). Additionally, several signaling pathways, namely, Notch and Shh, stimulate RPCs to remain proliferative over time. In mice, RGCs secrete Shh, which acts as a mitogen to regulate the rate of RPC cell division and multipotency (Amato, Boy, & Perron, 2004; Cwinn et al., 2011; Wall et al., 2009; Wang, Dakubo, Thurig, Mazerolle, & Wallace, 2005). In next sections we summarize the transcription factor hierarchies that regulate the development of each major retinal cell class.

8.1 Retinal ganglion cells

Retinal ganglion cells are the first neuronal cell type to differentiate. They initially appear in the dorsocentral optic cup but their subsequent neurogenesis spreads in a wave-like fashion to the ventral, peripheral retina (Masai, 2000; McCabe, Gunther, & Reh, 1999). A variety of patterning factors influence the onset and progression of RGC neurogenesis including *Sox2*, *Six3*, *Six6*, and *Pax6*. Among these, Pax6 directly activates the transcription of two proneural bHLH transcription factors, *Neurog2* and *Atoh7* (Brown, Patel, Brzezinski, & Glaser, 2001; Hufnagel, Le, Riesenberg, & Brown, 2010; Kay, Finger-Baier, Roeser, Staub, & Baier, 2001; Marquardt et al., 2001; Wang et al., 2001). In the mouse, *Neurog2* regulates the temporal progression of early neurogenesis (Hufnagel et al., 2010) while *Atoh7* is a competence factor for RGC fate, which is also cross-regulated by *Neurog2* (Skowronska-Krawczyk, Ballivet, Dynlacht, & Matter, 2004). These factors occupy the critical node for RGC development by activating a second wave of transcription factors that include *Sox4*, *Sox11*, *Neurod1*, *Pou4f2* and *Isl1* (Gan, Wang, Huang, & Klein, 1999; Jiang et al., 2013; Kiyama et al., 2011; Mu, Fu, Beremard, Thomas, & Klein, 2008; Pan, Deng, Xie, & Gan, 2008; Wu et al., 2015). In particular, Pou4f2 and Isl1 form heterodimeric protein complexes that lock-in the RGC fate by controlling a number of downstream genes that orchestrate cell cycle exit, RGC specification, neuronal differentiation and function. In the adult retina there are approximately 30 functional subtypes of RGCs (Sanes & Masland, 2015).

8.2 Cones, rods, and bipolar cells

Cones, rods, and bipolar cells differentiate from a tri-potential *Otx2*-expressing subpopulation of RPCs. Cones appear first, followed by rods and finally bipolar cells (Carter-Dawson & LaVail, 1979; Rapaport et al., 2004; Turner & Cepko, 1987; Turner et al., 1990; Young, 1985). In the *Otx2* lineage, Otx2 is expressed in terminally dividing RPCs (Muranishi et al., 2011). The importance of *Otx2* for maintaining the tri-potential precursors is illustrated by loss- and gain-of-function experiments. Removal of

Otx2 blocked the formation and overexpression stimulated the overproduction of all three cell classes (Koike et al., 2007; Nishida et al., 2003; Sato et al., 2007; Wang, Sengel, Emerson, & Cepko, 2014). Immediately downstream of *Otx2* are the activities of *Vsx2*, *Prdm1*, and *Crx*, which influence whether an *Otx2* positive cell will differentiate into a bipolar (*Vsx2*) or photoreceptor neuron (*Prdm1* and *Crx*). *Vsx2* is initially expressed by every RPC, but later restricted to the bipolar lineage, where it represses the molecular program for photoreceptor specification (Dorval, Bobechko, Ahmad, & Bremner, 2005; Dorval et al., 2006; Kim, Matsuda, & Cepko, 2008; Livne-Bar et al., 2006). In the mouse there are 14 functional classes of bipolar neurons (Euler, Haverkamp, Schubert, & Baden, 2014).

Prdm1 encodes a protein that promotes rod and represses bipolar fate (Brzezinski, Lamba, & Reh, 2010; Brzezinski, Uoon Park, & Reh, 2013; Katoh et al., 2010; Wang et al., 2014). *Crx* is also downstream of *Otx2* where it generally promotes both rod and cone fates, photoreceptor neuron outer segment growth and synapse formation (Chen et al., 1997; Freund et al., 1997; Furukawa, Morrow, & Cepko, 1997; Morrow, Furukawa, Raviola, & Cepko, 2005; Samuel, Housset, Fant, & Lamonerie, 2014). By itself, *Crx* cannot induce cone or rod development, but acts with the factors *Nrl* and *Rorβ,* which are essential for these fates (Fu et al., 2014; Mears et al., 2001; Swaroop, Kim, & Forrest, 2010). Genetic studies demonstrated that both Otx2 and Rorβ activate *Nrl* expression, which subsequently activates the transcription of the rod-specific factor *Nr2e3* (Akimoto et al., 2006; Chen, Rattner, & Nathans, 2005; Cheng et al., 2004; Fu et al., 2014; Kautzmann, Kim, Felder-Schmittbuhl, & Swaroop, 2011; Roger et al., 2014).

Unlike rods, there are multiple functional subtypes of cone photoreceptors in vertebrates. The subtypes, identifiable by their opsin expression, exhibit varied distribution frequencies among different species, related to relying on diurnal versus nocturnal vision. Generally, the most common cone subtypes include L (long wavelength), M (medium wavelength), and S (short wavelength). Both *Thrβ2* and *Rxrγ* are expressed in developing cones and required for proper subtype differentiation (Mori, Ghyselinck, Chambon, & Mark, 2001; Ng et al., 2001; Roberts, Hendrickson, McGuire, & Reh, 2005; Roberts, Srinivas, Forrest, de Escobar, & Reh, 2006). Overall, even though mutations to many different retina-specific transcription factors can cause a change in the distribution of photoreceptor and bipolar cells produced, *Otx2* is a key competence factor for the development of each cell type (Brzezinski & Reh, 2015; Cepko, 2015).

8.3 Amacrine and horizontal cells

Horizontal cells differentiate prior to birth whereas amacrine cells appear over a broader time period, both before and after birth in most vertebrates. These two cell classes originate from a bipotential lineage, which coexpress the transcription factors *Pax6*, *Foxn4*, and *Ptf1a*. Genetic studies have defined the individual roles for these transcription factors in amacrine and horizontal development. As mentioned previously, *Pax6* is broadly required by all RPCs whereas deletion of either *Foxn4* or *Ptf1a* completely abolishes horizontal cells and significantly reduces amacrine cells (Fujitani et al., 2006; Klimova & Kozmik, 2014; Li et al., 2004; Marquardt et al., 2001; Oron-Karni et al., 2008). Additionally, both *Foxn4* and *Ptf1a* overexpression can produce amacrine and horizontal cells, suggesting they are both necessary and sufficient for the development of these cell classes (Boije, Fard, Ring, & Hallbook, 2013; Dullin et al., 2007; Lelievre et al., 2011; Li et al., 2004). Based on genetic epistasis testing, *Ptf1a* acts downstream of *Foxn4* as it is missing in *Foxn4* mutants (Fujitani et al., 2006).

In addition to *Ptf1a*, *Onecut 1/2* encode transcription factors that act downstream of *Foxn4* in the horizontal cell lineage. Mutations in any of these three genes disrupt horizontal cell genesis, but rather than cross-regulating one another, these factors perform parallel functions (Fujitani et al., 2006; Klimova, Antosova, Kuzelova, Strnad, & Kozmik, 2015; Li et al., 2004; Wu et al., 2013). Downstream of both *Ptf1a* and *Onecut 1/2* are the transcription factors *Lim1*, *Prox1*, *Lhx1* and *Isl1* that confer global (*Lim1*, *Prox1*) or subtype-specific horizontal neuron identity (*Lhx1*, *Isl1*) (Edqvist, Myers, & Hallbook, 2006; Klimova et al., 2015; Liu, Wang, & Xiang, 2000; Poche et al., 2007; Wu et al., 2013). Additional details about horizontal neurogenesis and functional subtypes are reviewed in Boije, Fard, Edqvist, and Hallbook (2016).

Once the bipotential retinal precursor cells that express *Foxn4*, *Pft1a*, and *Pax6* have selected an amacrine fate, they require the functions of the bHLH factors *Neurod4* and *Neurod1*. Double *Neurod4/Neurod1* mutants completely lack amacrine neurons and both of these factors are still expressed in *Foxn4* and *Ptft1a* mutants, whereas *Neurod1* is still expressed in *Pax6* mutants (Fujitani et al., 2006; Inoue et al., 2002; Li et al., 2004; Marquardt et al., 2001; Morrow, Furukawa, Lee, & Cepko, 1999). Downstream of these main amacrine regulators are transcription factors controlling in amacrine cells subtype differentiation, including *Barhl2*, *Isl1*, *Bhlhb5*, *Neurod2*, *Neurod6*, and the *Ebf* family (Balasubramanian & Gan, 2014; Cherry et al., 2011; Ding et al., 2009; Duquette et al., 2010; Elshatory et al., 2007; Feng et al., 2006; Huang et al., 2014; Jin, Jiang, Mo, & Xiang, 2010).

8.4 Müller glia

Müller glia (MG) are the last cell type born in the retina, and their transcriptomic profiling suggests that they are the final, default state for RPCs. Terminally differentiated MG cells stretch across the retina from the outer to inner limiting membranes and their processes enclose all retinal neurons, thereby structurally supporting the retina (Bringmann et al., 2006). There is an important difference between retinal MG and glia in other regions of the CNS. In the eye, MG develop after most of the neurons and a laminated retinal architecture. However, in the cortex, glia are born much earlier and provide a scaffold on which neurons migrate (Barry, Pakan, & McDermott, 2014; Jadhav, Roesch, & Cepko, 2009). MG and RPCs have multiple shared characteristics, including in their molecular expression profiles. In the adult eye, MG specifically express several transcription factors present at the earliest stages of eye formation (Blackshaw et al., 2004; Jadhav et al., 2009; Livesey, Young, & Cepko, 2004; Roesch et al., 2008). This strongly suggests that MG retain stem cell-like properties (Bernardos, Barthel, Meyers, & Raymond, 2007; Fausett & Goldman, 2006; Fimbel, Montgomery, Burket, & Hyde, 2007). Some of the molecular markers of rodent Müller glia and RPCs include *Hes1*, *Hes5*, *Rax*, *Lhx2*, *Yap*, *Sox2, Sox8,* and *Sox9* (de Melo, Clark, & Blackshaw, 2016; de Melo, Zibetti, et al., 2016; Furukawa,Mukherjee, Bao, Morrow, & Cepko, 2000; Hamon et al., 2017; Hojo et al., 2000; Lin, Ouchi, Satoh, & Watanabe, 2009; Muto, Iida, Satoh, & Watanabe, 2009; Poche, Furuta, Chaboissier, Schedl, & Behringer, 2008; Satow et al., 2001). Of these factors, *Lhx2* is one example of a key regulator of MG genesis; its loss leads to proliferation defects in MG progenitors, downregulation of MG-specific genes, and an inability to obtain MG morphology (de Melo, Zibetti, et al., 2016). Some of the downstream MG genes linked genetically to *Lhx2* include Hes5, the bHLH anti-neurogenic transcription factor, and *Slc1a3* and *Car2* (de Melo, Zibetti, et al., 2016). Hes5, *Slc1a3* and *Car2* were all shown to be direct targets of Lhx2 (de Melo, Zibetti, et al., 2016). Once mature, MG express glial-specific genes and diversified functions (Jadhav et al., 2009).

9. Retina pigmented epithelium

The retina pigmented epithelium is a monolayer of cells located in the posterior eye. The apical surface of these cells contains microvilli that interdigitate between the photoreceptor outer segments. The photoreceptors continuously shed outer segments and the RPE phagocytose and degrade this waste material, which is redistributed to the basal side of the RPE, which abuts the choroidal vasculature. Therefore, the RPE serves as the retinal-blood barrier, since it derives from the eye field/optic vesicle simultaneously with the NR. Because RPE and NR specification and differentiation occur simultaneously, these two tissues are similarly regulated by the early eye gene hierarchy.

9.1 Specification

Mitf and *Otx2* are vital RPE specification transcription factors (Bharti et al., 2008; Lane & Lister, 2012; Zou & Levine, 2012). As discussed in the optic vesicle patterning section, *Mitf* (RPE) and *Vsx2* (NR) mutually repress each other's expression in the optic vesicle, to establish the early boundary between the RPE and NR (Bharti et al., 2008; Zou & Levine, 2012). All of the major signaling pathways (FGF and TGFβ, WNT, BMP, Notch and Hippo) play essential roles during RPE specification (Table 2) with several specifically regulating the expression of *Mitf* and *Otx2*. The WNT mediated transcription factor, *β-catenin*, directly activates *Otx2* and *Mitf* expression (Fujimura, Taketo, Mori, Korinek, & Kozmik, 2009; Westenskow, Piccolo, & Fuhrmann, 2009). Removal of WNT signaling from the optic vesicle blocks RPE specification resulting in *Vsx2* expansion in the RPE domain and RPE transdifferentiation to NR (Bharti et al., 2012; Hagglund, Berghard, & Carlsson, 2013). In addition, WNT and BMP signaling pathways from the surface ectoderm promote RPE specification through a β-catenin independent mechanism, where WNT inhibits GSK3β and BMP activates pSmad (Steinfeld et al., 2013). Later in development, loss of WNT from the surface ectoderm decreases β-catenin mediated *Mitf* and *Otx2* expression at the rim of the optic cup, which causes a decrease in RPE cell numbers (Carpenter et al., 2015; Westenskow et al., 2009). RPE cell autonomous signaling is also essential for fate specification and proper growth. In the both zebrafish and mice, the Hippo signaling pathway is active at the optic vesicle stage, with the transcriptional coactivator Yap, which is expressed throughout the optic cup (Kim et al., 2016; Miesfeld et al., 2015). In both species *Yap* is required for RPE specification with its role shown to be cell autonomous in zebrafish based on transplantation experiments of genetic mutants (Kim et al., 2016; Miesfeld et al., 2015). In mice, loss of *Yap* results in RPE to NR transdifferentiation (Kim et al., 2016). In complement to *Yap* in zebrafish, the other Hippo signaling coactivator, *Taz*, is not individually responsible for RPE specification but acts redundantly with Yap (Miesfeld et al., 2015). The direct downstream targets of Yap and Taz in the RPE are currently unknown. After RPE specification, an upstream Hippo signaling gene, *Nf2*, is expressed early in the optic vesicle, later in the RPE and required to maintain the proper balance of Yap/Taz activity in the developing RPE to prevent hyperplastic growth (Moon et al., 2018).

9.2 Differentiation

Pax6 and *Mitf* act reiteratively during both RPE specification and differentiation. Early in these processes, Pax6 activates the transcription of an RPE-specific *Mitf* isoform and subsequently acts in parallel with Mitf to regulate the transcription of the shared RPE downstream genes *Try, Dct, Tryp1* (Bharti et al., 2008; Cavodeassi & Bovolenta, 2014; Raviv et al., 2014; Tachibana, 2000; Tachibana et al., 1994). An important function of RPE cells is the deposition of melanin pigment, which absorbs stray light not captured by the photoreceptors. This critical function, along with removing photoreceptor cell waste products, prevents the accumulation of photo-oxidative stress that contributes to rod and cone cell death (Datta, Cano, Ebrahimi, Wang, & Handa, 2017). Best1, a Ca-dependent chloride channel, is a marker of differentiated RPE cells. *Best1* transcription is coordinately regulated by the *Mitf, Otx2,* and *Sox9* factors (Esumi, Kachi, Campochiaro, & Zack, 2007; Esumi et al., 2009; Masuda & Esumi, 2010). *Otx2*, *Sox9*, and *Lhx2* also act in concert to transcriptionally activate the RPE visual cycle genes *RPE65*, *RGR*, and *RLBP1* (Masuda et al., 2014). The Hedgehog (HH) signaling pathway is also important for RPE differentiation. In the frog RPE, the secreted ligands *Banded* and *Cephalic Hh*, downstream receptor *X-Smo* and transcription factors *Gli2*, *Gli3*, are expressed. Inhibition of HH signaling using dominant negative PKA, or the drug cyclopamine, reduces differentiation *Mitf* expression and RPE cell pigmentation (Perron et al., 2003). Additionally, IHH in the developing mouse choroid signals to the RPE to propagate RPE differentiation and pigmentation (Dakubo et al., 2008).

10. Summary

Organogenesis of the vertebrate eye begins early in gastrulation and requires intricate coordination between transcription factors, signaling pathways, and morphogenetic movements to produce the adult eye. Multiple ocular tissues arise from the eye field within the neuroectoderm, but even the tissues with other embryonic origins (cornea, lens) rely on optic cup/vesicle inductive signals for their proper formation. The development of all ocular structures described in this chapter highlights the reiterative use of a common set of transcription factors and signaling pathways from initial specification to terminal differentiation. For some tissues, rapid progress has been made in elucidating underlying developmental gene regulatory networks. With the advent of single cell techniques, however, new data sets will provide a more comprehensive view of genetic and molecular interactions that regulate all aspects of eye organogenesis.

References

Acampora, D., Mazan, S., Avantaggiato, V., Barone, P., Tuorto, F., Lallemand, Y., et al. (1996). Epilepsy and brain abnormalities in mice lacking the Otx1 gene. *Nature Genetics, 14*, 218–222.

Ahmad, N., Aslam, M., Muenster, D., Horsch, M., Khan, M. A., Carlsson, P., et al. (2013). Pitx3 directly regulates Foxe3 during early lens development. *International Journal of Developmental Biology, 57*, 741–751.

Akimoto, M., Cheng, H., Zhu, D. X., Brzezinski, J. A., Khanna, R., Filippova, E., et al. (2006). Targeting of GFP to newborn rods by Nrl promoter and temporal expression profiling of flow-sorted photoreceptors. *Proceedings of the National Academy of Sciences of the United States of America, 103*, 3890–3895.

Amato, M. A., Boy, S., & Perron, M. (2004). Hedgehog signaling in vertebrate eye development: A growing puzzle. *Cellular and Molecular Life Sciences, 61*, 899–910.

Bailey, T. J., El-Hodiri, H., Zhang, L., Shah, R., Mathers, P. H., & Jamrich, M. (2004). Regulation of vertebrate eye development by Rx genes. *The International Journal of Developmental Biology, 48*, 761–770.

Balasubramanian, R., & Gan, L. (2014). Development of retinal amacrine cells and their dendritic stratification. *Current Ophthalmology Reports, 2*, 100–106.

Barry, D. S., Pakan, J. M., & McDermott, K. W. (2014). Radial glial cells: Key organisers in CNS development. *The International Journal of Biochemistry & Cell Biology, 46*, 76–79.

Bassnett, S., Missey, H., & Vucemilo, I. (1999). Molecular architecture of the lens fiber cell basal membrane complex. *Journal of Cell Science, 112*(Pt. 13), 2155–2165.

Baulmann, D. C., Ohlmann, A., Flugel-Koch, C., Goswami, S., Cvekl, A., & Tamm, E. R. (2002). Pax6 heterozygous eyes show defects in chamber angle differentiation that are associated with a wide spectrum of other anterior eye segment abnormalities. *Mechanisms of Development, 118*, 3–17.

Baumer, N., Marquardt, T., Stoykova, A., Spieler, D., Treichel, D., Ashery-Padan, R., et al. (2003). Retinal pigmented epithelium determination requires the redundant activities of Pax2 and Pax6. *Development, 130*, 2903–2915.

Beebe, D. C. (1986). Development of the ciliary body: A brief review. *Transactions of the Ophthalmological Societies of the United Kingdom, 105*(Pt. 2), 123–130.

Belanger, M. C., Robert, B., & Cayouette, M. (2017). Msx1-positive progenitors in the retinal ciliary margin give rise to both neural and non-neural progenies in mammals. *Developmental Cell, 40*, 137–150.

Bernardos, R. L., Barthel, L. K., Meyers, J. R., & Raymond, P. A. (2007). Late-stage neuronal progenitors in the retina are radial Muller glia that function as retinal stem cells. *The Journal of Neuroscience, 27*, 7028–7040.

Berry, F. B., Lines, M. A., Oas, J. M., Footz, T., Underhill, D. A., Gage, P. J., et al. (2006). Functional interactions between FOXC1 and PITX2 underlie the sensitivity to

FOXC1 gene dose in Axenfeld-Rieger syndrome and anterior segment dysgenesis. *Human Molecular Genetics*, *15*, 905–919.

Bharti, K., Gasper, M., Ou, J., Brucato, M., Clore-Gronenborn, K., Pickel, J., et al. (2012). A regulatory loop involving PAX6, MITF, and WNT signaling controls retinal pigment epithelium development. *PLoS Genetics*, *8*, e1002757.

Bharti, K., Liu, W., Csermely, T., Bertuzzi, S., & Arnheiter, H. (2008). Alternative promoter use in eye development: The complex role and regulation of the transcription factor MITF. *Development*, *135*, 1169–1178.

Bhattacharyya, S., Bailey, A. P., Bronner-Fraser, M., & Streit, A. (2004). Segregation of lens and olfactory precursors from a common territory: Cell sorting and reciprocity of Dlx5 and Pax6 expression. *Developmental Biology*, *271*, 403–414.

Bielen, H., & Houart, C. (2012). BMP signaling protects telencephalic fate by repressing eye identity and its Cxcr4-dependent morphogenesis. *Developmental Cell*, *23*, 812–822.

Bishop, P. N., Takanosu, M., Le Goff, M., & Mayne, R. (2002). The role of the posterior ciliary body in the biosynthesis of vitreous humour. *Eye (London, England)*, *16*, 454–460.

Blackshaw, S., Harpavat, S., Trimarchi, J., Cai, L., Huang, H., Kuo, W. P., et al. (2004). Genomic analysis of mouse retinal development. *PLoS Biology*, *2*, E247.

Blixt, A., Landgren, H., Johansson, B. R., & Carlsson, P. (2007). Foxe3 is required for morphogenesis and differentiation of the anterior segment of the eye and is sensitive to Pax6 gene dosage. *Developmental Biology*, *302*, 218–229.

Blixt, A., Mahlapuu, M., Aitola, M., Pelto-Huikko, M., Enerback, S., & Carlsson, P. (2000). A forkhead gene, FoxE3, is essential for lens epithelial proliferation and closure of the lens vesicle. *Genes & Development*, *14*, 245–254.

Bohnsack, B. L., Kasprick, D. S., Kish, P. E., Goldman, D., & Kahana, A. (2012). A zebrafish model of Axenfeld-Rieger syndrome reveals that pitx2 regulation by retinoic acid is essential for ocular and craniofacial development. *Investigative Ophthalmology & Visual Science*, *53*, 7–22.

Boije, H., Fard, S. S., Edqvist, P. H., & Hallbook, F. (2016). Horizontal cells, the odd ones out in the retina, give insights into development and disease. *Frontiers in Neuroanatomy*, *10*, 77.

Boije, H., Fard, S. S., Ring, H., & Hallbook, F. (2013). Forkheadbox N4 (FoxN4) triggers context-dependent differentiation in the developing chick retina and neural tube. *Differentiation*, *85*, 11–19.

Bringmann, A., Pannicke, T., Grosche, J., Francke, M., Wiedemann, P., Skatchkov, S. N., et al. (2006). Muller cells in the healthy and diseased retina. *Progress in Retinal and Eye Research*, *25*, 397–424.

Brown, K. E., Keller, P. J., Ramialison, M., Rembold, M., Stelzer, E. H., Loosli, F., et al. (2010). Nlcam modulates midline convergence during anterior neural plate morphogenesis. *Developmental Biology*, *339*, 14–25.

Brown, N. L., Patel, S., Brzezinski, J., & Glaser, T. (2001). Math5 is required for retinal ganglion cell and optic nerve formation. *Development*, *128*, 2497–2508.

Brzezinski, J. A., Lamba, D. A., & Reh, T. A. (2010). Blimp1 controls photoreceptor versus bipolar cell fate choice during retinal development. *Development*, *137*, 619–629.

Brzezinski, J. A., & Reh, T. A. (2015). Photoreceptor cell fate specification in vertebrates. *Development*, *142*, 3263–3273.

Brzezinski, J. A., Uoon Park, K., & Reh, T. A. (2013). Blimp1 (Prdm1) prevents re-specification of photoreceptors into retinal bipolar cells by restricting competence. *Developmental Biology*, *384*, 194–204.

Burmeister, M., Novak, J., Liang, M. Y., Basu, S., Ploder, L., Hawes, N. L., et al. (1996). Ocular retardation mouse caused by Chx10 homeobox null allele: Impaired retinal progenitor proliferation and bipolar cell differentiation. *Nature Genetics*, *12*, 376–384.

Cammas, L., Wolfe, J., Choi, S. Y., Dedhar, S., & Beggs, H. E. (2012). Integrin-linked kinase deletion in the developing lens leads to capsule rupture, impaired fiber migration and non-apoptotic epithelial cell death. *Investigative Ophthalmology & Visual Science, 53*, 3067–3081.

Carbe, C., & Zhang, X. (2011). Lens induction requires attenuation of ERK signaling by Nf1. *Human Molecular Genetics, 20*, 1315–1323.

Carpenter, A. C., Smith, A. N., Wagner, H., Cohen-Tayar, Y., Rao, S., Wallace, V., et al. (2015). Wnt ligands from the embryonic surface ectoderm regulate 'bimetallic strip' optic cup morphogenesis in mouse. *Development, 142*, 972–982.

Carter-Dawson, L. D., & LaVail, M. M. (1979). Rods and cones in the mouse retina. II. Autoradiographic analysis of cell generation using tritiated thymidine. *The Journal of Comparative Neurology, 188*, 263–272.

Cavodeassi, F., & Bovolenta, P. (2014). New functions for old genes: Pax6 and Mitf in eye pigment biogenesis. *Pigment Cell & Melanoma Research, 27*, 1005–1007.

Cechmanek, P. B., & McFarlane, S. (2017). Retinal pigment epithelium expansion around the neural retina occurs in two separate phases with distinct mechanisms. *Developmental Dynamics, 246*, 598–609.

Cepko, C. L. (2015). The determination of rod and cone photoreceptor fate. *Annual Review of Vision Science, 1*, 211–234.

Chaloindufau, C., Sun, T. T., & Dhouailly, D. (1990). Appearance of the keratin pair K3/K12 during embryonic and adult corneal epithelial differentiation in the chick and in the rabbit. *Cell Differentiation and Development, 32*, 97–108.

Chang, B., Smith, R. S., Peters, M., Savinova, O. V., Hawes, N. L., Zabaleta, A., et al. (2001). Haploinsufficient Bmp4 ocular phenotypes include anterior segment dysgenesis with elevated intraocular pressure. *BMC Genetics, 2*, 18.

Chauhan, B. K., Disanza, A., Choi, S. Y., Faber, S. C., Lou, M., Beggs, H. E., et al. (2009). Cdc42- and IRSp53-dependent contractile filopodia tether presumptive lens and retina to coordinate epithelial invagination. *Development, 136*, 3657–3667.

Chauhan, B. K., Lou, M., Zheng, Y., & Lang, R. A. (2011). Balanced Rac1 and RhoA activities regulate cell shape and drive invagination morphogenesis in epithelia. *Proceedings of the National Academy of Sciences of the United States of America, 108*, 18289–18294.

Chauhan, B. K., Reed, N. A., Yang, Y., Cermak, L., Reneker, L., Duncan, M. K., et al. (2002). A comparative cDNA microarray analysis reveals a spectrum of genes regulated by Pax6 in mouse lens. *Genes to Cells, 7*, 1267–1283.

Chauhan, B. K., Reed, N. A., Zhang, W. J., Duncan, M. K., Kilimann, M. W., & Cvekl, A. (2002). Identification of genes downstream of Pax6 in the mouse lens using cDNA microarrays. *The Journal of Biological Chemistry, 277*, 11539–11548.

Chauhan, B. K., Zhang, W. Y., Cveklova, K., Kantorow, M., & Cvekl, A. (2002). Identification of differentially expressed genes in mouse Pax6 heterozygous lenses. *Investigative Ophthalmology & Visual Science, 43*, 1884–1890.

Chen, J. C., Rattner, A., & Nathans, J. (2005). The rod photoreceptor-specific nuclear receptor Nr2e3 represses transcription of multiple cone-specific genes. *Journal of Neuroscience, 25*, 118–129.

Chen, S. M., Wang, Q. L., Nie, Z. Q., Sun, H., Lennon, G., Copeland, N. G., et al. (1997). Crx, a novel Otx-like paired-homeodomain protein, binds to and transactivates photoreceptor cell-specific genes. *Neuron, 19*, 1017–1030.

Cheng, H., Khanna, H., Oh, E. C. T., Hicks, D., Mitton, K. P., & Swaroop, A. (2004). Photoreceptor-specific nuclear receptor NR2E3 functions as a transcriptional activator in rod photoreceptors. *Human Molecular Genetics, 13*, 1563–1575.

Cherry, T. J., Wang, S., Bormuth, I., Schwab, M., Olson, J., & Cepko, C. L. (2011). NeuroD factors regulate cell fate and neurite stratification in the developing retina. *Journal of Neuroscience, 31*, 7365–7379.

Chiang, C., Litingtung, Y., Lee, E., Young, K. E., Corden, J. L., Westphal, H., et al. (1996). Cyclopia and defective axial patterning in mice lacking Sonic hedgehog gene function. *Nature*, *383*, 407–413.
Cho, S. H., & Cepko, C. L. (2006). Wnt2b/beta-catenin-mediated canonical Wnt signaling determines the peripheral fates of the chick eye. *Development*, *133*, 3167–3177.
Cintron, C., Covington, H., & Kublin, C. L. (1983). Morphogenesis of rabbit corneal stroma. *Investigative Ophthalmology & Visual Science*, *24*, 543–556.
Coca-Prados, M., Escribano, J., & Ortego, J. (1999). Differential gene expression in the human ciliary epithelium. *Progress in Retinal and Eye Research*, *18*, 403–429.
Collomb, E., Yang, Y., Foriel, S., Cadau, S., Pearton, D. J., & Dhouailly, D. (2013). The corneal epithelium and lens develop independently from a common pool of precursors. *Developmental Dynamics*, *242*, 401–413.
Coulombre, J. L., & Coulombre, A. J. (1963). Lens development—Fiber elongation and lens orientation. *Science*, *142*, 1489–1490.
Creuzet, S., Vincent, C., & Couly, G. R. (2005). Neural crest derivatives in ocular and periocular structures. *International Journal of Developmental Biology*, *49*, 161–171.
Cui, W., Tomarev, S. I., Piatigorsky, J., Chepelinsky, A. B., & Duncan, M. K. (2004). Mafs, Prox1, and Pax6 can regulate chicken betaB1-crystallin gene expression. *The Journal of Biological Chemistry*, *279*, 11088–11095.
Cvekl, A., & Ashery-Padan, R. (2014). The cellular and molecular mechanisms of vertebrate lens development. *Development*, *141*, 4432–4447.
Cvekl, A., & Tamm, E. R. (2004). Anterior eye development and ocular mesenchyme: New insights from mouse models and human diseases. *BioEssays*, *26*, 374–386.
Cvekl, A., & Zhang, X. (2017). Signaling and gene regulatory networks in mammalian lens development. *Trends in Genetics*, *33*, 677–702.
Cwinn, M. A., Mazerolle, C., McNeill, B., Ringuette, R., Thurig, S., Hui, C. C., et al. (2011). Suppressor of fused is required to maintain the multipotency of neural progenitor cells in the retina. *The Journal of Neuroscience*, *31*, 5169–5180.
Dakubo, G. D., Mazerolle, C., Furimsky, M., Yu, C., St-Jacques, B., McMahon, A. P., et al. (2008). Indian hedgehog signaling from endothelial cells is required for sclera and retinal pigment epithelium development in the mouse eye. *Developmental Biology*, *320*, 242–255.
Danno, H., Michiue, T., Hitachi, K., Yukita, A., Ishiura, S., & Asashima, M. (2008). Molecular links among the causative genes for ocular malformation: Otx2 and Sox2 coregulate Rax expression. *Proceedings of the National Academy of Sciences of the United States of America*, *105*, 5408–5413.
Datta, S., Cano, M., Ebrahimi, K., Wang, L., & Handa, J. T. (2017). The impact of oxidative stress and inflammation on RPE degeneration in non-neovascular AMD. *Progress in Retinal and Eye Research*, *60*, 201–218.
Davis, J., Duncan, M. K., Robison, W. G., Jr., & Piatigorsky, J. (2003). Requirement for Pax6 in corneal morphogenesis: A role in adhesion. *Journal of Cell Science*, *116*, 2157–2167.
Davis, N., Yoffe, C., Raviv, S., Antes, R., Berger, J., Holzmann, S., et al. (2009). Pax6 dosage requirements in iris and ciliary body differentiation. *Developmental Biology*, *333*, 132–142.
Davis-Silberman, N., & Ashery-Padan, R. (2008). Iris development in vertebrates; genetic and molecular considerations. *Brain Research*, *1192*, 17–28.
Davis-Silberman, N., Kalich, T., Oron-Karni, V., Marquardt, T., Kroeber, M., Tamm, E. R., et al. (2005). Genetic dissection of Pax6 dosage requirements in the developing mouse eye. *Human Molecular Genetics*, *14*, 2265–2276.
Dawes, L. J., Shelley, E. J., McAvoy, J. W., & Lovicu, F. J. (2018). A role for Hippo/YAP-signaling in FGF-induced lens epithelial cell proliferation and fibre differentiation. *Experimental Eye Research*, *169*, 122–133.

de Melo, J., Clark, B. S., & Blackshaw, S. (2016). Multiple intrinsic factors act in concert with Lhx2 to direct retinal gliogenesis. *Scientific Reports*, *6*, 32757.

de Melo, J., Zibetti, C., Clark, B. S., Hwang, W., Miranda-Angulo, A. L., Qian, J., et al. (2016). Lhx2 is an essential factor for retinal gliogenesis and notch signaling. *The Journal of Neuroscience*, *36*, 2391–2405.

DelMonte, D. W., & Kim, T. (2011). Anatomy and physiology of the cornea. *Journal of Cataract and Refractive Surgery*, *37*, 588–598.

Ding, Q., Chen, H., Xie, X., Libby, R. T., Tian, N., & Gan, L. (2009). BARHL2 differentially regulates the development of retinal amacrine and ganglion neurons. *Journal of Neuroscience*, *29*, 3992–4003.

Dorval, K. M., Bobechko, B. P., Ahmad, K. F., & Bremner, R. (2005). Transcriptional activity of the paired-like homeodomain proteins CHX10 and VSX1. *The Journal of Biological Chemistry*, *280*, 10100–10108.

Dorval, K. M., Bobechko, B. P., Fujieda, H., Chen, S. M., Zack, D. J., & Bremner, R. (2006). CHX10 targets a subset of photoreceptor genes. *The Journal of Biological Chemistry*, *281*, 744–751.

Dullin, J. P., Locker, M., Robach, M., Henningfeld, K. A., Parain, K., Afelik, S., et al. (2007). Ptf1a triggers GABAergic neuronal cell fates in the retina. *BMC Developmental Biology*, 7, 110.

Duncan, M. K., Kozmik, Z., Cveklova, K., Piatigorsky, J., & Cvekl, A. (2000). Overexpression of PAX6(5a) in lens fiber cells results in cataract and upregulation of alpha 5 beta 1 integrin expression. *Journal of Cell Science*, *113*, 3173–3185.

Duquette, P. M., Zhou, X., Yap, N. L., MacLaren, E. J., Lu, J. J., Wallace, V. A., et al. (2010). Loss of LMO4 in the retina leads to reduction of GABAergic amacrine cells and functional deficits. *PLoS One*, *5*, e13232.

Ebert, A. M., Childs, S. J., Hehr, C. L., Cechmanek, P. B., & McFarlane, S. (2014). Sema6a and Plxna2 mediate spatially regulated repulsion within the developing eye to promote eye vesicle cohesion. *Development*, *141*, 2473–2482.

Edqvist, P. H. D., Myers, S. M., & Hallbook, F. (2006). Early identification of retinal subtypes in the developing, pre-laminated chick retina using the transcription factors Prox1, Lim1, Ap2 alpha, Pax6, Isl1, Isl2, Lim3 and Chx10. *European Journal of Histochemistry*, *50*, 147–154.

Eiraku, M., Takata, N., Ishibashi, H., Kawada, M., Sakakura, E., Okuda, S., et al. (2011). Self-organizing optic-cup morphogenesis in three-dimensional culture. *Nature*, *472*, 51–56.

Elliott, J., Jolicoeur, C., Ramamurthy, V., & Cayouette, M. (2008). Ikaros confers early temporal competence to mouse retinal progenitor cells. *Neuron*, *60*, 26–39.

Elshatory, Y., Everhart, D., Deng, M., Xie, X. L., Barlow, R. B., & Gan, L. (2007). Islet-1 controls the differentiation of retinal bipolar and cholinergic amacrine cells. *Journal of Neuroscience*, *27*, 12707–12720.

England, S. J., Blanchard, G. B., Mahadevan, L., & Adams, R. J. (2006). A dynamic fate map of the forebrain shows how vertebrate eyes form and explains two causes of cyclopia. *Development*, *133*, 4613–4617.

Esumi, N., Kachi, S., Campochiaro, P. A., & Zack, D. J. (2007). VMD2 promoter requires two proximal E-box sites for its activity in vivo and is regulated by the MITF-TFE family. *The Journal of Biological Chemistry*, *282*, 1838–1850.

Esumi, N., Kachi, S., Hackler, L., Jr., Masuda, T., Yang, Z., Campochiaro, P. A., et al. (2009). BEST1 expression in the retinal pigment epithelium is modulated by OTX family members. *Human Molecular Genetics*, *18*, 128–141.

Euler, T., Haverkamp, S., Schubert, T., & Baden, T. (2014). Retinal bipolar cells: Elementary building blocks of vision. *Nature Reviews. Neuroscience*, *15*, 507–519.

Evans, A. L., & Gage, P. J. (2005). Expression of the homeobox gene Pitx2 in neural crest is required for optic stalk and ocular anterior segment development. *Human Molecular Genetics*, *14*, 3347–3359.

Faber, S. C., Robinson, M. L., Makarenkova, H. P., & Lang, R. A. (2002). Bmp signaling is required for development of primary lens fiber cells. *Development*, *129*, 3727–3737.

Fausett, B. V., & Goldman, D. (2006). A role for alpha1 tubulin-expressing Muller glia in regeneration of the injured zebrafish retina. *The Journal of Neuroscience*, *26*, 6303–6313.

Feng, L., Xie, X. L., Joshi, P. S., Yang, Z. Y., Shibasaki, K., Chow, R. L., et al. (2006). Requirement for Bhlhb5 in the specification of amacrine and cone bipolar subtypes in mouse retina. *Development*, *133*, 4815–4825.

Fimbel, S. M., Montgomery, J. E., Burket, C. T., & Hyde, D. R. (2007). Regeneration of inner retinal neurons after intravitreal injection of ouabain in zebrafish. *The Journal of Neuroscience*, *27*, 1712–1724.

Freund, C. L., GregoryEvans, C. Y., Furukawa, T., Papaioannou, M., Looser, J., Ploder, L., et al. (1997). Cone-rod dystrophy due to mutations in a novel photoreceptor-specific homeobox gene (CRX) essential for maintenance of the photoreceptor. *Cell*, *91*, 543–553.

Fu, Y. L., Liu, H., Ng, L., Kim, J. W., Hao, H., Swaroop, A., et al. (2014). Feedback induction of a photoreceptor-specific isoform of retinoid-related orphan nuclear receptor beta by the rod transcription factor NRL. *The Journal of Biological Chemistry*, *289*, 32469–32480.

Fuhrmann, S., Levine, E. M., & Reh, T. A. (2000). Extraocular mesenchyme patterns the optic vesicle during early eye development in the embryonic chick. *Development*, *127*, 4599–4609.

Fujimura, N., Taketo, M. M., Mori, M., Korinek, V., & Kozmik, Z. (2009). Spatial and temporal regulation of Wnt/beta-catenin signaling is essential for development of the retinal pigment epithelium. *Developmental Biology*, *334*, 31–45.

Fujitani, Y., Fujitani, S., Luo, H. J., Qiu, F., Burlison, J., Long, Q. M., et al. (2006). Ptf1a determines horizontal and amacrine cell fates during mouse retinal development. *Development*, *133*, 4439–4450.

Furukawa, T., Morrow, E. M., & Cepko, C. L. (1997). Crx, a novel otx-like homeobox gene, shows photoreceptor-specific expression and regulates photoreceptor differentiation. *Cell*, *91*, 531–541.

Furukawa, T., Mukherjee, S., Bao, Z. Z., Morrow, E. M., & Cepko, C. L. (2000). rax, Hes1, and notch1 promote the formation of Muller glia by postnatal retinal progenitor cells. *Neuron*, *26*, 383–394.

Furuta, Y., & Hogan, B. L. M. (1998). BMP4 is essential for lens induction in the mouse embryo. *Genes & Development*, *12*, 3764–3775.

Gage, P. J., Suh, H., & Camper, S. A. (1999). Dosage requirement of Pitx2 for development of multiple organs. *Development*, *126*, 4643–4651.

Gan, L., Wang, S. W., Huang, Z., & Klein, W. H. (1999). POU domain factor Brn-3b is essential for retinal ganglion cell differentiation and survival but not for initial cell fate specification. *Developmental Biology*, *210*, 469–480.

Geng, X., Speirs, C., Lagutin, O., Inbal, A., Liu, W., Solnica-Krezel, L., et al. (2008). Haploinsufficiency of Six3 fails to activate Sonic hedgehog expression in the ventral forebrain and causes holoprosencephaly. *Developmental Cell*, *15*, 236–247.

Georgi, S. A., & Reh, T. A. (2011). Dicer is required for the maintenance of notch signaling and gliogenic competence during mouse retinal development. *Developmental Neurobiology*, *71*, 1153–1169.

Gerhart, J., Pfautz, J., Neely, C., Elder, J., DuPrey, K., Menko, A. S., et al. (2009). Noggin producing, MyoD-positive cells are crucial for eye development. *Developmental Biology*, *336*, 30–41.

Grocott, T., Johnson, S., Bailey, A. P., & Streit, A. (2011). Neural crest cells organize the eye via TGF-beta and canonical Wnt signalling. *Nature Communications*, *2*, 265.
Haddad, A., & Bennett, G. (1988). Synthesis of lens capsule and plasma membrane glycoproteins by lens epithelial cells and fibers in the rat. *The American Journal of Anatomy*, *183*, 212–225.
Hagglund, A. C., Berghard, A., & Carlsson, L. (2013). Canonical Wnt/beta-catenin signalling is essential for optic cup formation. *PLoS One*, *8*, e81158.
Hamon, A., Masson, C., Bitard, J., Gieser, L., Roger, J. E., & Perron, M. (2017). Retinal degeneration triggers the activation of YAP/TEAD in reactive Muller cells. *Investigative Ophthalmology & Visual Science*, *58*, 1941–1953.
Hanna, C., Bicknell, D. S., & Obrien, J. E. (1961). Cell turnover in adult human eye. *Archives of Ophthalmology*, *65*, 695–698.
Hanssen, E., Franc, S., & Garrone, R. (2001). Synthesis and structural organization of zonular fibers during development and aging. *Matrix Biology*, *20*, 77–85.
Hay, E. D. (1980). Development of the vertebrate cornea. *International Review of Cytology*, *63*, 263–322.
Heermann, S., Schutz, L., Lemke, S., Krieglstein, K., & Wittbrodt, J. (2015). Eye morphogenesis driven by epithelial flow into the optic cup facilitated by modulation of bone morphogenetic protein. *eLife*, *4*, e05216.
Hjalt, T. A., Semina, E. V., Amendt, B. A., & Murray, J. C. (2000). The Pitx2 protein in mouse development. *Developmental Dynamics*, *218*, 195–200.
Hojo, M., Ohtsuka, T., Hashimoto, N., Gradwohl, G., Guillemot, F., & Kageyama, R. (2000). Glial cell fate specification modulated by the bHLH gene Hes5 in mouse retina. *Development*, *127*, 2515–2522.
Hollemann, T., Bellefroid, E., & Pieler, T. (1998). The Xenopus homologue of the Drosophila gene tailless has a function in early eye development. *Development*, *125*, 2425–2432.
Houzelstein, D., Cohen, A., Buckingham, M. E., & Robert, B. (1997). Insertional mutation of the mouse Msx1 homeobox gene by an nlacZ reporter gene. *Mechanisms of Development*, *65*, 123–133.
Huang, L., Hu, F., Feng, L., Luo, X. J., Liang, G. Q., Zeng, X. Y., et al. (2014). Bhlhb5 is required for the subtype development of retinal amacrine and bipolar cells in mice. *Developmental Dynamics*, *243*, 279–289.
Huang, J., Liu, Y., Filas, B., Gunhaga, L., & Beebe, D. C. (2015). Negative and positive autoregulation of BMP expression in early eye development. *Developmental Biology*, *407*, 256–264.
Huang, J., Rajagopal, R., Liu, Y., Dattilo, L. K., Shaham, O., Ashery-Padan, R., et al. (2011). The mechanism of lens placode formation: A case of matrix-mediated morphogenesis. *Developmental Biology*, *355*, 32–42.
Hufnagel, R. B., Le, T. T., Riesenberg, A. L., & Brown, N. L. (2010). Neurog2 controls the leading edge of neurogenesis in the mammalian retina. *Developmental Biology*, *340*, 490–503.
Hyer, J., Kuhlman, J., Afif, E., & Mikawa, T. (2003). Optic cup morphogenesis requires prelens ectoderm but not lens differentiation. *Developmental Biology*, *259*, 351–363.
Inoue, T., Hojo, M., Bessho, Y., Tano, Y., Lee, J. E., & Kageyama, R. (2002). Math3 and NeuroD regulate amacrine cell fate specification in the retina. *Development*, *129*, 831–842.
Ittner, L. M., Wurdak, H., Schwerdtfeger, K., Kunz, T., Ille, F., Leveen, P., et al. (2005). Compound developmental eye disorders following inactivation of TGFbeta signaling in neural-crest stem cells. *Journal of Biology*, *4*, 11.
Ivanovitch, K., Cavodeassi, F., & Wilson, S. W. (2013). Precocious acquisition of neuroepithelial character in the eye field underlies the onset of eye morphogenesis. *Developmental Cell*, *27*, 293–305.

Jadhav, A. P., Roesch, K., & Cepko, C. L. (2009). Development and neurogenic potential of Muller glial cells in the vertebrate retina. *Progress in Retinal and Eye Research*, *28*, 249–262.
Jarrin, M., Pandit, T., & Gunhaga, L. (2012). A balance of FGF and BMP signals regulates cell cycle exit and *Equarin* expression in lens cells. *Molecular Biology of the Cell*, *23*, 3266–3274.
Jeong, Y., Leskow, F. C., El-Jaick, K., Roessler, E., Muenke, M., Yocum, A., et al. (2008). Regulation of a remote Shh forebrain enhancer by the Six3 homeoprotein. *Nature Genetics*, *40*, 1348–1353.
Jia, J. L., Lin, M., Zhang, L. N., York, J. P., & Zhang, P. M. (2007). The Notch signaling pathway controls the size of the ocular lens by directly suppressing p57(Kip2) expression. *Molecular and Cellular Biology*, *27*, 7236–7247.
Jiang, Y., Ding, Q., Xie, X. L., Libby, R. T., Lefebvre, V., & Gan, L. (2013). Transcription factors SOX4 and SOX11 function redundantly to regulate the development of mouse retinal ganglion cells. *The Journal of Biological Chemistry*, *288*, 18429–18438.
Jidigam, V. K., Srinivasan, R. C., Patthey, C., & Gunhaga, L. (2015). Apical constriction and epithelial invagination are regulated by BMP activity. *Biology Open*, *4*, 1782–1791.
Jin, K. X., Jiang, H. S., Mo, Z. Q., & Xiang, M. Q. (2010). Early B-cell factors are required for specifying multiple retinal cell types and subtypes from postmitotic precursors. *Journal of Neuroscience*, *30*, 11902–11916.
Johnston, M. C., Noden, D. M., Hazelton, R. D., Coulombre, J. L., & Coulombre, A. J. (1979). Origins of avian ocular and periocular tissues. *Experimental Eye Research*, *29*, 27–43.
Kallifatidis, G., Boros, J., Shin, E. H. H., McAvoy, J. W., & Lovicu, F. J. (2011). The fate of dividing cells during lens morphogenesis, differentiation and growth. *Experimental Eye Research*, *92*, 502–511.
Kamachi, Y., Uchikawa, M., Tanouchi, A., Sekido, R., & Kondoh, H. (2001). Pax6 and SOX2 form a co-DNA-binding partner complex that regulates initiation of lens development. *Genes & Development*, *15*, 1272–1286.
Katoh, K., Omori, Y., Onishi, A., Sato, S., Kondo, M., & Furukawa, T. (2010). Blimp1 suppresses Chx10 expression in differentiating retinal photoreceptor precursors to ensure proper photoreceptor development. *The Journal of Neuroscience*, *30*, 6515–6526.
Kautzmann, M. A. I., Kim, D. S., Felder-Schmittbuhl, M. P., & Swaroop, A. (2011). Combinatorial regulation of photoreceptor differentiation factor, neural retina leucine zipper gene NRL, revealed by in vivo promoter analysis. *The Journal of Biological Chemistry*, *286*, 28247–28255.
Kay, J. N., Finger-Baier, K. C., Roeser, T., Staub, W., & Baier, H. (2001). Retinal ganglion cell genesis requires lakritz, a zebrafish atonal homolog. *Neuron*, *30*, 725–736.
Kidson, S. H., Kume, T., Deng, K. Y., Winfrey, V., & Hogan, B. L. M. (1999). The forkhead/winged-helix gene, Mf1, is necessary for the normal development of the cornea and formation of the anterior chamber in the mouse eye. *Developmental Biology*, *211*, 306–322.
Kim, D. S., Matsuda, T., & Cepko, C. L. (2008). A core paired-type and POU homeodomain-containing transcription factor program drives retinal bipolar cell gene expression. *Journal of Neuroscience*, *28*, 7748–7764.
Kim, J. Y., Park, R., Lee, J. H., Shin, J., Nickas, J., Kim, S., et al. (2016). Yap is essential for retinal progenitor cell cycle progression and RPE cell fate acquisition in the developing mouse eye. *Developmental Biology*, *419*, 336–347.
Kiyama, T., Mao, C. A., Cho, J. H., Fu, X. Y., Pan, P., Mu, X. Q., et al. (2011). Overlapping spatiotemporal patterns of regulatory gene expression are required for neuronal progenitors to specify retinal ganglion cell fate. *Vision Research*, *51*, 251–259.
Klimova, L., Antosova, B., Kuzelova, A., Strnad, H., & Kozmik, Z. (2015). Onecut1 and Onecut2 transcription factors operate downstream of Pax6 to regulate horizontal cell development. *Developmental Biology*, *402*, 48–60.

Klimova, L., & Kozmik, Z. (2014). Stage-dependent requirement of neuroretinal Pax6 for lens and retina development. *Development, 141*, 1292–1302.

Koike, C., Nishida, A., Ueno, S., Saito, H., Sanuki, R., Sato, S., et al. (2007). Functional roles of Otx2 transcription factor in postnatal mouse retinal development. *Molecular and Cellular Biology, 27*, 8318–8329.

Koroma, B. M., Tseng, S. C. G., & Sundin, O. H. (1997). Expression of the PAX6 gene in the anterior segment in human aniridia and Sey mouse model. *Investigative Ophthalmology & Visual Science, 38*, 4405.

Kumar, S., & Duester, G. (2010). Retinoic acid signaling in perioptic mesenchyme represses Wnt signaling via induction of Pitx2 and Dkk2. *Developmental Biology, 340*, 67–74.

Kuracha, M. R., Burgess, D., Siefker, E., Cooper, J. T., Licht, J. D., Robinson, M. L., et al. (2011). Spry1 and Spry2 are necessary for lens vesicle separation and corneal differentiation. *Investigative Ophthalmology & Visual Science, 52*, 6887–6897.

Kwan, K. M., Otsuna, H., Kidokoro, H., Carney, K. R., Saijoh, Y., & Chien, C. B. (2012). A complex choreography of cell movements shapes the vertebrate eye. *Development, 139*, 359–372.

La Torre, A., Georgi, S., & Reh, T. A. (2013). Conserved microRNA pathway regulates developmental timing of retinal neurogenesis. *Proceedings of the National Academy of Sciences of the United States of America, 110*, E2362–E2370.

Lagutin, O., Zhu, C. C., Furuta, Y., Rowitch, D. H., McMahon, A. P., & Oliver, G. (2001). Six3 promotes the formation of ectopic optic vesicle-like structures in mouse embryos. *Developmental Dynamics, 221*, 342–349.

Lagutin, O. V., Zhu, C. C., Kobayashi, D., Topczewski, J., Shimamura, K., Puelles, L., et al. (2003). Six3 repression of Wnt signaling in the anterior neuroectoderm is essential for vertebrate forebrain development. *Genes & Development, 17*, 368–379.

Lane, B. M., & Lister, J. A. (2012). Otx but not Mitf transcription factors are required for zebrafish retinal pigment epithelium development. *PLoS One*, 7(11), e49357.

Lang, R. A., Herman, K., Reynolds, A. B., Hildebrand, J. D., & Plageman, T. F., Jr. (2014). p120-catenin-dependent junctional recruitment of Shroom3 is required for apical constriction during lens pit morphogenesis. *Development, 141*, 3177–3187.

Le, T. T., Conley, K. W., & Brown, N. L. (2009). Jagged 1 is necessary for normal mouse lens formation. *Developmental Biology, 328*, 118–126.

Le, T. T., Conley, K. W., Mead, T. J., Rowan, S., Yutzey, K. E., & Brown, N. L. (2012). Requirements for Jag1-Rbpj mediated Notch signaling during early mouse lens development. *Developmental Dynamics, 241*, 493–504.

Leconte, L., Lecoin, L., Martin, P., & Saule, S. (2004). Pax6 interacts with cVax and Tbx5 to establish the dorsoventral boundary of the developing eye. *The Journal of Biological Chemistry, 279*, 47272–47277.

Lelievre, E. C., Lek, M., Boije, H., Houille-Vernes, L., Brajeul, V., Slembrouck, A., et al. (2011). Ptf1a/Rbpj complex inhibits ganglion cell fate and drives the specification of all horizontal cell subtypes in the chick retina. *Developmental Biology, 358*, 296–308.

Lengler, J., Krausz, E., Tomarev, S., Prescott, A., Quinlan, R. A., & Graw, J. (2001). Antagonistic action of Six3 and Prox1 at the gamma-crystallin promoter. *Nucleic Acids Research, 29*, 515–526.

Li, S. G., Mo, Z. Q., Yang, X. J., Price, S. M., Shen, M. M., & Xiang, M. Q. (2004). Foxn4 controls the genesis of amacrine and horizontal cells by retinal progenitors. *Neuron, 43*, 795–807.

Li, X., Perissi, V., Liu, F., Rose, D. W., & Rosenfeld, M. G. (2002). Tissue-specific regulation of retinal and pituitary precursor cell proliferation. *Science, 297*, 1180–1183.

Li, G., Xu, F., Zhu, J., Krawczyk, M., Zhang, Y., Yuan, J., et al. (2015). Transcription factor PAX6 (paired box 6) controls limbal stem cell lineage in development and disease. *The Journal of Biological Chemistry, 290*, 20448–20454.

Lin, Y. P., Ouchi, Y., Satoh, S., & Watanabe, S. (2009). Sox2 plays a role in the induction of amacrine and Muller glial cells in mouse retinal progenitor cells. *Investigative Ophthalmology & Visual Science*, *50*, 68–74.

Litsiou, A., Hanson, S., & Streit, A. (2005). A balance of FGF, BMP and WNT signalling positions the future placode territory in the head. *Development*, *132*, 4051–4062.

Liu, P., & Johnson, R. L. (2010). Lmx1b is required for murine trabecular meshwork formation and for maintenance of corneal transparency. *Developmental Dynamics*, *239*, 2161–2171.

Liu, J. J., Kao, W. W., & Wilson, S. E. (1999). Corneal epithelium-specific mouse keratin K12 promoter. *Experimental Eye Research*, *68*, 295–301.

Liu, W., Lagutin, O., Swindell, E., Jamrich, M., & Oliver, G. (2010). Neuroretina specification in mouse embryos requires Six3-mediated suppression of Wnt8b in the anterior neural plate. *The Journal of Clinical Investigation*, *120*, 3568–3577.

Liu, W., Wang, J. H., & Xiang, M. Q. (2000). Specific expression of the LIM/homeodomain protein Lim-1 in horizontal cells during retinogenesis. *Developmental Dynamics*, *217*, 320–325.

Liu, H., Xu, S., Wang, Y., Mazerolle, C., Thurig, S., Coles, B. L., et al. (2007). Ciliary margin transdifferentiation from neural retina is controlled by canonical Wnt signaling. *Developmental Biology*, *308*, 54–67.

Livesey, F. J., Young, T. L., & Cepko, C. L. (2004). An analysis of the gene expression program of mammalian neural progenitor cells. *Proceedings of the National Academy of Sciences of the United States of America*, *101*, 1374–1379.

Livne-Bar, I., Pacal, M., Cheung, M. C., Hankin, M., Trogadis, J., Chen, D., et al. (2006). Chx10 is required to block photoreceptor differentiation but is dispensable for progenitor proliferation in the postnatal retina. *Proceedings of the National Academy of Sciences of the United States of America*, *103*, 4988–4993.

Lovicu, F. J., & McAvoy, J. W. (1999). Spatial and temporal expression of p57(KIP2) during murine lens development. *Mechanisms of Development*, *86*, 165–169.

Lupo, G., Gestri, G., O'Brien, M., Denton, R. M., Chandraratna, R. A., Ley, S. V., et al. (2011). Retinoic acid receptor signaling regulates choroid fissure closure through independent mechanisms in the ventral optic cup and periocular mesenchyme. *Proceedings of the National Academy of Sciences of the United States of America*, *108*, 8698–8703.

Lwigale, P. Y., Cressy, P. A., & Bronner-Fraser, M. (2005). Corneal keratocytes retain neural crest progenitor cell properties. *Developmental Biology*, *288*, 284–293.

Macdonald, R., Barth, K. A., Xu, Q., Holder, N., Mikkola, I., & Wilson, S. W. (1995). Midline signalling is required for Pax gene regulation and patterning of the eyes. *Development*, *121*, 3267–3278.

Maddala, R., Chauhan, B. K., Walker, C., Zheng, Y., Robinson, M. L., Lang, R. A., et al. (2011). Rac1 GTPase-deficient mouse lens exhibits defects in shape, suture formation, fiber cell migration and survival. *Developmental Biology*, *360*, 30–43.

Maeda, A., Moriguchi, T., Hamada, M., Kusakabe, M., Fujioka, Y., Nakano, T., et al. (2009). Transcription factor GATA-3 is essential for lens development. *Developmental Dynamics*, *238*, 2280–2291.

Marquardt, T., Ashery-Padan, R., Andrejewski, N., Scardigli, R., Guillemot, F., & Gruss, P. (2001). Pax6 is required for the multipotent state of retinal progenitor cells. *Cell*, *105*, 43–55.

Martinez-Morales, J. R., Signore, M., Acampora, D., Simeone, A., & Bovolenta, P. (2001). Otx genes are required for tissue specification in the developing eye. *Development*, *128*, 2019–2030.

Masai, I. (2000). Mechanisms underlying induction and progression of a neurogenic wave in the zebrafish developing retina. *Tanpakushitsu Kakusan Koso*, *45*, 2782–2790.

Masuda, T., & Esumi, N. (2010). SOX9, through interaction with microphthalmia-associated transcription factor (MITF) and OTX2, regulates BEST1 expression in the retinal pigment epithelium. *The Journal of Biological Chemistry*, *285*, 26933–26944.
Masuda, T., Wahlin, K., Wan, J., Hu, J., Maruotti, J., Yang, X., et al. (2014). Transcription factor SOX9 plays a key role in the regulation of visual cycle gene expression in the retinal pigment epithelium. *The Journal of Biological Chemistry*, *289*, 12908–12921.
Mathers, P. H., Grinberg, A., Mahon, K. A., & Jamrich, M. (1997). The Rx homeobox gene is essential for vertebrate eye development. *Nature*, *387*, 603–607.
Matt, N., Dupe, V., Garnier, J. M., Dennefeld, C., Chambon, P., Mark, M., et al. (2005). Retinoic acid-dependent eye morphogenesis is orchestrated by neural crest cells. *Development*, *132*, 4789–4800.
Matt, N., Ghyselinck, N. B., Pellerin, I., & Dupe, V. (2008). Impairing retinoic acid signalling in the neural crest cells is sufficient to alter entire eye morphogenesis. *Developmental Biology*, *320*, 140–148.
Mattar, P., Ericson, J., Blackshaw, S., & Cayouette, M. (2015). A conserved regulatory logic controls temporal identity in mouse neural progenitors. *Neuron*, *85*, 497–504.
Mcavoy, J. W. (1978a). Cell-division, cell elongation and coordination of crystallin gene-expression during lens morphogenesis in rat. *Journal of Embryology and Experimental Morphology*, *45*, 271–281.
Mcavoy, J. W. (1978b). Cell-division, cell elongation and distribution of alpha-crystallins, beta-crystallins and gamma-crystallins in rat lens. *Journal of Embryology and Experimental Morphology*, *44*, 149–165.
Mcavoy, J. W. (1980). Cytoplasmic processes interconnect lens placode and optic vesicle during eye morphogenesis. *Experimental Eye Research*, *31*, 527–534.
McCabe, K. L., Gunther, E. C., & Reh, T. A. (1999). The development of the pattern of retinal ganglion cells in the chick retina: Mechanisms that control differentiation. *Development*, *126*, 5713–5724.
McMahon, C., Gestri, G., Wilson, S. W., & Link, B. A. (2009). Lmx1b is essential for survival of periocular mesenchymal cells and influences Fgf-mediated retinal patterning in zebrafish. *Developmental Biology*, *332*, 287–298.
Mears, A. J., Kondo, M., Swain, P. K., Takada, Y., Bush, R. A., Saunders, T. L., et al. (2001). Nrl is required for rod photoreceptor development. *Nature Genetics*, *29*, 447–452.
Medina-Martinez, O., Shah, R., & Jamrich, M. (2009). Pitx3 controls multiple aspects of lens development. *Developmental Dynamics*, *238*, 2193–2201.
Miesfeld, J. B., Gestri, G., Clark, B. S., Flinn, M. A., Poole, R. J., Bader, J. R., et al. (2015). Yap and Taz regulate retinal pigment epithelial cell fate. *Development*, *142*, 3021–3032.
Modak, S. P., Morris, G., & Yamada, T. (1968). DNA synthesis and mitotic activity during early development of chick lens. *Developmental Biology*, *17*, 544–561.
Moon, K. H., Kim, H. T., Lee, D., Rao, M. B., Levine, E. M., Lim, D. S., et al. (2018). Differential expression of NF2 in neuroepithelial compartments is necessary for mammalian eye development. *Developmental Cell*, *44*, 13–28.e3.
Mori, M., Ghyselinck, N. B., Chambon, P., & Mark, M. (2001). Systematic immunolocalization of retinoid receptors in developing and adult mouse eyes. *Investigative Ophthalmology & Visual Science*, *42*, 1312–1318.
Morrow, E. M., Furukawa, T., Lee, J. E., & Cepko, C. L. (1999). NeuroD regulates multiple functions in the developing neural retina in rodent. *Development*, *126*, 23–36.
Morrow, E. M., Furukawa, T., Raviola, E., & Cepko, C. L. (2005). Synaptogenesis and outer segment formation are perturbed in the neural retina of Crx mutant mice. *BMC Neuroscience*, *6*, 5.

Mu, X. Q., Fu, X. Y., Beremard, P. D., Thomas, T. L., & Klein, W. H. (2008). Gene regulation logic in retinal ganglion cell development: Isl1 defines a critical branch distinct from but overlapping with Pou4f2. *Proceedings of the National Academy of Sciences of the United States of America*, *105*, 6942–6947.

Mukhopadhyay, M., Gorivodsky, M., Shtrom, S., Grinberg, A., Niehrs, C., Morasso, M. I., et al. (2006). Dkk2 plays an essential role in the corneal fate of the ocular surface epithelium. *Development*, *133*, 2149–2154.

Muranishi, Y., Terada, K., Inoue, T., Katoh, K., Tsujii, T., Sanuki, R., et al. (2011). An essential role for RAX homeoprotein and NOTCH-HES signaling in Otx2 expression in embryonic retinal photoreceptor cell fate determination. *Journal of Neuroscience*, *31*, 16792–16807.

Muto, A., Iida, A., Satoh, S., & Watanabe, S. (2009). The group E Sox genes Sox8 and Sox9 are regulated by Notch signaling and are required for Muller glial cell development in mouse retina. *Experimental Eye Research*, *89*, 549–558.

Napier, H. R., & Kidson, S. H. (2007). Molecular events in early development of the ciliary body: A question of folding. *Experimental Eye Research*, *84*, 615–625.

Ng, L., Hurley, L. B., Dierks, B., Srinivas, M., Salto, C., Vennstrom, B., et al. (2001). A thyroid hormone receptor that is required for the development of green cone photoreceptors. *Nature Genetics*, *27*, 94–98.

Nguyen, M., & Arnheiter, H. (2000). Signaling and transcriptional regulation in early mammalian eye development: A link between FGF and MITF. *Development*, *127*, 3581–3591.

Nishida, A., Furukawa, A., Koike, C., Tano, Y., Aizawa, S., Matsuo, I., et al. (2003). Otx2 homeobox gene controls retinal photoreceptor cell fate and pineal gland development. *Nature Neuroscience*, *6*, 1255–1263.

Nishiguchi, S., Wood, H., Kondoh, H., Lovell-Badge, R., & Episkopou, V. (1998). Sox1 directly regulates the gamma-crystallin genes and is essential for lens development in mice. *Genes & Development*, *12*, 776–781.

Oliver, G., Mailhos, A., Wehr, R., Copeland, N. G., Jenkins, N. A., & Gruss, P. (1995). Six3, a murine homologue of the sine oculis gene, demarcates the most anterior border of the developing neural plate and is expressed during eye development. *Development*, *121*, 4045–4055.

Oron-Karni, V., Farhy, C., Elgart, M., Marquardt, T., Remizova, L., Yaron, O., et al. (2008). Dual requirement for Pax6 in retinal progenitor cells. *Development*, *135*, 4037–4047.

Ozeki, H., Ogura, Y., Hirabayashi, Y., & Shimada, S. (2001). Suppression of lens stalk cell apoptosis by hyaluronic acid leads to faulty separation of the lens vesicle. *Experimental Eye Research*, *72*, 63–70.

Pan, L., Deng, M., Xie, X. L., & Gan, L. (2008). ISL1 and BRN3B co-regulate the differentiation of murine retinal ganglion cells. *Development*, *135*, 1981–1990.

Parmigiani, C., & McAvoy, J. (1984). Localisation of laminin and fibronectin during rat lens morphogenesis. *Differentiation*, *28*, 53–61.

Patel, A., & Sowden, J. C. (2017). Genes and pathways in optic fissure closure. *Seminars in Cell & Developmental Biology*. https://doi.org/10.1016/j.semcdb.2017.10.010.

Pei, Y. F., & Rhodin, J. A. G. (1970). The prenatal development of mouse eye. *The Anatomical Record*, *168*, 105–125.

Perron, M., Boy, S., Amato, M. A., Viczian, A., Koebernick, K., Pieler, T., et al. (2003). A novel function for Hedgehog signalling in retinal pigment epithelium differentiation. *Development*, *130*, 1565–1577.

Plageman, T. F., Chung, M. I., Lou, M., Smith, A. N., Hildebrand, J. D., Wallingford, J. B., et al. (2010). Pax6-dependent Shroom3 expression regulates apical constriction during lens placode invagination. *Development*, *137*, 405–415.

Poche, R. A., Furuta, Y., Chaboissier, M. C., Schedl, A., & Behringer, R. R. (2008). Sox9 is expressed in mouse multipotent retinal progenitor cells and functions in Muller glial cell development. *The Journal of Comparative Neurology*, *510*, 237–250.

Poche, R. A., Kwan, K. M., Raven, M. A., Furuta, Y., Reese, B. E., & Behringer, R. R. (2007). Lim1 is essential for the correct laminar positioning of retinal horizontal cells. *Journal of Neuroscience*, *27*, 14099–14107.

Pontoriero, G. F., Deschamps, P., Ashery-Padan, R., Wong, R., Yang, Y., Zavadil, J., et al. (2008). Cell autonomous roles for AP-2 alpha in lens vesicle separation and maintenance of the lens epithelial cell phenotype. *Developmental Dynamics*, *237*, 602–617.

Pontoriero, G. F., Smith, A. N., Miller, L. A. D., Radice, G. L., West-Mays, J. A., & Lang, R. A. (2009). Co-operative roles for E-cadherin and N-cadherin during lens vesicle separation and lens epithelial cell survival. *Developmental Biology*, *326*, 403–417.

Porazinski, S., Wang, H. J., Asaoka, Y., Behrndt, M., Miyamoto, T., Morita, H., et al. (2015). YAP is essential for tissue tension to ensure vertebrate 3D body shape. *Nature*, *521*, 217–221.

Porter, F. D., Drago, J., Xu, Y., Cheema, S. S., Wassif, C., Huang, S. P., et al. (1997). Lhx2, a LIM homeobox gene, is required for eye, forebrain, and definitive erythrocyte development. *Development*, *124*, 2935–2944.

Pressman, C. L., Chen, H. X., & Johnson, R. L. (2000). Lmx1b, a LIM homeodomain class transcription factor, is necessary for normal development of multiple tissues in the anterior segment of the murine eye. *Genesis*, *26*, 15–25.

Pressman, C. L., Martin, J., & Johnson, R. L. (2000). Identification of a genetic interaction between LMX1B and PITX2. *Developmental Biology*, *222*, 236.

Prosser, J., & van Heyningen, V. (1998). PAX6 mutations reviewed. *Human Mutation*, *11*, 93–108.

Rajagopal, R., Huang, J., Dattilo, L. K., Kaartinen, V., Mishina, Y., Deng, C. X., et al. (2009). The type I BMP receptors, Bmpr1a and Acvr1, activate multiple signaling pathways to regulate lens formation. *Developmental Biology*, *335*, 305–316.

Ramaesh, T., Collinson, J. M., Ramaesh, K., Kaufman, M. H., West, J. D., & Dhillon, B. (2003). Corneal abnormalities in Pax6 +/− small eye mice mimic human aniridia-related keratopathy. *Investigative Ophthalmology & Visual Science*, *44*, 1871–1878.

Rapaport, D. H., Wong, L. L., Wood, E. D., Yasumura, D., & LaVail, M. M. (2004). Timing and topography of cell genesis in the rat retina. *The Journal of Comparative Neurology*, *474*, 304–324.

Raviola, G. (1971). The fine structure of the ciliary zonule and ciliary epithelium. With special regard to the organization and insertion of the zonular fibrils. *Investigative Ophthalmology*, *10*, 851–869.

Raviv, S., Bharti, K., Rencus-Lazar, S., Cohen-Tayar, Y., Schyr, R., Evantal, N., et al. (2014). PAX6 regulates melanogenesis in the retinal pigmented epithelium through feed-forward regulatory interactions with MITF. *PLoS Genetics*, *10*, e1004360.

Rembold, M., Loosli, F., Adams, R. J., & Wittbrodt, J. (2006). Individual cell migration serves as the driving force for optic vesicle evagination. *Science*, *313*, 1130–1134.

Roberts, M. R., Hendrickson, A., McGuire, C. R., & Reh, T. A. (2005). Retinoid X receptor gamma is necessary to establish the S-opsin gradient in cone photoreceptors of the developing mouse retina. *Investigative Ophthalmology & Visual Science*, *46*, 2897–2904.

Roberts, M. R., Srinivas, M., Forrest, D., de Escobar, G. M., & Reh, T. A. (2006). Making the gradient: Thyroid hormone regulates cone opsin expression in the developing mouse retina. *Proceedings of the National Academy of Sciences of the United States of America*, *103*, 6218–6223.

Robinson, M. L., & Overbeek, P. A. (1996). Differential expression of alpha A- and alpha B-crystallin during murine ocular development. *Investigative Ophthalmology & Visual Science*, *37*, 2276–2284.

Roesch, K., Jadhav, A. P., Trimarchi, J. M., Stadler, M. B., Roska, B., Sun, B. B., et al. (2008). The transcriptome of retinal Muller glial cells. *The Journal of Comparative Neurology*, *509*, 225–238.

Roger, J. E., Hiriyanna, A., Gotoh, N., Hao, H., Cheng, D. F., Ratnapriya, R., et al. (2014). OTX2 loss causes rod differentiation defect in CRX-associated congenital blindness. *Journal of Clinical Investigation*, *124*, 631–643.

Rones, B. (1932). Development of the human cornea. *Archives of Ophthalmology*, *8*, 568–575.

Rowan, S., Conley, I. W., Le, T. T., Donner, A. L., Maas, R. L., & Brown, N. L. (2008). Notch signaling regulates growth and differentiation in the mammalian lens. *Developmental Biology*, *321*, 111–122.

Saika, S., Saika, S., Liu, C. Y., Azhar, M., Sanford, L. P., Doetschman, T., et al. (2001). TGFbeta2 in corneal morphogenesis during mouse embryonic development. *Developmental Biology*, *240*, 419–432.

Samuel, A., Housset, M., Fant, B., & Lamonerie, T. (2014). Otx2 ChIP-seq reveals unique and redundant functions in the mature mouse retina. *PLoS One*, *9*, e89110.

Sanes, J. R., & Masland, R. H. (2015). The types of retinal ganglion cells: Current status and implications for neuronal classification. *Annual Review of Neuroscience*, *38*, 221–246.

Saravanamuthu, S. S., Le, T. T., Gao, C. Y., Cojocaru, R. I., Pandiyan, P., Liu, C. Q., et al. (2012). Conditional ablation of the Notch2 receptor in the ocular lens. *Developmental Biology*, *362*, 219–229.

Sato, S., Inoue, T., Terada, K., Matsuo, I., Aizawa, S., Tano, Y., et al. (2007). Dkk3-Cre BAC transgenic mouse line: A tool for highly efficient gene deletion in retinal progenitor cells. *Genesis*, *45*, 502–507.

Satokata, I., & Maas, R. (1994). Msx1 deficient mice exhibit cleft palate and abnormalities of craniofacial and tooth development. *Nature Genetics*, *6*, 348–356.

Satow, T., Bae, S. K., Inoue, T., Inoue, C., Miyoshi, G., Tomita, K., et al. (2001). The basic helix-loop-helix gene hesr2 promotes gliogenesis in mouse retina. *Journal of Neuroscience*, *21*, 1265–1273.

Sawyer, J. M., Harrell, J. R., Shemer, G., Sullivan-Brown, J., Roh-Johnson, M., & Goldstein, B. (2010). Apical constriction: A cell shape change that can drive morphogenesis. *Developmental Biology*, *341*, 5–19.

Schedl, A., Ross, A., Lee, M., Engelkamp, D., Rashbass, P., van Heyningen, V., et al. (1996). Influence of PAX6 gene dosage on development: Overexpression causes severe eye abnormalities. *Cell*, *86*, 71–82.

Semina, E. V., Ferrell, R. E., Mintz-Hittner, H. A., Bitoun, P., Alward, W. L. M., Reiter, R. S., et al. (1998). A novel homeobox gene PITX3 is mutated in families with autosomal-dominant cataracts and ASMD. *Nature Genetics*, *19*, 167–170.

Semina, E. V., Murray, J. C., Reiter, R., Hrstka, R. F., & Graw, J. (2000). Deletion in the promoter region and altered expression of Pitx3 homeobox gene in aphakia mice. *Human Molecular Genetics*, *9*, 1575–1585.

Semina, E. V., Reiter, R., Leysens, N. J., Alward, W. L., Small, K. W., Datson, N. A., et al. (1996). Cloning and characterization of a novel bicoid-related homeobox transcription factor gene, RIEG, involved in Rieger syndrome. *Nature Genetics*, *14*, 392–399.

Semina, E. V., Reiter, R. S., & Murray, J. C. (1997). Isolation of a new homeobox gene belonging to the Pitx/Rieg family: Expression during lens development and mapping to the aphakia region on mouse chromosome 19. *Human Molecular Genetics*, *6*, 2109–2116.

Seo, S., Chen, L. S., Liu, W. Z., Zhao, D. M., Schultz, K. M., Sasman, A., et al. (2017). Foxc1 and Foxc2 in the neural crest are required for ocular anterior segment development. *Investigative Ophthalmology & Visual Science*, *58*, 1368–1377.

Seo, S., Singh, H. P., Lacal, P. M., Sasman, A., Fatima, A., Liu, T., et al. (2012). Forkhead box transcription factor FoxC1 preserves corneal transparency by regulating vascular growth. *Proceedings of the National Academy of Sciences of the United States of America*, *109*, 2015–2020.

Shiraishi, A., Converse, R. L., Liu, C. Y., Zhou, F., Kao, C. W., & Kao, W. W. (1998). Identification of the cornea-specific keratin 12 promoter by in vivo particle-mediated gene transfer. *Investigative Ophthalmology & Visual Science, 39*, 2554–2561.

Silla, Z. T., Naidoo, J., Kidson, S. H., & Sommer, P. (2014). Signals from the lens and Foxc1 regulate the expression of key genes during the onset of corneal endothelial development. *Experimental Cell Research, 322*, 381–388.

Sjodal, M., Edlund, T., & Gunhaga, L. (2007). Time of exposure to BMP signals plays a key role in the specification of the olfactory and lens placodes ex vivo. *Developmental Cell, 13*, 141–149.

Skowronska-Krawczyk, D., Ballivet, M., Dynlacht, B. D., & Matter, J. M. (2004). Highly specific interactions between bHLH transcription factors and chromatin during retina development. *Development, 131*, 4447–4454.

Smith, A. N., Miller, L. A. D., Song, N., Taketo, M. M., & Lang, R. A. (2005). The duality of beta-catenin function: A requirement in lens morphogenesis and signaling suppression of lens fate in periocular ectoderm. *Developmental Biology, 285*, 477–489.

Smith, R. S., Zabaleta, A., Kume, T., Savinova, O. V., Kidson, S. H., Martin, J. E., et al. (2000). Haploinsufficiency of the transcription factors FOXC1 and FOXC2 results in aberrant ocular development. *Human Molecular Genetics, 9*, 1021–1032.

Somasundaram, T., & Bhat, S. P. (2004). Developmentally dictated expression of heat shock factors: Exclusive expression of HSF4 in the postnatal lens and its specific interaction with alphaB-crystallin heat shock promoter. *The Journal of Biological Chemistry, 279*, 44497–44503.

Song, J. Y., Park, R., Kim, J. Y., Hughes, L., Lu, L., Kim, S., et al. (2014). Dual function of Yap in the regulation of lens progenitor cells and cellular polarity. *Developmental Biology, 386*, 281–290.

Spemann, H., & Mangold, H. (1924). The induction of embryonic predispositions by implantation of organizers foreign to the species. *Archiv für Mikroskopische Anatomie und Entwicklungsmechanik, 100*, 599–638.

Steinfeld, J., Steinfeld, I., Coronato, N., Hampel, M. L., Layer, P. G., Araki, M., et al. (2013). RPE specification in the chick is mediated by surface ectoderm-derived BMP and Wnt signalling. *Development, 140*, 4959–4969.

Steventon, B., Mayor, R., & Streit, A. (2016). Directional cell movements downstream of Gbx2 and Otx2 control the assembly of sensory placodes. *Biology Open, 5*, 1620–1624.

Swaroop, A., Kim, D., & Forrest, D. (2010). Transcriptional regulation of photoreceptor development and homeostasis in the mammalian retina. *Nature Reviews. Neuroscience, 11*, 563–576.

Tachibana, M. (2000). MITF: A stream flowing for pigment cells. *Pigment Cell Research, 13*, 230–240.

Tachibana, M., Perez-Jurado, L. A., Nakayama, A., Hodgkinson, C. A., Li, X., Schneider, M., et al. (1994). Cloning of MITF, the human homolog of the mouse microphthalmia gene and assignment to chromosome 3p14.1-p12.3. *Human Molecular Genetics, 3*, 553–557.

Takabatake, Y., Takabatake, T., Sasagawa, S., & Takeshima, K. (2002). Conserved expression control and shared activity between cognate T-box genes Tbx2 and Tbx3 in connection with Sonic hedgehog signaling during Xenopus eye development. *Development, Growth & Differentiation, 44*, 257–271.

Taranova, O. V., Magness, S. T., Fagan, B. M., Wu, Y., Surzenko, N., Hutton, S. R., et al. (2006). SOX2 is a dose-dependent regulator of retinal neural progenitor competence. *Genes & Development, 20*, 1187–1202.

Thein, T., de Melo, J., Zibetti, C., Clark, B. S., Juarez, F., Juarez, F., et al. (2016). Control of lens development by Lhx2-regulated neuroretinal FGFs. *Development, 143*, 3994–4002.

Thumann, G. (2001). Development and cellular functions of the iris pigment epithelium. *Survey of Ophthalmology*, *45*, 345–354.

Trimarchi, J. M., Cho, S. H., & Cepko, C. L. (2009). Identification of genes expressed preferentially in the developing peripheral margin of the optic cup. *Developmental Dynamics*, *238*, 2327–2329.

Turner, D. L., & Cepko, C. L. (1987). A common progenitor for neurons and glia persists in rat retina late in development. *Nature*, *328*, 131–136.

Turner, D. L., Snyder, E. Y., & Cepko, C. L. (1990). Lineage-independent determination of cell type in the embryonic mouse retina. *Neuron*, *4*, 833–845.

Tzahor, E., Kempf, H., Mootoosamy, R. C., Poon, A. C., Abzhanov, A., Tabin, C. J., et al. (2003). Antagonists of Wnt and BMP signaling promote the formation of vertebrate head muscle. *Genes & Development*, *17*, 3087–3099.

van Raamsdonk, C. D., & Tilghman, S. M. (2000). Dosage requirement and allelic expression of PAX6 during lens placode formation. *Development*, *127*, 5439–5448.

Varga, Z. M., Wegner, J., & Westerfield, M. (1999). Anterior movement of ventral diencephalic precursors separates the primordial eye field in the neural plate and requires cyclops. *Development*, *126*, 5533–5546.

Wall, D. S., Mears, A. J., McNeill, B., Mazerolle, C., Thurig, S., Wang, Y. P., et al. (2009). Progenitor cell proliferation in the retina is dependent on Notch-independent Sonic hedgehog/Hes1 activity. *The Journal of Cell Biology*, *184*, 101–112.

Wang, Y. P., Dakubo, G. D., Thurig, S., Mazerolle, C. J., & Wallace, V. A. (2005). Retinal ganglion cell-derived sonic hedgehog locally controls proliferation and the timing of RGC development in the embryonic mouse retina. *Development*, *132*, 5103–5113.

Wang, S. W., Kim, B. S., Ding, K., Wang, H., Sun, D., Johnson, R. L., et al. (2001). Requirement for math5 in the development of retinal ganglion cells. *Genes & Development*, *15*, 24–29.

Wang, S., Sengel, C., Emerson, M. M., & Cepko, C. L. (2014). A gene regulatory network controls the binary fate decision of rod and bipolar cells in the vertebrate retina. *Developmental Cell*, *30*, 513–527.

Wang, X., Shan, X., & Gregory-Evans, C. Y. (2017). A mouse model of aniridia reveals the in vivo downstream targets of Pax6 driving iris and ciliary body development in the eye. *Biochimica et Biophysica Acta*, *1863*, 60–67.

Westenskow, P., Piccolo, S., & Fuhrmann, S. (2009). Beta-catenin controls differentiation of the retinal pigment epithelium in the mouse optic cup by regulating Mitf and Otx2 expression. *Development*, *136*, 2505–2510.

Wigle, J. T., Chowdhury, K., Gruss, P., & Oliver, G. (1999). Prox1 function is crucial for mouse lens-fibre elongation. *Nature Genetics*, *21*, 318–322.

Wolf, L., Harrison, W., Huang, J., Xie, Q., Xiao, N. N., Sun, J., et al. (2013). Histone posttranslational modifications and cell fate determination: Lens induction requires the lysine acetyltransferases CBP and p300. *Nucleic Acids Research*, *41*, 10199–10214.

Wolf, L. V., Yang, Y., Wang, J. H., Xie, Q., Braunger, B., Tamm, E. R., et al. (2009). Identification of pax6-dependent gene regulatory networks in the mouse lens. *PLoS One*, *4*, e4159.

Wolf, L., Yang, Y., Wawrousek, E., & Cvekl, A. (2008). Transcriptional regulation of mouse alpha A-crystallin gene in a 148kb Cryaa BAC and its derivates. *BMC Developmental Biology*, *8*, 88.

Wolosin, J. M., Budak, M. T., & Akinci, M. A. (2004). Ocular surface epithelial and stem cell development. *The International Journal of Developmental Biology*, *48*, 981–991.

Wu, F. G., Kaczynski, T. J., Sethuramanujam, S., Li, R. Z., Jain, V., Slaughter, M., et al. (2015). Two transcription factors, Pou4f2 and Isl1, are sufficient to specify the retinal ganglion cell fate. *Proceedings of the National Academy of Sciences of the United States of America*, *112*, E1559–E1568.

Wu, F. G., Li, R. Z., Umino, Y., Kaczynski, T. J., Sapkota, D., Li, S. G., et al. (2013). Onecut1 is essential for horizontal cell genesis and retinal integrity. *Journal of Neuroscience, 33*, 13053–13065.

Wulle, K. G., & Lerche, W. (1969). Electron microscopic observations of early development of human corneal endothelium and descemets membrane. *Ophthalmologica, 157*, 451–461.

Wurm, A., Sock, E., Fuchshofer, R., Wegner, M., & Tamm, E. R. (2008). Anterior segment dysgenesis in the eyes of mice deficient for the high-mobility-group transcription factor Sox11. *Experimental Eye Research, 86*, 895–907.

Xie, Q., & Cvekl, A. (2011). The orchestration of mammalian tissue morphogenesis through a series of coherent feed-forward loops. *The Journal of Biological Chemistry, 286*, 43259–43271.

Xie, Q., McGreal, R., Harris, R., Gao, C. Y., Liu, W., Reneker, L. W., et al. (2016). Regulation of c-Maf and alpha A-crystallin in ocular lens by fibroblast growth factor signaling. *The Journal of Biological Chemistry, 291*, 3947–3958.

Yang, Y., Stopka, T., Golestaneh, N., Wang, Y., Wu, K., Li, A., et al. (2006). Regulation of alphaA-crystallin via Pax6, c-Maf, CREB and a broad domain of lens-specific chromatin. *The EMBO Journal, 25*, 2107–2118.

Yoshimoto, A., Saigou, Y., Higashi, Y., & Kondoh, H. (2005). Regulation of ocular lens development by Smad-interacting protein 1 involving Foxe3 activation. *Development, 132*, 4437–4448.

Young, R. W. (1985). Cell differentiation in the retina of the mouse. *The Anatomical Record, 212*, 199–205.

Young, R. W., & Ocumpaugh, D. E. (1996). Autoradiographic studies on the growth and development of the lens capsule in the rat. *Investigative Ophthalmology and Visual Science, 5*, 583–593.

Yun, S., Saijoh, Y., Hirokawa, K. E., Kopinke, D., Murtaugh, L. C., Monuki, E. S., et al. (2009). Lhx2 links the intrinsic and extrinsic factors that control optic cup formation. *Development, 136*, 3895–3906.

Zhao, S., Chen, Q., Hung, F. C., & Overbeek, P. A. (2002). BMP signaling is required for development of the ciliary body. *Development, 129*, 4435–4442.

Zou, C., & Levine, E. M. (2012). Vsx2 controls eye organogenesis and retinal progenitor identity via homeodomain and non-homeodomain residues required for high affinity DNA binding. *PLoS Genetics, 8*, e1002924.

Zuber, M. E., Gestri, G., Viczian, A. S., Barsacchi, G., & Harris, W. A. (2003). Specification of the vertebrate eye by a network of eye field transcription factors. *Development, 130*, 5155–5167.

CHAPTER ELEVEN

Fluid forces shape the embryonic heart: Insights from zebrafish

Pragya Sidhwani, Deborah Yelon*
Division of Biological Sciences, University of California, San Diego, CA, United States
*Corresponding author: e-mail address: dyelon@ucsd.edu

Contents

Abstract

Heart formation involves a complex series of tissue rearrangements, during which regions of the developing organ expand, bend, converge, and protrude in order to create the specific shapes of important cardiac components. Much of this morphogenesis takes place while cardiac function is underway, with blood flowing through the rapidly contracting chambers. Fluid forces are therefore likely to influence the regulation of cardiac morphogenesis, but it is not yet clear how these biomechanical cues direct specific cellular behaviors. In recent years, the optical accessibility and genetic amenability of zebrafish embryos have facilitated unique opportunities to integrate the analysis of flow parameters with the molecular and cellular dynamics underlying cardiogenesis. Consequently, we are making progress toward a comprehensive view of the biomechanical regulation of cardiac chamber emergence, atrioventricular canal differentiation, and ventricular trabeculation. In this review, we highlight a series of studies in zebrafish that have provided new insight into how cardiac function can shape cardiac morphology, with a particular focus on how hemodynamics can impact cardiac cell behavior. Over the long-term, this knowledge will undoubtedly guide our consideration of the potential causes of congenital heart disease.

Current Topics in Developmental Biology, Volume 132
ISSN 0070-2153
https://doi.org/10.1016/bs.ctdb.2018.12.009

1. Introduction

Biomechanical forces are important regulators of embryonic morphogenesis. As early as gastrulation, circumferential contractions and flow-frictional mechanisms drive the spreading of the enveloping cell layer over the zebrafish yolk (Behrndt et al., 2012). Compressing and stretching the apical surface of cells induces apoptosis and proliferation, respectively, to shape the overall structure of the Drosophila wing (Diaz de la Loza & Thompson, 2017). The periodic spacing of feather follicles in chick is governed by competing interactions between cellular contractility and substrate stiffness (Shyer et al., 2017). In many contexts, physical cues provoke specific cellular behaviors that ultimately give rise to tissue morphology.

Forces are especially crucial during cardiac development, since blood flow and cardiac contractility produce distinct physical inputs while heart formation is underway (Freund, Goetz, Hill, & Vermot, 2012; Haack & Abdelilah-Seyfried, 2016). It is particularly interesting to consider how flow-induced forces create spatiotemporally heterogeneous patterns that provide regionalized cues within the developing organ. Fluid frictional forces, for example, induce shear stress that deforms cells in the direction of blood flow. Blood pressure, on the other hand, causes cells to stretch circumferentially. Finally, the direction of blood flow can itself administer distinct mechanical cues to cells: reversing or retrograde flows, for instance, are thought to activate specific downstream pathways. Together, these fluid forces sculpt the heart, in a biological manifestation of "form follows function."

How do fluid forces prompt cardiac cells to enlarge, elongate, divide, converge, and protrude as they create the specific architecture of the embryonic heart? Despite awareness of several mechanosensitive pathways that operate in endothelial cells (Baratchi et al., 2017), we do not yet fully understand the cellular and molecular mechanisms that drive cardiac morphogenesis in response to hemodynamic cues. To gain insights into these mechanisms, we need to be able to integrate *in vivo* assessment of fluid forces, high-resolution analysis of cell behaviors, and precise manipulation of mechanosensitive genes. The zebrafish embryo serves as an ideal model organism in this regard: its optical clarity allows real-time visualization of fluid dynamics as well as live imaging of morphogenesis at cellular resolution, and its genetic tractability permits both classical genetic screens and cutting-edge genome editing (Collins & Stainier, 2016; Li, Zhao, Page-McCaw, & Chen, 2016). Importantly, zebrafish offer the unique advantage that they can survive throughout embryogenesis without convective oxygen, allowing the analysis of cardiovascular defects without confounding lethality. Finally, although the zebrafish heart is anatomically simpler than the mammalian heart, the mechanosensory pathways operating during cardiac development appear to be highly conserved (Goddard et al., 2017). Together, these benefits of zebrafish have facilitated a number of investigations into the ways in which function influences form in the developing heart.

Here, we review a number of recent advances in our understanding of the impact of fluid forces on cardiac morphogenesis, with a particular focus on work performed in zebrafish. We first outline several commonly used techniques for quantification of flow parameters in the zebrafish vasculature and heart. We then highlight a series of studies that have illuminated how biomechanical inputs regulate the characteristic curvatures of the cardiac chambers, the distinctive features of the atrioventricular canal, and the elaborate network of the ventricular trabeculae. Overall, these insights emphasize the important connections between the mechanical, cellular, and molecular pathways that collaborate to construct the embryonic heart. Importantly, these studies may provide guidance for understanding the etiology of cardiac birth defects in humans, since they imply that fluctuations in blood flow *in utero* could have significant consequences on cardiac morphogenesis.

2. Examination of fluid forces in zebrafish

2.1 Techniques for quantification of fluid forces in the vasculature

To evaluate the influence of fluid forces on cardiac development, it is essential to be able to quantify these forces *in vivo*. Parameters such as shear stress and retrograde flow can then be correlated with patterns of cell behavior in order to create models for how mechanics instruct morphology. A number of techniques have been found to be useful for the assessment of fluid forces in the zebrafish heart, and several of these were first developed in the context of the zebrafish vasculature, where the size and simplicity of the major vessels facilitate live imaging of blood flow.

Digital motion analysis was an initial approach employed to visualize blood flow in zebrafish vessels (Schwerte & Pelster, 2000). Using a simple dissecting microscope and a CCD camera for video imaging, "difference images" can be created to track the motion of erythrocytes in consecutive video frames. Analysis of these images allows assessment of erythrocyte distribution, erythrocyte velocity and vessel diameter (Schwerte, Uberbacher, & Pelster, 2003). Although easy to perform, a drawback of two-dimensional video microscopy is that slight variations in embryo orientation can significantly affect results; to some extent, this can be overcome by embedding larvae in tubular molds that are rotated to allow imaging from multiple angles (Bagatto & Burggren, 2006). Additionally, the use of high-speed cameras to follow erythrocytes expressing fluorescent reporter transgenes can substantially enhance the accuracy of cell tracking techniques (Fig. 1A–C) (Watkins et al., 2012).

With the advent of particle image velocimetry (PIV), methods for evaluating flow parameters became even more rigorous. Unlike cell tracking techniques that follow only individual features, PIV averages velocity fields of sub-regions or "interrogation windows" within an image with a high signal-to-noise ratio (Yalcin, Amindari, Butcher, Althani, & Yacoub, 2017). This information can be translated into estimates of wall shear stress when coupled with local area measurements, which can be obtained by simultaneously imaging vessel wall motion (Chen et al., 2011). PIV can be performed with both bright-field and fluorescent images, although confocal imaging of fluorescent dyes or transgene-expressing erythrocytes greatly facilitates subsequent image segmentation.

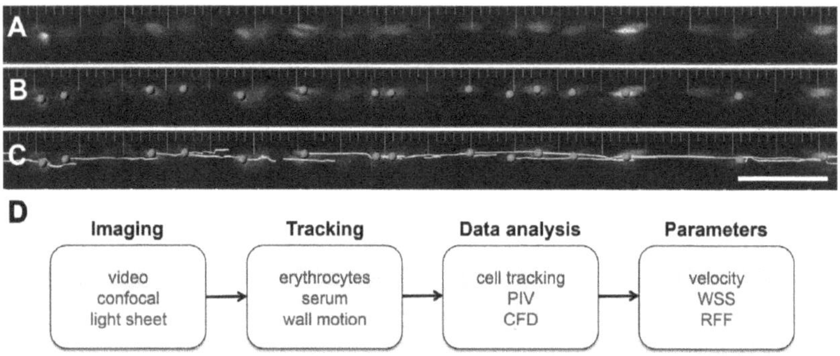

Fig. 1 Analysis of blood flow in the zebrafish. (A) Labeled erythrocytes in the dorsal aorta of an embryo carrying the transgene *Tg(gata1:dsRed)* are visualized at a single timepoint. (B) Image segmentation identifies individual erythrocytes (green dots). (C) A tracking algorithm reveals the trajectories (yellow lines) of the erythrocytes, allowing measurement of their displacement and estimation of their velocity. (D) Workflow options for analysis of blood flow in zebrafish. PIV, particle image velocimetry; CFD, computational fluid dynamics; WSS, wall shear stress; RFF, retrograde flow fraction. *Panels (A)–(C): Adapted from Watkins, S. C., Maniar, S., Mosher, M., Roman, B. L., Tsang, M., St Croix, C. M. (2012). High resolution imaging of vascular function in zebrafish.* PLoS One, *7, e44018.*

While PIV measurements can be used to estimate the distribution of physical forces, methods such as optical tweezing can directly measure the forces experienced by erythrocytes *in vivo* (Anton et al., 2013). In this technique, forces administered by a focused laser beam are used to hold an erythrocyte in position, such that the amount of force necessary allows assessment of the cell's mechanical milieu. Optical tweezing can be especially useful at branch points in the vasculature, where sudden changes in the direction of erythrocyte movement can be difficult to track.

Additional techniques have been utilized to image blood flow in the zebrafish vasculature, including fluorescence correlation microscopy (Pan, Yu, Shi, Korzh, & Wohland, 2007) and optical coherent tomography (Iftimia et al., 2008). Fluorescence correlation microscopy utilizes the inherent autofluorescence of erythrocytes and blood vessels to determine erythrocyte velocity, sidestepping the need to introduce fluorescent tracers. Similarly, optical coherent tomography measures backscattered and back-reflected light to image fluid flow in real time. Although these techniques are beneficial, they require specialized instruments that may not be readily accessible at all research institutions. Consequently, video imaging and confocal imaging, followed by cell tracking or PIV, remain the prevalent methods for assessing flow parameters in zebrafish vessels.

2.2 Techniques for analysis of flow parameters in the heart

The embryonic zebrafish heart is a morphologically elaborate structure that functions to propel blood from the atrium into the ventricle before finally supplying circulation to the aortic vessels. Owing to its particular architecture, imaging blood flow in the heart presents unique technical challenges distinct from those encountered in the vasculature. For example, the dimensions of the cardiac chambers provide many opportunities for erythrocytes to move in and out of the plane of focus. The motion of the chambers creates further complexity, since the chamber wall has to be properly segmented during image analysis. Moreover, the structure of the heart, coupled with the sequential contraction of the chambers, produces complex flow patterns featuring vortical and reversing flows. Nevertheless, techniques developed for the zebrafish vasculature have been successfully adapted for use in the embryonic heart.

As in the vasculature, PIV has been the method of choice to quantify wall shear stress and other flow parameters during cardiac development. Early applications of this technique used a fluorescent contrast agent to visualize serum motion and demonstrated vortical patterns of flow within the heart (Hove et al., 2003). Although still used widely, PIV has certain drawbacks when applied in the heart. For example, when utilizing video imaging to collect data for PIV analysis, three-dimensional information has to be projected in two dimensions, which results in underestimation of the distance traveled by erythrocytes (Hove et al., 2003). Techniques such as defocusing digital particle image velocimetry, which employ microscopes with two or more apertures to detect out-of-focus particles, can enable more precise tracking of intracardiac flows (Lu, Pereira, Fraser, & Gharib, 2008). More recently, the depth sectioning capabilities of light sheet microscopy have also been exploited for PIV analysis (Zickus & Taylor, 2018). Another complication associated with PIV measurements in the heart is the interference of cardiac wall motion with erythrocyte tracking, especially in bright-field images. To overcome this, a local signal-to-noise ratio can be used to achieve "cardiac-phase filtering," which effectively removes artifacts from cardiac wall motion post-image acquisition (Jamison, Samarage, Bryson-Richardson, & Fouras, 2013). Since this method identifies the position of the cardiac wall, it enables assessment of wall shear stress in addition to more accurate PIV calculations.

The development of faster imaging techniques has also been valuable for following the complex patterns of erythrocyte motion within the heart. Using a confocal laser slit scanning system, which executes line-wise illumination for faster imaging, significant retrograde flows were observed in the zebrafish heart, both during atrial contraction back into the sinus venosus, and during ventricular contraction back into the atrium (Liebling et al., 2006). Subsequent analysis with high-speed bright-field video microscopy confirmed that the region between the ventricle and the atrium, the atrioventricular canal, experiences a heightened retrograde flow fraction, which is defined as the fraction of the cardiac cycle exhibiting flow reversal (Vermot et al., 2009). These studies have motivated an interest in retrograde flow as a critical signal for atrioventricular canal differentiation, which we will discuss further in a subsequent section.

The emergence of additional strategies for image analysis, such as computational fluid dynamics (CFD), marks a new era for assessment of fluid forces in the zebrafish heart (Yalcin et al., 2017). In computational fluid dynamics, flow parameters are inferred from tracking both chamber wall motions and inlet velocities, which bypasses the requirement for high-resolution imaging as in PIV. Use of computational fluid dynamics in the zebrafish heart has allowed correlation of certain flow characteristics with specific morphogenetic processes (Lee et al., 2013). For example, mean peak shear stress increases significantly during chamber emergence and atrioventricular canal differentiation, suggesting an influence of shear forces on these steps of heart morphogenesis.

Altogether, several options are readily available for analysis of flow parameters in the zebrafish heart (Fig. 1D). Using video, confocal, or light sheet imaging to track erythrocytes, serum, or chamber walls, investigators can use cell tracking, PIV, or computational fluid dynamics approaches to assess erythrocyte velocity, wall shear stress, and retrograde flow fraction. This variety of strategies opens the door to connecting fluid forces to the regulation of a number of specific processes during cardiac morphogenesis, as discussed in Section 3.

3. Hemodynamics and cardiovascular morphogenesis

3.1 Fluid forces regulate multiple aspects of vessel development

When considering how fluid forces influence cardiac morphology, it is instructive to reflect upon our understanding of the influence of flow on vessel development. The impact of physical forces on endothelial cells in the vasculature is especially relevant to the potential roles of such forces on the endocardium, the specialized inner layer of endothelium that lines the muscular myocardial layer of the heart. Since the biomechanical regulation of vascular morphogenesis has been reviewed elsewhere (Baratchi et al., 2017; Boselli, Freund, & Vermot, 2015), this section will simply provide a few recent perspectives on how processes like lumen growth and vessel remodeling respond to fluid forces in zebrafish.

Blood flow has been shown to regulate vessel diameter through its influence on several types of endothelial cell behaviors. For example, Endoglin, a TGF-β receptor, modulates endothelial cell shape in response to hemodynamic cues, thereby restricting the diameters of the dorsal aorta and posterior cardinal vein (Sugden et al., 2017). In the caudal vein, shear forces lead to an increase in endothelial cell number *via* mechanosensation by Pkd2/Trpp2, a calcium channel present on endothelial cilia (Goetz et al., 2014). In the cranial vasculature, flow-induced expression of the TGF-β receptor Alk1 regulates the migration of endothelial cells in the direction opposite to blood flow, thereby limiting vessel caliber (Corti et al., 2011; Rochon, Menon, & Roman, 2016). Finally, Yap1, a Hippo pathway effector, localizes to the nucleus in a flow-dependent manner in the dorsal longitudinal anastomotic vessel, where it regulates the maintenance of lumen size (Nakajima et al., 2017).

Fluid forces are also critical for the morphogenetic processes that are involved in vessel remodeling. For instance, flow regulates the apical membrane invagination that is required for successful tube formation during vessel fusion (Herwig et al., 2011). Blood flow is also important for the establishment of planar cell polarity in endothelial cells (Kwon et al., 2016), which could influence vessel regression, as has been predicted in mice (Franco et al., 2015). Indeed, vessel regression in the zebrafish eye (Kochhan et al., 2013) and midbrain (Chen et al., 2012) has been shown to occur in response to blood flow, potentially *via* flow-mediated cell rearrangement and migration.

Together, these studies underscore the importance of fluid forces during multiple aspects of vascular morphogenesis and provide inspiration for the ways in which blood flow could influence cell shape, cell number, migration, and polarity during heart development. In the following sections, we will examine the impact of cardiac function and fluid forces on cell behavior during three essential phases of cardiac morphogenesis: chamber emergence, atrioventricular canal differentiation, and ventricular trabeculation.

3.2 Chamber emergence requires hemodynamic inputs

In the early embryo, the primitive heart is a simple, relatively linear cylinder positioned at the embryonic midline. As development proceeds, the heart tube enlarges and transforms into a morphologically distinct series of cardiac chambers, each expanding into its characteristic curvatures (Collins & Stainier, 2016). Since chamber emergence takes place while the heart is beating and the blood is flowing, biomechanical inputs generated by cardiac function have the potential to influence this process. Indeed, surgical obstruction of blood flow into the embryonic chick heart results in aberrant chamber morphology (Broekhuizen et al., 1999). Mouse mutants lacking atrial contractility due to a mutation in *atrial myosin light chain 2a* exhibit abnormal ventricular morphogenesis (Huang et al., 2003), further supporting a connection between blood flow and cardiac chamber development. Taking advantage of the opportunities for examining chamber emergence in the context of the zebrafish embryo, a number of studies have delved deeper into the cell behaviors that shape the chambers in response to biomechanical cues.

In zebrafish, as the ventricle and atrium emerge from the linear heart tube, distinct convex and concave surfaces form the outer and inner curvatures of each chamber. These tissue shape changes are associated with regional patterns of cell shape change: in the ventricle, for example, outer curvature cardiomocytes enlarge and elongate during chamber emergence, while inner curvature cardiomyocytes remain relatively small and round (Auman et al., 2007). Analysis of cardiomyocyte morphology in mutant embryos has suggested that cell shape change at the ventricular outer curvature depends upon inputs generated by cardiac function. In *weak atrium* (*amhc*) mutants, which lack atrial contractility and therefore have reduced blood flow into the ventricle, outer curvature cardiomyocytes are smaller and rounder than in the wild-type ventricle (Fig. 2A and B) (Auman et al., 2007); these cells also exhibit diminished myofibril maturation (Lin, Swinburne, & Yelon, 2012). In contrast, *half-hearted* (*vmhc*) mutants, which lack ventricular contractility, display excessively enlarged and elongated cells in the ventricular outer curvature (Fig. 2C and D) (Auman et al., 2007). These studies suggest that forces produced by both blood flow and contractility modulate cardiomyocyte shape and size and therefore contribute to the regulation of chamber emergence.

In conjunction with the outward expansion of the myocardium, the inner endocardial layer of the heart also grows as the chambers emerge. Cellular proliferation drives this endocardial expansion (Fig. 3A, E and F), while local differences in endocardial cell morphologies correlate with the curved chamber contours (Dietrich, Lombardo, Veerkamp, Priller, & Abdelilah-Seyfried, 2014). Interestingly, in conditions where fluid forces

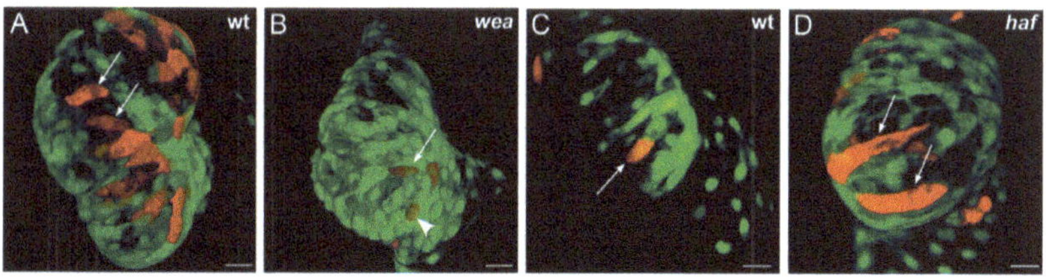

Fig. 2 Cardiac function modulates cardiomyocyte dimensions in the ventricle. Confocal projections show the ventricle in embryos carrying the transgene *Tg(myl7:egfp)*, with mosaic expression of *Tg(myl7:dsRedt4)*. (A and C) At 52h post-fertilization (hpf), cardiomyocytes in the outer curvature of the ventricle (arrows) have acquired a characteristic size and an elongated morphology. (B) In *weak atrium* (*wea*) mutants, blood flow into the ventricle is reduced, and the cardiomyocytes in the outer curvature are abnormally small (arrow) and circular (arrowhead). (D) In *half-hearted* (*haf*) mutants, the ventricle is non-contractile, and the outer curvature cardiomyocytes are abnormally large and distended (arrows). *Images adapted from Auman, H. J., Coleman, H., Riley, H. E., Olale, F., Tsai, H. J., Yelon, D. (2007). Functional modulation of cardiac form through regionally confined cell shape changes.* PLoS Biology, 5, *e53.*

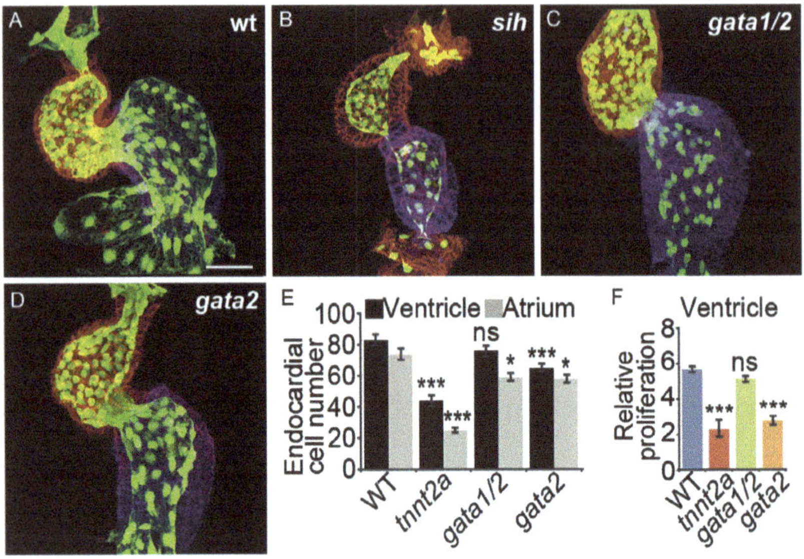

Fig. 3 Fluid forces influence endocardial cell number during chamber emergence. (A–D) Endocardial cells (green) are indicated by expression of the transgene *Tg*(*kdrl:GFP*); localization of Amhc (blue) and actin (red) are also shown. Whereas wild-type embryos have ~80 and ~70 endocardial cells in the ventricle and the atrium, respectively (A and E), the number of endocardial cells in both chambers is significantly reduced in *silent heart* (*sih; tnnt2a*) mutants (B and E), in which the heart is noncontractile. *gata1/2* morphants, which have diminished shear forces due to their reduced hematocrit, display fewer endocardial cells in the atrium (C and E). In contrast, *gata2* morphants, which have been shown to have reduced retrograde flow in addition to diminished shear forces, display fewer endocardial cells in both chambers (D and E), similar to *sih* mutants. (F) Consistent with this, endocardial proliferation in the ventricle is significantly reduced in *sih* mutants and *gata2* morphants, but not in *gata1/2* morphants. *Images adapted from Dietrich, A. C., Lombardo, V. A., Veerkamp, J., Priller, F., Abdelilah-Seyfried, S. (2014). Blood flow and bmp signaling control endocardial chamber morphogenesis.* Developmental Cell, 30, *367–377.*

are reduced, proliferation of the ventricular endocardium is markedly decreased (Fig. 3B–F) (Dietrich et al., 2014). Endocardial cells also fail to acquire normal morphologies under these circumstances. Together, these results suggest a model in which fluid forces trigger proliferation of endocardial cells, causing an overall enlargement of the endocardium that provokes cell shape changes within the myocardium, thereby facilitating synchronous expansion of both cardiac layers.

Which molecular pathways might drive chamber emergence in response to fluid forces? Exposure of endothelial cells to shear stress *in vitro* induces the expression of *KLF2*, which encodes a mechanoresponsive transcription factor (Dekker et al., 2002), and the expression of the zebrafish homolog *klf2a* also appears to be flow-responsive (Vermot et al., 2009). The *klf2* pathway is thought to contribute to the regulation of chamber emergence, since *klf2a* morphants exhibit aberrant endocardial morphology in the ventricle (Dietrich et al., 2014). Micro-RNAs are also interesting candidates for mechanoresponsive functions, since they can be regulated rapidly in response to flow (Banjo et al., 2013). Indeed, function-dependent expression of *miR-143* in both the myocardium and the endocardium impacts the process of ventricular emergence (Miyasaka et al., 2011). In the myocardium, *miR-143* targets *adducin3*, which regulates cytoskeletal actin dynamics and can thereby influence cardiomyocyte morphology (Deacon et al., 2010). In the endocardium, *miR-143* limits retinoic acid signaling by directly targeting *raldh2* and *rxrab* (Miyasaka et al., 2011). Intriguingly, *miR-143* morphants exhibit gaps in the ventricular endocardium (Miyasaka et al., 2011), hinting at defects in endocardial cell number. In future studies, it will be valuable to identify not only additional mechanosensitive genes that are relevant to chamber emergence but also additional downstream effectors responsible for executing the patterns of proliferation in the endocardium and cell shape change in the myocardium.

3.3 Retrograde flow drives atrioventricular canal differentiation

As the cardiac chambers emerge, the junction between the atrium and the ventricle constricts to form the atrioventricular canal (AVC). Within this region, distinct differentiation pathways create the endocardial cushions, specialized structures that will subsequently remodel to create the atrioventricular valve. Endocardial cushion formation has been examined with cellular precision in zebrafish, owing to the optical accessibility of the embryo. At early stages, endocardial cells accumulate both through proliferation and *via* convergence to the AVC region (Fig. 4A–C) (Boselli, Steed, Freund, & Vermot, 2017; Steed et al., 2016), while also acquiring distinctive cuboidal morphologies (Beis et al., 2005; Steed et al., 2016). Following this, the endocardial cells invaginate as a sheet into the extracellular matrix, ultimately forming valve leaflets (Pestel et al., 2016).

Early work in chick had shown that impediment of mechanical forces in the heart, *via* surgical alteration, can result in abnormal morphology of the atrioventricular valve (Sedmera, Pexieder, Rychterova, Hu, & Clark, 1999). Following this, several studies in zebrafish have indicated that cardiac

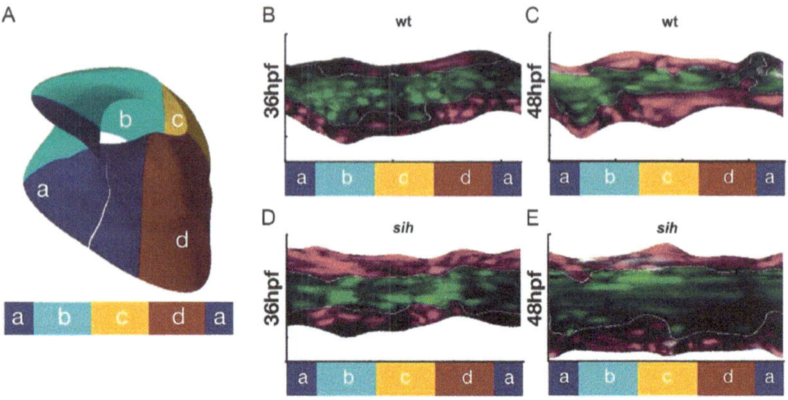

Fig. 4 Cardiac function regulates endocardial convergence at the AVC. (A) The AVC endocardium can be subdivided into four regions: superior (a), exterior (b), inferior (c) and interior (d). (B–E) Representations of the unfolded endocardium in *Tg(fli:kaede)* embryos in which the AVC endocardium has been labeled *via* photoconversion. In wild-type embryos, the AVC endocardial cells converge between 36 hpf (B) and 48 hpf (C). In contrast, in *sih* mutants, the AVC endocardium widens between 36 hpf (D) and 48 hpf (E). *Images adapted from Boselli, F., Steed, E., Freund, J. B., Vermot, J. (2017). Anisotropic shear stress patterns predict the orientation of convergent tissue movements in the embryonic heart.* Development, 144, *4322–4327.*

function provides essential cues for endocardial cushion formation. Endocardial cushions fail to form in *silent heart* mutants and *cardiofunk* mutants, both of which have defects in cardiac contractility (Bartman et al., 2004; Beis et al., 2005). Similarly, physical occlusion of blood flow into the zebrafish heart impairs atrioventricular valve development (Hove et al., 2003).

The oscillatory component of wall shear stress, or retrograde flow, is known to be particularly elevated within the AVC (Heckel et al., 2015), suggesting that the magnitude and the direction of fluid forces could influence AVC differentiation. To examine the effects of these types of mechanical inputs, the phenotypes generated by knocking down *gata1* and *gata2*, genes involved in hematopoiesis, were examined (Vermot et al., 2009). Since both genes contribute to erythrocyte development, knockdown of either should reduce shear forces in the heart. In addition, *gata2* morphants display a reduction in retrograde flow fraction (RFF) in the AVC, whereas *gata1* morphants display an increased RFF. Furthermore, *gata2* morphants fail to develop valve leaflets, whereas *gata1* morphants appear strikingly normal. Likewise, valve formation fails in *weak atrium* mutants, which also exhibit a reduced RFF in the AVC (Kalogirou et al., 2014). Notably, the valve defects in *gata2* morphants are preceded by a failure to accumulate endocardial cells at the AVC, as well as irregular cellular morphologies (Vermot et al., 2009).

The former is consistent with the observation that endocardial convergence at the AVC fails in *silent heart* mutants (Fig. 4B–E) (Boselli et al., 2017). Together, these findings imply that retrograde flow, and not shear, is the driving force for atrioventricular valve formation, potentially through its influence on cellular rearrangement during AVC differentiation.

Investigation of the molecular pathways triggered by retrograde flow in the AVC has primarily centered on *klf2a*. Expression of *klf2a* in the AVC endocardium is reduced in *gata2* morphants (Heckel et al., 2015; Vermot et al., 2009), and *klf2a* morphants resemble *gata2* morphants in that their AVC endocardial cells are reduced in number as well as abnormally elongated and flat (Vermot et al., 2009). Moreover, the calcium channels Trpp2 and Trpv4 appear to regulate *klf2a* expression in response to oscillatory flow patterns within the AVC endocardium (Heckel et al., 2015). Downstream effectors of Klf2 in the AVC endocardium include the extracellular matrix component Fibronectin: *fibronectin1b* morphants have defective cell clustering at the AVC, similar to *klf2a* morphants and *gata2* morphants (Steed et al., 2016). In addition, *wnt9b* acts a paracrine factor downstream of *klf2a* during AVC differentiation (Goddard et al., 2017).

Given the importance of *klf2a* to mechanosensation at the AVC, it is reasonable to imagine that the levels of *klf2a* expression must be carefully controlled. In accordance with this, the cerebral cavernous malformation (CCM) proteins Ccm1 and Ccm2 have been shown to restrict *klf2a* expression, with Ccm2 acting in a manner independent of cardiac function (Renz et al., 2015). Interestingly, the expression of Heg1, which acts to stabilize Ccm1, appears to be dependent on both cardiac function and *klf2a*, suggesting that *klf2a* acts in a negative feedback loop to influence its own expression (Donat et al., 2018). Altogether, studies of the factors upstream and downstream of *klf2a* place it as a key node in a retrograde flow-regulated pathway that drives AVC differentiation.

When considering additional factors that may act in or parallel to the *klf2a* pathway in this context, it is again interesting to examine the possible roles of micro-RNAs. For example, *miR-21* is expressed in the AVC in a function-dependent manner, and *miR-21* morphants fail to exhibit normal atrioventricular valve formation (Banjo et al., 2013). *miR-21* is thought to activate the MAP kinase cascade by targeting *sprouty2*, suggesting a possible mechanism for its influence on endocardial cell behavior during AVC differentiation. Further work will be necessary to evaluate whether these and/or other players interface with the *klf2a* pathway in executing the biomechanical regulation of AVC differentiation by retrograde flow.

3.4 Functional regulation of ventricular trabeculation

Following cardiac chamber emergence, the architecture of the ventricle becomes more elaborate through the formation of trabeculae—myocardial ridges that extend into the ventricular lumen. Analyses in zebrafish have shown that trabeculation is driven by the directional migration of cells that delaminate from the compact layer of the myocardium (Liu et al., 2010; Staudt et al., 2014). This delamination process involves constriction of the abluminal cardiomyocyte surface and is dependent upon the proper apicobasal polarization of the myocardial tissue (Jimenez-Amilburu et al., 2016; Staudt et al., 2014). In a noncontractile zebrafish heart, trabeculation fails: in *silent heart* morphants, for example, cardiomyocytes still extend protrusions into the ventricular lumen, but these protrusions frequently retract and stable trabeculae do not form (Staudt et al., 2014). These findings indicate that cardiac function is important for regulating trabeculation, consistent with prior observations in chick, where ventricular afterload has been implicated in regulating the thickening of the compact and trabecular layers (Sedmera et al., 1999).

Do fluid forces play a key role in controlling the cell behaviors that underlie trabeculation? Indeed, reduction of blood flow in both *weak atrium* mutants and *gata1* morphants inhibits trabeculation (Fig. 5A, C, and D) (Lee et al., 2016; Peshkovsky, Totong, & Yelon, 2011; Staudt et al., 2014; Vedula et al., 2017). Furthermore, studies in *weak atrium* mutants and morphants suggest that both the initial formation of luminal protrusions and their progression into stable trabeculae could be sensitive to the patterns of blood flow through the ventricle (Peshkovsky et al., 2011; Staudt et al., 2014). Flow parameters at the onset of ventricular trabeculation have not yet been carefully examined; however, the oscillatory shear index has been shown to be higher in trabecular grooves and lower in trabecular ridges at 4 days post-fertilization, after mature trabeculae are in place, and this distinct pattern is perturbed when trabeculation is inhibited (Fig. 5) (Vedula et al., 2017). In future studies, it will be valuable to investigate flow patterns in the ventricle at earlier stages in order to discern whether parameters like the oscillatory shear index can predict where trabecular protrusions will form.

Several signaling pathways that are known to be important regulators of trabeculation are also responsive to biomechanical cues. For example, the Neuregulin signaling pathway, operating through the Erbb2 receptor tyrosine kinase, is required for trabeculation (Liu et al., 2010), and expression of both *erbb2* (Lee et al., 2016) and its ligand *nrg2a* (Rasouli & Stainier, 2017)

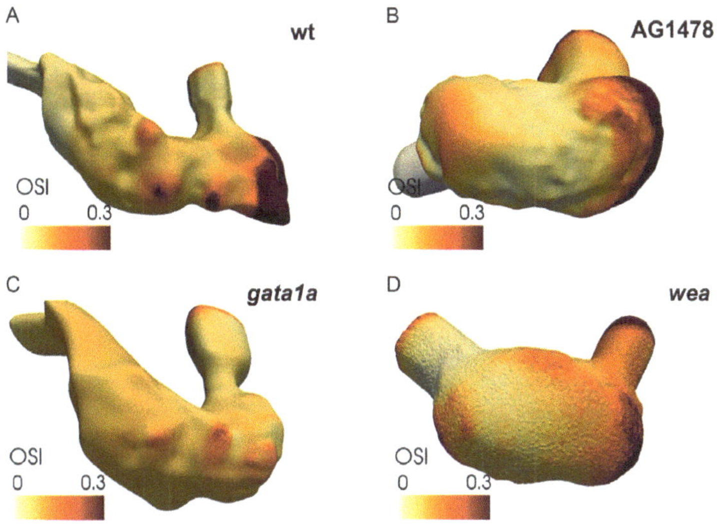

Fig. 5 Interplay between fluid forces and ventricular trabeculation. Representations of the inner surface of the ventricle, overlaid with a depiction of the oscillatory shear index (OSI) at 4 dpf. (A) In wild-type, the OSI is relatively high in trabecular grooves and relatively low in trabecular ridges. (B) In AG1478-treated embryos, the Neuregulin signaling pathway is inhibited, resulting in reduced trabeculation and a smoother inner surface of the ventricle. A similar reduction of trabeculation is observed in *gata1a* morphants (C) and *wea* mutants (D). *Images adapted from Vedula, V., Lee, J., Xu, H., Kuo, C.J., Hsiai, T.K., Marsden, A. L. (2017). A method to quantify mechanobiologic forces during zebrafish cardiac development using 4-d light sheet imaging and computational modeling.* PLoS Computational Biology, 13, *e1005828.*

has been shown to be dependent on cardiac function. Function-mediated Notch signaling activity in the endocardium has also been implicated in trabeculation, potentially *via* its role in regulating expression of the Erbb2 ligand *nrg1* (Samsa et al., 2015), as has been suggested in mice (Grego-Bessa et al., 2007). Wwtr1, a Hippo pathway effector that localizes to the nucleus in response to function, is an essential regulator of the compact wall architecture supporting trabeculation, where it can also influence myocardial Notch signaling (Lai et al., 2018). Furthermore, Erbb2 signaling can direct Wwtr1 localization (Lai et al., 2018) and Notch signaling in the myocardium (Jimenez-Amilburu et al., 2016). Finally, cardiac function, Erbb2, Wwtr1, and Notch have all independently been shown to be important for the relocalization of N-cadherin in delaminating cardiomyocytes (Cherian, Fukuda, Augustine, Maischein, & Stainier, 2016; Han et al., 2016; Lai et al., 2018). Collectively, these data signify complex spatiotemporal interactions between cardiac function and the Erbb2, Notch, and Hippo signaling pathways during the modulation of trabeculation. Ongoing studies will continue to illuminate the precise nature of these interactions, while also deciphering how these pathways instruct cell behaviors as trabeculae form.

4. Summary and future directions

Recent studies in zebrafish have amply demonstrated the broad utility of this model organism for analysis of the biomechanical regulation of cardiac development. The exceptional optical access to the developing heart, coupled with a variety of options for manipulation of gene function, has illuminated multiple morphogenetic processes that are dependent upon cardiac function. Moreover, scenarios in which blood flow is altered have provided intriguing evidence linking fluid forces to several key steps in cardiac maturation. Finally, further genetic analyses have created an emerging framework of the mechanoresponsive pathways that orchestrate important patterns of cell behavior in the embryonic heart (Fig. 6).

Despite these recent advances, there are still substantial challenges inherent to studying the relationship between fluid forces and cardiac morphogenesis. Importantly, common strategies for interfering with cardiac function often alter several biomechanical variables simultaneously. Inhibition of cardiac contractility, for example, changes both the stretch of cardiac tissues and the various forces associated with blood flow. In addition, manipulations that alter circulation while still maintaining contractility can perturb both the velocity and the direction of flow. Hence, most conclusions reached thus far are based on correlating particular functional scenarios with interesting phenotypes, rather than demonstrating which specific types of fluid forces cause which cellular outcomes. To evaluate the effects of distinct biomechanical cues, it will be instrumental for future studies to employ techniques that administer specific forces in a targeted fashion. Applications of single molecule spectroscopy could be adapted for this purpose: optical tweezing, for instance, could be used to manipulate erythrocyte velocity or direction in a confined region (Anton et al., 2013). Similarly, magnetic tweezers could be employed to stretch cells coated with ligand-conjugated magnetic beads, where the specificity of the ligand determines spatial control (Marjoram, Guilluy, & Burridge, 2016). These types of directed manipulations will help to delineate causative relationships between distinct fluid forces and specific cellular responses.

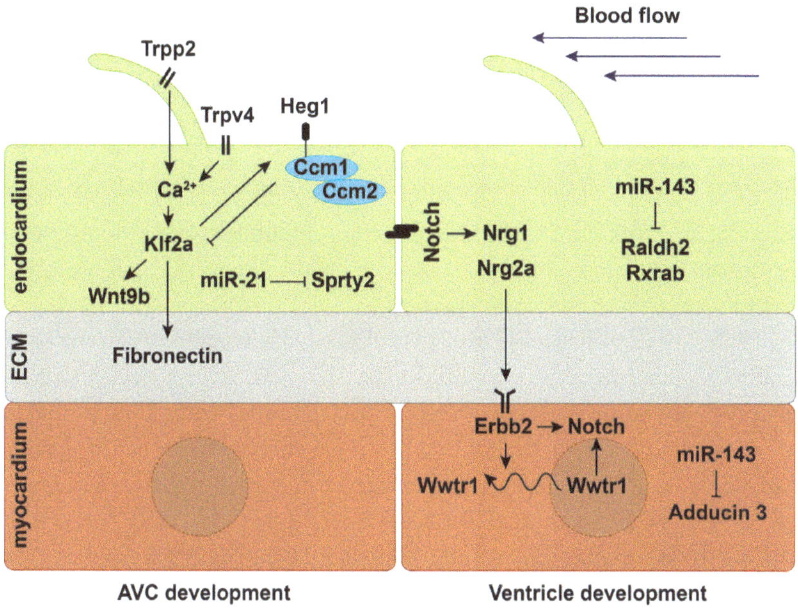

Fig. 6 Mechanoresponsive pathways regulating cardiac development in zebrafish. During AVC development, oscillatory flow activates mechanosensitive calcium channels Trpp2 and Trpv4 in the endocardium, which then promote *klf2a* expression (Heckel et al., 2015). In addition, flow-induced Heg1 represses *klf2a* expression in conjunction with the CCM complex, establishing a feedback loop for *klf2a* (Donat et al., 2018; Renz et al., 2015). Klf2a, in turn, regulates *wnt9b* expression in the endocardium (Goddard et al., 2017) and Fibronectin deposition in the extracellular matrix (ECM) (Steed et al., 2016). At the same time, flow-induced *miR-21* represses receptor tyrosine kinase signaling by targeting *sprty2* (Banjo et al., 2013). Meanwhile, during ventricle emergence, function-dependent *miR-143* inhibits retinoic acid signaling in the endocardium (Miyasaka et al., 2011) and targets Adducin3 in the myocardium (Deacon et al., 2010). Mechanical forces also activate Notch signaling in the endocardium, which further induces expression of the Erbb2 ligand *nrg1* (Samsa et al., 2015). During trabeculation, cardiac function regulates *nrg2a* (Rasouli & Stainier, 2017) and *erbb2* expression (Lee et al., 2016), as well as Wwtr1 localization (Lai et al., 2018). The Nrg/Erbb2 pathway can further affect Notch signaling in cardiomyocytes (Jimenez-Amilburu et al., 2016) and regulate Wwtr1 localization, which can also activate Notch signaling (Lai et al., 2018).

Even with experiments that uncouple the effects of specific flow parameters, limitations remain in our ability to interpret how particular forces exert their impact on both the endocardium and the myocardium. These two layers of the heart appear to respond to fluid forces in a harmonious fashion, yet it is unknown how they coordinate their morphogenesis. Does the pressure created by blood flow induce stretch of both the endocardium and the myocardium simultaneously? Or do fluid forces act primarily on the endocardium, with signals relayed to the myocardium only secondarily? Does the ECM that resides between the endocardium and myocardium transmit biomechanical cues between the layers, in addition to regulating biochemical cross-talk between them? To address these questions, it will be exciting for future studies to employ tools like tissue-specific, FRET-based tension sensors (Lagendijk et al., 2017) or tissue-embedded, force-sensitive oil microdroplets (Campas et al., 2014) to elucidate the precise spatiotemporal dynamics of cardiac mechanosensation.

Finally, we cannot resolve the mechanical interactions between the endocardium and the myocardium without knowing the full roster of molecular pathways responding to forces in each of the two layers. Interesting candidates for future exploration include Piezo1, a mechanosensitive calcium channel (Ranade et al., 2014), Yap1, a nuclear target of Hippo signaling (Nakajima et al., 2017), TGF-β receptors like Endoglin (Sugden et al., 2017), and junctional components like VE-cadherin (Baratchi et al., 2017), all of which have been shown to have mechanoresponsive roles in the endothelium. It will also be important to consider possible functional overlap between paralogs of genes that have already been implicated in cardiac mechanobiology, especially since discrepancies between mutant and morphant phenotypes suggest compensatory interactions between related factors (Novodvorsky et al., 2015).

Altogether, the body of work examining the influence of fluid forces during zebrafish heart development has opened up a number of exciting avenues for further investigation into the mechanosensitive pathways that control essential morphogenetic mechanisms. Going forward, we foresee that future studies in this area will actively bridge the gap between biomechanical inputs and cellular outcomes. By measuring and manipulating specific flow parameters and monitoring their impacts, it will be feasible to connect the fluid mechanics, signal transduction, downstream effector genes, and cellular dynamics that are responsible for shaping essential features of the developing heart. Ultimately, a mechanistic understanding of the roles of fluid forces during cardiac morphogenesis is likely to provide a valuable foundation for new insights regarding how hemodynamic conditions *in utero* affect the incidence of congenital heart disease.

References

Anton, H., Harlepp, S., Ramspacher, C., Wu, D., Monduc, F., Bhat, S., et al. (2013). Pulse propagation by a capacitive mechanism drives embryonic blood flow. *Development*, *140*, 4426–4434.

Auman, H. J., Coleman, H., Riley, H. E., Olale, F., Tsai, H. J., & Yelon, D. (2007). Functional modulation of cardiac form through regionally confined cell shape changes. *PLoS Biology*, *5*, e53.

Bagatto, B., & Burggren, W. (2006). A three-dimensional functional assessment of heart and vessel development in the larva of the zebrafish (Danio rerio). *Physiological and Biochemical Zoology*, *79*, 194–201.

Banjo, T., Grajcarek, J., Yoshino, D., Osada, H., Miyasaka, K. Y., Kida, Y. S., et al. (2013). Haemodynamically dependent valvulogenesis of zebrafish heart is mediated by flow-dependent expression of miR-21. *Nature Communications*, *4*, 1978.

Baratchi, S., Khoshmanesh, K., Woodman, O. L., Potocnik, S., Peter, K., & McIntyre, P. (2017). Molecular sensors of blood flow in endothelial cells. *Trends in Molecular Medicine*, *23*, 850–868.

Bartman, T., Walsh, E. C., Wen, K. K., McKane, M., Ren, J., Alexander, J., et al. (2004). Early myocardial function affects endocardial cushion development in zebrafish. *PLoS Biology*, *2*, e129.

Behrndt, M., Salbreux, G., Campinho, P., Hauschild, R., Oswald, F., Roensch, J., et al. (2012). Forces driving epithelial spreading in zebrafish gastrulation. *Science*, *338*, 257–260.

Beis, D., Bartman, T., Jin, S. W., Scott, I. C., D'Amico, L. A., Ober, E. A., et al. (2005). Genetic and cellular analyses of zebrafish atrioventricular cushion and valve development. *Development*, *132*, 4193–4204.

Boselli, F., Freund, J. B., & Vermot, J. (2015). Blood flow mechanics in cardiovascular development. *Cellular and Molecular Life Sciences*, *72*, 2545–2559.

Boselli, F., Steed, E., Freund, J. B., & Vermot, J. (2017). Anisotropic shear stress patterns predict the orientation of convergent tissue movements in the embryonic heart. *Development*, *144*, 4322–4327.

Broekhuizen, M. L., Hogers, B., DeRuiter, M. C., Poelmann, R. E., Gittenberger-de Groot, A. C., & Wladimiroff, J. W. (1999). Altered hemodynamics in chick embryos after extraembryonic venous obstruction. *Ultrasound in Obstetrics & Gynecology*, *13*, 437–445.

Campas, O., Mammoto, T., Hasso, S., Sperling, R. A., O'Connell, D., Bischof, A. G., et al. (2014). Quantifying cell-generated mechanical forces within living embryonic tissues. *Nature Methods*, *11*, 183–189.

Chen, Q., Jiang, L., Li, C., Hu, D., Bu, J. W., Cai, D., et al. (2012). Haemodynamics-driven developmental pruning of brain vasculature in Zebrafish. *PLoS Biology*, *10*, e1001374.

Chen, C. Y., Patrick, M. J., Corti, P., Kowalski, W., Roman, B. L., & Pekkan, K. (2011). Analysis of early embryonic great-vessel microcirculation in zebrafish using high-speed confocal muPIV. *Biorheology*, *48*, 305–321.

Cherian, A. V., Fukuda, R., Augustine, S. M., Maischein, H. M., & Stainier, D. Y. (2016). N-cadherin relocalization during cardiac trabeculation. *Proceedings of the National Academy of Sciences of the United States of America*, *113*, 7569–7574.

Collins, M. M., & Stainier, D. Y. (2016). Organ function as a modulator of organ formation: Lessons from zebrafish. *Current Topics in Developmental Biology*, *117*, 417–433.

Corti, P., Young, S., Chen, C. Y., Patrick, M. J., Rochon, E. R., Pekkan, K., et al. (2011). Interaction between Alk1 and blood flow in the development of arteriovenous malformations. *Development*, *138*, 1573–1582.

Deacon, D. C., Nevis, K. R., Cashman, T. J., Zhou, Y., Zhao, L., Washko, D., et al. (2010). The miR-143-adducin3 pathway is essential for cardiac chamber morphogenesis. *Development*, *137*, 1887–1896.

Dekker, R. J., van Soest, S., Fontijn, R. D., Salamanca, S., de Groot, P. G., VanBavel, E., et al. (2002). Prolonged fluid shear stress induces a distinct set of endothelial cell genes, most specifically lung Kruppel-like factor (Klf2). *Blood*, *100*, 1689–1698.

Diaz de la Loza, M. C., & Thompson, B. J. (2017). Forces shaping the Drosophila wing. *Mechanisms of Development*, *144*, 23–32.

Dietrich, A. C., Lombardo, V. A., Veerkamp, J., Priller, F., & Abdelilah-Seyfried, S. (2014). Blood flow and Bmp signaling control endocardial chamber morphogenesis. *Developmental Cell*, *30*, 367–377

Donat, S., Lourenco, M., Paolini, A., Otten, C., Renz, M., & Abdelilah-Seyfried, S. (2018). Hegl and Ccm1/2 proteins control endocardial mechanosensitivity during zebrafish valvulogenesis. *eLife*, *7*, e28939.

Franco, C. A., Jones, M. L., Bernabeu, M. O., Geudens, I., Mathivet, T., Rosa, A., et al. (2015). Dynamic endothelial cell rearrangements drive developmental vessel regression. *PLoS Biology*, *13*, e1002125.

Freund, J. B., Goetz, J. G., Hill, K. L., & Vermot, J. (2012). Fluid flows and forces in development: Functions, features and biophysical principles. *Development*, *139*, 1229–1245.

Goddard, L. M., Duchemin, A. L., Ramalingan, H., Wu, B., Chen, M., Bamezai, S., et al. (2017). Hemodynamic forces sculpt developing heart valves through a Klf2-Wnt9b paracrine signaling axis. *Developmental Cell*, *43*, 274–289.

Goetz, J. G., Steed, E., Ferreira, R. R., Roth, S., Ramspacher, C., Boselli, F., et al. (2014). Endothelial cilia mediate low flow sensing during zebrafish vascular development. *Cell Reports*, *6*, 799–808.

Grego-Bessa, J., Luna-Zurita, L., del Monte, G., Bolos, V., Melgar, P., Arandilla, A., et al. (2007). Notch signaling is essential for ventricular chamber development. *Developmental Cell*, *12*, 415–429.

Haack, T., & Abdelilah-Seyfried, S. (2016). The force within: Endocardial development, mechanotransduction and signalling during cardiac morphogenesis. *Development*, *143*, 373–386.

Han, P., Bloomekatz, J., Ren, J., Zhang, R., Grinstein, J. D., Zhao, L., et al. (2016). Coordinating cardiomyocyte interactions to direct ventricular chamber morphogenesis. *Nature*, *534*, 700–704.

Heckel, E., Boselli, F., Roth, S., Krudewig, A., Belting, H. G., Charvin, G., et al. (2015). Oscillatory flow modulates mechanosensitive Klf2a expression through Trpv4 and Trpp2 during heart valve development. *Current Biology*, *25*, 1354–1361.

Herwig, L., Blum, Y., Krudewig, A., Ellertsdottir, E., Lenard, A., Belting, H. G., et al. (2011). Distinct cellular mechanisms of blood vessel fusion in the zebrafish embryo. *Current Biology*, *21*, 1942–1948.

Hove, J. R., Koster, R. W., Forouhar, A. S., Acevedo-Bolton, G., Fraser, S. E., & Gharib, M. (2003). Intracardiac fluid forces are an essential epigenetic factor for embryonic cardiogenesis. *Nature*, *421*, 172–177.

Huang, C., Sheikh, F., Hollander, M., Cai, C., Becker, D., Chu, P. H., et al. (2003). Embryonic atrial function is essential for mouse embryogenesis, cardiac morphogenesis and angiogenesis. *Development*, *130*, 6111–6119.

Iftimia, N. V., Hammer, D. X., Ferguson, R. D., Mujat, M., Vu, D., & Ferrante, A. A. (2008). Dual-beam Fourier domain optical Doppler tomography of zebrafish. *Optics Express*, *16*, 13624–13636.

Jamison, R. A., Samarage, C. R., Bryson-Richardson, R. J., & Fouras, A. (2013). In vivo wall shear measurements within the developing zebrafish heart. *PLoS One*, *8*, e75722.

Jimenez-Amilburu, V., Rasouli, S. J., Staudt, D. W., Nakajima, H., Chiba, A., Mochizuki, N., et al. (2016). In vivo visualization of cardiomyocyte apicobasal polarity reveals epithelial to mesenchymal-like transition during cardiac trabeculation. *Cell Reports*, *17*, 2687–2699.

Kalogirou, S., Malissovas, N., Moro, E., Argenton, F., Stainier, D. Y., & Beis, D. (2014). Intracardiac flow dynamics regulate atrioventricular valve morphogenesis. *Cardiovascular Research*, *104*, 49–60.

Kochhan, E., Lenard, A., Ellertsdottir, E., Herwig, L., Affolter, M., Belting, H. G., et al. (2013). Blood flow changes coincide with cellular rearrangements during blood vessel pruning in zebrafish embryos. *PLoS One*, *8*, e75060.

Kwon, H. B., Wang, S., Helker, C. S., Rasouli, S. J., Maischein, H. M., Offermanns, S., et al. (2016). In vivo modulation of endothelial polarization by Apelin receptor signalling. *Nature Communications*, *7*, 11805.

Lagendijk, A. K., Gomez, G. A., Baek, S., Hesselson, D., Hughes, W. E., Paterson, S., et al. (2017). Live imaging molecular changes in junctional tension upon Ve-cadherin in zebrafish. *Nature Communications*, *8*, 1402.

Lai, J. K. H., Collins, M. M., Uribe, V., Jimenez-Amilburu, V., Gunther, S., Maischein, H. M., et al. (2018). The Hippo pathway effector Wwtr1 regulates cardiac wall maturation in zebrafish. *Development*, *145*, pii: dev159210.

Lee, J., Fei, P., Sevag Packard, R. R., Kang, H., Xu, H., Baek, K. I., et al. (2016). 4-Dimensional light-sheet microscopy to elucidate shear stress modulation of cardiac trabeculation. *The Journal of Clinical Investigation*, *126*, 3158.

Lee, J., Moghadam, M. E., Kung, E., Cao, H., Beebe, T., Miller, Y., et al. (2013). Moving domain computational fluid dynamics to interface with an embryonic model of cardiac morphogenesis. *PLoS One*, *8*, e72924.

Li, M., Zhao, L., Page-McCaw, P. S., & Chen, W. (2016). Zebrafish genome engineering using the CRISPR-Cas9 system. *Trends in Genetics*, *32*, 815–827.

Liebling, M., Forouhar, A. S., Wolleschensky, R., Zimmermann, B., Ankerhold, R., Fraser, S. E., et al. (2006). Rapid three-dimensional imaging and analysis of the beating embryonic heart reveals functional changes during development. *Developmental Dynamics*, *235*, 2940–2948.

Lin, Y. F., Swinburne, I., & Yelon, D. (2012). Multiple influences of blood flow on cardiomyocyte hypertrophy in the embryonic zebrafish heart. *Developmental Biology*, *362*, 242–253.

Liu, J., Bressan, M., Hassel, D., Huisken, J., Staudt, D., Kikuchi, K., et al. (2010). A dual role for Erbb2 signaling in cardiac trabeculation. *Development*, *137*, 3867–3875.

Lu, J., Pereira, F., Fraser, S. E., & Gharib, M. (2008). Three-dimensional real-time imaging of cardiac cell motions in living embryos. *Journal of Biomedical Optics*, *13*, 014006.

Marjoram, R. J., Guilluy, C., & Burridge, K. (2016). Using magnets and magnetic beads to dissect signaling pathways activated by mechanical tension applied to cells. *Methods*, *94*, 19–26.

Miyasaka, K. Y., Kida, Y. S., Banjo, T., Ueki, Y., Nagayama, K., Matsumoto, T., et al. (2011). Heartbeat regulates cardiogenesis by suppressing retinoic acid signaling via expression of miR-143. *Mechanisms of Development*, *128*, 18–28.

Nakajima, H., Yamamoto, K., Agarwala, S., Terai, K., Fukui, H., Fukuhara, S., et al. (2017). Flow-dependent endothelial YAP regulation contributes to vessel maintenance. *Developmental Cell*, *40*. 523–536.e6.

Novodvorsky, P., Watson, O., Gray, C., Wilkinson, R. N., Reeve, S., Smythe, C., et al. (2015). Klf2ash317 mutant zebrafish do not recapitulate morpholino-induced vascular and haematopoietic phenotypes. *PLoS One*, *10*, e0141611.

Pan, X., Yu, H., Shi, X., Korzh, V., & Wohland, T. (2007). Characterization of flow direction in microchannels and zebrafish blood vessels by scanning fluorescence correlation spectroscopy. *Journal of Biomedical Optics*, *12*, 014034.

Peshkovsky, C., Totong, R., & Yelon, D. (2011). Dependence of cardiac trabeculation on neuregulin signaling and blood flow in zebrafish. *Developmental Dynamics*, *240*, 446–456.

Pestel, J., Ramadass, R., Gauvrit, S., Helker, C., Herzog, W., & Stainier, D. Y. (2016). Real-time 3d visualization of cellular rearrangements during cardiac valve formation. *Development*, *143*, 2217–2227.

Ranade, S. S., Qiu, Z., Woo, S. H., Hur, S. S., Murthy, S. E., Cahalan, S. M., et al. (2014). Piezo1, a mechanically activated ion channel, is required for vascular development in mice. *Proceedings of the National Academy of Sciences of the United States of America*, *111*, 10347–10352.

Rasouli, S. J., & Stainier, D. Y. R. (2017). Regulation of cardiomyocyte behavior in zebrafish trabeculation by neuregulin 2a signaling. *Nature Communications*, *8*, 15281.

Renz, M., Otten, C., Faurobert, E., Rudolph, F., Zhu, Y., Boulday, G., et al. (2015). Regulation of β1 integrin-Klf2-mediated angiogenesis by CCM proteins. *Developmental Cell*, *32*, 181–190.

Rochon, E. R., Menon, P. G., & Roman, B. L. (2016). Alk1 controls arterial endothelial cell migration in lumenized vessels. *Development*, *143*, 2593–2602.

Samsa, L. A., Givens, C., Tzima, E., Stainier, D. Y., Qian, L., & Liu, J. (2015). Cardiac contraction activates endocardial Notch signaling to modulate chamber maturation in zebrafish. *Development*, *142*, 4080–4091.

Schwerte, T., & Pelster, B. (2000). Digital motion analysis as a tool for analysing the shape and performance of the circulatory system in transparent animals. *The Journal of Experimental Biology*, *203*, 1659–1669.

Schwerte, T., Uberbacher, D., & Pelster, B. (2003). Non-invasive imaging of blood cell concentration and blood distribution in zebrafish Danio rerio incubated in hypoxic conditions in vivo. *The Journal of Experimental Biology*, *206*, 1299–1307.

Sedmera, D., Pexieder, T., Rychterova, V., Hu, N., & Clark, E. B. (1999). Remodeling of chick embryonic ventricular myoarchitecture under experimentally changed loading conditions. *The Anatomical Record*, *254*, 238–252.

Shyer, A. E., Rodrigues, A. R., Schroeder, G. G., Kassianidou, E., Kumar, S., & Harland, R. M. (2017). Emergent cellular self-organization and mechanosensation initiate follicle pattern in the avian skin. *Science*, *357*, 811–815.

Staudt, D. W., Liu, J., Thorn, K. S., Stuurman, N., Liebling, M., & Stainier, D. Y. (2014). High-resolution imaging of cardiomyocyte behavior reveals two distinct steps in ventricular trabeculation. *Development*, *141*, 585–593.

Steed, E., Faggianelli, N., Roth, S., Ramspacher, C., Concordet, J. P., & Vermot, J. (2016). Klf2a couples mechanotransduction and zebrafish valve morphogenesis through fibronectin synthesis. *Nature Communications*, *7*, 11646.

Sugden, W. W., Meissner, R., Aegerter-Wilmsen, T., Tsaryk, R., Leonard, E. V., Bussmann, J., et al. (2017). Endoglin controls blood vessel diameter through endothelial cell shape changes in response to haemodynamic cues. *Nature Cell Biology*, *19*, 653–665.

Vedula, V., Lee, J., Xu, H., Kuo, C. J., Hsiai, T. K., & Marsden, A. L. (2017). A method to quantify mechanobiologic forces during zebrafish cardiac development using 4-D light sheet imaging and computational modeling. *PLoS Computational Biology*, *13*, e1005828.

Vermot, J., Forouhar, A. S., Liebling, M., Wu, D., Plummer, D., Gharib, M., et al. (2009). Reversing blood flows act through Klf2a to ensure normal valvulogenesis in the developing heart. *PLoS Biology*, *7*, e1000246.

Watkins, S. C., Maniar, S., Mosher, M., Roman, B. L., Tsang, M., & St Croix, C. M. (2012). High resolution imaging of vascular function in zebrafish. *PLoS One*, *7*, e44018.

Yalcin, H. C., Amindari, A., Butcher, J. T., Althani, A., & Yacoub, M. (2017). Heart function and hemodynamic analysis for zebrafish embryos. *Developmental Dynamics*, *246*, 868–880.

Zickus, V., & Taylor, J. M. (2018). 3d + time blood flow mapping using SPIM-microPIV in the developing zebrafish heart. *Biomedical Optics Express*, *9*, 2418–2435.

CHAPTER TWELVE

New perspectives on the mechanisms establishing the dorsal-ventral axis of the spinal cord

Madeline G. Andrews[a,b], Jennifer Kong[a,b], Bennett G. Novitch[a,c], Samantha J. Butler[a,c,]*

[a]Department of Neurobiology, University of California, Los Angeles, CA, United States
[b]Neuroscience Graduate Program, University of California, Los Angeles, CA, United States
[c]Eli and Edythe Broad Center of Regenerative Medicine and Stem Cell Research, University of California, Los Angeles, CA, United States
*Corresponding author: e-mail address: butlersj@ucla.edu

Contents

Abstract

Distinct classes of neurons arise at different positions along the dorsal-ventral axis of the spinal cord leading to spinal neurons being segregated along this axis according to their physiological properties and functions. Thus, the neurons associated with motor control

Current Topics in Developmental Biology, Volume 132
ISSN 0070-2153
https://doi.org/10.1016/bs.ctdb.2018.12.010

are generally located in, or adjacent to, the ventral horn whereas the interneurons (INs) that mediate sensory activities are present within the dorsal horn. Here, we review classic and recent studies examining the developmental mechanisms that establish the dorsal-ventral axis in the embryonic spinal cord. Intriguingly, while the cellular organization of the dorsal and ventral halves of the spinal cord looks superficially similar during early development, the underlying molecular mechanisms that establish dorsal vs ventral patterning are markedly distinct. For example, the ventral spinal cord is patterned by the actions of a single growth factor, sonic hedgehog (Shh) acting as a morphogen, i.e., concentration-dependent signal. Recent studies have shed light on the mechanisms by which the spatial and temporal gradient of Shh is transduced by cells to elicit the generation of different classes of ventral INs, and motor neurons (MNs). In contrast, the dorsal spinal cord is patterned by the action of multiple factors, most notably by members of the bone morphogenetic protein (BMP) and Wnt families. While less is known about dorsal patterning, recent studies have suggested that the BMPs do not act as morphogens to specify dorsal IN identities as previously proposed, rather each BMP has signal-specific activities. Finally, we consider the promise that elucidation of these mechanisms holds for neural repair.

1. Functional organization of the adult spinal cord

The central nervous system (CNS) is comprised of the brain and spinal cord, and is collectively responsible for much of what we consider to be vital for life and the experience of being human: cognition, movement, emotion, sensation, respiration and learning. While the spinal cord is less structurally complex than the brain, it nonetheless performs an essential task, receiving and processing incoming sensory information from the peripheral nervous system and then modulating and/or executing appropriate motor outputs (Andersson et al., 2012; Goetz, Pivetta, & Arber, 2015; Lai, Seal, & Johnson, 2016) (Fig. 1A). The spinal cord coordinates movement, orients our bodies in space, quickly processes harmful stimuli, and allows us to respond to our environment. The cells that process these sensory and motor functions are organized along the dorsal-ventral axis. Motor neurons (MNs) in the ventral horn control and coordinate motor output. Motor circuits are comprised of MNs, whose axons exit the spinal cord through the ventral root to synapse onto specific muscles, and spinal interneurons (INs), which modulate the activity of MNs. In contrast, sensory circuits relay sensory information from the periphery to the brain and motor systems. Different populations of peripheral sensory neurons in the dorsal root ganglion (DRG) detect distinct types of sensory information and send afferent projections either directly onto MNs or into the dorsal horn of spinal cord, where they synapse onto spinal INs. These dorsal INs then process and relay

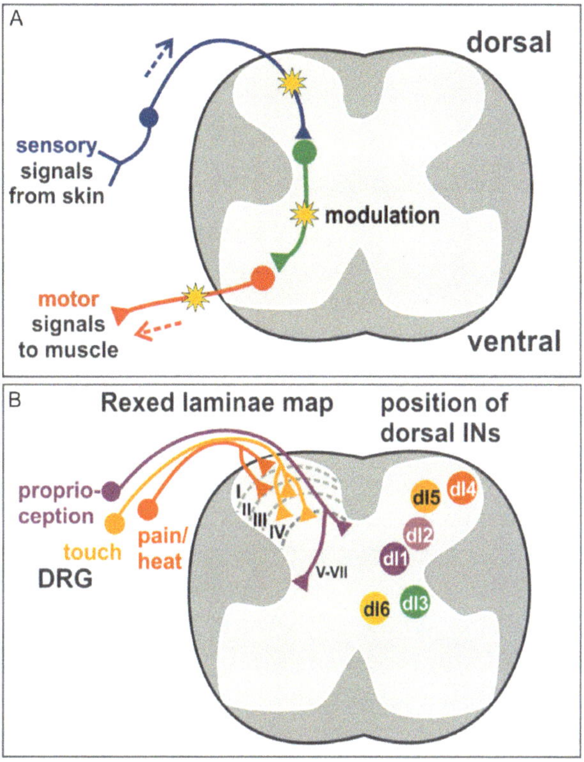

Fig. 1 Organization of the adult spinal cord. Schematics depicting the organization and function of the adult spinal cord. (A) Sensory information from the peripheral nervous system, is relayed into the dorsal horn of the spinal cord, processed by interneurons in the dorsal and ventral spinal cord and then relayed to higher centers in the brains, or modulate motor function. (B) The adult spinal cord is organized with a laminar architecture comprising seven distinct layers. Afferent sensory information is decoded in specific layers. Through fate mapping, genetic and connectivity studies, the general function and location of the dorsal IN populations have been identified in the adult spinal cord.

the relevant categories of sensory information either to the brain (Fig. 1B) or to MNs to modulate reflex-specific MN output (Arber, 2012).

In the adult spinal cord, the dorsal horn is organized with a layered, or laminar, architecture (Rexed, 1954). Seven laminae contain neurons with distinct physiological properties; some neurons locally process afferent sensory information while others relay it to higher order centers in the brain including the brainstem, thalamus and cerebellum (Bermingham et al., 2001; Brown & Fyffe, 1981; Lai et al., 2016; Yasaka, Tiong, Hughes, Riddell, & Todd, 2010). Information about nociception (pain) (Villeda, Akopians, Babayan, Basbaum, & Phelps, 2006; Xu et al., 2013;

Yasaka et al., 2010) or thermosensation (temperature) is processed in layers I-II, the most superficial layers of the horn (Bautista et al., 2007; Szabo et al., 2015). Somatosensation (touch) is processed in the intermediate layers, III-IV (Bourane, Grossmann, et al., 2015), and finally, proprioception, or perception of where your body resides in space, is processed in the deep horn, layers V-VI (Miesegaes et al., 2009; Yang, Bermingham, Finegold, & Zoghbi, 2001; Yuengert et al., 2015) (Fig. 1B). While the type of information processed in each laminae is generally understood, the transcriptional identities and functional characteristics of the IN subpopulations present in each laminae are still being characterized (Abraira et al., 2017; Bourane, Duan, et al., 2015; Bourane, Grossmann, et al., 2015).

In the ventral horn of the spinal cord (layer IX), MNs are segregated into distinct clusters, called pools, which innervate individual or groups of muscles. Motor pools are topographically arranged according to their muscle targets. They form columns along the rostral-caudal axis, settling at the appropriate axial position to innervate adjacent muscle targets (Dasen, De Camilli, Wang, Tucker, & Jessell, 2008; Jessell, 2000; Jessell, Surmeli, & Kelly, 2011; Rousso, Gaber, Wellik, Morrisey, & Novitch, 2008). The most medial MNs become the medial motor column (MMC), which innervates back and vertebral muscles. In the cervical and thoracic spinal cord, the hypaxial motor column (HMC) lies slightly medial to the MMC and innervates trunk muscles, many of which are associated with respiration. The thoracic spinal cord also contains the more laterally positioned preganglionic motor column (PGC) MNs which innervate sympathetic targets. At forelimb and hindlimb levels are two lateral motor columns (LMC) which are divided into medial and lateral sub-columns, LMCm and LMCl, to, respectively, control flexor and extensor muscles. The remaining layers of the ventral horn contain a multitude of interneurons that coordinate the activities of MNs (Goulding, 2009).

1.1 The organization of the embryonic spinal cord along the dorsal-ventral axis

Over the past 30 years, the embryonic spinal cord has emerged as an exceptional model system for assessing the mechanisms that direct the specification of neuronal and glial cell fate. It has a relatively simple architecture, arising when the neural plate folds to form the neural tube from a pseudostratified epithelium, which is only one cell thick at the earliest stages (Fig. 2A). The neural tube is comprised of rapidly dividing cells, which are the progenitors for all of the neurons and glia in the spinal cord. As the progenitors differentiate, the cells migrate laterally such that a simple laminar structure is

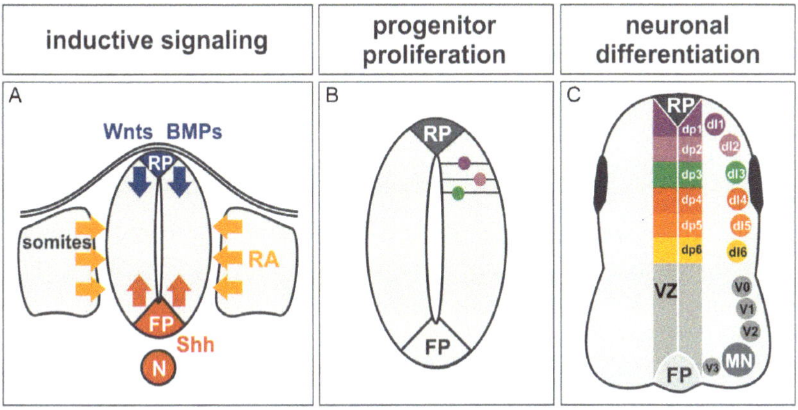

Fig. 2 Spinal cord development. Schematics describing the stages of neuronal development in the spinal cord. (A) Inductive signaling centers signal to neural tube progenitors to pattern neural identity. Sonic hedgehog (Shh) is secreted from the notochord and floor plate (FP) to pattern the ventral spinal cord. BMPs and Wnts are expressed from the roof plate (RP) to pattern the dorsal spinal cord. Retinoic acid (RA) is secreted from the somites in the paraxial mesoderm to specify axial level and promote neuronal differentiation. (B) During early spinal cord development, the nuclei of proliferating progenitors oscillate along the medial-lateral axis as they undergo cell division. (C) In response to inductive growth factors, progenitors are patterned into distinct domains within the ventricular zone (VZ) along the dorsal-ventral axis. As progenitors differentiate they exit the cell cycle and migrate to the lateral edge of the spinal cord. The dorsal spinal cord has six distinct interneurons (IN) populations, the dI1–6, while the ventral spinal cord has five distinct neuronal populations, the V0–V3 IN populations and the spinal MNs. *Panel A modified with permission from Tanabe, Y., & Jessell, T. M. (1996). Diversity and pattern in the developing spinal cord.* Science (New York, N.Y.), 274, *1115–1123.*

formed: the inner ventricular zone contains the nuclei of the proliferating progenitors; the mantle layer contains the post-mitotic neural cell bodies; and the outer marginal layer is comprised of processes. At the earliest stages, the marginal layer contains the thin processes of the neuroepithelial cells; however, it expands over time to accommodate the fiber tracts, becoming better known as the white matter of the adult spinal cord. The mantle layer ultimately becomes the gray matter.

The identities of the progenitors within the ventricular zone are patterned by inductive signaling molecules released from adjacent structures. These signals include the bone morphogenetic proteins (BMPs) and Wnt family from the roof plate (RP) at the dorsal midline, sonic hedgehog (Shh) from the floor plate (FP) at the ventral midline, and retinoic acid (RA) from somites in the adjacent paraxial mesoderm (Briscoe & Novitch, 2008; Tanabe & Jessell, 1996) (Fig. 2B). These signals compete and cooperate to specify distinct domains of progenitors along the dorsal-ventral axis. Domains of progenitors become distinct from each other following the expression of unique complements of transcription factors specifying their identities. Upon neurogenesis, patterned progenitor cells exit the cell cycle and migrate laterally into the mantle layer (Helms & Johnson, 2003; Le Dreau & Marti, 2013) (Fig. 2B and C). This process in turn results in discrete populations of neurons arrayed along the dorsal-ventral axis at the lateral edge of the spinal cord. The dorsal spinal cord contains at least six dorsal progenitor (dP) domains, dP1–dP6, that differentiate into the dorsal interneuron (dI) populations 1–6 (Fig. 2C) (Helms & Johnson, 2003). Each neuronal population is distinguished by its expression of a distinct subset of transcription factors which appear to regulate their position, development and later connectivity and function (Lai et al., 2016). The dIs mature into sensory relay INs. The function of many of these neural populations has only recently been identified. The dI1s mediate proprioception (Bermingham et al., 2001; Yuengert et al., 2015), the dI3s facilitate some mechanosensory behaviors including grasping control (Bui et al., 2013; Goetz et al., 2015), the dI4s and dI5s have multiple functions including itch/touch/pain (Abraira et al., 2017; Bourane, Duan, et al., 2015; Bourane, Grossmann, et al., 2015) and the dI6s regulate motor output (Lai et al., 2016) (Fig. 1).

The ventral spinal cord encompasses at least five progenitor (p) domains, p0–p3 and the pMN, which give rise to five distinct cell types, the ventral (V) IN, V0–V3 and MNs (Briscoe & Novitch, 2008; Goulding, 2009; Jessell, 2000). The MNs control movement throughout the body axis, while the ventral INs, V0–V3, function to modulate motor activities. The V0–V3 IN populations are a mix of ipsilaterally and contralaterally projecting populations that regulate MN output (Gosgnach et al., 2017). These populations are further subdivided based on their dorsal-ventral position, axon projection type and function. Broadly, V0 and V2 are a mix of excitatory and inhibitory INs, while V1s are inhibitory and V3 are excitatory (Gosgnach et al., 2017; Goulding, 2009). V0 and V2 neurons serve to alternate left/right motor activities (Crone et al., 2008; Griener, Zhang, Kao, Wagner, & Gosgnach, 2015), while V1 and V2 coordinate flexor-extensor activities (Zhang et al., 2014). V1 neurons also influence motor speed (Gosgnach et al., 2006), while V3 modulate the quality of motor bursts as well as overall motor performance (Zhang et al., 2008). As with the dorsal neural populations, considerable work has defined the transcription factor code that specifies the identity of the motor pools and ventral INs (Catela, Shin, & Dasen, 2015; Gosgnach et al., 2017).

2. The specification of dorsal and ventral patterning in the spinal cord

The cellular mechanisms that establish the embryonic dorsal and ventral spinal cord are remarkably similar. Many of the neural populations are specified by inductive signaling from clusters of glial-like cells at the opposing ventral and dorsal midlines. These cells have distinct origins: the dorsal RP is formed in the region where the neural plate seals to form the neural tube, whereas the ventral FP is induced by signals from the underlying notochord, a transient mesodermal structure (Yamada, Placzek, Tanaka, Dodd, & Jessell, 1991). These structures are considered to be analogous signaling centers, secreting inductive factors that pattern the identity of the adjacent tissue. However, the identity of these factors and the mechanisms used to pattern cell fate have recently been shown to be strikingly different. We will begin by reviewing the mechanisms of ventral patterning, which has been extensively studied as a model for morphogen signaling, and then discuss how the mechanisms of dorsal patterning appear to use a signal-specific strategy.

2.1 The Hh signaling pathway

In vertebrates, the hedgehog (Hh) signaling pathway is activated when extracellular Hh ligands bind to the transmembrane receptor patched (Ptch) (Chen & Struhl, 1996; Ingham, Taylor, & Nakano, 1991; Marigo, Davey, Zuo, Cunningham, & Tabin, 1996; Stone et al., 1996). This interaction relieves Ptch-mediated inhibition of the transmembrane protein Smo, allowing for the activation and translocation of Smo into the primary cilium, a single nonmotile microtubule-based projection present on nearly all vertebrate cells (Corbit et al., 2005; Rohatgi, Milenkovic, & Scott, 2007). The presence of activated Smo within the primary cilium triggers a cascade of downstream signaling events that culminate in the conversion of Gli proteins into transcriptional activators (GliA) that translocate into the nucleus and directly turn on downstream genes including *Gli1* and *Ptch* (Goodrich, Johnson, Milenkovic, McMahon, & Scott, 1996; Lee, Platt, Censullo, & Ruiz i Altaba, 1997) (Fig. 3B). In the absence of Hh ligands, Ptch represses both the activation and ciliary localization of Smo, resulting in the truncation of Gli proteins into transcriptional repressors (GliR) (Hui, Slusarski, Platt, Holmgren, & Joyner, 1994; Hynes et al., 1997; Ruiz i Altaba, 1998; Sasaki, Nishizaki, Hui, Nakafuku, & Kondoh, 1999) (Fig. 3A). While the link between Smo activation and Gli processing has been well established, the mechanistic details explaining this connection remain unclear.

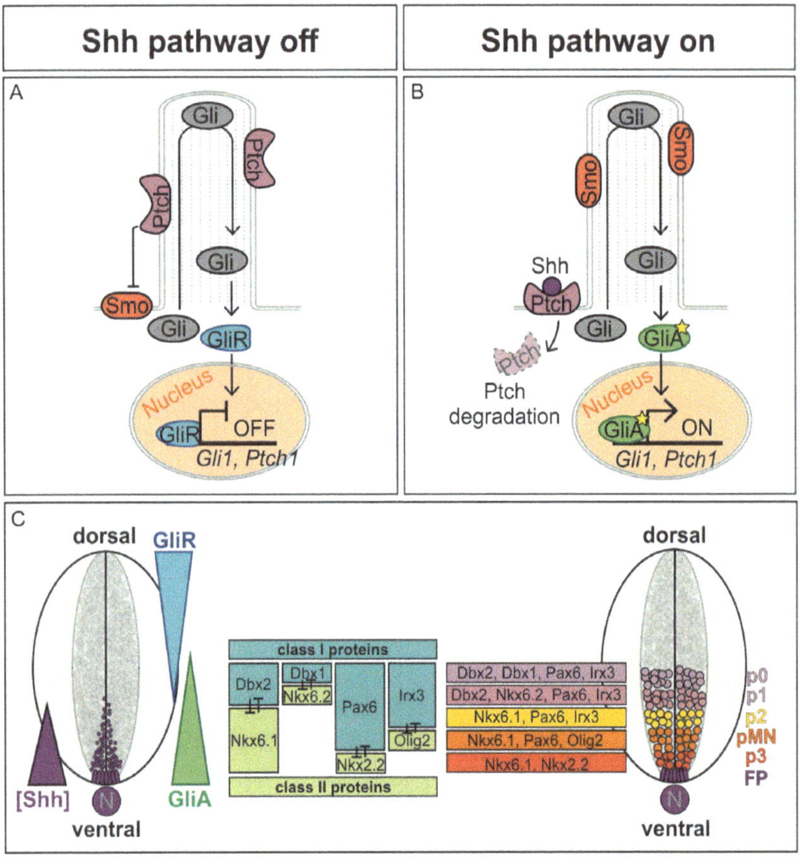

Fig. 3 Role of the Shh signaling pathway. (A) Shh signal transduction. In the absence of Shh ligand, the patched (Ptch) accumulates at the base of, and within, the primary cilium. Through an indirect mechanism, the presence of Ptch represses the activation and movement of smoothened (Smo) into the primary cilium. In the absence of Smo, full-length Gli proteins (Gli) are cleaved and processed into transcriptional repressors (GliR), which then translocate to the nucleus and repress downstream gene targets that include *Gli1* and *Ptch1*. (B) In the presence of Shh, Shh ligand binds to Ptch, which is then is endocytosed and degraded. In the absence of Ptch, Smo is activated and translocates into the primary cilium. In the presence of Smo, Gli proteins are processed into transcriptional activators (GliA). (C) Ventral spinal cord patterning. Shh ligand is secreted by the notochord (N) and the floor plate (FP) of the neural tube. Shh acts in a dose-dependent manner to alter the processing of Gli proteins from transcriptional repressors (GliR) to transcriptional activators (GliA). The net balance between GliA and GliR influences the expression of various homeodomain transcription factors along the dorsal-ventral axis. These transcription factors can be grouped into two classes: class I proteins (Dbx2, Dbx1, Pax6, and Irx3) that are influenced by RA signaling and present only in the absence of Shh and class II proteins (Nkx6.1, Nkx6.2, Nkx2.2, and Olig2) that require Shh for their activation. Ultimately, unique combinations of class I and class II

Due to genome duplication events in early vertebrates (Wada & Makabe, 2006) there are three mammalian Hh genes: *sonic hedgehog* (*Shh*); *indian hedgehog* (*Ihh*); and *desert hedgehog* (*Dhh*) (Hammerschmidt, Brook, & McMahon, 1997). The three mammalian Hh proteins are remarkably similar, as they are all processed by the same mechanisms, have the capacity to bind to the Hh receptors patched1 and patched2 (Ptch1 and Ptch2), and ultimately upregulate the expression of known Hh pathway target genes (Carpenter et al., 1998). However, each Hh protein has a unique expression pattern in the developing embryo. Shh is the most broadly expressed member of the mammalian Hh family and is the key signal that patterns the ventral spinal cord (Echelard et al., 1993; Varjosalo & Taipale, 2008).

2.2 Shh acts as a canonical morphogen in ventral patterning

Ventral patterning commences in the spinal cord when signals from the notochord induce the formation of the FP. This inductive interaction was first observed by Johannes Holtfreter in the 1930s (Gerhart, 1996), but the molecule responsible was not identified until many decades later. It took an unexpected convergence of molecular and evolutionary genetics in the late 1980s for researchers to realize that the genes responsible for patterning segments in Drosophila (Lee, von Kessler, Parks, & Beachy, 1992; Mohler & Vani, 1992; Nusslein-Volhard & Wieschaus, 1980; Tabata, Eaton, & Kornberg, 1992; Tashiro et al., 1993) were also critically involved in the organization of cellular identity in the vertebrate embryo.

Ventral patterning was found to be induced by a single signaling molecule, Shh, identified by its homology with the Drosophila gene, hedgehog (Roelink et al., 1994). Shh is secreted from the notochord and induces the formation of the FP at the ventral midline (Roelink et al., 1994).

transcription factors subdivide the ventral spinal cord into five molecularly distinct progenitor domains (p0, p1, p2, pMN, and p3). Over time, each of these progenitor domains gives rise to distinct types of neurons and then glia. *Panel A modified with permission from Briscoe, J., & Therond, P. P. (2013). The mechanisms of Hedgehog signalling and its roles in development and disease.* Nature Reviews. Molecular Cell Biology, 14, *416–429. Panel C modified with permission from Briscoe, J., & Novitch, B. G. (2008). Regulatory pathways linking progenitor patterning, cell fates and neurogenesis in the ventral neural tube.* Philosophical Transactions of the Royal Society of London. Series B, Biological Sciences, 363, *57–70; Jessell, T. M. (2000). Neuronal specification in the spinal cord: Inductive signals and transcriptional codes.* Nature Reviews. Genetics, 1, *20–29; Oosterveen, T., Kurdija, S., Alekseenko, Z., Uhde, C. W., Bergsland, M., Sandberg, M., et al. (2012). Mechanistic differences in the transcriptional interpretation of local and long-range Shh morphogen signaling.* Developmental Cell, 23, *1006–1019.*

The FP itself then also secretes Shh, which specifies neuronal identity by acting as a morphogen (Ericson, Briscoe, Rashbass, van Heyningen, & Jessell, 1997; Yamada, Pfaff, Edlund, & Jessell, 1993) (Fig. 2A). Morphogens were first hypothesized to exist as diffusible factors that act in a concentration-dependent manner to direct cellular identity (Bier & De Robertis, 2015; Briscoe et al., 1999; Wolpert, 1969). Shh is one of relatively few molecules that have been credibly shown to act as a canonical morphogen during development. In a classic series of experiments, Jessell and co-workers used spinal tissue culture assays to demonstrate that high concentrations of solubilized Shh specify the ventral-most identities in the spinal cord, FP and p3, while progressively lower concentrations specify more intermediate cell types, i.e., the pMN, and then V2–V0 (Ericson, Briscoe, et al., 1997; Roelink et al., 1995). Subsequent analyses showed that loss of Shh *in vivo* resulted in the profound loss of ventral spinal identities, far beyond the FP, the source of *Shh* expression (Chiang et al., 1996). These experiments are among the some of the most convincing to demonstrate that morphogens exist in vertebrates, suggesting a model whereby a localized, diffusible signal at the FP patterns the adjacent tissue into discrete domains of specific neuronal identities (Tanabe & Jessell, 1996). The morphogen model of patterning has been highly influential, with other localized sources of inductive signals being hypothesized to pattern other areas of the CNS, including the dorsal spinal cord (Liem, Tremml, & Jessell, 1997).

2.3 The Shh gradient is interpreted by a code of Gli transcription factors

The Hh signal transduction pathway converges on Gli proteins, an evolutionarily conserved family of zinc finger transcription factors. Vertebrates have three Gli homologs: Gli1, Gli2, and Gli3. Gli1 only exists as a transcriptional activator and primarily serves as a secondary amplifier of Hh signaling activity (Bai, Auerbach, Lee, Stephen, & Joyner, 2002; Lee et al., 1997). In contrast, Gli2 and Gli3 can become either transcriptional activators or repressors because they possess both a C-terminal activator domain and a N-terminal repressor domain. Despite this bifunctional potential, Gli2 is typically processed into a transcriptional activator and Gli3 into a repressor through differential phosphorylation of the proteins at specific sites that modulate transcriptional activities as well as proteolytic cleavage of their C-terminal activator domains (Aza-Blanc, Lin, Ruiz i Altaba, & Kornberg, 2000; Hui & Angers, 2011; Niewiadomski et al., 2014; Persson et al., 2002; Ruiz i Altaba, 1998; Sasaki et al., 1999).

Within the developing spinal cord, Shh is initially synthesized and secreted from the notochord and later the FP of the neural tube (Echelard et al., 1993; Krauss, Concordet, & Ingham, 1993; Roelink et al., 1994). Over time, Shh diffuses dorsally away from this fixed source, generating a high-to-low Shh concentration gradient along the ventral-to-dorsal axis of the neural tube (Burke et al., 1999; Chamberlain, Jeong, Guo, Allen, & McMahon, 2008; Goetz, Singh, Suber, Kull, & Robbins, 2006; Zeng et al., 2001) (Fig. 3C). This extracellular Shh concentration gradient is translated into an intracellular gradient of Gli activity, whereby GliA levels are elevated in the ventral spinal cord and GliR is elevated in the dorsal spinal cord (Fig. 3C). The balance between GliA and GliR, defined as net Gli activity, ultimately regulates the expression of transcription factors that will define the progenitor identity of the cells (Stamataki, Ulloa, Tsoni, Mynett, & Briscoe, 2005). Within the spinal cord, the transcription factors regulated by net Gli activity are divided into two classes: the dorsally expressed class I proteins (Dbx2, Dbx1, Pax6, and Irx3), which are repressed by Shh signaling activity, and the ventrally expressed class II proteins (Nkx6.1, Nkx6.2, Olig2, and Nkx2.2) that are activated by Shh signaling (Briscoe, Pierani, Jessell, & Ericson, 2000; Ericson, Rashbass, et al., 1997). The dorsal/ventral limits of transcription factor expression are a consequence of Shh signaling activity, as the well-defined boundaries between progenitor domains are established by cross repressive interactions between complementary class I/II protein pairs. Ultimately, unique combinations of class I and class II proteins subdivide the ventral neural tube into five molecularly distinct progenitor domains (p3, pMN, p2, p1, and p0) (Fig. 3C). Each of these progenitor domains subsequently gives rise to a distinct subset of neurons (V3, MN, V2, V1, and V0) and glial cells later in development (Rowitch & Kriegstein, 2010).

2.4 Shh also functions as a temporal morphogen

In addition to the specific extracellular concentration of Shh, the transcriptional profile of a ventral neural progenitor cell is dependent on the duration of time it is exposed to Shh. This temporal component of Shh signaling is evident in the early embryonic spinal cord, where ventral progenitor domains gradually appear over time as a result of prolonged Shh signaling activity driving the sequential expression of increasingly ventral class II transcription factors (Dessaud et al., 2010, 2007). Remarkably, this progressive patterning process can be reproduced *in vitro*, by exposing chick neural tube explants and mouse neural progenitor cells to Shh ligands and Smo agonists for extended durations of time (Dessaud et al., 2010, 2007; Kutejova, Sasai, Shah, Gouti, & Briscoe, 2016; Sagner et al., 2018).

When Shh is absent from these cultures, neural progenitors express class I transcription factors, such as Pax6 and Pax7. The addition of Shh turns on the class II transcription factors. Initially, Nkx6.1 and Olig2 (markers of p2 and pMN state) are induced by low levels of Shh signaling activity and reductions in GliR levels. However, sustained exposure to Shh over time results in the accumulation of GliA and expression of the p3-associated transcription factor Nkx2.2, which itself serves as a repressor of Olig2 expression and pMN fate. This shift in patterning occurs due to temporal adaptation to the morphogen. Prolonged exposure to Shh causes progenitor cells to upregulate Ptch, which functions as a negative feedback inhibitor of the Shh signaling pathway. Increased Ptch levels cause progenitors to become insensitive to extracellular Shh over time (Balaskas et al., 2012; Dessaud et al., 2010, 2007; Jeong & McMahon, 2005). Only the ventral-most progenitor cells, exposed to the highest concentrations of extracellular Shh, maintain the elevated levels of Shh signaling activity needed to turn on the transcription factors such as Nkx2.2 that define the ventral-most domains.

While exposure to Shh causes progenitors to become desensitized to Shh, it was recently shown that Notch signaling can sustain a progenitor's responsivity to Shh (Kong et al., 2015; Stasiulewicz et al., 2015). Experiments using both transgenic mice and neural progenitor cultures demonstrated that elevating Notch signaling activity extends the duration of time a cell is maintained as a Shh-responsive progenitor. This increase in Shh signaling activity shifts the fate of the progenitors to a more ventral identity. Interactions between Notch and Shh signaling have been described in other systems (Casso, Biehs, & Kornberg, 2011; Hallahan et al., 2004; Huang, Xiong, Megason, & Schier, 2012; Lopez et al., 2003). Shh pathway components are directly regulated by the Rbpj transcriptional complex (Hamidi, Gustafason, Pellegrini, & Gasson, 2011; Li, Hibbs, Gard, Shylo, & Yun, 2012), and elevated Notch signaling promotes the Shh driven accumulation of Smo to the primary cilia (Kong et al., 2015). However, the precise means through which Notch signaling interacts with Shh signaling remains largely unknown.

2.5 The RA signaling pathway

Retinoic acid (RA) signaling is a core developmental signaling pathway that also contributes to dorsal-ventral patterning in the spinal cord. Derived from dietary vitamin A, the importance of RA in development is perhaps best illustrated from studies on vitamin A deficiencies. When pregnant female rats were maintained on a vitamin A-deficient diet over a short gestational window, the resulting pups exhibited wide spread developmental defects to the eyes, skull, trachea, esophagus, limbs, heart, and CNS (Dickman, Thaller, & Smith, 1997).

When vitamin A (all-trans retinol, at-retinol) is consumed, it is absorbed into the blood stream via the small intestine, where it binds to retinol binding protein 4 (RBP4). RBP4 serves as a tag, binding to the membrane receptor STRA6 and allowing for at-retinol to identify and enter the cytoplasm of target cells. Once in the cytoplasm, at-retinol is converted into its biologically active form, all-trans retinoic acid (RA), via a two-step oxidation process involving the enzymes retinol or alcohol dehydrogenases (RoDHs or ADHs) and retinaldehyde dehydrogenases (RALDHs) (Maden, 2002). RA is then shuttled into the nucleus by binding to cellular retinoic acid binding protein 2 (CRABP2) where it binds to either the RA receptors (RARs) or the retinoid X receptors (RXRs). These RA bound receptors (RA-RAR and RA-RXR) then heterodimerize and bind to retinoic acid response elements (RAREs) on target genes to activate their transcription (Kam, Deng, Chen, & Zhao, 2012; Maden, 2007). Through the use of ChIP-on-chip (chromatin immunoprecipitation combined with microarray hybridization) and ChIP-seq technology, over 300 potential RA-regulated genes have been identified (Delacroix et al., 2010). Among the targets were Hox genes and components of the TGFβ (transforming growth factor β) signaling pathway, a pathway known to interact with RA signaling to regulate cell growth and differentiation.

In the developing CNS, RA signaling plays a major role in both anteroposterior and dorsal-ventral progenitor patterning. Multiple studies have shown that RA signaling drives progenitor cells toward more posterior/rostral cell identities by directly binding to and manipulating the expression of certain Hox genes (Conlon & Rossant, 1992; Delacroix et al., 2010; Durston et al., 1989; Huang, Chen, & Gudas, 2002; Koop et al., 2010). Additional studies conducted in mouse embryonic cell culture, avian embryos, and chick neural plate explants have also shown that RA drives spinal cord progenitors toward more dorsal and intermediate identities (Novitch, Wichterle, Jessell, & Sockanathan, 2003; Okada, Shimazaki, Sobue, & Okano, 2004; Wilson, Gale, Chambers, & Maden, 2004). One of the ways in which RA contributes to fate specification is by positively regulating the expression of class I transcription factors, thereby opposing the ventral patterning actions of the Shh pathway (Briscoe & Novitch, 2008;

Diez del Corral et al., 2003; Novitch et al., 2003). Beyond its role in patterning, RA signaling can also drive progenitors toward neuronal differentiation (Diez del Corral et al., 2003; Novitch et al., 2003; Sharpe & Goldstone, 1997).

2.6 The specification of dorsal patterning in the dorsal spinal cord

The similarity between the organization of early dorsal and ventral spinal cord led to the hypothesis that the molecular mechanisms that assembled cellular identity were also similar. However, while the ventral spinal cord is patterned largely by the action of one molecule, multiple members of the BMP and Wnt families are critical for dorsal patterning. It was thus unclear how >5 growth factors could cooperate to form a morphogen gradient. Indeed, recent studies have found no evidence that the BMPs act as morphogens to pattern the spinal cord, rather each BMP ligand directs a unique range of dorsal cellular identities.

2.7 The BMP signaling pathway

There are ~30 BMP family members expressed during embryonic development (Ducy & Karsenty, 2000). Although they were first discovered by their ability to direct the formation of bone and cartilage (Urist, Dowell, Hay, & Strates, 1968), they are also required for the development of the dorsal spinal cord, limb, eye, forebrain and the kidney (Ahn, Mishina, Hanks, Behringer, & Crenshaw, 2001; Dudley, Lyons, & Robertson, 1995; Lyons, Hogan, & Robertson, 1995; Wyatt, Osborne, Stewart, & Ragge, 2010). BMPs can regulate many processes including cell division, apoptosis, proliferation, cell migration, cell survival and differentiation (Bier & De Robertis, 2015; Ducy & Karsenty, 2000; Hogan, 1996; Shi & Massague, 2003; Wu & Hill, 2009). The BMPs signal through a complex of type I and type II serine/threonine kinase receptors (Balemans & Van Hul, 2002) (Fig. 4A). Signal specificity is thought to reside in the identity of the BMP type I receptor (BmprI), which include the BmprIa (also known as Alk3) and BmprIb (Alk6) (Moustakas & Heldin, 2009; ten Dijke et al., 1994). Upon BMP binding, BmprII transphosphorylates BmprI, which in turn phosphorylates and activates the Smad proteins, the intracellular mediators of canonical BMP signaling (Shi & Massague, 2003) (Fig. 4A).

There are three known classes of Smads: the receptor (R) activated, inhibitory (I) and common mediator (Co) Smads (Balemans & Van Hul, 2002).

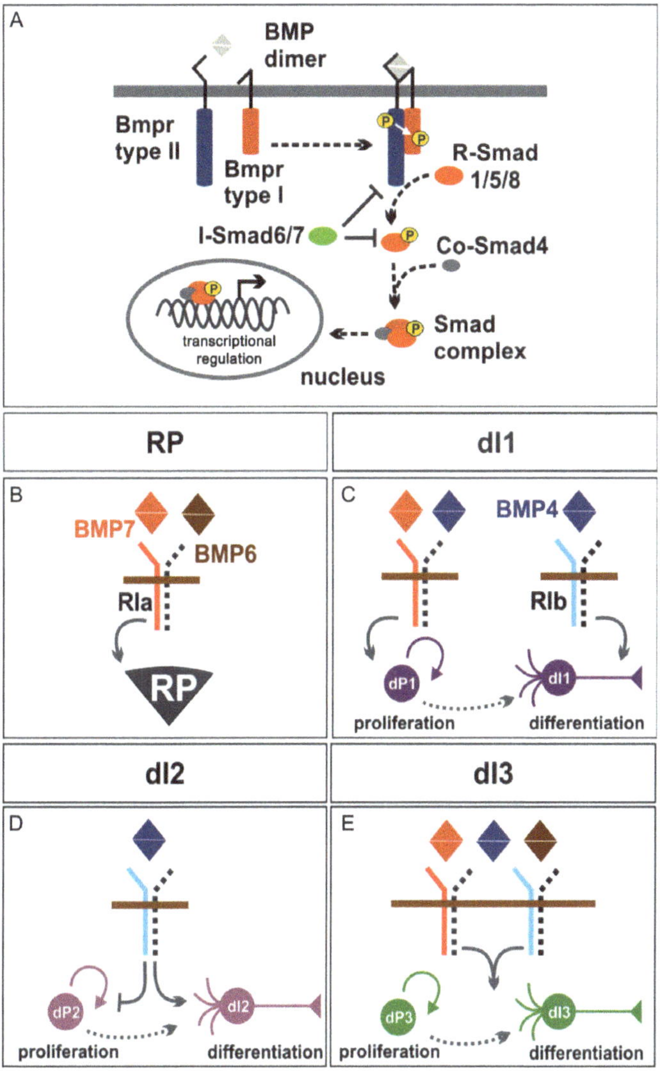

Fig. 4 BMP signaling pathway. (A) Schematic of BMP signaling. BMP dimers bind to a complex of type I and type II receptors which phosphorylate receptor (R) activated Smads 1, 5 and 8. R-Smads complex with Co-Smad4 and enter the nucleus to regulate transcription. Inhibitory (I) Smads 6 and 7 negatively regulate the signaling pathway by competitively binding with other components of the signaling pathway. (B–E) Model for a "mix and match" code for the specification of the RP, dI1s, dI2s and dI3s. In this model (B) BMP6 (mouse) and BMP7 (chicken) are the most effective at directing RP identity through the BmprIa receptor (mouse). (C) Both BMP4 and BMP7 can promote dP1 patterning through BmprIa or BmprIb (chicken), but only BMP4 directs progenitors to differentiate as dI1s through BmprIb (mouse and chicken). (D) BMP4 also specifically directs dI2 differentiation in chicken, thereby depleting the pool of dP2s. (E) All BMPs tested in both species, including BMP4, BMP5, BMP6 and BMP7, can act though either BmprIa or BmprIb to promote the dI3 fate. *Panel A modified with permission from Balemans, W., & Van Hul, W. (2002). Extracellular regulation of BMP signaling in vertebrates: A cocktail of modulators.* Developmental Biology, 250, *231–250.*

After phosphorylation by BmprI, the R-Smads complex with Co-Smad4 and translocate into the nucleus to regulate transcription (Moustakas & Heldin, 2009). Three of the R-Smads, Smad1, Smad5 and Smad8, are considered BMP specific, while two, Smad2 and Smad3, are activated by the TGFβ pathway. The I-Smads, Smad6 and Smad7, are involved in the negative regulation of BMP signaling (Fig. 3) (Massague, Seoane, & Wotton, 2005; Wu & Hill, 2009). Smad6 competitively binds to phosphorylated R-Smads to inhibit the formation of the complex with Co-Smad4 (Fig. 4A). Smad7 competes with the R-Smads for binding to activated BmprIs. Additionally, both I-Smads, Smad6 and Smad7, can regulate BMP receptor degradation by promoting ubiquitination in conjunction with another inhibitory molecule, Smufl (Massague et al., 2005).

2.8 BMPs were first proposed to signal as morphogens to pattern the dorsal spinal cord

During neurulation, BMPs in the non-neural ectoderm promote the formation of the RP at the dorsal midline of the spinal cord (Liem, Tremml, Roelink, & Jessell, 1995). Subsequently, the RP itself expresses a range of BMPs in nested domains (Butler & Dodd, 2003; Lee, Mendelsohn, & Jessell, 1998; Liem et al., 1997). The rodent RP expresses at least three members of the BMP family, BMP6, BMP7 and Growth/Differentiation Factor (GDF) 7. The complement of BMPs expressed in the chicken RP is larger, comprising at least four members of the BMP family, BMP4, BMP5, BMP7 and BMP9 (dorsalin1) (Liem et al., 1997). At later stages, the expression of the BMPs extends more broadly into the dorsal spinal cord (Andrews et al., 2017; Le Dreau & Marti, 2013). Studies examining the loss (Hazen et al., 2012; Le Dreau et al., 2012; Lee et al., 1998; Nguyen et al., 2000; Wine-Lee et al., 2004) or gain (Andrews et al., 2017; Chizhikov & Millen, 2004; Timmer, Chesnutt, & Niswander, 2005; Timmer, Wang, & Niswander, 2002; Yamauchi, Phan, & Butler, 2008) of BMP signaling have shown that BMP signaling is critical *in vivo* for the formation of the dI1, dI2 and dI3 classes of sensory neurons. The I-Smads, Smad6 and Smad7, appear to act to confine the activities of the BMPs to the dorsal-most spinal cord (Hazen, Phan, Hudiburgh, & Butler, 2011).

The mechanism by which the BMPs promote dorsal cell fate had remained unresolved until recently. The BMPs were first proposed to act collectively to establish a morphogen gradient (Lee & Jessell, 1999), by analogy with the Shh morphogen model in the ventral spinal cord (Ericson, Rashbass, et al., 1997; Yamada et al., 1993). In the canonical spatial morphogen model, high levels of BMP signaling should induce the dorsal-most fates, i.e., the RP and dI1s, whereas lower levels of BMPs should induce more ventral-dorsal cell types, such as dI3s. Studies in the Jessell laboratory demonstrated that conditioned medium from either BMP4- or GDF7-tranfected COS cells was sufficient to specify dI1s in naïve chicken neural plate tissue explants (Lee et al., 1998; Liem et al., 1997). However, treatment of these explants with diluted BMP4-conditioned medium was not particularly effective at inducing dI3s, the predicted result if BMP4 functioned as a canonical spatial morphogen (Liem et al., 1997).

A more recent study has suggested that BMP4 functions as a "temporal" morphogen (Tozer, Le Dreau, Marti, & Briscoe, 2013), again by analogy with Shh signaling (Dessaud et al., 2007). Dorsal progenitor identities within chicken neural plate explants were observed to change when cultured with BMP4 for different durations of time. The explants were first directed toward the dP2 identity, but then assumed the more dorsal dP1 fate as time progressed, suggesting that progenitors can integrate BMP signaling over time to assume a more dorsal spinal identity. Collectively, these experiments provided some evidence that there is a concentration- and/or timing-dependent component to activity of the BMPs. However, none of these studies addressed a key difference between signaling in the RP and FP: how the multiple members of the BMP family present in the dorsal spinal cord, might cooperate to establish a unified morphogen gradient.

2.9 BMPs have distinct roles directing dorsal spinal fates

An alternative model is that the different BMPs have unique activities specifying the induction of particular neural fates (Lee & Jessell, 1999). Early support for this hypothesis came when loss of the *Gdf7* gene in mice was found to specifically ablate a subpopulation of dI1s without affecting the identity of the other populations of dorsal INs (Lee et al., 1998). More recently, knocking down the expression of *Bmp4* and *Bmp7* in chicken embryos was shown to have qualitatively different effects: decreasing *Bmp4* expression reduces the number of dI1s, while depleting *Bmp7* expression reduces the number of dI1s, dI3s and dI5s (Le Dreau & Marti, 2013). These findings support the hypothesis that BMP ligands have unique activities specifying dorsal cell fates.

We have recently resolved the mode by which the different BMP ligands direct dorsal spinal identity. The ability of RP-derived BMPs to direct dorsal spinal fates was methodically tested by either manipulating *Bmp* expression in chicken embryos *in vivo* (Andrews et al., 2017), or culturing mouse (m) or human (h) embryonic stem cells (ESC) with different BMP recombinant proteins *in vitro* (Andrews et al., 2017; Gupta et al., 2018). These studies demonstrated unambiguously that BMPs do not act as concentration-dependent morphogens. Dose-response experiments in both the chicken and mouse systems showed that while altering the level of BMPs changed the effectiveness of the response, i.e., the number of cells converted to a specific fate, there were none of the changes in the dorsal IN identity predicted by either the spatial or temporal morphogen models.

Rather, BMPs were observed to be distinct in their abilities to direct the four dorsal-most cell types: the RP, dI1, dI2 and dI3. Each BMP has a complex range of activities, with some species-specific variation: they have overlapping and distinct functions directing progenitor patterning and the specification of dorsal IN identities. Thus, BMP6 (mouse) and BMP7 (chicken) are the most effective at directing the RP fates (Fig. 4B). BMP4 and BMP7 both promote dP1 proliferation (chicken), but only BMP4 can direct progenitors to differentiate as dI1s (mouse and chicken) (Fig. 4C). BMP4 is the only BMP that can direct dP2 to differentiate into dI2s in chicken (Fig. 4D). All BMPs tested in both species, including BMP4, BMP5, BMP6 and BMP7, can promote modest levels of dP3 proliferation and their differentiation into dI3s (Fig. 4E). No BMP was identified that direct the dI4–dI6 fates, rather BMP signaling tends to suppress the ventral-dorsal fates, consistent with earlier findings (Hazen et al., 2011). Thus, the BMPs have an evolutionarily conserved role specifying the dorsal spinal cord, with a species-specific code of BMP signals that directs dorsal IN formation.

BMPs mediate their signal-specific activities patterning the dorsal spinal cord by activating different type I Bmprs. In both chicken and mouse, the activities of BMP4 are mediated through BmprIb. In contrast, the activities of BMP7 (chicken) and BMP6 (mouse) are mediated, at least in part, by BmprIa (Andrews et al., 2017). However, the mechanisms by which different BMP receptor complexes then promote distinct dorsal cell identities remain unclear. Previous studies have differed in their conclusions as to whether both Smad1 and Smad5 (Le Dreau et al., 2012) or Smad5 alone (Hazen et al., 2012) mediate the specification of dorsal IN fate. R-Smads (Smad1/5/8) are phosphorylated (p), i.e., activated, in a gradient in the dorsal-most spinal cord (Hazen et al., 2012). This distribution suggested that specific levels of Smad activation might encode specific cell fates. However, this hypothesis has not been supported by recent studies: misexpression of *Bmp4* and *Bmp7* can direct similar levels of pSmad, yet results in different cell fate outcomes (Andrews et al., 2017). Thus, it appears that the R-Smads can be activated in a BMP-specific manner to result in distinct patterning activities. Future studies need to evaluate how the type I Bmprs differentially regulate the ability of the R-Smads to turn on the specific genes involved the assumption of specific dorsal cell fates.

2.10 The specification of dorsal spinal fate may be discontinuous

A curious, and mechanistically unresolved, feature of patterning along the dorsal spinal cord is the apparently discontinuous specification of dorsal IN identity, akin the expression of pair rule genes in the specification of segments in the Drosophila embryo (Pankratz & Jäckle, 1990). This finding was first reported by Elisa Martí and co-workers when they reported that BMP7 signaling is required for the generation of dI1, dI3 and dI5 subpopulations (Le Dreau et al., 2012). They found the depletion of *Bmp7* expression reduces the number of dI1, dI3 and dI5 by ~50% in chicken embryos using RNA knockdown approaches, and by ~25% in *Bmp7* mutant mouse embryos. However, the identity of the BMP mediating these activities remains unclear, as this phenotype was not previously observed in *Bmp7* mutants (Butler & Dodd, 2003), and gain of function studies in mESC cultures using BMP7 recombinant protein do not have the same effects on dorsal IN identity (Andrews et al., 2017). However, the ability of individual BMPs to induce discrete populations of dorsal INs is consistent across these studies. BMP4 is sufficient to specify dI1 and dI3 both *in vivo* and *in vitro*. *Bmp4* misexpression does not induce additional dI5s in chicken embryos *in vivo*, but its effect on dI5s has not yet been assessed in mESC or hESC cultures *in vitro*.

Moreover, the dI2, dI4 and dI6 subpopulations may be formed in a similarly discrete manner. BMP signaling is both necessary (Wine-Lee et al., 2004) and sufficient (Andrews et al., 2017) for the differentiation of dI2s *in vivo*. Recent chicken misexpression studies have shown that *Bmp4* elevation drives the $Ngn1^+$ domain of dP2s to rapidly differentiate into dI2s, thereby depleting the dP2 population (Fig. 4D). While low levels of *Bmp4* expression solely deplete dP2, higher *Bmp4* expression levels can additionally deplete a more ventral $Ngn1^+$ domain, which includes dP6, resulting in the formation of a large mass of dI2-like cells (Andrews et al., 2017). Unexpectedly, we were unable to derive BMP-dependent dI2s *in vitro* in either mESC or hESC cultures. However, these cultures are also treated with RA, to drive the ESCs toward a spinal neural identity (Wichterle, Lieberam, Porter, & Jessell, 2002). dI2s and neurons with shared dI4/dI6 identities were rather found to be arising in the RA-alone "control" condition in the hESC directed differentiation protocol (Gupta et al., 2018).

These disparate findings are complex to interpret but are generally consistent with a model in which the dorsal progenitor domains are first subdivided into either a dP1/dP3/(dP5?) or dP2/(dP4/dP6?) common identity. Temporal morphogen signaling by BMPs may distinguish between these progenitor identities (Tozer et al., 2013), i.e., the dP2/dP4/dP6 state may arise from cells with limited exposure to BMP signaling, while the dP1/dP3/(dP5?) state arises from cells with more sustained exposure to BMP signaling. Discrete series of transcription factors are expressed in these domains and over time subdivide progenitor, and later neuronal, populations. For example, the basic-helix-loop-helix transcription factor, *Olig3*, under the regulation of BMP and Wnt signaling activity, is first expressed a broad dP1–dP3 domain (Zechner et al., 2007). Subsequently, *Atoh1* and *Ngn1* are expressed in the dP1 and dP2 domain, respectively, where they cross-repress one another. *Ngn1* also represses *Ascl1* which is expressed in the dP3–5 region (Gowan et al., 2001). The sequential activation/repression of these transcription factors appear to be determined by early BMP patterning events, but then later affects the competency of the cells to respond to differentiation signals by the BMPs. Thus, following these early patterning events, cells in the dP1/dP3/(dP5?) state can respond to BMP4 to become either a dI1 or dI3, while the dP2/dP4/dP6 progenitor can only respond to BMP4 to become a dI2. Thus, prolonged addition of BMP4/RA to ESC cultures suppresses the dP2/dP4/dP6 state, such that no dI2s arise *in vitro*. In contrast, RA alone condition induces the dP2/dP4/dP6 state, permitting dI2s and a small number of dI4/dI6s to be generated.

While these studies have shown that the BMPs are reiteratively used to pattern distinct progenitor and neuronal domains in the developing dorsal spinal cord, many mysteries still remain; BMP4 is not expressed in the mouse or human spinal cord, thus another BMP must be directing dorsal spinal identity in these species. Markers to distinguish between the dP1/dP3/dP5 vs dP2/dP4/dP6 state have not been identified. It is also unresolved how different types of dorsal INs arise from a common progenitor fate. BMP4 can act as both a patterning and differentiation signal: for example, it specifies dP1/dP3 progenitors and then reiteratively directs them to form either dI1s or dI3s. What are signals allow dP1/dP3 cells to determine whether to respond to BMP4 to become either a dI1 or dI3? Finally, how are the dI4–dI6s, which are responsible for many important sensory modalities, specified? These cells appear to be dependent on RA signaling, as well as the transcription factor *Lbx1*, which is expressed in these populations. *Lbx1* is both necessary and sufficient to prevent dI4–dI6s from becoming dI2–dI4s (Gross, Dottori, & Goulding, 2002; Muller et al., 2002). Thus, like the I-Smads, Lbx1 may prevent the intermediate populations from interpreting the dorsal BMP signaling and thereby becoming more dorsalized.

2.11 Wnt signaling controls dorsal progenitor proliferation and patterning

Like the BMPs, the Wnt family of signaling molecules is also large; 19 *Wnt* genes have been identified in the human genome and are reiteratively used throughout development (Clevers & Nusse, 2012). Wnts regulate many cellular processes, aided by their ability to interact with other signaling pathways, including the mTOR, AKT CREB and PKC pathways, the core pathways that control cell proliferation, migration and survival as well as transcription (Logan & Nusse, 2004). Wnts signal through three known pathways: the canonical signaling pathway, where Wnt binding leads to the stabilization of the β-catenin second messenger, and two non-canonical pathways, the planar cell polarity (PCP) pathway and Wnt/calcium pathway, which are both β-catenin-independent. In the canonical Wnt pathway, the binding of Wnts to the frizzled (Fz) and LRP5/6 co-receptors, recruits disheveled (Dvl), and subsequently axin, to the signaling complex at the membrane (Cong, Schweizer, & Varmus, 2004; Hart, de los Santos, Albert, Rubinfeld, & Polakis, 1998). In the absence of Wnt signaling, axin is bound to a complex of proteins that degrades β-catenin. However, once axin is recruited to the membrane, β-catenin is stabilized and can translocate to the nucleus to act as transcriptional activator for Wnt target genes (Behrens et al., 1996; Tolwinski & Wieschaus, 2004). In contrast, the PCP pathway mediates cytoskeleton modification by activating actin regulators such as profilin and the Rho-associated protein kinase (ROCK) (Shulman, Perrimon, & Axelrod, 1998), while the calcium pathway modulates G protein-regulated calcium release from the endoplasmic reticulum (Slusarski, Yang-Snyder, Busa, & Moon, 1997).

While Wnt ligands are critical for inductive patterning in the brain (McMahon & Bradley, 1990) and have a key role in the specification of the anterior-posterior axis (Agalliu, Takada, Agalliu, McMahon, & Jessell, 2009; Green, Whitener, Mohanty, & Lekven, 2015), relatively little is known about their role in patterning the dorsal-ventral axis of the spinal cord. At least four Wnts are expressed in the RP and dorsal spinal cord: Wnt1, Wnt3, Wnt3a and Wnt4 (Hollyday, McMahon, & McMahon, 1995; Megason & McMahon, 2002; Parr, Shea, Vassileva, & McMahon, 1993) (Fig. 2A). Of these Wnts, Wnt1 and Wnt3a can act as mitogens to regulate cell proliferation during development of chicken spinal cord. Mis-expression of either gene results in a ~twofold increase in the number of cells in S-phase in the cell cycle (Megason & McMahon, 2002) and a general increase in the expression of dorsal markers (Alvarez-Medina, Cayuso, Okubo, Takada, & Marti, 2008). The loss-of-function experiments were performed in mice (Muroyama, Fujihara, Ikeya, Kondoh, & Takada, 2002) and were consistent with the role of Wnt1 and Wnt3a as mitogens. They are genetically redundant; a single mutant in either Wnt had no effect on dorsal spinal identity. However, both dP1 and dP2 were depleted in *Wnt1; Wnt3a* double mutants, and there was profound reduction in the numbers of dI1s and dI3s. The concomitant increase in the dI4s suggests that Wnts might also have patterning activities; however, this hypothesis requires further study. The mechanism that transduces these activities also remains unresolved. Wnt mediates its role as a mitogen through the canonical β-catenin pathway (Alvarez-Medina et al., 2008; Megason & McMahon, 2002). Inhibiting the Wnt transcriptional response resulting in an expansion of ventral gene expression at the expense of dorsal identities. This response was shown to be Gli-dependent, suggesting that Wnt signaling is also required to restrict the activity of Shh to the ventral spinal cord (Alvarez-Medina et al., 2008).

Taken together, these studies most strongly support the hypothesis that Wnts act as proliferative signals. Wnts are required for the appropriate proliferation of dorsal progenitors; without this mitogenic activity, fewer differentiated dorsal IN neurons are produced. Nevertheless, it remains unresolved whether the Wnts directly specify neuronal identity in the dorsal spinal cord.

3. Looking ahead: What does knowledge of these manipulations mean for spinal cord regeneration?

Spinal cord injuries (SCI) are a major public health concern. The spinal cord is particularly vulnerable to damage from either injury or disease because it has a relatively exposed position in the body and shares no functional redundancy with other structures in the CNS. Damage to the spinal cord can affect both sensory and motor systems below the injury site and can be debilitating emotionally and physically for patients. SCI patients often have reduced quality of life. The loss of motor circuits results in patients being unable to move. When the ability to process and relay sensory input to the brain is affected, patients can lose the capacity to sense and respond to external stimuli, like pain or temperature. They are at risk for serious burn injuries, from being unable to detect noxious heat, and suffer the emotional consequences of being unable to receive reassuring touch. SCIs are estimated to affect over 1 million people in the United States, and cost ~$40 billion annually in health care (Abraira et al., 2017; Armour, Courtney-Long, Fox, Fredine, & Cahill, 2016). SCI patients have significantly lower life expectancy than uninjured individuals, regardless of injury severity; their average life expectancy has not improved since the 1980s.

There is currently no treatment that can reverse damage to the spinal cord. While important progress has been made on neuromodulatory mechanisms that can improve the function of the nerves remaining after SCI (AuYong & Lu, 2014), there is currently no means of reversing congenital or acute damage to the spinal cord. Such a treatment will require the regeneration of substantial amounts of tissue to concomitantly reestablish sensory and motor spinal circuitry. A tractable path toward that objective is use pluripotent stem cells (PSCs) to generate the necessary component neurons, which can then be used in cellular replacement therapies to replace lost tissue (Kadoya et al., 2016; Pawar et al., 2015; Piltti et al., 2015). PSCs are also a powerful reagent for understanding disease mechanisms and serving as a platform for drug screening.

An important milestone toward a treatment for injured or diseased spinal cords has been the establishment of protocols to derive spinal MNs from both mouse and human ESCs (Li et al., 2005; Wichterle et al., 2002) and induced pluripotent stem cells (iPSCs) (Dimos et al., 2008; Karumbayaram et al., 2009). ESCs are directed toward a neurogenic lineage and caudalized by the addition of RA. Finally, they are ventralized by the addition of a Shh signaling agonist, such as purmorphamine, to specify spinal MN identity (Wichterle & Peljto, 2008). Stem cell-derived MNs have provided a critical window into understanding of the pathology of motor diseases, including amyotrophic lateral sclerosis (ALS) (Sances et al., 2016) and spinal muscular atrophy (SMA) (Ebert et al., 2009), revealing, for example, changes in the excitability state of MNs as either a potential cause or immediate outcome of these diseases (Wainger et al., 2014). The ability to readily generate MNs *in vitro* permits detailed studies of the underlying disease mechanisms in the context of human cells and has opened the door to high-throughput drug screening strategies.

Stem cell-derived MNs have shown further value as cellular replacement therapies in animal models of SCI (Harper et al., 2004; Thomsen, Gowing, Svendsen, & Svendsen, 2014). While these advances raise the possibility of repairing motor functions in SCI patients, this objective also requires the regeneration of the sensory circuits that modulate motor output. However, progress has been slowed by the lack of a directed differentiation protocol to generate sensory spinal relay INs from PSCs. We recently published the first directed differentiation protocols for *in vitro* derived sensory spinal INs for both mESCs (Andrews et al., 2017) and hESCs and hiPSCs (Gupta et al., 2018). While BMP4 was used in the place of purmorphamine to dorsalize the identity of the RA-dependent spinal neurons, critical to the human protocol was the realization that has finding that the progenitors had an early window of competence to respond to BMP signaling to become dorsal neurons. These directed differentiation protocols can generate four key populations of dorsal sensory INs: the dI1s (proprioception), dI2s (unknown function), dI3s (touch-activated motor behaviors) and dI4/dI6s (pain/itch/touch) (Lai et al., 2016). These studies also established that hiPSCs can be directed to differentiate into dorsal sensory INs with comparable efficiency with hESCs.

Multiple challenges remain. While PSC-derived MNs have been shown to physiologically mirror their endogenous counterparts (Umbach, Adams, Gundersen, & Novitch, 2012), this outcome has yet to be demonstrated for PSC-derived dorsal sensory INs. Both PSC-derived MNs and dorsal INs arise as mixed populations, with varied efficiencies in the current protocols. It is critical to continue to work toward protocols that have the highest efficiency and purity of generating specific populations of spinal neurons. More studies also need to be performed to evaluate the ability to incorporate into the injured spinal cord of adult animal, form synaptic connections and provide functional recovery. However, there is now genuine hope that cellular replacement therapies can be successfully used to regenerate complex tissues such as the spinal cord. Recent studies have shown that human-derived spinal neural precursors can successfully integrate into an injured non-human primate spinal cord resulting in some functional recovery (Rosenzweig et al., 2018). The ability to reverse congenital and acute damage to the spinal cord has the potential to impact millions of patients worldwide.

References

Abraira, V. E., Kuehn, E. D., Chirila, A. M., Springel, M. W., Toliver, A. A., Zimmerman, A. L., et al. (2017). The cellular and synaptic architecture of the mechanosensory dorsal horn. *Cell*, *168*, 295–310 e219.

Agalliu, D., Takada, S., Agalliu, I., McMahon, A. P., & Jessell, T. M. (2009). Motor neurons with axial muscle projections specified by Wnt4/5 signaling. *Neuron*, *61*, 708–720.

Ahn, K., Mishina, Y., Hanks, M. C., Behringer, R. R., & Crenshaw, E. B., 3rd. (2001). BMPR-IA signaling is required for the formation of the apical ectodermal ridge and dorsal-ventral patterning of the limb. *Development (Cambridge, England)*, *128*, 4449–4461.

Alvarez-Medina, R., Cayuso, J., Okubo, T., Takada, S., & Marti, E. (2008). Wnt canonical pathway restricts graded Shh/Gli patterning activity through the regulation of Gli3 expression. *Development (Cambridge, England)*, *135*, 237–247.

Andersson, L. S., Larhammar, M., Memic, F., Wootz, H., Schwochow, D., Rubin, C. J., et al. (2012). Mutations in DMRT3 affect locomotion in horses and spinal circuit function in mice. *Nature*, *488*, 642–646.

Andrews, M. G., Del Castillo, L. M., Ochoa-Bolton, E., Yamauchi, K., Smogorzewski, J., & Butler, S. J. (2017). BMPs direct sensory interneuron identity in the developing spinal cord using signal-specific not morphogenic activities. *eLife*, *6*e30647.

Arber, S. (2012). Motor circuits in action: Specification, connectivity, and function. *Neuron*, *74*, 975–989.

Armour, B. S., Courtney-Long, E. A., Fox, M. H., Fredine, H., & Cahill, A. (2016). Prevalence and causes of paralysis-United States, 2013. *American Journal of Public Health*, *106*, 1855–1857.

AuYong, N., & Lu, D. C. (2014). Neuromodulation of the lumbar spinal locomotor circuit. *Neurosurgery Clinics of North America*, *25*, 15–23.

Aza-Blanc, P., Lin, H. Y., Ruiz i Altaba, A., & Kornberg, T. B. (2000). Expression of the vertebrate Gli proteins in Drosophila reveals a distribution of activator and repressor activities. *Development (Cambridge, England)*, *127*, 4293–4301.

Bai, C. B., Auerbach, W., Lee, J. S., Stephen, D., & Joyner, A. L. (2002). Gli2, but not Gli1, is required for initial Shh signaling and ectopic activation of the Shh pathway. *Development (Cambridge, England)*, *129*, 4753–4761.

Balaskas, N., Ribeiro, A., Panovska, J., Dessaud, E., Sasai, N., Page, K. M., et al. (2012). Gene regulatory logic for reading the Sonic Hedgehog signaling gradient in the vertebrate neural tube. *Cell*, *148*, 273–284.

Balemans, W., & Van Hul, W. (2002). Extracellular regulation of BMP signaling in vertebrates: A cocktail of modulators. *Developmental Biology*, *250*, 231–250.

Bautista, D. M., Siemens, J., Glazer, J. M., Tsuruda, P. R., Basbaum, A. I., Stucky, C. L., et al. (2007). The menthol receptor TRPM8 is the principal detector of environmental cold. *Nature*, *448*, 204–208.

Behrens, J., von Kries, J. P., Kuhl, M., Bruhn, L., Wedlich, D., Grosschedl, R., et al. (1996). Functional interaction of beta-catenin with the transcription factor LEF-1. *Nature*, *382*, 638–642.

Bermingham, N. A., Hassan, B. A., Wang, V. Y., Fernandez, M., Banfi, S., Bellen, H. J., et al. (2001). Proprioceptor pathway development is dependent on Math1. *Neuron*, *30*, 411–422.

Bier, E., & De Robertis, E. M. (2015). EMBRYO DEVELOPMENT. BMP gradients: A paradigm for morphogen-mediated developmental patterning. *Science (New York, N.Y.)*, *348*, aaa5838.

Bourane, S., Duan, B., Koch, S. C., Dalet, A., Britz, O., Garcia-Campmany, L., et al. (2015). Gate control of mechanical itch by a subpopulation of spinal cord interneurons. *Science (New York, N.Y.)*, *350*, 550–554.

Bourane, S., Grossmann, K. S., Britz, O., Dalet, A., Del Barrio, M. G., Stam, F. J., et al. (2015). Identification of a spinal circuit for light touch and fine motor control. *Cell*, *160*, 503–515.

Briscoe, J., & Novitch, B. G. (2008). Regulatory pathways linking progenitor patterning, cell fates and neurogenesis in the ventral neural tube. *Philosophical Transactions of the Royal Society of London. Series B, Biological Sciences*, *363*, 57–70.

Briscoe, J., Pierani, A., Jessell, T. M., & Ericson, J. (2000). A homeodomain protein code specifics progenitor cell identity and neuronal fate in the ventral neural tube. *Cell*, *101*, 435–445.

Briscoe, J., Sussel, L., Serup, P., Hartigan-O'Connor, D., Jessell, T. M., Rubenstein, J. L., et al. (1999). Homeobox gene Nkx2.2 and specification of neuronal identity by graded Sonic hedgehog signalling. *Nature*, *398*, 622–627.

Brown, A. G., & Fyffe, R. E. W. (1981). Form and function of dorsal horn neurons with axons ascending the dorsal columns in cat. *The Journal of Physiology*, *321*, 31–47.

Bui, T. V., Akay, T., Loubani, O., Hnasko, T. S., Jessell, T. M., & Brownstone, R. M. (2013). Circuits for grasping: Spinal dI3 interneurons mediate cutaneous control of motor behavior. *Neuron*, *78*, 191–204.

Burke, R., Nellen, D., Bellotto, M., Hafen, E., Senti, K. A., Dickson, B. J., et al. (1999). Dispatched, a novel sterol-sensing domain protein dedicated to the release of cholesterol-modified hedgehog from signaling cells. *Cell*, *99*, 803–815.

Butler, S. J., & Dodd, J. (2003). A role for BMP heterodimers in roof plate-mediated repulsion of commissural axons. *Neuron*, *38*, 389–401.

Carpenter, D., Stone, D. M., Brush, J., Ryan, A., Armanini, M., Frantz, G., et al. (1998). Characterization of two patched receptors for the vertebrate hedgehog protein family. *Proceedings of the National Academy of Sciences of the United States of America*, *95*, 13630–13634.

Casso, D. J., Biehs, B., & Kornberg, T. B. (2011). A novel interaction between hedgehog and Notch promotes proliferation at the anterior-posterior organizer of the Drosophila wing. *Genetics*, *187*, 485–499.

Catela, C., Shin, M. M., & Dasen, J. S. (2015). Assembly and function of spinal circuits for motor control. *Annual Review of Cell and Developmental Biology*, *31*, 669–698.

Chamberlain, C. E., Jeong, J., Guo, C., Allen, B. L., & McMahon, A. P. (2008). Notochord-derived Shh concentrates in close association with the apically positioned basal body in neural target cells and forms a dynamic gradient during neural patterning. *Development (Cambridge, England)*, *135*, 1097–1106.

Chen, Y., & Struhl, G. (1996). Dual roles for patched in sequestering and transducing Hedgehog. *Cell*, *87*, 553–563.

Chiang, C., Litingtung, Y., Lee, E., Young, K. E., Corden, J. L., Westphal, H., et al. (1996). Cyclopia and defective axial patterning in mice lacking Sonic hedgehog gene function. *Nature*, *383*, 407–413.

Chizhikov, V. V., & Millen, K. J. (2004). Control of roof plate formation by Lmx1a in the developing spinal cord. *Development (Cambridge, England)*, *131*, 2693–2705.

Clevers, H., & Nusse, R. (2012). Wnt/beta-catenin signaling and disease. *Cell*, *149*, 1192–1205.

Cong, F., Schweizer, L., & Varmus, H. (2004). Wnt signals across the plasma membrane to activate the beta-catenin pathway by forming oligomers containing its receptors, frizzled and LRP. *Development (Cambridge, England)*, *131*, 5103–5115.

Conlon, R. A., & Rossant, J. (1992). Exogenous retinoic acid rapidly induces anterior ectopic expression of murine Hox-2 genes in vivo. *Development (Cambridge, England)*, *116*, 357–368.

Corbit, K. C., Aanstad, P., Singla, V., Norman, A. R., Stainier, D. Y., & Reiter, J. F. (2005). Vertebrate smoothened functions at the primary cilium. *Nature*, *437*, 1018–1021.

Crone, S. A., Quinlan, K. A., Zagoraiou, L., Droho, S., Restrepo, C. E., Lundfald, L., et al. (2008). Genetic ablation of V2a ipsilateral interneurons disrupts left-right locomotor coordination in mammalian spinal cord. *Neuron*, *60*, 70–83.

Dasen, J. S., De Camilli, A., Wang, B., Tucker, P. W., & Jessell, T. M. (2008). Hox repertoires for motor neuron diversity and connectivity gated by a single accessory factor, FoxP1. *Cell*, *134*, 304–316.

Delacroix, L., Moutier, E., Altobelli, G., Legras, S., Poch, O., Choukrallah, M. A., et al. (2010). Cell-specific interaction of retinoic acid receptors with target genes in mouse embryonic fibroblasts and embryonic stem cells. *Molecular and Cellular Biology*, *30*, 231–244.

Dessaud, E., Ribes, V., Balaskas, N., Yang, L. L., Pierani, A., Kicheva, A., et al. (2010). Dynamic assignment and maintenance of positional identity in the ventral neural tube by the morphogen sonic hedgehog. *PLoS Biology*, *8*, e1000382.

Dessaud, E., Yang, L. L., Hill, K., Cox, B., Ulloa, F., Ribeiro, A., et al. (2007). Interpretation of the sonic hedgehog morphogen gradient by a temporal adaptation mechanism. *Nature*, *450*, 717–720.

Dickman, E. D., Thaller, C., & Smith, S. M. (1997). Temporally-regulated retinoic acid depletion produces specific neural crest, ocular and nervous system defects. *Development (Cambridge, England)*, *124*, 3111–3121.

Diez del Corral, R., Olivera-Martinez, I., Goriely, A., Gale, E., Maden, M., & Storey, K. (2003). Opposing FGF and retinoid pathways control ventral neural pattern, neuronal differentiation, and segmentation during body axis extension. *Neuron*, *40*, 65–79.

Dimos, J. T., Rodolfa, K. T., Niakan, K. K., Weisenthal, L. M., Mitsumoto, H., Chung, W., et al. (2008). Induced pluripotent stem cells generated from patients with ALS can be differentiated into motor neurons. *Science (New York, N.Y.)*, *321*, 1218–1221.

Ducy, P., & Karsenty, G. (2000). The family of bone morphogenetic proteins. *Kidney International*, *57*, 2207–2214.

Dudley, A. T., Lyons, K. M., & Robertson, E. J. (1995). A requirement for bone morphogenetic protein-7 during development of the mammalian kidney and eye. *Genes & Development*, *9*, 2795–2807.

Durston, A. J., Timmermans, J. P., Hage, W. J., Hendriks, H. F., de Vries, N. J., Heideveld, M., et al. (1989). Retinoic acid causes an anteroposterior transformation in the developing central nervous system. *Nature*, *340*, 140–144.

Ebert, A. D., Yu, J., Rose, F. F., Jr., Mattis, V. B., Lorson, C. L., Thomson, J. A., et al. (2009). Induced pluripotent stem cells from a spinal muscular atrophy patient. *Nature*, *457*, 277–280.

Echelard, Y., Epstein, D. J., St-Jacques, B., Shen, L., Mohler, J., McMahon, J. A., et al. (1993). Sonic hedgehog, a member of a family of putative signaling molecules, is implicated in the regulation of CNS polarity. *Cell*, *75*, 1417–1430.

Ericson, J., Briscoe, J., Rashbass, P., van Heyningen, V., & Jessell, T. M. (1997). Graded sonic hedgehog signaling and the specification of cell fate in the ventral neural tube. *Cold Spring Harbor Symposia on Quantitative Biology*, *62*, 451–466.

Ericson, J., Rashbass, P., Schedl, A., Brenner-Morton, S., Kawakami, A., van Heyningen, V., et al. (1997). Pax6 controls progenitor cell identity and neuronal fate in response to graded Shh signaling. *Cell*, *90*, 169–180.

Gerhart, J. (1996). Johannes Holtfreter's contributions to ongoing studies of the organizer. *Developmental Dynamics*, *205*, 245–256.

Goetz, C., Pivetta, C., & Arber, S. (2015). Distinct limb and trunk premotor circuits establish laterality in the spinal cord. *Neuron*, *85*, 131–144.

Goetz, J. A., Singh, S., Suber, L. M., Kull, F. J., & Robbins, D. J. (2006). A highly conserved amino-terminal region of sonic hedgehog is required for the formation of its freely diffusible multimeric form. *The Journal of Biological Chemistry*, *281*, 4087–4093.

Goodrich, L. V., Johnson, R. L., Milenkovic, L., McMahon, J. A., & Scott, M. P. (1996). Conservation of the hedgehog/patched signaling pathway from flies to mice: Induction of a mouse patched gene by Hedgehog. *Genes & Development*, *10*, 301–312.

Gosgnach, S., Bikoff, J. B., Dougherty, K. J., El Manira, A., Lanuza, G. M., & Zhang, Y. (2017). Delineating the diversity of spinal interneurons in locomotor circuits. *The Journal of Neuroscience*, *37*, 10835–10841.

Gosgnach, S., Lanuza, G. M., Butt, S. J., Saueressig, H., Zhang, Y., Velasquez, T., et al. (2006). V1 spinal neurons regulate the speed of vertebrate locomotor outputs. *Nature*, *440*, 215–219.

Goulding, M. (2009). Circuits controlling vertebrate locomotion: Moving in a new direction. *Nature Reviews. Neuroscience*, *10*, 507–518

Gowan, K., Helms, A. W., Hunsaker, T. L., Collisson, T., Ebert, P. J., Odom, R., et al. (2001). Crossinhibitory activities of Ngn1 and Math1 allow specification of distinct dorsal interneurons. *Neuron*, *31*, 219–232.

Green, D., Whitener, A. E., Mohanty, S., & Lekven, A. C. (2015). Vertebrate nervous system posteriorization: Grading the function of Wnt signaling. *Developmental Dynamics*, *244*, 507–512.

Griener, A., Zhang, W., Kao, H., Wagner, C., & Gosgnach, S. (2015). Probing diversity within subpopulations of locomotor-related V0 interneurons. *Developmental Neurobiology*, *75*, 1189–1203.

Gross, M. K., Dottori, M., & Goulding, M. (2002). Lbx1 specifies somatosensory association interneurons in the dorsal spinal cord. *Neuron*, *34*, 535–549.

Gupta, S., Sivalingam, D., Hain, S., Makkar, C., Sosa, E., Clark, A., et al. (2018). Deriving dorsal spinal sensory interneurons from human pluripotent stem cells. *Stem Cell Reports*, *10*, 390–405.

Hallahan, A. R., Pritchard, J. I., Hansen, S., Benson, M., Stoeck, J., Hatton, B. A., et al. (2004). The SmoA1 mouse model reveals that notch signaling is critical for the growth and survival of sonic hedgehog-induced medulloblastomas. *Cancer Research*, *64*, 7794–7800.

Hamidi, H., Gustafason, D., Pellegrini, M., & Gasson, J. (2011). Identification of novel targets of CSL-dependent Notch signaling in hematopoiesis. *PLoS One*, *6*, e20022.

Hammerschmidt, M., Brook, A., & McMahon, A. P. (1997). The world according to hedgehog. *Trends in Genetics*, *13*, 14–21.

Harper, J. M., Krishnan, C., Darman, J. S., Deshpande, D. M., Peck, S., Shats, I., et al. (2004). Axonal growth of embryonic stem cell-derived motoneurons in vitro and in motoneuron-injured adult rats. *Proceedings of the National Academy of Sciences of the United States of America*, *101*, 7123–7128.

Hart, M. J., de los Santos, R., Albert, I. N., Rubinfeld, B., & Polakis, P. (1998). Downregulation of beta-catenin by human Axin and its association with the APC tumor suppressor, beta-catenin and GSK3 beta. *Current Biology*, *8*, 573–581.

Hazen, V. M., Andrews, M. A., Umans, L., Crenshaw, E. B., 3rd, Zwijsen, A., & Butler, S. J. (2012). BMP receptor-activated Smads direct diverse functions during the development of the dorsal spinal cord. *Developmental Biology*, *367*, 216–227.

Hazen, V. M., Phan, K. D., Hudiburgh, S., & Butler, S. J. (2011). Inhibitory Smads differentially regulate cell fate specification and axon dynamics in the dorsal spinal cord. *Developmental Biology*, *356*, 566–575.

Helms, A. W., & Johnson, J. E. (2003). Specification of dorsal spinal cord interneurons. *Current Opinion in Neurobiology*, *13*, 42–49.

Hogan, B. L. (1996). Bone morphogenetic proteins in development. *Current Opinion in Genetics & Development*, *6*, 432–438.

Hollyday, M., McMahon, J. A., & McMahon, A. P. (1995). Wnt expression patterns in chick embryo nervous system. *Mechanisms of Development*, *52*, 9–25.

Huang, D., Chen, S. W., & Gudas, L. J. (2002). Analysis of two distinct retinoic acid response elements in the homeobox gene Hoxb1 in transgenic mice. *Developmental Dynamics*, *223*, 353–370.

Huang, P., Xiong, F., Megason, S. G., & Schier, A. F. (2012). Attenuation of Notch and Hedgehog signaling is required for fate specification in the spinal cord. *PLoS Genetics*, *8*, e1002762.

Hui, C. C., & Angers, S. (2011). Gli proteins in development and disease. *Annual Review of Cell and Developmental Biology*, *27*, 513–537.

Hui, C. C., Slusarski, D., Platt, K. A., Holmgren, R., & Joyner, A. L. (1994). Expression of three mouse homologs of the Drosophila segment polarity gene cubitus interruptus, Gli, Gli-2, and Gli-3, in ectoderm- and mesoderm-derived tissues suggests multiple roles during postimplantation development. *Developmental Biology*, *162*, 402–413.

Hynes, M., Stone, D. M., Dowd, M., Pitts-Meek, S., Goddard, A., Gurney, A., et al. (1997). Control of cell pattern in the neural tube by the zinc finger transcription factor and oncogene Gli-1. *Neuron, 19*, 15–26.
Ingham, P. W., Taylor, A. M., & Nakano, Y. (1991). Role of the Drosophila patched gene in positional signalling. *Nature, 353*, 184–187.
Jeong, J., & McMahon, A. P. (2005). Growth and pattern of the mammalian neural tube are governed by partially overlapping feedback activities of the hedgehog antagonists patched 1 and Hhip1. *Development (Cambridge, England), 132*, 143–154.
Jessell, T. M. (2000). Neuronal specification in the spinal cord: Inductive signals and transcriptional codes. *Nature Reviews. Genetics, 1*, 20–29.
Jessell, T. M., Surmeli, G., & Kelly, J. S. (2011). Motor neurons and the sense of place. *Neuron, 72*, 419–424.
Kadoya, K., Lu, P., Nguyen, K., Lee-Kubli, C., Kumamaru, H., Yao, L., et al. (2016). Spinal cord reconstitution with homologous neural grafts enables robust corticospinal regeneration. *Nature Medicine, 22*, 479–487.
Kam, R. K., Deng, Y., Chen, Y., & Zhao, H. (2012). Retinoic acid synthesis and functions in early embryonic development. *Cell & Bioscience, 2*, 11.
Karumbayaram, S., Novitch, B. G., Patterson, M., Umbach, J. A., Richter, L., Lindgren, A., et al. (2009). Directed differentiation of human-induced pluripotent stem cells generates active motor neurons. *Stem Cells, 27*, 806–811.
Kong, J. H., Yang, L., Dessaud, E., Chuang, K., Moore, D. M., Rohatgi, R., et al. (2015). Notch activity modulates the responsiveness of neural progenitors to sonic hedgehog signaling. *Developmental Cell, 33*, 373–387.
Koop, D., Holland, N. D., Semon, M., Alvarez, S., de Lera, A. R., Laudet, V., et al. (2010). Retinoic acid signaling targets Hox genes during the amphioxus gastrula stage: Insights into early anterior-posterior patterning of the chordate body plan. *Developmental Biology, 338*, 98–106.
Krauss, S., Concordet, J. P., & Ingham, P. W. (1993). A functionally conserved homolog of the Drosophila segment polarity gene hh is expressed in tissues with polarizing activity in zebrafish embryos. *Cell, 75*, 1431–1444.
Kutejova, E., Sasai, N., Shah, A., Gouti, M., & Briscoe, J. (2016). Neural progenitors adopt specific identities by directly repressing all alternative progenitor transcriptional programs. *Developmental Cell, 36*, 639–653.
Lai, H. C., Seal, R. P., & Johnson, J. E. (2016). Making sense out of spinal cord somatosensory development. *Development (Cambridge, England), 143*, 3434–3448.
Le Dreau, G., Garcia-Campmany, L., Rabadan, M. A., Ferronha, T., Tozer, S., Briscoe, J., et al. (2012). Canonical BMP7 activity is required for the generation of discrete neuronal populations in the dorsal spinal cord. *Development (Cambridge, England), 139*, 259–268.
Le Dreau, G., & Marti, E. (2013). The multiple activities of BMPs during spinal cord development. *Cellular and Molecular Life Sciences, 70*, 4293–4305.
Lee, K. J., & Jessell, T. M. (1999). The specification of dorsal cell fates in the vertebrate central nervous system. *Annual Review of Neuroscience, 22*, 261–294.
Lee, K. J., Mendelsohn, M., & Jessell, T. M. (1998). Neuronal patterning by BMPs: A requirement for GDF7 in the generation of a discrete class of commissural interneurons in the mouse spinal cord. *Genes & Development, 12*, 3394–3407.
Lee, J., Platt, K. A., Censullo, P., & Ruiz i Altaba, A. (1997). Gli1 is a target of Sonic hedgehog that induces ventral neural tube development. *Development (Cambridge, England), 124*, 2537–2552.
Lee, J. J., von Kessler, D. P., Parks, S., & Beachy, P. A. (1992). Secretion and localized transcription suggest a role in positional signaling for products of the segmentation gene hedgehog. *Cell, 71*, 33–50.

Li, X. J., Du, Z. W., Zarnowska, E. D., Pankratz, M., Hansen, L. O., Pearce, R. A., et al. (2005). Specification of motoneurons from human embryonic stem cells. *Nature Biotechnology*, *23*, 215–221.

Li, Y., Hibbs, M. A., Gard, A. L., Shylo, N. A., & Yun, K. (2012). Genome-wide analysis of N1ICD/RBPJ targets in vivo reveals direct transcriptional regulation of Wnt, SHH, and hippo pathway effectors by Notch1. *Stem Cells*, *30*, 741–752.

Liem, K. F., Jr., Tremml, G., & Jessell, T. M. (1997). A role for the roof plate and its resident TGFbeta-related proteins in neuronal patterning in the dorsal spinal cord. *Cell*, *91*, 127–138.

Liem, K. F., Tremml, G., Roelink, H., & Jessell, T. M. (1995). Dorsal differentiation of neural plate cells induced by BMP-mediated signals from epidermal ectoderm. *Cell*, *82*, 969–979.

Logan, C. Y., & Nusse, R. (2004). The Wnt signaling pathway in development and disease. *Annual Review of Cell and Developmental Biology*, *20*, 781–810.

Lopez, S. L., Paganelli, A. R., Siri, M. V., Ocana, O. H., Franco, P. G., & Carrasco, A. E. (2003). Notch activates sonic hedgehog and both are involved in the specification of dorsal midline cell-fates in Xenopus. *Development (Cambridge, England)*, *130*, 2225–2238.

Lyons, K. M., Hogan, B. L., & Robertson, E. J. (1995). Colocalization of BMP 7 and BMP 2 RNAs suggests that these factors cooperatively mediate tissue interactions during murine development. *Mechanisms of Development*, *50*, 71–83.

Maden, M. (2002). Retinoid signalling in the development of the central nervous system. *Nature Reviews. Neuroscience*, *3*, 843–853.

Maden, M. (2007). Retinoic acid in the development, regeneration and maintenance of the nervous system. *Nature Reviews. Neuroscience*, *8*, 755–765.

Marigo, V., Davey, R. A., Zuo, Y., Cunningham, J. M., & Tabin, C. J. (1996). Biochemical evidence that patched is the Hedgehog receptor. *Nature*, *384*, 176–179.

Massague, J., Seoane, J., & Wotton, D. (2005). Smad transcription factors. *Genes & Development*, *19*, 2783–2810.

McMahon, A. P., & Bradley, A. (1990). The Wnt-1 (int-1) proto-oncogene is required for development of a large region of the mouse brain. *Cell*, *62*, 1073–1085.

Megason, S. G., & McMahon, A. P. (2002). A mitogen gradient of dorsal midline Wnts organizes growth in the CNS. *Development (Cambridge, England)*, *129*, 2087–2098.

Miesegaes, G. R., Klisch, T. J., Thaller, C., Ahmad, K. A., Atkinson, R. C., & Zoghbi, H. Y. (2009). Identification and subclassification of new Atoh1 derived cell populations during mouse spinal cord development. *Developmental Biology*, *327*, 339–351.

Mohler, J., & Vani, K. (1992). Molecular organization and embryonic expression of the hedgehog gene involved in cell-cell communication in segmental patterning of Drosophila. *Development (Cambridge, England)*, *115*, 957–971.

Moustakas, A., & Heldin, C. H. (2009). The regulation of TGFbeta signal transduction. *Development (Cambridge, England)*, *136*, 3699–3714.

Muller, T., Brohmann, H., Pierani, A., Heppenstall, P. A., Lewin, G. R., Jessell, T. M., et al. (2002). The homeodomain factor lbx1 distinguishes two major programs of neuronal differentiation in the dorsal spinal cord. *Neuron*, *34*, 551–562.

Muroyama, Y., Fujihara, M., Ikeya, M., Kondoh, H., & Takada, S. (2002). Wnt signaling plays an essential role in neuronal specification of the dorsal spinal cord. *Genes & Development*, *16*, 548–553.

Nguyen, V. H., Trout, J., Connors, S. A., Andermann, P., Weinberg, E., & Mullins, M. C. (2000). Dorsal and intermediate neuronal cell types of the spinal cord are established by a BMP signaling pathway. *Development (Cambridge, England)*, *127*, 1209–1220.

Niewiadomski, P., Kong, J. H., Ahrends, R., Ma, Y., Humke, E. W., Khan, S., et al. (2014). Gli protein activity is controlled by multisite phosphorylation in vertebrate Hedgehog signaling. *Cell Reports*, *6*, 168–181.

Novitch, B. G., Wichterle, H., Jessell, T. M., & Sockanathan, S. (2003). A requirement for retinoic acid-mediated transcriptional activation in ventral neural patterning and motor neuron specification. *Neuron, 40*, 81–95.
Nusslein-Volhard, C., & Wieschaus, E. (1980). Mutations affecting segment number and polarity in Drosophila. *Nature, 287*, 795–801.
Okada, Y., Shimazaki, T., Sobue, G., & Okano, H. (2004). Retinoic-acid-concentration-dependent acquisition of neural cell identity during in vitro differentiation of mouse embryonic stem cells. *Developmental Biology, 275*, 124–142.
Pankratz, M. J., & Jäckle, H. (1990). Making stripes in the Drosophila embryo. *Trends in Genetics, 6*, 287–292.
Parr, B. A., Shea, M. J., Vassileva, G., & McMahon, A. P. (1993). Mouse Wnt genes exhibit discrete domains of expression in the early embryonic CNS and limb buds. *Development (Cambridge, England), 119*, 247–261.
Pawar, K., Cummings, B. J., Thomas, A., Shea, L. D., Levine, A., Pfaff, S., et al. (2015). Biomaterial bridges enable regeneration and re-entry of corticospinal tract axons into the caudal spinal cord after SCI: Association with recovery of forelimb function. *Biomaterials, 65*, 1–12.
Persson, M., Stamataki, D., te Welscher, P., Andersson, E., Bose, J., Ruther, U., et al. (2002). Dorsal-ventral patterning of the spinal cord requires Gli3 transcriptional repressor activity. *Genes & Development, 16*, 2865–2878.
Piltti, K. M., Avakian, S. N., Funes, G. M., Hu, A., Uchida, N., Anderson, A. J., et al. (2015). Transplantation dose alters the dynamics of human neural stem cell engraftment, proliferation and migration after spinal cord injury. *Stem Cell Research, 15*, 341–353.
Rexed, B. (1954). A cytoarchitectonic atlas of the spinal cord in the cat. *The Journal of Comparative Neurology, 100*, 297–379.
Roelink, H., Augsburger, A., Heemskerk, J., Korzh, V., Norlin, S., Ruiz i Altaba, A., et al. (1994). Floor plate and motor neuron induction by vhh-1, a vertebrate homolog of hedgehog expressed by the notochord. *Cell, 76*, 761–775.
Roelink, H., Porter, J. A., Chiang, C., Tanabe, Y., Chang, D. T., Beachy, P. A., et al. (1995). Floor plate and motor neuron induction by different concentrations of the amino-terminal cleavage product of sonic hedgehog autoproteolysis. *Cell, 81*, 445–455.
Rohatgi, R., Milenkovic, L., & Scott, M. P. (2007). Patched1 regulates hedgehog signaling at the primary cilium. *Science (New York, N.Y.), 317*, 372–376.
Rosenzweig, E. S., Brock, J. H., Lu, P., Kumamaru, H., Salegio, E. A., Kadoya, K., et al. (2018). Restorative effects of human neural stem cell grafts on the primate spinal cord. *Nature Medicine, 24*, 484–490.
Rousso, D. L., Gaber, Z. B., Wellik, D., Morrisey, E. E., & Novitch, B. G. (2008). Coordinated actions of the forkhead protein Foxp1 and Hox proteins in the columnar organization of spinal motor neurons. *Neuron, 59*, 226–240.
Rowitch, D. H., & Kriegstein, A. R. (2010). Developmental genetics of vertebrate glial-cell specification. *Nature, 468*, 214–222.
Ruiz i Altaba, A. (1998). Combinatorial Gli gene function in floor plate and neuronal inductions by Sonic hedgehog. *Development (Cambridge, England), 125*, 2203–2212.
Sagner, A., Gaber, Z. B., Delile, J., Kong, J. H., Rousso, D. L., Pearson, C. A., et al. (2018). Olig2 and Hes regulatory dynamics during motor neuron differentiation revealed by single cell transcriptomics. *PLoS Biology, 16*, e2003127.
Sances, S., Bruijn, L. I., Chandran, S., Eggan, K., Ho, R., Klim, J. R., et al. (2016). Modeling ALS with motor neurons derived from human induced pluripotent stem cells. *Nature Neuroscience, 19*, 542–553.
Sasaki, H., Nishizaki, Y., Hui, C., Nakafuku, M., & Kondoh, H. (1999). Regulation of Gli2 and Gli3 activities by an amino-terminal repression domain: Implication of Gli2 and Gli3 as primary mediators of Shh signaling. *Development (Cambridge, England), 126*, 3915–3924.

Sharpe, C. R., & Goldstone, K. (1997). Retinoid receptors promote primary neurogenesis in Xenopus. *Development (Cambridge, England)*, *124*, 515–523.
Shi, Y., & Massague, J. (2003). Mechanisms of TGF-beta signaling from cell membrane to the nucleus. *Cell*, *113*, 685–700.
Shulman, J. M., Perrimon, N., & Axelrod, J. D. (1998). Frizzled signaling and the developmental control of cell polarity. *Trends in Genetics*, *14*, 452–458.
Slusarski, D. C., Yang-Snyder, J., Busa, W. B., & Moon, R. T. (1997). Modulation of embryonic intracellular Ca2+ signaling by Wnt-5A. *Developmental Biology*, *182*, 114–120.
Stamataki, D., Ulloa, F., Tsoni, S. V., Mynett, A., & Briscoe, J. (2005). A gradient of Gli activity mediates graded Sonic Hedgehog signaling in the neural tube. *Genes & Development*, *19*, 626–641.
Stasiulewicz, M., Gray, S. D., Mastromina, I., Silva, J. C., Bjorklund, M., Seymour, P. A., et al. (2015). A conserved role for Notch signaling in priming the cellular response to Shh through ciliary localisation of the key Shh transducer Smo. *Development (Cambridge, England)*, *142*, 2291–2303.
Stone, D. M., Hynes, M., Armanini, M., Swanson, T. A., Gu, Q., Johnson, R. L., et al. (1996). The tumour-suppressor gene patched encodes a candidate receptor for Sonic hedgehog. *Nature*, *384*, 129–134.
Szabo, N. E., da Silva, R. V., Sotocinal, S. G., Zeilhofer, H. U., Mogil, J. S., & Kania, A. (2015). Hoxb8 intersection defines a role for Lmx1b in excitatory dorsal horn neuron development, spinofugal connectivity, and nociception. *The Journal of Neuroscience*, *35*, 5233–5246.
Tabata, T., Eaton, S., & Kornberg, T. B. (1992). The Drosophila hedgehog gene is expressed specifically in posterior compartment cells and is a target of engrailed regulation. *Genes & Development*, *6*, 2635–2645.
Tanabe, Y., & Jessell, T. M. (1996). Diversity and pattern in the developing spinal cord. *Science (New York, N.Y.)*, *274*, 1115–1123.
Tashiro, S., Michiue, T., Higashijima, S., Zenno, S., Ishimaru, S., Takahashi, F., et al. (1993). Structure and expression of hedgehog, a Drosophila segment-polarity gene required for cell-cell communication. *Gene*, *124*, 183–189.
ten Dijke, P., Yamashita, H., Sampath, T. K., Reddi, A. H., Estevez, M., Riddle, D. L., et al. (1994). Identification of type I receptors for osteogenic protein-1 and bone morphogenetic protein-4. *The Journal of Biological Chemistry*, *269*, 16985–16988.
Thomsen, G. M., Gowing, G., Svendsen, S., & Svendsen, C. N. (2014). The past, present and future of stem cell clinical trials for ALS. *Experimental Neurology*, *262*(Pt. B), 127–137.
Timmer, J., Chesnutt, C., & Niswander, L. (2005). The activin signaling pathway promotes differentiation of dI3 interneurons in the spinal neural tube. *Developmental Biology*, *285*, 1–10.
Timmer, J. R., Wang, C., & Niswander, L. (2002). BMP signaling patterns the dorsal and intermediate neural tube via regulation of homeobox and helix-loop-helix transcription factors. *Development (Cambridge, England)*, *129*, 2459–2472.
Tolwinski, N. S., & Wieschaus, E. (2004). A nuclear function for armadillo/beta-catenin. *PLoS Biology*, *2*, E95.
Tozer, S., Le Dreau, G., Marti, E., & Briscoe, J. (2013). Temporal control of BMP signalling determines neuronal subtype identity in the dorsal neural tube. *Development (Cambridge, England)*, *140*, 1467–1474.
Umbach, J. A., Adams, K. L., Gundersen, C. B., & Novitch, B. G. (2012). Functional neuromuscular junctions formed by embryonic stem cell-derived motor neurons. *PLoS One*, *7*, e36049.
Urist, M. R., Dowell, T. A., Hay, P. H., & Strates, B. S. (1968). Inductive substrates for bone formation. *Clinical Orthopaedics and Related Research*, *59*, 59–96.

Varjosalo, M., & Taipale, J. (2008). Hedgehog: Functions and mechanisms. *Genes & Development, 22*, 2454–2472.
Villeda, S. A., Akopians, A. L., Babayan, A. H., Basbaum, A. I., & Phelps, P. E. (2006). Absence of Reelin results in altered nociception and aberrant neuronal positioning in the dorsal spinal cord. *Neuroscience, 139*, 1385–1396.
Wada, H., & Makabe, K. (2006). Genome duplications of early vertebrates as a possible chronicle of the evolutionary history of the neural crest. *International Journal of Biological Sciences, 2*, 133–141.
Wainger, B. J., Kiskinis, E., Mellin, C., Wiskow, O., Han, S. S., Sandoe, J., et al. (2014). Intrinsic membrane hyperexcitability of amyotrophic lateral sclerosis patient-derived motor neurons. *Cell Reports, 7*, 1–11.
Wichterle, H., Lieberam, I., Porter, J. A., & Jessell, T. M. (2002). Directed differentiation of embryonic stem cells into motor neurons. *Cell, 110*, 385–397.
Wichterle, H., & Peljto, M. (2008). Differentiation of mouse embryonic stem cells to spinal motor neurons. *Current Protocols in Stem Cell Biology*Chapter 1, 1H.1.1–1H.1.9.
Wilson, L., Gale, E., Chambers, D., & Maden, M. (2004). Retinoic acid and the control of dorsoventral patterning in the avian spinal cord. *Developmental Biology, 269*, 433–446.
Wine-Lee, L., Ahn, K. J., Richardson, R. D., Mishina, Y., Lyons, K. M., & Crenshaw, E. B., 3rd. (2004). Signaling through BMP type 1 receptors is required for development of interneuron cell types in the dorsal spinal cord. *Development (Cambridge, England), 131*, 5393–5403.
Wolpert, L. (1969). Positional information and the spatial pattern of cellular differentiation. *Journal of Theoretical Biology, 25*, 1–47.
Wu, M. Y., & Hill, C. S. (2009). Tgf-beta superfamily signaling in embryonic development and homeostasis. *Developmental Cell, 16*, 329–343.
Wyatt, A. W., Osborne, R. J., Stewart, H., & Ragge, N. K. (2010). Bone morphogenetic protein 7 (BMP7) mutations are associated with variable ocular, brain, ear, palate, and skeletal anomalies. *Human Mutation, 31*, 781–787.
Xu, Y., Lopes, C., Wende, H., Guo, Z., Cheng, L., Birchmeier, C., et al. (2013). Ontogeny of excitatory spinal neurons processing distinct somatic sensory modalities. *The Journal of Neuroscience, 33*, 14738–14748.
Yamada, T., Pfaff, S. L., Edlund, T., & Jessell, T. M. (1993). Control of cell pattern in the neural tube: Motor neuron induction by diffusible factors from notochord and floor plate. *Cell, 73*, 673–686.
Yamada, T., Placzek, M., Tanaka, H., Dodd, J., & Jessell, T. M. (1991). Control of cell pattern in the developing nervous system: Polarizing activity of the floor plate and notochord. *Cell, 64*, 635–647.
Yamauchi, K., Phan, K. D., & Butler, S. J. (2008). BMP type I receptor complexes have distinct activities mediating cell fate and axon guidance decisions. *Development (Cambridge, England), 135*, 1119–1128.
Yang, Q., Bermingham, N. A., Finegold, M. J., & Zoghbi, H. Y. (2001). Requirement of Math1 for secretory cell lineage commitment in the mouse intestine. *Science (New York, NY), 294*, 2155–2158.
Yasaka, T., Tiong, S. Y., Hughes, D. I., Riddell, J. S., & Todd, A. J. (2010). Populations of inhibitory and excitatory interneurons in lamina II of the adult rat spinal dorsal horn revealed by a combined electrophysiological and anatomical approach. *Pain, 151*, 475–488.
Yuengert, R., Hori, K., Kibodeaux, E. E., McClellan, J. X., Morales, J. E., Huang, T. W., et al. (2015). Origin of a non-Clarke's column division of the dorsal spinocerebellar tract and the role of caudal proprioceptive neurons in motor function. *Cell Reports, 13*, 1258–1271.
Zechner, D., Müller, T., Wende, H., Walther, I., Taketo, M. M., Crenshaw, E. B., 3rd, et al. (2007). Bmp and Wnt/beta-catenin signals control expression of the transcription factor Olig3 and the specification of spinal cord neurons. *Developmental Biology, 303*, 181–190.

Zeng, X., Goetz, J. A., Suber, L. M., Scott, W. J., Jr., Schreiner, C. M., & Robbins, D. J. (2001). A freely diffusible form of Sonic hedgehog mediates long-range signalling. *Nature*, *411*, 716–720.

Zhang, J., Lanuza, G. M., Britz, O., Wang, Z., Siembab, V. C., Zhang, Y., et al. (2014). V1 and v2b interneurons secure the alternating flexor-extensor motor activity mice require for limbed locomotion. *Neuron*, *82*, 138–150.

Zhang, Y., Narayan, S., Geiman, E., Lanuza, G. M., Velasquez, T., Shanks, B., et al. (2008). V3 spinal neurons establish a robust and balanced locomotor rhythm during walking. *Neuron*, *60*, 84–96.

CHAPTER THIRTEEN

Development, repair, and regeneration of the limb musculoskeletal system

Jane Y. Song[a], Kyriel M. Pineault[b], Deneen M. Wellik[b,*]
[a]Program in Cell and Molecular Biology Program, University of Michigan, Ann Arbor, MI, United States
[b]Department of Cell & Regenerative Biology, University of Wisconsin, Madison, WI, United States
*Corresponding author: e-mail address: wellik@wisc.edu

Contents

Abstract

The limb musculoskeletal system provides a primary means for locomotion, manipulation of objects and protection for most vertebrate organisms. Intricate integration of the bone, tendon and muscle tissues are required for function. These three tissues arise largely independent of one another, but the connections formed during later development are maintained throughout life and are re-established following injury. Each of these tissues also have mesenchymal stem/progenitor cells that function in maintenance and repair. Here in, we will review the major events in the development of limb skeleton, tendon, and muscle tissues, their response to injury, and discuss current knowledge regarding resident progenitor/stem cells within each tissue that participate in development, repair, and regeneration *in vivo*.

Current Topics in Developmental Biology, Volume 132
ISSN 0070-2153
https://doi.org/10.1016/bs.ctdb.2018.12.011

1. Introduction

Many early studies of limb development capitalized on the accessibility and ease of manipulation of the developing chick (and/or quail) embryo. With the advent of genetic engineering technology, studies have expanded to include targeted manipulation of the mouse genome to uncover the molecular underpinnings of limb development (Mansour, Thomas, & Capecchi, 1988; Thomas & Capecchi, 1987). Mouse embryonic fore- and hindlimb buds are first observed on the lateral sides of the embryo as bulges of lateral plate mesoderm at embryonic day (E) 9.25 and E9.75, respectively. While the appearance and development of the forelimb precedes that of the hindlimb by approximately 12 h, much of the developmental program is conserved between both pairs of limbs (Martin, 1990). The musculoskeletal unit is composed of bone, tendon, and muscle, which functions together to produce locomotion. Therefore, all three tissues must develop in concert and precisely integrate to generate a functional musculoskeletal system.

Populations of tissue resident progenitor/stem cells have been identified in bone, tendon, and muscle that participate in the development, adult homeostasis, and injury repair/regeneration of each of these tissues. The bone and the muscle tissues, in particular, have a remarkable capacity for regeneration. In the adult animal, many of the processes that regulate the development of each tissue are re-deployed during injury repair (Howell et al., 2017; Murphy, Lawson, Mathew, Hutcheson, & Kardon, 2011; Relaix & Zammit, 2012; Sambasivan et al., 2011; Voleti, Buckley, & Soslowsky, 2012). Significant ongoing efforts are being pursued to fully characterize these progenitor/stem cell populations and harness their therapeutic potential. In this review, we will discuss the development and the repair/regeneration processes of bone, muscle, and tendon tissues individually. We will additionally discuss what is known about the progenitor/stem cell populations that have been characterized in each tissue.

2. Development

2.1 Skeleton

Mesenchymal condensations that preconfigure the limb skeletal elements initiate formation in a proximal to distal fashion beginning at E10.5 in the forelimb. These pre-chondrogenic mesenchymal cells derive from the lateral plate mesoderm and are first observed as morphologically dense, compacted groups of mesenchymal cells within the limb bud (Thorogood & Hinchliffe, 1975). The earliest marker of pre-chondrogenic cells is the transcription factor, *Sox9* (Akiyama, Chaboissier, Martin, Schedl, & De Crombrugghe, 2002; Bi, Deng, Zhang, Behringer, & De Crombrugghe, 1999). *Sox9*-expressing cells are found scattered in the limb bud mesenchyme prior to condensation. These chondrogenic precursors subsequently aggregate in a proximal (stylopod, ~E10.5) to distal (autopod, ~E11.5) fashion to form the cartilage anlage that will become the skeletal elements of the limb (Akiyama et al., 2002, 2005; Bi et al., 1999; Bi et al. 2001). The nuclear factor kappa B (NF-κB) family member, RelA, has been shown to directly promote *Sox9* expression and to induce chondrogenic differentiation in HeLa and ATDC5 cells (a chondrogenic cell line derived from teratocarcinoma) cells (Ushita et al., 2009). While the upstream regulation of *Sox9* expression *in vivo* is less clear, input from multiple signaling pathways have been implicated in the regulation of chondrogenesis, including Shh, Ihh, Wnt/β-catenin, Notch, TGFβ, BMP, NF-κB, and FGF (Akiyama, 2008; Akiyama et al., 2007; Deschaseaux, Pontikoglou, & Sensébé, 2010; Govindarajan & Overbeek, 2006; Mead & Yutzey, 2009; Mori-Akiyama & Zhang, 2004; Pan et al., 2008; Rockel, Kudirka, Guzi, & Bernier, 2008; St-Jacques, Hammerschmidt, & McMahon, 1999; Ushita et al., 2009).

Unlike *Sox9, Sox5* and *Sox6* are not required for chondrogenesis, however, global inactivation of both genes leads to short, cartilage primordia with matrix defects and immature growth plates. Results from molecular analyses strongly support the notion that *Sox5* and *Sox6* increases the efficiency of *Sox9* binding to DNA to potentiate *Sox9*'s ability to activate cartilage gene enhancers (Han & Lefebvre, 2008; Lefebvre, Li, & De Crombrugghe, 1998). A recent genome-wide approach revealed that *Sox9* and *Sox5/6* bind thousands of sites but preferably bind near each other by primarily targeting super-enhancers, which are extremely active enhancers or clusters of enhancers used as a hub by master transcription factors to regulate the expression of genes that impart cell identity in stem cells (Liu & Lefebvre, 2015). The binding of *Sox9* and *Sox5/6* are responsible for coding critical cartilage extracellular matrix components as well as the SOX trio proteins (Liu & Lefebvre, 2015).

Longitudinal bone growth is fueled by the regulated proliferation and differentiation of chondrocytes at the distal ends of the bone. Chondrocytes within the growth plate are organized into distinct cellular zones with specific proliferative and maturity characteristics (Fig. 1). The reserve zone,

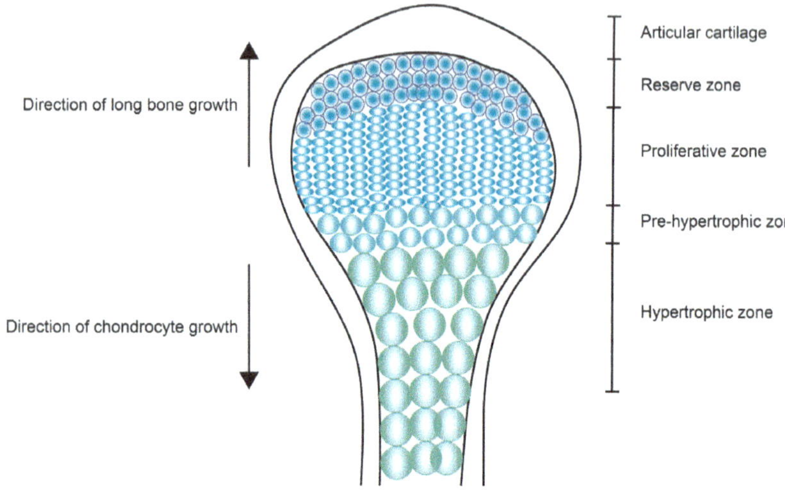

Fig. 1 Chondrocytes within the growth plate fuel longitudinal growth of long bones. Depicted is a representative cartoon of an E14.5 vertebrate long bone. The reserve zone is comprised of round chondrocytes that slowly cycle, which transitions into proliferative chondrocytes that have a distinct columnar appearance. The chondrocytes begin to hypertrophy in the pre-hypertrophic zone and eventually exit the cell cycle as they enter the hypertrophic zone. Of note, the direction of chondrocyte growth and the direction of long bone growth are in opposite directions.

at the distal-most ends of the bone, contains slow dividing chondrocyte progenitors, which give rise to all growth plate chondrocytes (Abad et al., 2002). Chondrocytes in the proliferative zone are oriented into columns of cells, parallel to the long axis of the bone, and undergo significant proliferation to increase the number of cells within the growth plate (Hunziker, 1994). The proliferative zone is characterized by expression of a type II collagen matrix. At the central metaphyseal side of each growth plate, the hypertrophic zone is observed. Chondrocytes exit the cell cycle undergo hypertrophy, and transition to secreting a type X collagen matrix (Beier, 2005). The hypertrophic transition zone provides the bulk of the longitudinal expansion of the bone.

Concomitant to condensation of the pre-chondrogenic skeletal mesenchyme, a dense connective tissue layer forms surrounding the skeletal anlage termed the perichondrium. It has been suggested that the perichondrium functions to constrain the radial growth and promote longitudinal growth of the developing skeletal anlage (Rooney & Archer, 1992). The perichondrium and, later periosteum, additionally serves as a primary site of osteogenesis in the central bone shaft region. Osteogenesis initiates at E13.5 in the mouse, and it can be visualized at its earliest stages by expression of the osteogenic markers *Runx2* and *Osterix*. Osteoblasts, defined by expression of *Runx2* and Osterix (*Osx*), are first observed at E14.5 at the medial aspect of the anlage (Akiyama et al., 2005). Starting around E14.5, Osx-expressing pre-osteoblasts enter the bone marrow space concurrent with vascular invasion into the medial region of the developing bone, which is mostly made up of hypertrophic chondrocytes (Maes et al., 2010). Osteoclasts erode the hypertrophic matrix and allow invasion of endothelial cells into the cartilage (Karsenty & Wagner, 2002; Kronenberg, 2007). Osteoblasts and other stromal populations invade the developing bone marrow space concurrent with the vasculature and begin depositing bony matrix that make up the primary ossification center (Maes et al., 2010).

2.2 Tendon

Tendons and the remaining connective tissue elements of the limb are also derived from the lateral plate mesoderm (Kieny & Chevallier, 1979). The bHLH transcription factor, *Scleraxis* (*Scx*) is the earliest known marker for tendon and is first detected in the limb bud at ~E10.5 (Brent, Braun, & Tabin, 2005; Schweitzer et al., 2001). Analogous to the formation of the skeletal anlage, induction of tendon progenitors in the limb occurs in a proximal to distal fashion as *Scx*-positive tendon progenitors arise within the limb bud mesenchyme in the dorsal and ventral subectodermal domains adjacent to the condensing skeletal anlage (Murchison et al., 2007; Schweitzer et al., 2001). Ectodermal signals have been implicated in the induction of tendon progenitors as ectodermal removal prior to onset of *Scx* expression results in a complete abrogation of *Scx* expression. The ectodermal signal is not required for the maintenance of *Scx* expression, however, as removal of the ectoderm after the onset of expression had no effect on tendon development (Schweitzer et al., 2001). There is also evidence that tendon progenitor induction and skeletal condensation are tightly linked as increases or decreases in the number of digits results in a corresponding increase or decrease in formation of the associated tendons (Huang, Riordan, et al., 2015). Following induction, TGFβ signaling has been found to be critical for tendon progenitor maintenance. In *Tgfb2*$^{-/-}$;*Tgfb3*$^{-/-}$ mutant mouse embryo limbs, *Scx*-positive tendon progenitors appear normally at E11.5, but are lost over the next 24 h, resulting in complete abrogation of tendon development (Pryce et al., 2009). In addition to *Scx* and TGFβ signaling, roles for *Mohawk* (*Mkx*), *tenomodulin* (*Tnmd*), *Thrombospondin 2* and *4*

(*Thbs-2, 4*), growth differentiation factor-5 (*Gdf-5*) have been demonstrated (Alberton et al., 2015; Brent, Schweitzer, & Tabin, 2003; Docheva, Hunziker, Fassler, & Brandau, 2005; Frolova et al., 2014; Ito et al., 2010; Koda et al., 2017; Kyriakides et al., 1998; Liu et al., 2010, 2014; Mendias, Bakhurin, & Faulkner, 2008; Mikic, 2004; Mikic, Schalet, Clark, Gaschen, & Hunziker, 2001; Onizuka et al., 2014).

Initial tendon specification and early patterning is tissue autonomous, however continued maintenance of tendon progenitors requires interactions with the muscle and the skeleton (Bonnin et al., 2005; Havis et al., 2016; Kardon, 1998). Interestingly, the requirement for tendon-muscle and tendon-cartilage interactions differs along the proximal to distal axis of the limb (Bonnin et al., 2005; Havis et al., 2016; Huang, Lu, & Schweitzer, 2015; Kardon, 1998). In muscle-less limb models in both chick and mouse, maintenance of proximal tendon progenitors requires the presence of muscle progenitors while distal tendon development is unaffected by the absence of muscle (Bonnin et al., 2005; Kardon, 1998; Schweitzer et al., 2001). In contrast, induction and maintenance of autopod tendons requires the presence of the digit skeletal anlages (Huang, Riordan, et al., 2015). Curiously, there are several long tendons in the limb that originate within the zeugopod but anchor to the skeleton within the autopod suggesting that, within a single tendon, unique developmental programs regulate progenitor differentiation along the proximal to distal axis. Careful examination of tendon development during the transition from the muscle-independent phase (before E12.5) to the muscle-dependent phase (after E12.5) revealed a spatial, modular mechanism of tendon development. Initially, short tendon progenitors integrate with the distal end of the individual muscle masses in the zeugopod. As limb outgrowth continues, the short tendons form connections with their counterparts that arise within the autopod, modularly forming a differentiated linear tendon (Huang, Riordan, et al., 2015). Following the integration of zeugopod and autopod tendons at the wrist, muscle contractions appear to be necessary for further splitting of the autopod tendons.

2.3 Muscle

In contrast to the lateral plate mesoderm-derived skeleton and tendon tissues, limb muscle precursors arise outside the limb field from the dermomyotome compartment of the axial somites adjacent to the limb bud regions. Somites are segmented mesodermal masses of tissue that form along the anterior to posterior axis of the embryo adjacent to the notochord during gastrulation. The muscle precursor myotomal cells delaminate from somites and migrate into the developing limb bud (Chevallier, Kieny, & Mauger, 1977; Christ, Jacob, & Jacob, 1977; Christ & Ordahl, 1995; Ordahl & Le Douarin, 1992; Wachtler, 1984). Muscle precursors that delaminate from this compartment express *Pax3* and migrate into the limb field in two streams dorsal and ventral to the limb skeleton anlage (Chevallier et al., 1977). The muscle precursors stream into the limb field and are initially intermixed with Scx-expressing tendon progenitors (Schweitzer et al., 2001). However, lineage-tracing analysis using *Pax7Cre;Rosa TdTomato;ScxGFP* revealed that the muscle precursors and tendon precursors remain completely distinct populations and give rise to either muscle fibers or tendons, respectively (Bonnin et al., 2005; Huang et al., 2013). Migration of the muscle precursors into the forelimbs is largely complete by E10.5, and limb muscle progenitors subsequently undergo significant proliferation and differentiation within the developing limbs (Murphy & Kardon, 2011).

Information that leads to the establishment of muscle pattern is contained within the connective tissue compartment of the limb. Muscle precursors themselves have no intrinsic or pre-determined patterning information (Chevallier et al., 1977; Kardon, Campbell, & Tabin, 2002). Prior to muscle invasion, muscle connective tissue fibroblasts form muscle pre-patterns that sub-divide the dorsal and ventral muscle streams into individual muscle groups (Kardon, Harfe, & Tabin, 2003). Muscle connective tissue fibroblasts, marked by the transcription factor *Tcf4*, are observed as early as E12.5 in the developing limb bud (Kardon et al., 2003; Mathew et al., 2011). Tcf4-positive fibroblasts are intimately associated with the myofibers of the limb musculature during development and remain associated throughout life (Mathew et al., 2011). There is evidence that Tcf4-positive connective tissue fibroblasts function as an extrinsic source of signals to regulate muscle myofiber type and myofiber maturation (Mathew et al., 2011). In addition to being a marker of the muscle connective tissue, Tcf4 also functions in regulating muscle patterning. Loss-of-function of *Tcf4* leads to disruption of several aspects of muscle patterning, including abnormal muscle splitting, truncation of several muscles, and mis-patterning of the sites of muscle integration with the bone (Kardon et al., 2003; Mathew et al., 2011).

3. Establishing limb pattern and morphology

As the limb bud emerges, it is initially polarized along three axes; proximal to distal (PD; from shoulder to fingers), dorsal to ventral (DV; back of the hand to the palm), and anterior to posterior (AP; from digit I to V). Mature bone, muscle, and tendon architecture is not symmetrical along these axes and establishing appropriate polarity within the limb bud is critical for later development. During limb development, the PD axis is mainly under the control of the apical ectodermal ridge (AER) that maintain undifferentiated, underlying mesenchymal cells to continue to proliferate (Summerbell, Lewis, & Wolpert, 1973; reviewed in Fernandez-Teran & Ros, 2008). The DV identity is controlled by the expression of Engrailed-1 in the ventral ectoderm that restricts *Wnt7a* to the dorsal ectoderm. Concurrently, Wnt7a prompts the expression of *Lmx1b* in the dorsal mesoderm (Chen & Johnson, 2002; Chen et al., 1998; Riddle et al., 1995). Finally, the AP axis is established when *Hox9* in the forelimb and *Isl1* in the hindlimb is expressed in the flank, inducing the expression of *Hand2* in the posterior limb bud (Tahara et al., 2017; Xu & Wellik, 2011). *Hand2* promotes *Shh* expression in the zone of polarizing activity (ZPA) in the posterior limb bud and this establishes a *Shh* morphogen gradient along the AP axis, thereby imparting AP identity and growth (Martí, Takada, Bumcrot, Sasaki, & McMahon, 1995; Riddle, Johnson, Laufer, & Tabin, 1993).

After the initial establishment of limb pattern of the stylopod (humerus, femur), zeugopod (radius/ulna, tibia/fibula) and autopod (fingers, toes), *Hox* genes play fundamental roles in the regional development of morphology of the skeleton. *Hox* genes are highly conserved and are arranged as tightly linked, colinear arrangement of genes such that their position along the chromosome reflects their spatial and temporal expression along the anteroposterior axis of the body (reviewed by Mallo & Alonso, 2013; Mallo, Vinagre, & Carapuço, 2009; Wellik, 2007, 2009). The posterior subset of *Hox* genes, *Hox9-Hox13*, additionally function to pattern the proximal to distal axis of the appendicular limb skeleton (Davis, Witte, Hsieh-Li, Potter, & Capecchi, 1995; Fromental-Ramain et al., 1996, 1996; Wellik & Capecchi, 2003).

The vertebrate limb skeleton can be divided into three regions along the proximal to distal axis; the stylopod (humerus or femur), the zeugopod (radius/ulna or tibia/fibula), and the autopod (carpals/metacarpals or tarsals/metatarsals). In the early limb bud the 5′ Hox genes are expressed in progressively distally restricted nested and overlapping domains

(Haack & Gruss, 1993; Nelson et al., 1996; Tarchini & Duboule, 2006; Zakany & Duboule, 2007). As cartilage anlage for the limb skeletal elements appear, the expression domains of the *Hox* paralogs resolve such that *Hox9* and *Hox10* genes are largely restricted to the stylopod, *Hox11* genes to the zeugopod and *Hox13* genes to the autopod (Davis et al., 1995; Fromental-Ramain, Warot, Lakkaraju, et al., 1996; Fromental-Ramain, Warot, Messadecq, et al., 1996; Wellik & Capecchi, 2003). Unlike the axial skeleton where loss of *Hox* function results in homeotic transformation of vertebrate morphology, in the limb, *Hox* mutants display regional defects in skeletal morphology (Davis et al., 1995; Fromental-Ramain, Warot, Lakkaraju, et al., 1996; Fromental-Ramain, Warot, Messadecq, et al., 1996; Wellik & Capecchi, 2003). For example, *Hox11* triple mutants (*Hoxa11−/−;Hoxc11−/−;Hoxd11−/−*) exhibit dramatically shortened and malformed zeugopod skeletal elements while the stylopod and autopod elements develop largely normally (Davis et al., 1995; Wellik & Capecchi, 2003). Similar regional patterning defects in the proximal elements are observed in *Hox9* or *Hox10* paralog mutants, and complete agenesis of the autopod elements is observed in *Hoxa13/Hoxd13* mutants (Fromental-Ramain, Warot, Lakkaraju, et al., 1996; Fromental-Ramain, Warot, Messadecq, et al., 1996).

The cellular mechanisms for Hox function in determining limb morphology have remained largely a mystery. Despite their fundamental roles in establishing limb shape, *Hox* expression is excluded from all differentiated cell types of the developing limb skeleton including chondrocytes and osteoblasts (Nelson, Rakshit, Sun, & Wellik, 2008; Swinehart, Schlientz, Quintanilla, Mortlock, & Wellik, 2013). Instead, *Hox* expression is observed throughout the stromal components of the limb, including the perichondrium/periosteum surrounding the skeletal elements, the tendons, and the muscle connective tissue (Fig. 2).

The continuum of *Hox* expression throughout the musculoskeletal connective tissue elements led to a re-examination of the *Hox11* mutant phenotype. Muscle and tendon development in *Hox11* compound mutants where one wild-type *Hoxa11* or *Hoxd11* allele remains (*Hoxa11+/−; Hoxd11−/−* or *Hoxa11−/−;Hoxd11+/−*) revealed that mis-patterning of the muscles and tendons occur even in the absence of skeletal defects (Swinehart et al., 2013). These data provide evidence that *Hox* gene function in connective tissue stroma is likely a key regulator for the patterning of each these tissues.

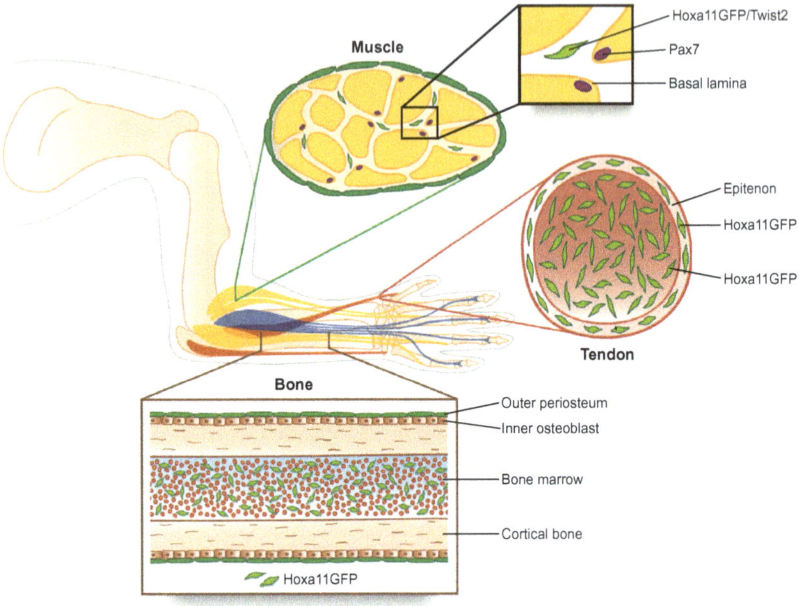

Fig. 2 Regional expression of Hoxa11eGFP in the mesenchymal tissues of the musculoskeletal system. In the adult, *Hoxa11eGFP* is expressed in undifferentiated mesenchymal tissues in the bone, tendon, and muscle. *Hoxa11*-expressing cells (green) are observed in the connective tissues surrounding the muscle bundles (epimysium and perimysium) and within the muscle bundles outside the basal lamina distinct from Pax7-positive satellite cells (purple). Hoxa11eGFP expression is observed in tenocytes throughout the tendon. In the skeleton, Hoxa11eGFP-expressing bone marrow MSCs are observed on the periosteal and endosteal (not indicated) bone surfaces and throughout the bone marrow space.

4. Injury repair and regeneration

4.1 Skeleton

The skeleton is one of the few organs in the body that is capable of regenerating damaged tissue and re-establishing original structure and biomechanical function throughout life. Many of these repair events closely mirror the processes of embryonic development (reviewed in Gerstenfeld, Cullinane, Barnes, Graves, & Einhorn, 2003; Vortkamp et al., 1998). The fracture healing process proceeds through three phases: anabolic, catabolic, and remodeling. At the onset of fracture injury and subsequent disruption of the bone microenvironment, the anabolic phase is characterized by the recruitment to and expansion of multi-potent progenitors at the injury site to form a soft, cartilaginous callus that stabilizes the fracture. During the subsequent catabolic phase, the soft cartilaginous callus is resorbed and replaced with an ossified bony callus. Finally, remodeling of the callus is achieved through the activity of osteoblasts and osteoclasts to re-establish the marrow space and re-shaping the skeleton back to its original pre-injury structure (Hilton, 2014).

Mesenchymal stem cells (MSCs, described in more detail below) from both the bone marrow and the periosteal compartments participate in fracture healing (Ono & Kronenberg, 2016). However, there is emerging evidence that periosteal progenitors and bone marrow progenitors may have distinct functions during fracture repair. Bone injuries that only damage the periosteum or only damage the bone marrow/endosteum (without damage to the other compartment) combined with genetic lineage-tracing approaches, demonstrated that while both compartments are capable of generating osteoblasts, the periosteal progenitors contribute to the majority of the cartilage present within the fracture callus while bone marrow cells demonstrate minimal cartilage contribution (Colnot, 2009). Additionally, transplantation of periosteal cells marked by *Prx1Cre;mTmG* into the fracture callus demonstrate that periosteal MSCs display a higher regenerative capacity than bone marrow MSCs in this experimental setting (Duchamp de Lageneste et al., 2018). Cell extrinsic factors and/or cues within the bone marrow microenvironment appear to have a negative effect on the chondrogenic differentiation capacity of bone marrow/endosteal MSCs. Grafting of bone marrow MSCs onto the periosteal surface led to an increase in their chondrogenic potential (Colnot, 2009). Microarray analysis of periosteal and bone marrow stromal cells that were isolated based solely on their plastic adherence characteristics how that at both pre-injury and post-injury stages, the two populations of cells display distinct expression profiles (Duchamp de Lageneste et al., 2018). The functional, *in vivo* relevance of these differences in molecular profiles remains to be explored. Taken together, these studies support the idea that there are likely functionally distinct stem and progenitor cell populations within the skeletal compartment.

4.2 Tendon

Mature tendon is composed of a hierarchical structure of extracellular matrix proteins, mainly type I collagen molecules, that assemble into fibrils, that make up fibers and further organize into fascicle that, together, compose the tendon unit (Nourissat, Berenbaum, & Duprez, 2015). There are two main types of tendon-specific pathologies: acute tendon injuries and tendinopathies (chronic tendon injuries). Acute tendon injury involves a partial or complete rupture of a tendon that disrupts the continuity of the structure (Sharma & Maffulli, 2006). Chronic tendon disorders are usually caused by a combination of both intrinsic and extrinsic factors that lead to a disruption of normal architecture. For example, tendinopathy is histologically characterized by disorganized collagen fibrils, increased glycosaminoglycans and proteoglycan content, increased non-collagenous ECM, neovascularization and hypercellularity of the tendons (Jelinsky et al., 2011; Sharma & Maffulli, 2006; Walden et al., 2017; Xu & Murrell, 2008). Excessive loading and continuous physical training are a major extrinsic stimulus that results in chronic tendon injury. Similar to skeletal repair, tendon injury repair occurs in three overlapping phases (Hope & Saxby, 2007). The inflammatory phase takes place at the onset of tendon rupture with infiltration of red blood cells and immune cells. A fibrin clot is formed to provide temporary stability and stiffness. During the repair/proliferative phase, tenocytes begin depositing new matrix composed mainly of type III collagen. Finally, in the remodeling phase, the temporary collagen III matrix is replaced by collagen I and the extracellular matrix becomes more organized and structured (Nourissat et al., 2015; Walden et al., 2017). Tendon injuries are generally less efficient in repair compared to the bone or muscle, largely due to the higher tendency of tendon to form scar tissue during healing. An injured, adult tendon will never completely regain the biomechanical properties it possessed prior to injury (Docheva, Müller, Majewski, & Evans, 2015; Voleti et al., 2012). Interestingly, tendon healing in neonatal animals heals by forming Scx-positive "neotendon" while adult tendon repair *via* Scx-negative scar. Unlike with adult injuries, neonatal tendon healing leads to restored functional gait and biomechanical properties (Howell et al., 2017).

Cells from the outer epitenon/paratenon (a thin layer of cells oriented circumferentially around the tendon) are the main source of cells that contribute to tendon regeneration (Fig. 2; Dyment et al., 2013). Paratenon cells are normally quiescent and do not express tenogenic markers such as Scx, however, in response to patellar tendon injury, Scx expression is activated in these cells. Following injury, the cells within the paratenon begin to proliferate and produce tenascin-C and fibromodulin, which are important assembly genes that organizes the new matrix (Dyment et al., 2013). The proliferating paratenon cells migrate towards the defect, begin to express

Scx, and produce circumferential collagen fibers (Dyment et al., 2013). Concomitantly, the proliferation of paratenon cells results in a thickening of this layer. The migration of the paratenon cells as well as the thickening of the paratenon, together, ultimately bridges the tendon defect. Additionally, it has been shown that cells that lie in the periphery of the tendon defect that co-express Scx and smooth muscle actin (SMA) also contribute to healing (Dyment et al., 2013). While cells from the paratenon seems to be the major source of cells that participate during tendon healing, existing tendon progenitor markers and genetic lineage-tracing tools are insufficiently specific to exclude contribution of other cell populations to healing within the tendon. Efforts are being made to identify better markers for resident tendon progenitors that contribute to tendon injury repair.

During postnatal tendon growth, smooth muscle actin (SMA)-expressing cells label paratenon/sheath/retinacular cells on the tendon surface and tendon fibroblasts within the center of the developing tendon. At adult stages, these cells greatly expand in response to tendon injury, spanning the defect space, and are main contributors to healing (Dyment et al., 2014). An *SMA-CreERT2*; *Scx-GFP* animal model demonstrated that SMA-lineage marked cells expand during injury repair and give rise to Scx-positive tendon cells during the healing process (Dyment et al., 2014). The broad nature of SMA expression limits the conclusions that can be reached regarding this model, but shows that non-Scx cells can gain Scx expression.

Hoxa11eGFP is highly expressed both in the periosteum of the zeugopod skeletal elements and continuously into the zeugopod-attached tendon structures, including the paratenon (Swinehart et al., 2013). Given that Hox-expressing cells have been identified as skeletal stem/progenitor cells (MSCs, Rux et al., 2016), it will be important to determine whether this population additionally serves as a tendon progenitor as well. Closer investigation of the similarities or differences between the tendon *vs.* periosteal cells that express Hox, SMA and other markers would aid in beginning to understand whether these populations represent equivalent or unique mesenchymal populations.

4.3 Muscle

Similar to the skeleton, the skeletal muscle also boasts a regenerative capacity and is capable of fully restoring injured tissue back to its previous uninjured state (Charge & Rudnicki, 2004). Minor muscle lesions occur from daily activities, and acute muscle injuries can result from rigorous physical activities that exceed the load-bearing capacity of the muscle, or from lacerations, freezing, burning, contusions, or exposure to toxins (Charge & Rudnicki, 2004; Tidball, 2011). Skeletal muscle repair process involves two phases; a degeneration/necrosis phase followed by a regeneration/myogenesis phase. At the onset of muscle injury, disruption of myofiber integrity triggers the damaged myofibers to undergo localized autolysis (Hurme, Kalimo, Lehto, & Järvinen, 1991; Järvinen, Järvinen, & Kalimo, 2013). During regeneration, there is massive proliferation of myogenic cells and the newly generated muscle fibers fuse to existing damaged myofibers and to each other to restore muscle mass and muscle integrity (Charge & Rudnicki, 2004).

Numerous cell populations both within the muscle milieu as well as extrinsic cells, such as immune cells, have been implicated in contributing to the muscle repair process (Dey, Goldhamer, & Yu, 2015; Wosczyna & Rando, 2018). Satellite cells are widely accepted as the resident muscle stem cell for regeneration and repair. Satellite cells are located beneath the surrounding basal lamina of each muscle fiber (Yin, Price, & Rudnicki, 2013). Satellite cells satisfy the pre-requisites for adult stem cells including self-renewal and multi-lineage capacity (Collins & Partridge, 2005). In *ex vivo* experiments, transplantation of a limited number of satellite cells into irradiated, regeneration-defective mice resulted in the generation of both new muscle fibers and the establishment of a new satellite cell pool. Further, the transplanted satellite cells were sufficient to support subsequent rounds of muscle regeneration (Collins et al., 2005; Yin et al., 2013). The presence of satellite cells are necessary for muscle regeneration as the genetic ablation of these cells results in complete failure of new muscle fiber formation following injury (Fry et al., 2015; Lepper, Conway, & Fan, 2009; Murphy et al., 2011; Sambasivan et al., 2011).

5. Musculoskeletal stromal progenitor and stem cells

5.1 Skeletal stem cells/bone marrow MSCs

The existence of multi-potent mesenchymal stromal/stem cells (MSCs) that can serve as skeletal progenitors was first demonstrated by Friedenstein and colleagues when they established that plastic adherent cells isolated from bone marrow could form fibroblastic colonies (CFU-F) in cell culture, differentiate into bone, and reconstitute an ectopic hematopoietic microenvironment when transplanted *in vivo* (Friedenstein, Chailakhjan, & Lalykina, 1970; Friedenstein, Gorskaja, & Kulagina, 1976; Friedenstein, Latzinik,

Grosheva, & Gorskaya, 1982; Friedenstein, Petrakova, Kurolesova, & Frolova, 1968). Two main functions have been attributed to skeletal MSCs. The first function is to serve as a reservoir of new cellular material to maintain skeletal integrity throughout the life of the animal (Bianco et al., 2013; Ono & Kronenberg, 2015). The second function is as a component of the HSC niche (Kfoury & Scadden, 2015; Kunisaki et al., 2013; Méndez-Ferrer et al., 2010; Morrison & Scadden, 2014). The niche refers to the physical environment where HSCs reside within the bone marrow space that functions to preserve the stem cell state both through direct cell-cell contact and through paracrine production of specific signaling factors (Wilson & Trumpp, 2006). Bone marrow MSCs express several important HSC niche factors including *C-X-C motif chemokine 12* (*CXCL12*), *stem cell factor* (*Scf*), and *angiopoietin 1* (*Ang1*) that are necessary for HSC maintenance (Sacchetti et al., 2007). Elimination of these bone marrow stromal populations or deletion of HSC niche factor expression in these cells through genetic or pharmaceutical methods leads to reductions of HSCs (Ding & Morrison, 2013; Ding, Saunders, Enikolopov, & Morrison, 2012; Isern et al., 2014; Kunisaki et al., 2013; Méndez-Ferrer et al., 2010; Omatsu et al., 2010). Today, several overlapping mesenchymal stem/stromal populations of cells have been defined that enrich, to varying degrees, for cells with mesenchymal stem cell capacity [reviewed in extensive detail (Chen, Johnson, & Robey, 2017; Morrison & Scadden, 2014; Ono & Kronenberg, 2016)]. However, unlike the hematopoietic system where HSCs have been unequivocally defined, at least in *ex vivo* experiments, an analogous cell has not been defined for the skeletal stromal compartment. It is important to highlight that the existence of a self-renewing stem cells capable of giving rise to bone, cartilage, and fat has, thus far, only been demonstrated clonally *ex vivo* through *in vitro* differentiation assays or *via* transplantation of a previously cultured, clonal population. Single-cell clonal tracing to demonstrate both multi-lineage contribution and self-renewal has yet to be performed *in vivo* and may be impossible to demonstrate due to the spatial and temporal differences in cartilage, bone and adipocyte production within the skeleton.

The earliest genetically marked population to be rigorously tested to enrich for stem/progenitor activity was defined by GFP expression driven transgenically by the *Nestin* promoter (Méndez-Ferrer et al., 2010). *Nestin*-GFP-positive cells were shown to be rare, perivascularly-localized stromal cells within the bone marrow space. This *Nestin*-GFP positive population was shown to contain all the CFU-F capacity in the bone marrow and to undergo multi-lineage differentiation *in vitro* into the skeletal mesenchymal lineages; osteoblasts, chondrocytes, and adipocytes (Méndez-Ferrer et al., 2010). Self-renewal was demonstrated clonally through serial transplantation supporting that this population enriches for cells with multi-potent progenitor and self-renewal activity (Kunisaki et al., 2013; Méndez-Ferrer et al., 2010). When the lineage capacity of the *Nestin*-GFP positive population was tested *in vivo* using an inducible lineage-tracing model, *Nestin*-CreER, the *Nestin*-lineage marked cells were shown to contribute to bone and cartilage lineages *in vivo* [Fig. 3 and (Méndez-Ferrer et al., 2010; Ono, Ono,

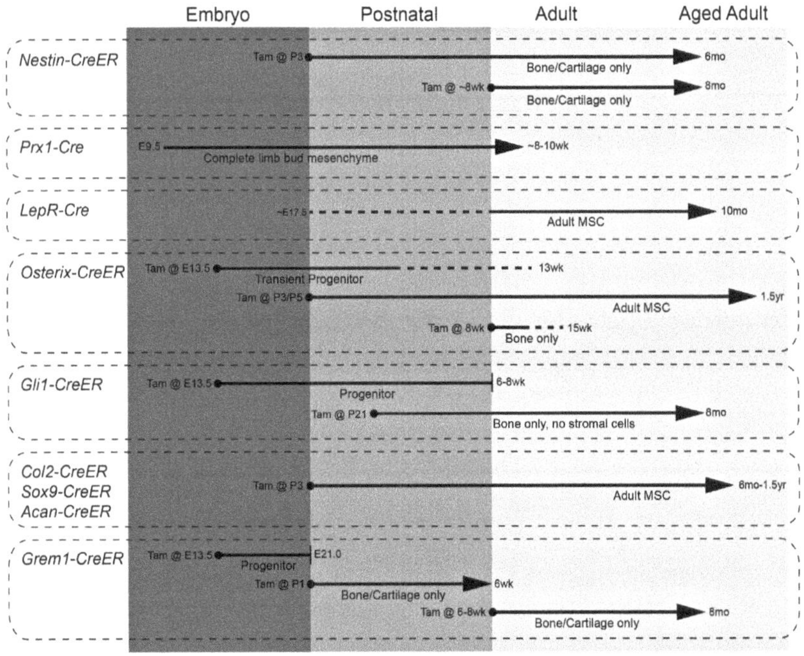

Fig. 3 Comparison of genetic MSC lineage populations. MSC populations genetically marked by various Cre/CreER models display different lineage dynamics over time. The appearance of each population or the time point of tamoxifen induction (black circle) and the analysis end point (arrow) for each model is noted. Lineage contribution not reported beyond a certain stage is indicated by vertical bar. Brief summary of the skeletal contribution and MSC-like behavior is indicated. References listed: *Nestin-CreER* (Méndez-Ferrer et al., 2010; Ono, Ono, Mizoguchi, et al., 2014; Ono, Ono, Nagasawa, & Kronenberg, 2014), *Prx1-Cre* (Ding et al., 2012; Greenbaum et al., 2013), *LepR-Cre* (Zhou, Yue, Murphy, Peyer, & Morrison, 2014), *Osterix-CreER* (Liu et al., 2013; Mizoguchi et al., 2014; Ono, Ono, Mizoguchi, et al., 2014; Ono, Ono, Nagasawa, et al., 2014), *Gli1-CreER* (Shi et al., 2017), *Col2-CreER; Sox9-CreER; Acan-CreER* (Ono, Ono, Mizoguchi, et al., 2014; Ono, Ono, Nagasawa, et al., 2014), *Grem1-CreER* (Greenbaum et al., 2013).

Mizoguchi, et al., 2014; Ono, Ono, Nagasawa, et al., 2014)]. Interestingly, contribution to adipocytes was not reported following adult-lineage induction, and no contribution to any skeletal, mesenchymal compartments was observed from postnatal *Nestin*-lineage marked cells (Ono, Ono, Mizoguchi, et al., 2014; Ono, Ono, Nagasawa, et al., 2014). This apparent contradiction in multi-lineage capacity of *Nestin*-GFP expressing stromal cells and *Nestin*-lineage marked cells has not been resolved, but the assays in which cells have been placed *in vitro* and subsequently transplanted may not accurately predict *in vivo* functions. It is important to note that *Nestin*-GFP is also expressed in endothelial cells within bone marrow, and *Nestin*-CreER lineage tracing reveals substantial contributions to the endothelial lineage particularly from fetal and early postnatal stages (Ono, Ono, Nagasawa, et al., 2014). Curiously, several *Nestin* transgenes have been generated and they do not identify the same populations. Cells expressing the *Nestin*-GFP transgene reported in Ding et al. express little endogenous *Nestin*, *Nestin*-CreER, or *Nestin*-mCherry, obfuscating the identity of the marked population (Ding et al., 2012; Ono, Ono, Nagasawa, et al., 2014). *Nestin*-CreER was additionally shown to preferentially label *Nestin*-GFP expressing endothelial cells during developmental stages (Ono, Ono, Nagasawa, et al., 2014). These differences in transgene expression could also explain some of the discrepancies between the clonal *in vivo/in vitro* analyses and the genetic models.

Bone marrow stromal cells expressing cell surface markers PDGFRα/CD51 or Leptin Receptor (LepR) identify a largely overlapping population within the cells marked by *Nestin*-GFP (Pinho et al., 2013; Zhou et al., 2014). The stromal cell population identified by these markers also encompasses the majority of CFU-F in the bone marrow, undergoes tri-lineage differentiation into bone, cartilage and fat *in vitro*, and is capable of self-renewal by CFU-F culture analyses and serial clonal transplantation (Pinho et al., 2013; Zhou et al., 2014). It is important to note that only ~10% of LepR cells form CFU-Fs and only ~10% of those clonal colonies undergo tri-lineage differentiation. The remaining clones show a bias toward osteogenic-forming colonies, highlighting the heterogeneity within this population, at least *in vitro* (Zhou et al., 2014). While CFU-F capacity is not a direct measure of stem cell number, and clonal *in vitro* differentiation experiments are technically challenging as well as potentially complicated by changes that occur in cell culture, the cumulative data suggest that the true skeletal stem cell is only a fraction of the identified population and it is critical that further work resolve these issues *in vivo*.

Additional bone marrow, stromal sub-populations have been identified that enrich for progenitor-enriched MSC activity with varying characteristics. Cells co-expressing PDGFRα and Sca1 represent a very small fraction of the PDGFRα/CD51 population that is highly enriched for CFU-Fs, however they do not contain all CFU-F in the bone marrow (Morikawa et al., 2009; Zhou et al., 2014). PDGFRα/Sca1 cells are clonally capable of tri-lineage differentiation *in vitro* (Morikawa et al., 2009). Stromal cells expressing CXCL12, otherwise known as CAR cells, are bi-potent osteo-adipo progenitors *in vitro* (Ding & Morrison, 2013). CAR cells appear to be mutually exclusive from PDGFRα/Sca1 cells as no CXCL12 expression was measured within the PDGFRα/Sca1 population (Morikawa et al., 2009). Through extensive fractionation of the bone marrow stroma using markers CD51 (AlphaV integrin), CD105, Thy, 6C3 and CD200, cells co-expressing CD51 and CD200 were proposed to sit at the top of a stromal hierarchy giving rise to all seven of the other identified sub-populations (Chan et al., 2015). For these experiments, each sub-population of stromal cells was either clonally cultured *in vitro* or transplanted into the renal capsule *in vivo* and the lineage hierarchy was determined by flow cytometry analysis of the expanded population following a period of colony expansion (Chan et al., 2015). Of note, all of these experiments were performed after culturing cells *ex vivo* prior to further tests. CD51/CD200 cells, termed mouse skeletal stem cells (mSSCs), are multi-potent in culture and in transplantation experiments, and self-renewal was demonstrated through serial transplantation after *ex vivo* cultures (Chan et al., 2015). During development, PDGFRα/Sca1 cells and CD200-positive mSSCs arise as two mutually exclusive sub-populations from within the PDGFRα/CD51 population (Nusspaumer et al., 2017). Most recently, *Hoxa11eGFP*-expressing cells were found to be exclusively identified as a sub-population of PDGFRα/CD51- and LepR-positive bone marrow stromal cells (Rux et al., 2016). The Hoxa11eGFP-positive cells enrich for CFU-F potential over and above PDGFRα/CD51 and LepR populations, suggesting higher progenitor potential, and Hox-positive cells undergo tri-lineage differentiation *in vitro* (Rux et al., 2016). As mentioned above, Hox expression is regionally restricted within the skeleton during development. Using *LepR-Cre* lineage-marked MSCs isolated from different regions of the skeleton, Rux and colleagues demonstrated that Hox expression is exclusive to the *LepR*-lineage marked bone marrow stromal fraction, and that *LepR*-lineage marked cells isolated from different regions of the body express a Hox profile consistent with the Hox expression domains established during embryonic development (Rux et al., 2016). Thus, postnatal and adult MSCs can be regionally defined by their retained Hox expression.

Ubiquitous and inducible lineage-tracing alleles have been used to interrogate the multi-lineage potential of progenitor-enriched MSC populations *in vivo*. Adult bone marrow stromal cells marked by a transgenic Cre driven by the LepR promoter (*LepR-Cre*) correlate with the PDGFRα/CD51 population, and are the major source of bone and fat in the adult mouse (Zhou et al., 2014). *LepR*-lineage marked cells do not give rise to chondrocytes during homeostasis but are capable of giving rise to cartilage within the callus following fracture injury (Mizoguchi et al., 2014; Zhou et al., 2014). Interestingly, *LepR*-lineage marked cells begin to appear around E17.5 but do not make substantial contribution to osteoblasts and osteocytes until at least 2 months of age. The authors conclude that a distinct cell population supports embryonic development and postnatal growth (Mizoguchi et al., 2014; Zhou et al., 2014). It is important to note that *LepR*-Cre is not temporally controlled and, therefore, new cells can be continuously marked by this allele the transgene turns on. This precludes unequivocal identification of a single MSC population or a given time point after initial activation. *Prx1*-Cre, also a non-inducible lineage allele, labels the entirety of the early limb bud mesoderm, and at later stages, gives rise to CAR cells, periosteal cells, and LepR-positive stromal cells allowing the conclusion that adult MSCs arise from cells within the early limb bud mesoderm [Fig. 3 (Ding & Morrison, 2013; Greenbaum et al., 2013)].

Probably the most rigorously investigated, inducible genetic model is *Osterix*-CreER (Liu et al., 2013; Mizoguchi et al., 2014; Ono, Ono, Mizoguchi, et al., 2014; Ono, Ono, Nagasawa, et al., 2014). *Osterix* (*Osx*) is a transcription factor required for early osteoblast differentiation. *Osx*-lineage marked cells behave differently depending on the timing of lineage induction although this lineage, expectedly, predominantly gives rise to osteoblasts at all stages (Fig. 3). Cells marked by *Osx*-CreER at embryonic stages contribute substantially to bone formation during development and postnatal growth, but the lineage-marked population is depleted during adult stages (Liu et al., 2013; Ono, Ono, Mizoguchi, et al., 2014; Ono, Ono, Nagasawa, et al., 2014). Cells lineage-marked by *Gli1*-CreER at E13.5 exhibit similar multi-lineage contribution to the skeleton out to 8 weeks of age, but the long-term persistence of this population was not reported (Shi et al., 2017). When the same *Gli1*-CreER is induced at 1 month of age, lineage-marked cells are solely osteo-lineage restricted progenitors with no contribution to other cell types (Shi et al., 2017).

In contrast, early postnatal *Osx*-lineage marked cells contribute to osteoblasts during growth and adult homeostasis, but unlike induction from embryonic stages, a small number of these cells also persist as skeletal progenitors past 1 year of age (Mizoguchi et al., 2014; Ono, Ono, Nagasawa, et al., 2014). Interestingly, the contribution of postnatal *Osx*-lineage marked cells to bone decreases as the animal ages despite the persistence of a marked stromal population (Ono, Ono, Nagasawa, et al., 2014). Long-lived *Osx*-lineage marked stromal cells are *Nes*tin-GFP-, LepR-, and PDGFRα-positive (Mizoguchi et al., 2014; Ono, Ono, Nagasawa, et al., 2014).

Lineage-tracing of perichondrial/periosteal cells by *Col2*-CreER or chondrocyte lineage cells by *Sox9*-CreER or *Aggrecan*-CreER demonstrate similar capacities to give rise to long-lived bone marrow stromal cell populations from postnatal stages, but not from embryonic or adult induction stages (Ono, Ono, Nagasawa, et al., 2014). Postnatal and adult cells labeled by *Gremlin 1*-CreER (*Grem1*-CreER) also exhibit long-term labeling of bone marrow stroma, but function as lineage-restricted osteochondro progenitors with no contribution to bone marrow adipose cells (Worthley et al., 2015). The embryonic *Grem1*-lineage was not reported beyond birth.

Finally, adult *Osx*-CreER lineage-marked cells are osteo-lineage restricted and only contribute to bone for a short period of time before they are depleted (Ono, Ono, Nagasawa, et al., 2014). Collectively, these data that show different potentials depending on the timing of Cre activation have supported a model for skeletal progenitors whereby embryonic development is supported by early skeletal progenitors that are replaced by definitive skeletal MSCs that arise during postnatal stages (Kfoury & Scadden, 2015; Ono & Kronenberg, 2016).

However, a major consideration with regards to the "two-waves-of-progenitors" model is the nature of the genetic models used in these analyses. The majority of the mouse models analyzed employ inducible CreERs driven by promoters of genes known to be involved in early chondrogenic (Sox9 and Aggrecan) or osteogenic (Osterix) differentiation. As has been shown for many stem cell populations, skeletal stem cells are primed prior to differentiate into multiple lineages, therefore, early commitment factors are expressed at low levels in non-committed progenitors (Méndez-Ferrer et al., 2015). An alternative interpretation of these results is that while most of the cells marked by these inducible CreER alleles will be committed to the cell fates toward which these factors drive differentiation, a few of the most primitive progenitors at some stages could be labeled, leading to a small number of lineage-labeled, self-renewing progenitors.

Recently, we generated a *Hoxa11*-CreER^{T2} model that, in contrast to the *Waves of Progenitors* model, provides data that supports the existence of a lineage-continuous skeletal stem cell population originating from embryonic development. Similar to *Osx*-CreER experiments, the *Hoxa11* lineage was marked from embryonic, postnatal stages or adult stages out to at least 1 year of age and lineage-marked cells contribute to all three skeletal mesenchymal lineages from all times points and also give rise to PDGFRα/CD51 and LepR-positive bone marrow stromal cells that persist (self-renew) throughout the life of the animal (unpublished data). The previous demonstration at embryonic and adult stages that Hox11-expressing cells do not mark any differentiated population (Rux et al., 2016; Swinehart et al., 2013) the continuous contribution of the *Hoxa11CreERT*-lineage to differentiated skeletal cells, and the persistence of lineage-marked cells that remain progenitor-enriched MSCs throughout the life of the animal provide strong evidence that Hox-expressing cells are a regionally restricted, long-lived, self-renewing skeletal stem/progenitor population *in vivo*.

5.2 Tendon stem/progenitor cells

Tendon is a less regenerative tissue and injury often results in the formation of scar tissue. Thus, it was originally believed that tendons were devoid of progenitor cells, however recently, tendon stem/progenitor cells (TSCPs) have been isolated and characterized. Putative tendon stem cells, harvested from patellar tendon, were isolated based on their ability to adhere to plastic, form CFU-Fs and multi-lineage differentiation into bone, cartilage, and fat *in vivo* (Bi et al., 2007). TSCPs express several common progenitor/stem cell markers in common with progenitor-enriched bone marrow MSCs including Sca1, CD44, and CD105 (Bi et al., 2007). In contrast, semi-quantitative PCR demonstrated that unlike bone marrow stromal cells, TSPCs expressed genes associated with tendon identity as well, such as *Scx*, *Tenomodulin*, and *Mohawk* (Liu et al., 2015; Tan, Lui, Rui, & Wong, 2012). It is important to note that these data do not rule out a shared stem/progenitor population for tendon and skeleton; the niche environment may influence the gene expression profiles in the tissue-resident stem/progenitor cells. The authors of this study used cultured cells as opposed to freshly isolated cells for these comparative studies and it is known that cell surface expression can change substantially during cell culture, potentially masking important differences between these populations (Mark et al., 2013). Thus, direct comparison of freshly isolated TSPCs and bone marrow stromal cells may yield different results pertaining to the comparisons of the two populations. Further characterization of these cells and comparisons to skeletal MSCs are critical.

The lack of a unique surface markers for TSPCs prohibits unequivocal identification of this population *in vivo*. Recently, cells marked by αSMA-CreER in the juvenile tendon were found to contribute to the tendon during postnatal tendon growth (Dyment et al., 2014). This population was also demonstrated to be the main contributor to tendon injury repair. *In vitro*, progenitor/stem cells from both skeletal and tendon tissue exhibit tri-lineage, supporting similarities between the two progenitor populations (Rux et al., 2016; Zhang & Wang, 2010). Using a *Hoxa11eGFP* real-time reporter allele, Hoxa11 expression was shown to overlap with tenacin C within the developing tendons in the limb and this expression continues well into adulthood (Fig. 2; Swinehart et al., 2013). As Hox11 expression also identifies skeletal MSCs, it will be interesting to investigate the relationship of Hox-expressing cells and identified tendon progenitor markers.

5.3 Muscle: Satellite cells and stromal progenitors

Skeletal muscle has a remarkable capacity for regeneration. Various studies have established that *Pax3* and *Pax7* mark muscle progenitors in the developing embryo (Buckingham, 2007; Relaix, Rocancourt, Mansouri, & Buckingham, 2004). *Pax3* expression is first detected at E8.0 in the presomitic mesoderm as the somites form, but becomes progressively restricted to the dermomyotomal lip of the somite. *Pax3* expression is maintained in early migrating muscle progenitors, however expression is lost by E13.5 (Relaix et al., 2006). *Pax7* expression initiates in myoblasts around E11.5 and, at adult stages, Pax7 labels all satellite cells (Relaix et al., 2004; Seale et al., 2000). Extensive evidence supports Pax7-positive satellite cells as adult muscle stem cells that contribute to homeostasis and muscle regeneration. Following injury, satellite cells undergo rapid proliferation and differentiation into new muscle fibers and also self-renew to maintain a stem cell pool (Collins & Partridge, 2005). These cell are located under the basal lamina, adjacent to the myofibrils (Mauro, 1961). Both genetic lineage-tracing studies and transplantation models establish the ability of these cells to give rise to regenerating myofibers *in vivo* (Chen, Lin, & Slack, 2006; Collins & Partridge, 2005; Lepper et al., 2009; Lepper & Fan, 2010; Shea et al., 2010). Genetic ablation of Pax7-positive satellite cells using *Pax7*-CreER-mediated, diphtheria toxin expression (R26R-DTA) results in complete failure of muscle regeneration in response to injury, confirming that satellite cells are necessary for muscle repair (Lepper, Partridge, & Fan, 2011; Murphy et al., 2011; Sambasivan et al., 2011). That satellite cells are the sole muscle stem cell is not supported, however, by the demonstration that genetic ablation of satellite cells does not lead to accelerated muscle sarcopenia in aging mice, suggesting that, while satellite cells are required for muscle regeneration, they may not contribute to muscle homeostasis (Fry et al., 2015; Keefe et al., 2015; McCarthy et al., 2011).

Numerous subpopulations of mesenchymal cells within the muscle that are distinct from satellite cells have been found capable of contributing to muscle culture experiments. One function of these cells *in vivo* is to support muscle function and regeneration in a non-myogenic capacity. Fibroadipogenic progenitors (FAPs), named for their role as the main mediator of fibrotic and fatty tissue accumulation in the muscles of aged or muscular dystrophy patients, reside within the interstitial space of the muscle (Joe et al., 2010; Uezumi et al., 2010). FAPs greatly expand following muscle injury and express factors that are known to favor myogenic differentiation such as IGF-1, IL-6, Wnt1, Wnt3A and Wnt5A (Bodine et al., 2001; Brack et al., 2007; Serrano et al., 2008). It is presumed that these cells do not contribute to muscle cells directly, but this has not been tested due to lack of genetic tools to specifically mark these population. Thus, it is also unknown how muscle fiber regeneration would be affected during regeneration in the absence of FAPs. The vast majority of FAPs express the marker PDGFRα, which is also expressed by *Tcf4*-expressing fibroblasts within the muscle (Joe et al., 2010; Uezumi et al., 2010). From this, it can be speculated that FAPs could be an overlapping population with *Tcf4*-expressing cells. Genetic ablation of the muscle connective tissue population marked by *Tcf4* has been performed and the presence of this population was demonstrated to be crucial to effectively undergo muscle regeneration (Murphy et al., 2011). In the absence of Tcf4-positive fibroblasts, satellite cells prematurely differentiated into myoblasts resulting in depletion of the stem cell pool and smaller regenerated fibers (Murphy et al., 2011). These data demonstrate that adult muscle connective tissue fibroblasts function in analogous mechanism to muscle connective tissue cells during development to promote muscle fiber organization and fiber type identity during muscle regeneration.

Recent and more controversial evidence for a role for non-satellite mesenchymal cells contributing to new myofibers, challenges the dogma of the satellite cell as the only muscle stem cell. The first report to challenge this concept *in vitro* showed that PWI+/PAX7− interstitial cells (PICs) were muscle-resident, Pax7-negative cell localized to the muscle interstitium, capable of myogenic differentiation *in vitro* and contributing to new myofibers in transplantation following injury (Kuang, Gillespie, & Rudnicki, 2008; Mitchell et al., 2010).

In a more recent study, Twist2 reporter expression was shown to identify a non-satellite cell, mesenchymal, interstitial population with myogenic potential *in vivo* (Liu et al., 2017). It is unclear if this population is distinct from PW1 cells as no enrichment for PW1 was observed by flow cytometry and instead high levels of Sca1 and PDGFRα were observed. Lineage-tracing experiments using an inducible *Twist2*-CreER model provided clear genetic evidence of contribution of a non-satellite mesenchymal cell to mature muscle fibers at homeostasis *in vivo* (Liu et al., 2017). Interestingly, *Twist2*-lineage positive cells were reported to only contribute to type II muscle fibers, specifically type IIb and type IIx, suggesting that this population may represent a fiber-type specific progenitor in muscle. Ablation of the *Twist2*-lineage marked cells using a *Twist2-CreER;Rosa-DTA* system resulted in the atrophy of only type II myofibers, supporting this assertion. Open questions remain on whether there are additional progenitor sub-populations present within the adult muscle that are critical for the support of other types of myofibers. Importantly, Twist2+ cells are not descendants of Pax7-positive satellite cells, as linage-tracing studies revealed that the Pax7-lineage cells were distinct from Twist2-expressing cells (Liu et al., 2017). Myogenic contribution by non-satellite cells is a surprising area of new investigation with many questions remaining, including how the identified mesenchymal populations are related and if specific sub-populations have unique functional roles at homeostasis and following injury.

6. Conclusion/perspective

Recent work using more sophisticated genetic alleles that allow real-time reporting, conditional loss-of-function as well as temporally controlled lineage-tracing and cell ablation studies have greatly expanded our knowledge of musculoskeletal biology after initial developmental stages. In all three tissues, bone, tendon and muscle, many of the same key, developmental regulators are important in tissue maintenance and in repair/regeneration after injury, however, there are some critical differences that should be areas of focus for continuing investigation. The ability of the tendon to heal at postnatal stages as well as identification of potential stem/progenitor populations suggests that the inability to heal at adult stages might be overcome by prevention of scar tissue formation. The apparent differences in cells that can contribute to adult muscle regeneration from embryonic to adult stages is even more unclear. Whether the changes in potential of the cell populations at adult stages is functionally relevant is one key question.

With regards to the progenitor populations in the skeleton, the similarities and differences between stromal populations in the periosteum and bone marrow requires further investigation. It is known that the bone marrow cells originate from the perichondrium, however, how and when the perichondrial cells infiltrate the bone marrow space is unknown. Also, there is a lack of knowledge regarding which compartment is responsible for skeletal homeostasis at adult stages and whether these cells differ in their expression profiles upon spatial separation. Future experiments examining the compartment-specific contribution to development, homeostasis, and injury repair/regeneration and investigation into potential cellular distinctions would be imperative in progressing or knowledge in skeletal biology. Layering on top of that, is the relatively new information that there are regional differences in MSC populations, at least with respect to Hox expression, leads to more questions regarding differential potentials in these sub-populations.

Probably the single most important area that is in need of further exploration is a comparison of the various "stem/progenitor" cell populations from all three tissues—bone, tendon and muscle. Careful bulk analyses as well as single cell analyses, more specific genetic tools, and critically, studies that directly compare populations that have been identified *via* different methods are critically needed. This will require investigators sharing tools and knowledge with the ultimate goal of placing current pieces of knowledge into a more coherent whole. It seems the time is ripe for program level activities that include investigators with expertise in each of the three tissue types to work collaboratively to define the similarities and differences between various stem/progenitor cell populations.

References

Abad, V., Meyers, J. L., Weise, M., Gafni, R. I., Barnes, K. M., Nilsson, O., et al. (2002). The role of the resting zone in growth plate chondrogenesis. *Endocrinology*, *143*(5), 1851–1857. https://doi.org/10.1210/endo.143.5.8776.

Akiyama, H. (2008). Control of chondrogenesis by the transcription factor Sox9. *Modern Rheumatology*, *18*(3), 213–219. https://doi.org/10.1007/s10165-008-0048-x.

Akiyama, H., Chaboissier, M. C., Martin, J. F., Schedl, A., & De Crombrugghe, B. (2002). The transcription factor *Sox9* has essential roles in successive steps of the chondrocyte differentiation pathway and is required for expression of *Sox5* and *Sox6*. *Genes & Development*, *16*, 2813–2828. https://doi.org/10.1101/gad.1017802.

Akiyama, H., Kim, J.-E., Nakashima, K., Balmes, G., Iwai, N., Deng, J. M., et al. (2005). Osteo-chondroprogenitor cells are derived from Sox9 expressing precursors. *Proceedings of the National Academy of Sciences of the United States of America*, *102*(41), 14665–14670. https://doi.org/10.1073/pnas.0504750102.

Akiyama, H., Stadler, H. S., Martin, J. F., Ishii, T. M., Beachy, P. A., Nakamura, T., et al. (2007). Misexpression of Sox9 in mouse limb bud mesenchyme induces polydactyly and rescues hypodactyly mice. *Matrix Biology*, *26*(4), 224–233. https://doi.org/10.1016/j.matbio.2006.12.002.

Alberton, P., Dex, S., Popov, C., Shukunami, C., Schieker, M., & Docheva, D. (2015). Loss of tenomodulin results in reduced self-renewal and augmented senescence of tendon stem/progenitor cells. *Stem Cells and Development*, *24*, 597–609. https://doi.org/10.1089/scd.2014.0314.

Beier, F. (2005). Cell-cycle control and the cartilage growth plate. *Journal of Cellular Physiology*, *202*(1), 1–8. https://doi.org/10.1002/jcp.20111.

Bi, W., Deng, J. M., Zhang, Z., Behringer, R. R., & De Crombrugghe, B. (1999). *Sox9* is required for cartilage formation. *Nature Genetics*, *22*(1), 85–89. https://doi.org/10.1038/8792.

Bi, Y., Ehirchiou, D., Kilts, T. M., Inkson, C. A., Embree, M. C., Sonoyama, W., et al. (2007). Identification of tendon stem/progenitor cells and the role of the extracellular matrix in their niche. *Nature Medicine*, *13*(10), 1219–1227. https://doi.org/10.1038/nm1630.

Bi, W., Huang, W., Whitworth, D. J., Deng, J. M., Zhang, Z., Behringer, R. R., et al. (2001). Haploinsufficiency of Sox9 results in defective cartilage primordia and premature skeletal mineralization. *Proceedings of the National Academy of Sciences*, *98*(12), 6698–6703. http://doi.org/10.1073/pnas.111092198.

Bianco, P., Cao, X., Frenette, P. S., Mao, J. J., Robey, P. G., Simmons, P. J., et al. (2013). The meaning, the sense and the significance: Translating the science of mesenchymal stem cells into medicine. *Nature Medicine*, *19*(1), 35–42. https://doi.org/10.1038/nm.3028.

Bodine, S. C., Stitt, T. N., Gonzalez, M., Kline, W. O., Stover, G. L., Bauerlein, R., et al. (2001). Akt/mTOR pathway is a crucial regulator of skeletal muscle hypertrophy and can prevent muscle atrophy in vivo. *Nature Cell Biology*, *3*(11), 1014–1019. https://doi.org/10.1038/ncb1101-1014.

Bonnin, M. A., Laclef, C., Blaise, R., Eloy-Trinquet, S., Relaix, F., Maire, P., et al. (2005). Six1 is not involved in limb tendon development, but is expressed in limb connective tissue under Shh regulation. *Mechanisms of Development*, *122*(4), 573–585. https://doi.org/10.1016/j.mod.2004.11.005.

Brack, A. S., Conboy, M. J., Roy, S., Lee, M., Kuo, C. J., Keller, C., et al. (2007). Increased Wnt signaling during aging alters muscle stem cell fate and increases fibrosis. *Science*, *317*, 807–810. https://doi.org/10.1038/020493a0.

Brent, A. E., Braun, T., & Tabin, C. J. (2005). Genetic analysis of interactions between the somitic muscle, cartilage and tendon cell lineages during mouse development. *Development*, *132*(3), 515–528. https://doi.org/10.1242/dev.01605.

Brent, A. E., Schweitzer, R., & Tabin, C. J. (2003). A somitic compartment of tendon progenitors. *Cell*, *113*(2), 235–248. https://doi.org/10.1016/S0092-8674(03)00268-X.

Buckingham, M. (2007). Skeletal muscle progenitor cells and the role of Pax genes. *Comptes Rendus Biologies*, *330*(6–7), 530–533. https://doi.org/10.1016/j.crvi.2007.03.015.

Chan, C. K. F., Seo, E. Y., Chen, J. Y., Lo, D., Mcardle, A., Sinha, R., et al. (2015). Identification and specification of the mouse skeletal stem cell. *Cell*, *160*(1–2), 285–298. https://doi.org/10.1016/j.cell.2014.12.002.

Charge, S. B. P., & Rudnicki, M. A. (2004). Cellular and molecular regulation of muscle regeneration. *Physiological Reviews*, *84*(1), 209. https://doi.org/10.1152/physrev.00019.2003.
Chen, H., & Johnson, R. L. (2002). Interactions between dorsal-ventral patterning genes lmx1b, engrailed-1 and wnt-7a in the vertebrate limb. *International Journal of Developmental Biology*, *46*(7), 937–941. https://doi.org/10.1387/IJDB.12455631.
Chen, K. G., Johnson, K. R., & Robey, P. G. (2017). Mouse genetic analysis of bone marrow stem cell niches: Technological pitfalls, challenges, and translational considerations. *Stem Cell Reports*, *9*(5), 1343–1358. https://doi.org/10.1016/j.stemcr.2017.09.014.
Chen, Y., Lin, G., & Slack, J. M. W. (2006). Control of muscle regeneration in the Xenopus tadpole tail by Pax7. *Development*, *133*(12), 2303–2313. https://doi.org/10.1242/dev.02397.
Chen, H., Lun, Y., Ovchinnikov, D., Kokubo, H., Oberg, K. C., Pepicelli, C. V., et al. (1998). Limb and kidney defects in Lmx1b mutant mice suggest an involvement of LMX1B in human nail patella syndrome. *Nature Genetics*, *19*(1), 51–55. https://doi.org/10.1038/ng0598-51.
Chevallier, A., Kieny, M., & Mauger, A. (1977). Limb-somite relationship: Origin of the limb musculature. *Journal of Embryology and Experimental Morphology*, *41*(1951), 245–258.
Christ, B., Jacob, H. J., & Jacob, M. (1977). Experimental analysis of the origin of the wing musculature in avian embryos. *Anatomy and Embryology*, *150*(2), 171–186. https://doi.org/10.1007/BF00316649.
Christ, B., & Ordahl, C. P. (1995). Early stages of chick somite development. *Anatomy and Embryology*, *191*(5), 381–396. https://doi.org/10.1007/BF00304424.
Collins, C. A., Olsen, I., Zammit, P. S., Heslop, L., Petrie, A., Partridge, T. A., et al. (2005). Stem cell function, self-renewal, and behavioral heterogeneity of cells from the adult muscle satellite cell niche. *Cell*, *122*(2), 289–301. https://doi.org/10.1016/j.cell.2005.05.010.
Collins, C. A., & Partridge, T. A. (2005). Self-renewal of the adult skeletal muscle satellite cell. *Cell Cycle*, *4*(10), 1338–1341. https://doi.org/10.4161/cc.4.10.2114.
Colnot, C. (2009). Skeletal cell fate decisions within periosteum and bone marrow during bone regeneration. *Journal of Bone and Mineral Research*, *24*(2), 274–282. https://doi.org/10.1359/jbmr.081003.
Davis, A. P., Witte, D. P., Hsieh-Li, H. M., Potter, S. S., & Capecchi, M. R. (1995). Absence of radius and ulna in mice lacking hoxa-11 and hoxd-11. *Nature*, *375*(6534), 791–795. https://doi.org/10.1038/375791a0.
Deschaseaux, F., Pontikoglou, C., & Sensébé, L. (2010). Bone regeneration: The stem/progenitor cells point of view. *Journal of Cellular and Molecular Medicine*, *14*(1–2), 103–115. https://doi.org/10.1111/j.1582-4934.2009.00878.x.
Dey, D., Goldhamer, D. J., & Yu, P. B. (2015). Contributions of muscle-resident progenitor cells to homeostasis and disease. *Current Molecular Biology Reports*, *1*(4), 175–188. https://doi.org/10.1007/s40610-015-0025-z.
Ding, L., & Morrison, S. J. (2013). Haematopoietic stem cells and early lymphoid progenitors occupy distinct bone marrow niches. *Nature*, *495*(7440), 231–235. https://doi.org/10.1038/nature11885.
Ding, L., Saunders, T. L., Enikolopov, G., & Morrison, S. J. (2012). Endothelial and perivascular cells maintain haematopoietic stem cells. *Nature*, *481*(7382), 457–462. https://doi.org/10.1038/nature10783.
Docheva, D., Hunziker, E. B., Fassler, R., & Brandau, O. (2005). Tenomodulin is necessary for tenocyte proliferation and tendon maturation. *Molecular and Cellular Biology*, *25*(2), 699–705. https://doi.org/10.1128/MCB.25.2.699-705.2005.
Docheva, D., Müller, S. A., Majewski, M., & Evans, C. H. (2015). Biologics for tendon repair. *Advanced Drug Delivery Reviews*, *84*, 222–239. https://doi.org/10.1016/j.addr.2014.11.015.

Duchamp de Lageneste, O., Julien, A., Abou-Khalil, R., Frangi, G., Carvalho, C., Cagnard, N., et al. (2018). Periosteum contains skeletal stem cells with high bone regenerative potential controlled by Periostin. *Nature Communications*, *9*(1), 773. https://doi.org/10.1038/s41467-018-03124-z.

Dyment, N. A., Hagiwara, Y., Matthews, B. G., Li, Y., Kalajzic, I., & Rowe, D. W. (2014). Lineage tracing of resident tendon progenitor cells during growth and natural healing. *PLoS One*, *9*(4), e96113. https://doi.org/10.1371/journal.pone.0096113.

Dyment, N. A., Liu, C. F., Kazemi, N., Aschbacher-Smith, L. E., Kenter, K., Breidenbach, A. P., et al. (2013). The paratenon contributes to scleraxis-expressing cells during patellar tendon healing. *PLoS One*, *8*(3). https://doi.org/10.1371/journal.pone.0059944.

Fernandez-Teran, M., & Ros, M. A. (2008). The apical ectodermal ridge: Morphological aspects and signaling pathways. *International Journal of Developmental Biology*, *52*, 857–871. https://doi.org/10.1387/ijdb.072416mf.

Friedenstein, A. J., Chailakhjan, R. K., & Lalykina, K. S. (1970). The development of fibroblast colonies in monolayer cultures of guinea-pig bone marrow and spleen cells. *Cell and Tissue Kinetics*, *3*, 393–403.

Friedenstein, A. J., Gorskaja, J. F., & Kulagina, N. N. (1976). Fibroblast precursors in normal and irradiated mouse hematopoietic organs. *Experimental Hematology*, *4*(5), 267–274.

Friedenstein, A. J., Latzinik, N. W., Grosheva, A. G., & Gorskaya, U. F. (1982). Marrow microenvironment transfer by heterotopic transplantation of freshly isolated and cultured cells in porous sponges. *Experimental Hematology*, *10*(2), 217–227.

Friedenstein, A. J., Petrakova, K. V., Kurolesova, A. I., & Frolova, G. P. (1968). Heterotopic transplants of bone marrow. *Transplantation*, *10*(5), 557–567. https://doi.org/10.1097/00007890-196803000-00009.

Frolova, E. G., Drazba, J., Krukovets, I., Kostenko, V., Blech, L., Harry, C., et al. (2014). Control of organization and function of muscle and tendon by thrombospondin-4. *Matrix Biology*, *37*, 35–48. https://doi.org/10.1016/j.matbio.2014.02.003.

Fromental-Ramain, C., Warot, X., Lakkaraju, S., Favier, B., Haack, H., Birling, C., et al. (1996). Specific and redundant functions of the paralogous Hoxa-9 and Hoxd-9 genes in forelimb and axial skeleton patterning. *Development (Cambridge, England)*, *122*(2), 461–472.

Fromental-Ramain, C., Warot, X., Messadecq, N., LeMeur, M., Dollé, P., & Chambon, P. (1996). Hoxa-13 and Hoxd-13 play a crucial role in the patterning of the limb autopod. *Development (Cambridge, England)*, *122*(10), 2997–3011.

Fry, C. S., Lee, J. D., Mula, J., Kirby, T. J., Jackson, J. R., Liu, F., et al. (2015). Inducible depletion of satellite cells in adult, sedentary mice impairs muscle regenerative capacity without affecting sarcopenia. *Nature Medicine*, *21*(1), 76–80. https://doi.org/10.1038/nm.3710.

Gerstenfeld, L. C., Cullinane, D. M., Barnes, G. L., Graves, D. T., & Einhorn, T. A. (2003). Fracture healing as a post-natal developmental process: Molecular, spatial, and temporal aspects of its regulation. *Journal of Cellular Biochemistry*, *88*(5), 873–884. https://doi.org/10.1002/jcb.10435.

Govindarajan, V., & Overbeek, P. A. (2006). FGF9 can induce endochondral ossification in cranial mesenchyme. *BMC Developmental Biology*, *6*, 1–14. https://doi.org/10.1186/1471-213X-6-7.

Greenbaum, A., Hsu, Y.-M. S., Day, R. B., Schuettpelz, L. G., Christopher, M. J., Borgerding, J. N., et al. (2013). CXCL12 in early mesenchymal progenitors is required for haematopoietic stem-cell maintenance. *Nature*, *495*(7440), 227–230. https://doi.org/10.1038/nature11926.

Haack, H., & Gruss, P. (1993). The establishment of murine Hox-1 expression domains during patterning of the limb. *Developmental Biology*, *157*(2), 410–422. https://doi.org/10.1006/dbio.1993.1145.

Han, Y., & Lefebvre, V. (2008). L-Sox5 and Sox6 drive expression of the aggrecan gene in cartilage by securing binding of Sox9 to a far-upstream enhancer. *Molecular and Cellular Biology*, *28*(16), 4999–5013. https://doi.org/10.1128/MCB.00695-08.

Havis, E., Bonnin, M.-A., Esteves de Lima, J., Charvet, B., Milet, C., & Duprez, D. (2016). TGFβ and FGF promote tendon progenitor fate and act downstream of muscle contraction to regulate tendon differentiation during chick limb development. *Development*, *143*(20), 3839–3851. http://doi.org/10.1242/dev.136242.

Hilton, M. J. (2014). Skeletal development and repair: Methods and protocols. *Vol. 1130*. Humana Press. https://doi.org/10.1007/978-1-62703-989-5.

Hope, M., & Saxby, T. S. (2007). Tendon healing. *Foot and Ankle Clinics*, *12*(4), 553–567. https://doi.org/10.1016/j.fcl.2007.07.003.

Howell, K., Chien, C., Bell, R., Laudier, D., Tufa, S. F., Keene, D. R., et al. (2017). Novel model of tendon regeneration reveals distinct cell mechanisms underlying regenerative and fibrotic tendon healing. *Scientific Reports*, 7(February), 1–14. https://doi.org/10.1038/srep45238.

Huang, A. H., Lu, H. H., & Schweitzer, R. (2015). Molecular regulation of tendon cell fate during development. *Journal of Orthopaedic Research*, *33*(6), 800–812. http://doi.org/10.1002/jor.22834.

Huang, A. H., Riordan, T. J., Pryce, B., Weibel, J. L., Watson, S. S., Long, F., et al. (2015). Musculoskeletal integration at the wrist underlies modular development of limb tendons. *Development*, *142*(14), 2431–2441. https://doi.org/10.1242/dev.122374.

Huang, A. H., Riordan, T. J., Wang, L., Eyal, S., Zelzer, E., Brigande, J. V., et al. (2013). Repositioning forelimb superficialis muscles: Tendon attachment and muscle activity enable active relocation of functional myofibers. *Developmental Cell*, *26*(5), 544–551. https://doi.org/10.1016/ej.devcel.2013.08.007.

Hunziker, E. B. (1994). Mechanism of longitudinal bone growth and its regulation by growth plate chondrocytes. *Microscopy Research and Technique*, *28*(6), 505–519. https://doi.org/10.1002/jemt.1070280606.

Hurme, T., Kalimo, H., Lehto, M., & Järvinen, M. (1991). Healing of skeletal muscle injury: An ultrastructural and immunohistochemical study. *Medicine and Science in Sports and Exercise*, *23*(7), 801–810. Retrieved from http://www.ncbi.nlm.nih.gov/pubmed/1921672.

Isern, J., García-García, A., Martín, A. M., Arranz, L., Martín-Pérez, D., Torroja, C., et al. (2014). The neural crest is a source of mesenchymal stem cells with specialized hematopoietic stem cell niche function. *eLife*, *3*(e03696), 1–28. https://doi.org/10.7554/eLife.03696.

Ito, Y., Toriuchi, N., Yoshitaka, T., Ueno-Kudoh, H., Sato, T., Yokoyama, S., et al. (2010). The Mohawk homeobox gene is a critical regulator of tendon differentiation. *Proceedings of the National Academy of Sciences of the United States of America*, *107*(23), 10538–10542. https://doi.org/10.1073/pnas.1000525107.

Järvinen, T. A. H., Järvinen, M., & Kalimo, H. (2013). Regeneration of injured skeletal muscle after the injury. *Muscles, Ligaments and Tendons Journal*, *3*(4), 337–345. http://doi.org/10.11138/mltj/2013.3.4.337.

Jelinsky, S. A., Rodeo, S. A., Li, J., Gulotta, L. V., Archambault, J. M., & Seeherman, H. J. (2011). Regulation of gene expression in human tendinopathy. *BMC Musculoskeletal Disorders*, *12*, 86. https://doi.org/10.1186/1471-2474-12-86.

Joe, A. W. B., Yi, L., Natarajan, A., Le Grand, F., So, L., Wang, J., et al. (2010). Muscle injury activates resident fibro/adipogenic progenitors that facilitate myogenesis. *Nature Cell Biology*, *12*(2), 153–163. https://doi.org/10.1038/ncb2015.

Kardon, G. (1998). Muscle and tendon morphogenesis in the avian hind limb. *Development (Cambridge, England)*, *125*(20), 4019–4032.

Kardon, G., Campbell, J. K., & Tabin, C. J. (2002). Local extrinsic signals determine muscle and endothelial cell fate and patterning in the vertebrate limb. *Developmental Cell*, *3*(4), 533–545. https://doi.org/10.1016/S1534-5807(02)00291-5.

Kardon, G., Harfe, B. D., & Tabin, C. J. (2003). A Tcf4-positive mesodermal population provides a prepattern for vertebrate limb muscle patterning. *Developmental Cell*, *5*(6), 937–944. https://doi.org/10.1016/S1534-5807(03)00360-5.

Karsenty, G., & Wagner, E. F. (2002). Reaching a genetic and molecular understanding of skeletal development. *Developmental Cell*, *2*(4), 389–406. https://doi.org/10.1016/S1534-5807(02)00157-0.

Keefe, A. C., Lawson, J. A., Flygare, S. D., Fox, Z. D., Colasanto, M. P., Mathew, S. J., et al. (2015). Muscle stem cells contribute to myofibres in sedentary adult mice. *Nature Communications*, *6*, 7087. https://doi.org/10.1038/ncomms8087.

Kfoury, Y., & Scadden, D. T. (2015). Mesenchymal cell contributions to the stem cell niche. *Cell Stem Cell*, *16*(3), 239–253. https://doi.org/10.1016/j.stem.2015.02.019.

Kieny, M., & Chevallier, A. (1979). Autonomy of tendon development in the embryonic chick wing. *Journal of Embryology and Experimental Morphology*, *49*, 153–165.

Koda, N., Sato, T., Shinohara, M., Ichinose, S., Ito, Y., Nakamichi, R., et al. (2017). The transcription factor Mohawk homeobox regulates homeostasis of the periodontal ligament. *Development*, *144*(2), 313–320. https://doi.org/10.1242/dev.135798.

Kronenberg, H. M. (2007). The role of the perichondrium in fetal bone development. *Annals of the New York Academy of Sciences*, *1116*, 59–64. https://doi.org/10.1196/annals.1402.059.

Kuang, S., Gillespie, M. A., & Rudnicki, M. A. (2008). Niche regulation of muscle satellite cell self-renewal and differentiation. *Cell Stem Cell*, *2*, 22–31. https://doi.org/10.1016/j.stem.2007.12.012.

Kunisaki, Y., Bruns, I., Scheiermann, C., Ahmed, J., Pinho, S., Zhang, D., et al. (2013). Arteriolar niches maintain haematopoietic stem cell quiescence. *Nature*, *502*(7473), 637–643. https://doi.org/10.1038/nature12612.

Kyriakides, T. R., Zhu, Y. H., Smith, L. T., Bain, S. D., Yang, Z., Lin, M. T., et al. (1998). Mice that lack thrombospondin 2 display connective tissue abnormalities that are associated with disordered collagen fibrillogenesis, an increased vascular density, and a bleeding diathesis. *Journal of Cell Biology*, *140*, 419–430. https://doi.org/10.1083/jcb.140.2.419.

Lefebvre, V., Li, P., & De Crombrugghe, B. (1998). A new long form of Sox5 (L-Sox5), Sox6 and Sox9 are coexpressed in chondrogenesis and cooperatively activate the type II collagen gene. *The EMBO Journal*, *17*(19), 5718–5733. https://doi.org/10.1093/emboj/17.19.5718.

Lepper, C., Conway, S. J., & Fan, C. M. (2009). Adult satellite cells and embryonic muscle progenitors have distinct genetic requirements. *Nature*, *460*(7255), 627–631. https://doi.org/10.1038/nature08209.

Lepper, C., & Fan, C. M. (2010). Inducible lineage tracing of Pax7-descendant cells reveals embryonic origin of adult satellite cells. *Genesis*, *48*(7), 424–436. https://doi.org/10.1002/dvg.20630.

Lepper, C., Partridge, T. A., & Fan, C.-M. (2011). An absolute requirement for Pax7-positive satellite cells in acute injury-induced skeletal muscle regeneration. *Development*, *138*(17), 3639–3646. https://doi.org/10.1242/dev.067595.

Liu, N., Garry, G. A., Li, S., Bezprozvannaya, S., Sanchez-Ortiz, E., Chen, B., et al. (2017). A Twist2-dependent progenitor cell contributes to adult skeletal muscle. *Nature Cell Biology*, *19*, 202–213. https://doi.org/10.1038/ncb3477.

Liu, C. F., & Lefebvre, V. (2015). The transcription factors SOX9 and SOX5/SOX6 cooperate genome-wide through super-enhancers to drive chondrogenesis. *Nucleic Acids Research*, *43*(17), 8183–8203. https://doi.org/10.1093/nar/gkv688.

Liu, Y., Strecker, S., Wang, L., Kronenberg, M. S., Wang, W., Rowe, D. W., et al. (2013). Osterix-Cre labeled progenitor cells contribute to the formation and maintenance of the bone marrow stroma. *PLoS One*, *8*(8), e71318. https://doi.org/10.1371/journal.pone.0071318.

Liu, W., Watson, S. S., Lan, Y., Keene, D. R., Ovitt, C. E., Liu, H., et al. (2010). The atypical homeodomain transcription factor Mohawk controls tendon morphogenesis. *Molecular and Cellular Biology*, *30*, 4797–4807. https://doi.org/10.1128/MCB.00207-10.

Liu, H., Zhang, C., Zhu, S., Lu, P., Zhu, T., Gong, X., et al. (2015). Mohawk promotes the tenogenesis of mesenchymal stem cells through activation of the TGFβ signaling pathway. *Stem Cells (Dayton, Ohio)*, *33*(2), 443–455. https://doi.org/10.1002/stem.1866.

Liu, H., Zhu, S., Zhang, C., Lu, P., Hu, J., Yin, Z., et al. (2014). Crucial transcription factors in tendon development and differentiation: Their potential for tendon regeneration. *Cell and Tissue Research*, *356*(2), 287–298. https://doi.org/10.1007/s00441-014-1834-8.

Maes, C., Kobayashi, T., Selig, M. K., Torrekens, S., Roth, S. I., Mackem, S., et al. (2010). Osteoblast precursors, but not mature osteoblasts, move into developing and fractured bones along with invading blood vessels. *Developmental Cell*, *19*(2), 329–344. https://doi.org/10.1016/j.devcel.2010.07.010.

Mallo, M., & Alonso, C. R. (2013). The regulation of Hox gene expression during animal development. *Development*, *140*(19), 3951–3963. https://doi.org/10.1242/dev.068346.

Mallo, M., Vinagre, T., & Carapuço, M. (2009). The road to the vertebral formula. *International Journal of Developmental Biology*, *53*(8–10), 1469–1481. https://doi.org/10.1387/ijdb.072276mm.

Mansour, S. L., Thomas, K. R., & Capecchi, M. R. (1988). Disruption of the proto-oncogene in int-2 in mouse embryo-derived stem cells: A general strategy for targeting mutations to non-selectable genes. *Nature*, *336*, 403–405. https://doi.org/10.1038/332141a0.

Mark, P., Kleinsorge, M., Gaebel, R., Lux, C. A., Toelk, A., Pittermann, E., et al. (2013). Human mesenchymal stem cells display reduced expression of CD105 after culture in serum-free medium. *Stem Cells International*, *2013*, 698076.

Martí, E., Takada, R., Bumcrot, D. A., Sasaki, H., & McMahon, A. P. (1995). Distribution of Sonic hedgehog peptides in the developing chick and mouse embryo. *Development (Cambridge, England)*, *121*, 2537–2547. https://doi.org/10.1016/0092-8674(94)90522-3.

Martin, P. (1990). Tissue patterning in the developing mouse limb. *International Journal of Developmental Biology*, *34*(3), 323–336.

Mathew, S. J., Hansen, J. M., Merrell, A. J., Murphy, M. M., Lawson, J. A., Hutcheson, D. A., et al. (2011). Connective tissue fibroblasts and Tcf4 regulate myogenesis. *Development*, *138*(2), 371–384. https://doi.org/10.1242/dev.057463.

Mauro, A. (1961). Satellite cell of skeletal muscle fibers. *Journal of Biophysical and Biochemical Cytology*, *9*, 493–495. https://doi.org/10.1083/jcb.9.2.493.

McCarthy, J. J., Mula, J., Miyazaki, M., Erfani, R., Garrison, K., Farooqui, A. B., et al. (2011). Effective fiber hypertrophy in satellite cell-depleted skeletal muscle. *Development*, *138*(17), 3657–3666. https://doi.org/10.1242/dev.068858.

Mead, T. J., & Yutzey, K. E. (2009). Notch pathway regulation of chondrocyte differentiation and proliferation during appendicular and axial skeleton development. *Proceedings of the National Academy of Sciences of the United States of America*, *106*(34), 14420–14425. https://doi.org/10.1073/pnas.0902306106.

Méndez-Ferrer, S., Michurina, T. V., Ferraro, F., Mazloom, A. R., Macarthur, B. D., Lira, S. A., et al. (2010). Mesenchymal and haematopoietic stem cells form a unique bone marrow niche. *Nature*, *466*(7308), 829–834. https://doi.org/10.1038/nature09262.

Méndez-Ferrer, S., Scadden, D. T., & Sánchez-Aguilera, A. (2015). Bone marrow stem cells: Current and emerging concepts. *Annals of the New York Academy of Sciences*, *1335*(1), 32–44. http://doi.org/10.1111/nyas.12641.

Mendias, C. L., Bakhurin, K. I., & Faulkner, J. A. (2008). Tendons of myostatin-deficient mice are small, brittle, and hypocellular. *Proceedings of the National Academy of Sciences of the United States of America*, *105*(1), 388–393. https://doi.org/10.1073/pnas.0707069105.

Mikic, B. (2004). Multiple effects of GDF-5 deficiency on skeletal tissues: Implications for therapeutic bioengineering. *Annals of Biomedical Engineering*, *32*(3), 466–476. https://doi.org/10.1023/B:ABME.0000017549.57126.51.

Mikic, B., Schalet, B. J., Clark, R. T., Gaschen, V., & Hunziker, E. B. (2001). GDF-5 deficiency in mice alters the ultrastructure, mechanical properties and composition of the Achilles tendon. *Journal of Orthopaedic Research*, *19*(3), 365–371. https://doi.org/10.1016/S0736-0266(00)90018-4.

Mitchell, K. J., Pannérec, A., Cadot, B., Parlakian, A., Besson, V., Gomes, E. R., et al. (2010). Identification and characterization of a non-satellite cell muscle resident progenitor during postnatal development. *Nature Cell Biology*, *12*(3), 257–266. https://doi.org/10.1038/ncb2025.

Mizoguchi, T., Pinho, S., Ahmed, J., Kunisaki, Y., Hanoun, M., Mendelson, A., et al. (2014). Osterix marks distinct waves of primitive and definitive stromal progenitors during bone marrow development. *Developmental Cell*, *29*(3), 340–349. https://doi.org/10.1016/j.devcel.2014.03.013.

Mori-Akiyama, A., & Zhang, Y. (2004). Interactions between Sox9 and beta-catenin control chondrocyte differentiation. *Genes & Development*, 1072–1087. https://doi.org/10.1101/gad.1171104.Wagner.

Morikawa, S., Mabuchi, Y., Kubota, Y., Nagai, Y., Niibe, K., Hiratsu, E., et al. (2009). Prospective identification, isolation, and systemic transplantation of multipotent mesenchymal stem cells in murine bone marrow. *The Journal of Experimental Medicine*, *206*(11), 2483–2496. https://doi.org/10.1084/jem.20091046.

Morrison, S. J., & Scadden, D. T. (2014). The bone marrow niche for haematopoietic stem cells. *Nature Reviews. Molecular Cell Biology*, *505*, 327–3334. https://doi.org/10.1038/nature12984.

Murchison, N. D., Price, B. A., Conner, D. A., Keene, D. R., Olson, E. N., Tabin, C. J., et al. (2007). Regulation of tendon differentiation by scleraxis distinguishes force-transmitting tendons from muscle-anchoring tendons. *Development*, *134*(14), 2697–2708. https://doi.org/10.1242/dev.001933.

Murphy, M., & Kardon, G. (2011). Origin of vertebrate limb muscle: The role of progenitor and myoblast populations. *Current Topics in Developmental Biology*, *96*, 1–32. https://doi.org/10.1016/B978-0-12-385940-2.00001-2.

Murphy, M. M., Lawson, J. A., Mathew, S. J., Hutcheson, D. A., & Kardon, G. (2011). Satellite cells, connective tissue fibroblasts and their interactions are crucial for muscle regeneration. *Development*, *138*(17), 3625–3637. https://doi.org/10.1242/dev.064162.

Nelson, C. E., Morgan, B. A., Burke, A. C., Laufer, E., DiMambro, E., Murtaugh, L. C., et al. (1996). Analysis of Hox gene expression in the chick limb bud. *Development (Cambridge, England)*, *122*(5), 1449–1466. https://doi.org/10.1038/342767a0.

Nelson, L. T., Rakshit, S., Sun, H., & Wellik, D. M. (2008). Generation and expression of a Hoxa11eGFP targeted allele in mice. *Developmental Dynamics*, *237*(11), 3410–3416. https://doi.org/10.1002/dvdy.21756.

Nourissat, G., Berenbaum, F., & Duprez, D. (2015). Tendon injury: From biology to tendon repair. *Nature Reviews Rheumatology*, *11*(4), 223–233. https://doi.org/10.1038/nrrheum.2015.26.

Nusspaumer, G., Jaiswal, S., Barbero, A., Reinhardt, R., Ishay Ronen, D., Haumer, A., et al. (2017). Ontogenic identification and analysis of mesenchymal stromal cell populations during mouse limb and long bone development. *Stem Cell Reports*, *9*, 1–15. https://doi.org/10.1016/j.stemcr.2017.08.007.

Omatsu, Y., Sugiyama, T., Kohara, H., Kondoh, G., Fujii, N., Kohno, K., et al. (2010). The essential functions of Adipo-osteogenic progenitors as the hematopoietic stem and progenitor cell niche. *Immunity*, *33*(3), 387–399. https://doi.org/10.1016/j.immuni.2010.08.017.

Onizuka, N., Ito, Y., Inagawa, M., Nakahara, H., Takada, S., Lotz, M., et al. (2014). The Mohawk homeobox transcription factor regulates the differentiation of tendons and volar plates. *Journal of Orthopaedic Science*, *19*(1), 172–180. https://doi.org/10.1007/s00776-013-0485-z.

Ono, N., & Kronenberg, H. M. (2015). Mesenchymal progenitor cells for the osteogenic lineage. *Current Molecular Biology Reports*, 95–100. https://doi.org/10.1007/s40610-015-0017-z.

Ono, N., & Kronenberg, H. M. (2016). Bone repair and stem cells. *Current Opinion in Genetics & Development*, *40*, 103–107. https://doi.org/10.1016/j.gde.2016.06.012.

Ono, N., Ono, W., Mizoguchi, T., Nagasawa, T., Frenette, P. S., & Kronenberg, H. M. (2014). Vasculature-associated cells expressing nestin in developing bones encompass early cells in the osteoblast and endothelial lineage. *Developmental Cell*, *29*(3), 330–339. https://doi.org/10.1016/j.devcel.2014.03.014.

Ono, N., Ono, W., Nagasawa, T., & Kronenberg, H. M. (2014). A subset of chondrogenic cells provides early mesenchymal progenitors in growing bones. *Nature Cell Biology*, *16*(12), 1157–1167. https://doi.org/10.1038/ncb3067.

Ordahl, C. P., & Le Douarin, N. M. (1992). Two myogenic lineages within the developing somite. *Development*, *114*(2), 339–353. Retrieved from http://www.ncbi.nlm.nih.gov/pubmed/1591996.

Pan, Q., Yu, Y., Chen, Q., Li, C., Wu, H., Wan, Y., et al. (2008). Sox9, a key transcription factor of bone morphogenetic protein-2-induced chondrogenesis, is activated through BMP pathway and a CCAAT box in the proximal promoter. *Journal of Cellular Physiology*, *217*(1), 228–241. https://doi.org/10.1002/jcp.21496.

Pinho, S., Lacombe, J., Hanoun, M., Mizoguchi, T., Bruns, I., Kunisaki, Y., et al. (2013). PDGFRα and CD51 mark human nestin+ sphere-forming mesenchymal stem cells capable of hematopoietic progenitor cell expansion. *The Journal of Experimental Medicine*, *210*(7), 1351–1367. https://doi.org/10.1084/jem.20122252.

Pryce, B. A., Watson, S. S., Murchison, N. D., Staverosky, J. A., Dunker, N., & Schweitzer, R. (2009). Recruitment and maintenance of tendon progenitors by TGF signaling are essential for tendon formation. *Development*, *136*(8), 1351–1361. https://doi.org/10.1242/dev.027342.

Relaix, F., Montarras, D., Zaffran, S., Gayraud-Morel, B., Rocancourt, D., Tajbakhsh, S., et al. (2006). Pax3 and Pax7 have distinct and overlapping functions in adult muscle progenitor cells. *Journal of Cell Biology*, *172*(1), 91–102. https://doi.org/10.1083/jcb.200508044.

Relaix, F., Rocancourt, D., Mansouri, A., & Buckingham, M. (2004). Divergent functions of murine Pax3 and Pax7 in limb muscle development. *Genes and Development*, *18*(9), 1088–1105. https://doi.org/10.1101/gad.301004.

Relaix, F., & Zammit, P. S. (2012). Satellite cells are essential for skeletal muscle regeneration: The cell on the edge returns centre stage. *Development*, *139*(16), 2845–2856. https://doi.org/10.1242/dev.069088.

Riddle, R. D., Ensini, M., Nelson, C., Tsuchida, T., Jessell, T. M., & Tabin, C. (1995). Induction of the LIM homeobox gene Lmx1 by WNT6a establishes dorsoventral pattern in the vertebrate limb. *Cell*, *83*, 631–640. https://doi.org/10.1016/0092-8674(95)90103-5.

Riddle, R. D., Johnson, R. L., Laufer, E., & Tabin, C. (1993). Sonic hedgehog mediates the polarizing activity of the ZPA. *Cell*, *75*(7), 1401–1416. https://doi.org/10.1016/0092-8674(93)90626-2.

Rockel, J. S., Kudirka, J. C., Guzi, A. J., & Bernier, S. M. (2008). Regulation of Sox9 activity by crosstalk with nuclear factor-κB and retinoic acid receptors. *Arthritis Research and Therapy*, *10*(1), 1–12. https://doi.org/10.1186/ar2349.

Rooney, P., & Archer, C. W. (1992). The development of the perichondrium in the avian ulna. *Journal of Anatomy*, *181*(Pt. 3), 393–401.

Rux, D. R., Song, J. Y., Swinehart, I. T., Pineault, K. M., Schlientz, A. J., Trulik, K. G., et al. (2016). Regionally restricted Hox function in adult bone marrow multipotent mesenchymal stem/stromal cells. *Developmental Cell*, *39*(6), 653–666. https://doi.org/10.1016/j.devcel.2016.11.008.

Sacchetti, B., Funari, A., Michienzi, S., Di Cesare, S., Piersanti, S., Saggio, I., et al. (2007). Self-renewing osteoprogenitors in bone marrow sinusoids can organize a hematopoietic microenvironment. *Cell*, *131*(2), 324–336. https://doi.org/10.1016/j.cell.2007.08.025.

Sambasivan, R., Yao, R., Kissenpfennig, A., Van Wittenberghe, L., Paldi, A., Gayraud-Morel, B., et al. (2011). Pax7-expressing satellite cells are indispensable for adult skeletal muscle regeneration. *Development*, *138*(19), 3647–3656. https://doi.org/10.1242/dev.073601.

Schweitzer, R., Chyung, J. H., Murtaugh, L. C., Brent, A. E., Rosen, V., Olson, E. N., et al. (2001). Analysis of the tendon cell fate using Scleraxis, a specific marker for tendons and ligaments. *Development (Cambridge, England)*, *128*(19), 3855–3866. https://doi.org/10.1242/dev.001933.

Seale, P., Sabourin, L. A., Girgis-Gabardo, A., Mansouri, A., Gruss, P., & Rudnicki, M. A. (2000). Pax7 is required for the specification of myogenic satellite cells skeletal muscle are mitotically quiescent and are activated in response to diverse stimuli, including stretching, exercise, injury, and electrical stimulation (Schultz). *Cell*, *102*, 777–786. https://doi.org/10.1016/S0092-8674(00)00066-0.

Serrano, A. L., Baeza-Raja, B., Perdiguero, E., Jardí, M., & Muñoz-Cánoves, P. (2008). Interleukin-6 is an essential regulator of satellite cell-mediated skeletal muscle hypertrophy. *Cell Metabolism*, 7(1), 33–44. https://doi.org/10.1016/j.cmet.2007.11.011.

Sharma, P., & Maffulli, N. (2006). Biology of tendon injury: Healing, modeling and remodeling. *Journal of Musculoskeletal & Neuronal Interactions*, *6*(2), 181–190. https://doi.org/10.1016/j.math.2014.02.002.

Shea, K. L., Xiang, W., LaPorta, V. S., Licht, J. D., Keller, C., Basson, M. A., et al. (2010). Sprouty1 regulates reversible quiescence of a self-renewing adult muscle stem cell pool during regeneration. *Cell Stem Cell*, *6*(2), 117–129. https://doi.org/10.1016/j.stem.2009.12.015.

Shi, Y., He, G., Lee, W.-C., McKenzie, J. A., Silva, M. J., & Long, F. (2017). Gli1 identifies osteogenic progenitors for bone formation and fracture repair. *Nature Communications*, *8*(1), 2043. https://doi.org/10.1038/s41467-017-02171-2.

St-Jacques, B., Hammerschmidt, M., & McMahon, A. P. (1999). Indian hedgehog signaling regulates proliferation and differentiation of chondrocytes and is essential for bone formation. *Genes & Development*, *13*(16), 2072–2086. https://doi.org/10.1101/gad.13.16.2072.

Summerbell, D., Lewis, J. H., & Wolpert, L. (1973). Positional information in chick limb morphogenesis. *Nature*, *244*(5417), 492–496. https://doi.org/10.1038/244492a0.

Swinehart, I. T., Schlientz, A. J., Quintanilla, C. A., Mortlock, D. P., & Wellik, D. M. (2013). Hox11 genes are required for regional patterning and integration of muscle, tendon and bone. *Development (Cambridge, England)*, *140*(22), 4574–4582. https://doi.org/10.1242/dev.096693.

Tahara, N., Akiyama, R., Theisen, J. W. M., Kawakami, H., Wong, J., Garry, D. J., et al. (2017). Gata6 restricts Isl1 to the posterior of nascent hindlimb buds through Isl1 cis-regulatory modules. *Developmental Biology*, *434*(1), 74–83. https://doi.org/10.1016/j.ydbio.2017.11.013.

Tan, Q., Lui, P. P. Y., Rui, Y. F., & Wong, Y. M. (2012). Comparison of potentials of stem cells isolated from tendon and bone marrow for musculoskeletal tissue engineering. *Tissue Engineering Part A*, *18*(7–8), 840–851. https://doi.org/10.1089/ten.tea.2011.0362.

Tarchini, B., & Duboule, D. (2006). Control of Hoxd genes' collinearity during early limb development. *Developmental Cell*, *10*(1), 93–103. https://doi.org/10.1016/j.devcel.2005.11.014.

Thomas, K. R., & Capecchi, M. R. (1987). Site-directed mutagenesis by gene targeting in mouse embryo-derived stem cells. *Cell*, *51*(3), 503–512. https://doi.org/10.1016/0092-8674(87)90646-5.

Thorogood, P. V., & Hinchliffe, J. R. (1975). An analysis of the condensation process during chondrogenesis in the embryonic chick hind limb. *Journal of Embryology and Experimental Morphology*, *33*(3), 581–606. Retrieved from http://www.ncbi.nlm.nih.gov/pubmed/1176861.

Tidball, J. G. (2011). Mechanisms of muscle injury, repair, and regeneration. *Comprehensive Physiology*, *1*(4), 2029–2062. https://doi.org/10.1002/cphy.c100092.

Uezumi, A., Fukada, S. I., Yamamoto, N., Takeda, S., & Tsuchida, K. (2010). Mesenchymal progenitors distinct from satellite cells contribute to ectopic fat cell formation in skeletal muscle. *Nature Cell Biology*, *12*(2), 143–152. https://doi.org/10.1038/ncb2014.

Ushita, M., Saito, T., Ikeda, T., Yano, F., Higashikawa, A., Ogata, N., et al. (2009). Transcriptional induction of SOX9 by NF-κB family member RelA in chondrogenic cells. *Osteoarthritis and Cartilage*, *17*(8), 1065–1075. https://doi.org/10.1016/j.joca.2009.02.003.

Voleti, P. B., Buckley, M. R., & Soslowsky, L. J. (2012). Tendon healing: Repair and regeneration. *Annual Review of Biomedical Engineering*, *14*(1), 47–71. https://doi.org/10.1146/annurev-bioeng-071811-150122.

Vortkamp, A., Pathi, S., Peretti, G. M., Caruso, E. M., Zaleske, D. J., & Tabin, C. J. (1998). Recapitulation of signals regulating embryonic bone formation during postnatal growth and in fracture repair. *Mechanisms of Development*, *71*(1–2), 65–76. https://doi.org/10.1016/S0925-4773(97)00203-7.

Wachtler, F. (1984). On the migration of epidermal melanoblasts in the avian embryonic wing bud. *Anatomy and Embryology*, *170*(3), 307–312. https://doi.org/10.1007/BF00318734.

Walden, G., Liao, X., Donell, S., Raxworthy, M. J., Riley, G. P., & Saeed, A. (2017). A clinical, biological, and biomaterials perspective into tendon injuries and regeneration. *Tissue Engineering Part B: Reviews*, *23*(1), 44–58. https://doi.org/10.1089/ten.teb.2016.0181.

Wellik, D. M. (2007). Hox patterning of the vertebrate axial skeleton. *Developmental Dynamics*, *236*(9), 2454–2463. https://doi.org/10.1002/dvdy.21286.

Wellik, D. M. (2009). Chapter 9 Hox genes and vertebrate axial pattern. In P. M. Wassarman & O. Pourquié (Eds.), *Current topics in developmental biology*. (1st ed.)., Vol. 88. Elsevier Inc. https://doi.org/10.1016/S0070-2153(09)88009-5.

Wellik, D. M., & Capecchi, M. R. (2003). Hox10 and Hox11 genes are required to globally pattern the mammalian skeleton. *Science (New York, N.Y.)*, *301*(5631), 363–367. https://doi.org/10.1126/science.1085672.

Wilson, A., & Trumpp, A. (2006). Bone-marrow haematopoietic-stem-cell niches. *Nature Reviews. Immunology*, *6*(2), 93–106. https://doi.org/10.1038/nri1779.

Worthley, D. L., Churchill, M., Compton, J. T., Tailor, Y., Rao, M., Si, Y., et al. (2015). Gremlin 1 identifies a skeletal stem cell with bone, cartilage and reticular stromal potential. *Cell*, *160*(1–2), 269–284. https://doi.org/10.1016/j.cell.2014.11.042.

Wosczyna, M. N., & Rando, T. A. (2018). A muscle stem cell support group: Coordinated cellular responses in muscle regeneration. *Developmental Cell*, *46*(2), 135–143. Figure 1. https://doi.org/10.1016/j.devcel.2018.06.018.

Xu, Y., & Murrell, G. A. C. (2008). The basic science of tendinopathy. *Clinical Orthopaedics and Related Research*, *466*(7), 1528–1538. https://doi.org/10.1007/s11999-008-0286-4.

Xu, B., & Wellik, D. M. (2011). Axial Hox9 activity establishes the posterior field in the developing forelimb. *Proceedings of the National Academy of Sciences of the United States of America*, *108*(12), 4888–4891. https://doi.org/10.1073/pnas.1018161108.

Yin, H., Price, F., & Rudnicki, M. A. (2013). Satellite cells and the muscle stem cell niche. *Physiological Reviews*, *93*(1), 23–67. https://doi.org/10.1152/physrev.00043.2011.

Zakany, J., & Duboule, D. (2007). The role of Hox genes during vertebrate limb development. *Current Opinion in Genetics & Development*, *17*(4), 359–366. https://doi.org/10.1016/j.gde.2007.05.011.

Zhang, J., & Wang, J. H. C. (2010). Characterization of differential properties of rabbit tendon stem cells and tenocytes. *BMC Musculoskeletal Disorders*, *11*, 10. https://doi.org/10.1186/1471-2474-11-10.

Zhou, B. O., Yue, R., Murphy, M. M., Peyer, J. G., & Morrison, S. J. (2014). Leptin-receptor-expressing mesenchymal stromal cells represent the main source of bone formed by adult bone marrow. *Cell Stem Cell*, *15*(2), 154–168. https://doi.org/10.1016/j.stem.2014.06.008.

FSC
www.fsc.org
MIX
Papier | Fördert
gute Waldnutzung
FSC® C083411

Kolibri 360 GmbH
Ferdinand-Jühlke-Straße 7
99095 Erfurt, Deutschland
produktsicherheit@kolibri360.de